ARANA

Basque Diaspora Series No. 14

Arana

William A. Douglass

Center for Basque Studies
University of Nevada, Reno

Basque Diaspora Series No. 14
Series Editor: Xabier Irujo

William A. Douglass Center for Basque Studies
University of Nevada, Reno
Reno, Nevada 89557
http://basque.unr.edu

Copyright © 2019 William A. Douglass
Edition copyright © 2019 Center for Basque Studies, University of Nevada, Reno
All rights reserved. Printed in the United States of America

Library of Congress Cataloging-in-Publication Data
Names: Douglass, William A, author.
Title: Arana / William A Douglass.
Description: Reno : Center for Basque Studies, University of Nevada, Reno,
  2019. | Series: Basque diaspora series; no. 14 | Includes
  bibliographical references.
Identifiers: LCCN 2019026906 | ISBN 9781935709978 (paperback)
Subjects: LCSH: Indians of South America--Violence against--Putumayo River
  Valley--History--20th century. | Indians of South America--Putumayo
  River Valley--Social conditions--20th century. | Putumayo River
  Valley--History--20th century. | Rubber industry and trade--Peru. |
  Peruvian Amazon Company--History. | Arana, Julio César. |
  Industrialists--Peru--Biography.
Classification: LCC F3451.P94 D68 2019 | DDC 986.1/63--dc23
LC record available at https://lccn.loc.gov/2019026906

Contents

Introduction 1

Part 1
    Narratives 9

Part 2 179
    The Scandal

Part 3 285
    The Hearings

Part 4 341
    The Aftermath

Part 5 449
    Configuring New Indigenist Policy

Part 6 509
    Analysis

Appendix 1 555
    Memorandum 25: Sketch of the Putumayo
Appendix 2 559
    Translation of Testimony of Felipe Cabrera M.
Appendix 3 565
    The Flower of the Selva: A Tale
    of the Upper Amazon
Appendix 4 573
    Letter of Julio César Arana
Appendix 5 579
    A Criminal's Life Story: The Career
    of Armando Normand
Appendix 6 587
    Poem by Rómulo Paredes (1918)
Appendix 7 593
    Words of the Elders of Tobacco, Coca,
    and Sweet Yucca
Appendix 8 597
    Configuring and Comparing Rubber Barons

Acknowledgments 601
Bibliography 603
Index 617

Introduction

I write of the "Basque" Arana. Not Sabino de Arana Goiri, the tragic dreamer and founder of the modernBasque nationalist movement who died before turning forty (but whose legacy continues to reverberate in contemporary Basque politics). Nor do I mean Beatriz Enríquez de Arana, the Andalusia-born Converso and Araban-Basque-descended mistress (and heiress) of Christopher Columbus and mother of his son Hernando. Rather, I refer to the aptly named Julio César Arana del Águila Hidalgo—a Peruvian-born person of Basque descent who, as a rubber baron, established a personal empire in a remote region (sometimes referred to as "Huitocia") of Northwest Amazonia in the middle stretches of the coterminous watersheds of three parallel rivers—the Napo, Putumayo, and Caquetá—that flowed from the Andes to the Marañón, main tributary of the Amazon River.

Beginning in the mid-nineteenth century, the planet's economy experienced what is typically called the rubber "boom," fed initially by the collection of wild latex in the Congo and Amazon river basins. While it transformed both areas profoundly, it proved highly circumscribed and episodic, being eclipsed by the more efficient production of rubber under industrial conditions on Asian and African rubber plantations (and the subsequent development of synthetic rubber during World War II).

In both of its wild-rubber venues, the boom would eventuate in international scandal. The two bête noirs of the story were King Leopold of Belgium and the Colombian rubber baron Julio César Arana. If Leopold wrote the introduction and early chapters of the book of rubber scandals, Arana would be the author of its final chapter and conclusion. For both men, their nemesis and key antagonist was the same larger-than-life individual, Sir Roger Casement—one of the twentieth century's truly tragic figures. It all made for drama of epic proportions.

It was in the Middle Putumayo/Caquetá,[1] or what I shall call "Aranalandia," that Julio César was allegedly responsible for the deaths of thousands of Huitoto, Andoke, Bora, and Ocaina Indians.[2]

Eventually, the atrocities provoked major international concern over what came to be known as the "Putumayo Scandal"—leading to months of public "Hearings" in London (1912–1913) conducted by a Select Committee of the British House of Commons.[3] Arana has passed into history as one of the iconic perpetrators of the Age of Genocide, otherwise known as the twentieth century.

I was also drawn to Julio César's story by some of my other professional and personal interests. I have never been to the Putumayo (so far at least), but I feel that I know it well. I am an inveterate globe-trotting fly fisherman and have spent about 25 weeks fishing in tropical Brazil, Paraguay, Argentina, Bolivia, Venezuela, and Guyana. I have camped on and boated down many rivers like the Putumayo and am well acquainted with their feel. Furthermore, some of my angling has been within Indian reserves, and I have more than a passing interest in the interplay among the economics of sports fishing, national policy, indigenous rights, and environmental conservation.[4] Reading books about Amazonian exploration and ecology is one of my life-long passions.

---

1. I opt for this designation, reserving *Bajo Putumayo* for the lower part of the river from about the area that would later come to be called the Leticia Trapezoid to where the Brazilian portion of the stream, called the Içá River, joins the Marañon. *Alto Putumayo*, or Upper Putumayo, refers to the stretch from its birth near Mocoa to about the future town of Puerto Leguízamo. Some authors limit their geographical designations to the Upper and Lower Putumayo, and would include Arana's domain in the latter.
2. I am aware that the term "Indian" is no longer regarded as politically correct. However, I have decided to alternate it according to context with "indigene" in the interest of variety. Certainly, the voices and documents of most of the protagonists of this account employed *indios* rather than *indígenas*, although the latter form was not entirely absent. From a stylistic standpoint, it is handy to have two terms of reference rather than a single one when the invocations run into the thousands. I would emphasize at the outset, however, that my use of the terms interchangeably at times is meant to be value neutral. I would further note that, except when indicated otherwise, all translations are mine.
3. British Parliament. House of Commons, *Report and* Special *Report from the Select Committee on the Putumayo together with the Proceedings of the Committee, Minutes of Evidence and Appendices* (London: Wyman and Sons, Limited, 1913).
4. I have published a fishing book with some chapters on South American tropical destinations (see William A. Douglass, *Casting About in the Reel World.* [Oakland: RDR Books, 2002]) and am currently working on lengthy essays on fishing operations in the Securé River drainage in Bolivia and on the Rewa River in Guyana—both of which are Indian reserves.

Furthermore, when my daughter Ana co-edited a book on witnessing and asked me to contribute, I elected to compare the irreconcilable visions of the Brazilian Highlands published by former American president Theodore Roosevelt and famed French anthropologist Claude Lévi-Strauss.[5] I concluded that different perspectives of Roosevelt's overly optimistic description and Levi-Strauss's extremely pessimistic one were a function of the distinct personalities of these two witnesses as they went into the wilderness. Ana's book was an exercise in the literary criticism that has become so influential recently in my social anthropological discipline's exercise in deconstruction. And here I can close the circle, as it were, since arguably one of the most influential books in that trend has been Michael Taussig's *Shamanism, Colonialism, and the Wild Man*,[6] which devotes considerable analysis to the Putumayo scandal and the role of Arana within it, particularly the perceptions and prejudices of white and Indian "witnesses" regarding one another. As we shall see, the nature of witnessing—ranging through the disparate perceptions and interpretations of different eyewitnesses of events to their processing of hearsay—is of fundamental concern in the following text.

This is an oft-told tale, so there is the legitimate question of why tell it again. There is, of course, the obvious answer that each retelling differs from the others and supposedly brings new perspective, if only in the form of author bias. As we shall see, the story is enormously complex and replete with a bewildering array of players—persons, governments, and religious denominations. Consequently, there has been a tendency on the part of most interpreters to summarize the arguments. I have opted instead to employ generous quotes from original documents and statements of the protagonists themselves. Hopefully, my selection allows them to craft much of this narrative in their own voices. There is also the fact that the amount of information expands as new research progresses. The most recent data is obviously absent in the earlier treatments.[7] Then, too, I try

---

5. William A. Douglass, "Witness in the Wilderness: The Tropical Tryst of Claude Lévi-Strauss and Theodore Roosevelt," in Ana Douglass and Thomas A. Vogler (eds.), *Witness and Memory: The Discourse of Trauma* (New York and London: Routledge, 2003).

6. Michael Taussig, *Shamanism, Colonialism, and the Wild Man* (Chicago: The University of Chicago Press, 1987) [*Shamanism*].

7. A particularly telling example in this regard is the extensive work curated and edited by anthropologist Augusto Javier Gómez López based on the exhaustive combing of the records in the Colombian National Archive in Bogotá, as well as regional ones in places like Pasto, assisted by a team of Colombian anthropology graduate students. They gleaned documents submitted to their governmental agencies (1900–1910) by irate and desperate Colombians who felt themselves to be under assault in the Putumayo/Caquetá by Peruvians spearheaded by Arana. Augusto Javier Gómez López (compiler and

to situate the story within the broad framework of half a millennium of South American civil and ecclesiastical history, while elaborating it down to the present day (many treatments are limited to its late nineteenth and early twentieth century chapter exclusively). I also focus more on Julio César than, as is usually the case, his nemesis Roger Casement.

It is my unique concern, as underscored by the quotation marks around the fifth word of this introduction, to highlight and assess the extent to which Arana's Basque ethnic heritage possibly influenced his behavior. I am co-author of the book *Amerikanuak: Basques in the New World*.[8] One of its key premises is that historians of Latin America have typically treated the Iberian settlers as culturally uniform Spaniards (and Portuguese), whereas some of the actions and attitudes of the so-called *peninsulares* and their creole descendants can only be understood in terms of their Iberian regional ethnic differences.

There are numerous New World and Old World examples of Basques acting in concert, predicated upon their shared ethnic identity. In effect, these are histories underscoring the existence and activities of a Basque ethnic network throughout the Spanish colonial realm. In what follows, I shall identify every instance of a Basque-surnamed protagonist before subsequently assessing Arana's Basqueness as one of the possible factors informing his behavior.

In sum, whether dealing with scholarly analyses or the testimony of witnesses, the following text is not a protracted concinnity that reconciles the many accounts while reconfiguring them into a single overarching narrative. Unfortunately, in my view, that is the approach that characterizes far too much of historical writing, not to mention fiction. If this work fails to provide a story's tidy resolution—the beginning, middle, and end of the skilled novelist or playwright—so be it. No doubt my text's open-

---

annotator), *Putumayo: La vorágine de las caucherías. Memoria y testimonio* (Bogotá: Centro Nacional de Memoria Histórica, 2014).

8. William A. Douglass and Jon Bilbao, *Amerikanuak: Basques in the New World* (Reno and Las Vegas: University of Nevada Press, 1975). It is fair to say that *Amerikanuak* has become an iconic cornerstone of today's fairly vibrant sub-field of diaspora research within Basque Studies. Dozens of investigators have followed its lead in perusing New World history from a Basque perspective—usually by pursuing the trail of (distinctively) Basque-surnamed individuals through the records, as well as their clustering in particular events and initiatives. The year 2015 marked the fortieth anniversary of the book's appearance in English (there was a Spanish edition in 1986). Throughout the year commemorative conferences were organized in Cuba, Iceland, Nevada, and Idaho and at five university and museum venues in the Basque Country. There has also been a spate of publications regarding various aspects of Basque diaspora studies. The conferences and publications all display a unifying logo.

ended nature will frustrate that reader who expects closure (certainty), but I believe that my approach captures far more accurately the chaotic realities of life as lived by real human beings.

Avanti!

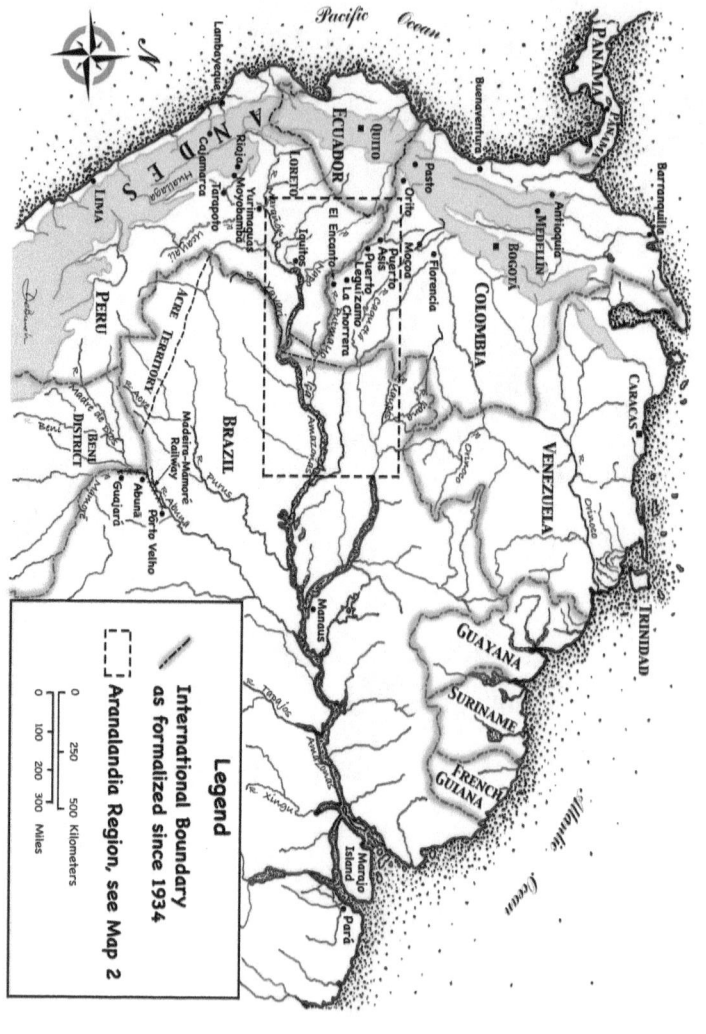

Map 1. Aranalandia, general map. Drawn by Patti DeBunch

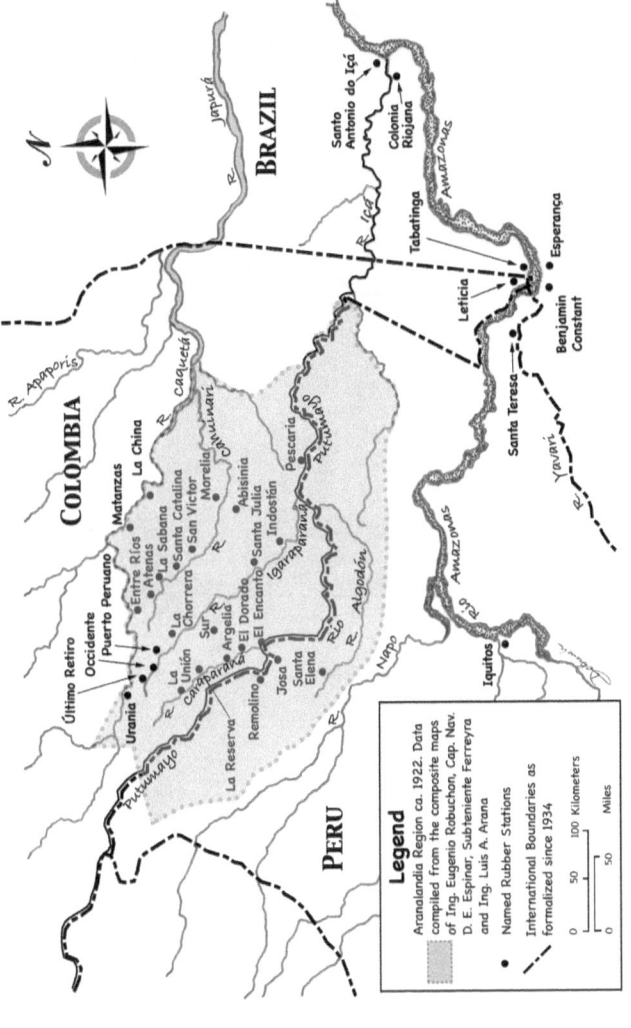

Map 2. Aranalandia, detail map. Drawn by Patti DeBunch

Part 1
Narratives

The River That God Forgot[1]

The "discovery" of the New World in 1492 by Christopher Columbus intensified an existing rivalry between Portugal and Castile. During the fifteenth century, Portugal had explored the West African coast, and, finally (in 1488), managed to circumnavigate the Cape of Good Hope. By century's end, the Portuguese were establishing both East African and Indian enclaves and were probing Malaysia and the Spice Islands—the latter the source of the lucrative spice trade. Portugal had effectively managed an end-run around the fabled Silk Route that had heretofore been an Arab and Venetian monopoly. Columbus's proposal (twice-rejected by Portuguese monarchs and once by Castile's queen on the accurate grounds that he was underestimating the circumference of the globe) was to sail west in order to open a new sea route to the Orient—particularly to China. Indeed, Columbus believed that his landfall was there.

The potential of open conflict between the two Catholic powers was obvious. Consequently, in 1494, Pope Alexander VI brokered the Treaty of Tordesillas that divided the world between the two Iberian competitors. Presumably, Africa and most of the Orient were within the Portuguese sphere—although the inability at the time to calculate longitude accurately meant that both nations would continue to claim and contest the Spice Islands for the next three decades. As a result, most of the Pacific region and Caribbean, as well as much of the two American continents, fell within the Spanish orbit. As it became increasingly evident that Columbus had discovered a continental barrier of immense proportions, the

---

1. This is the main title of the work by Richard Collier, *The River that God Forgot: The Dramatic Story of the Rise and Fall of the Despotic Amazon Rubber Barons* (New York: E.P. Dutton, 1968).

dream of establishing a sea route to the Orient triggered exploration of the New World coastline and the search for either a southern or northern passage. Ironically, it would be the expedition (1519–1522) of Ferdinand Magellan, a Portuguese native sailing under the Spanish flag, that would finally circumvent the Americas through the Strait of Magellan more than a quarter of a century after the Genoese's discovery of the New World.[2]

By then, however, it was clear that Brazil fell on the Portuguese side of the Tordesillas Line. First visted by Cabral in 1500, for the next three decades the Portuguese did little more than log the area for Brazilwood (employing indigenes in the task). In 1531, there was a concerted effort to colonize Brazil through a system of awarding captaincies over particular regions to private investors. Most failed, but the seeds of permanent European settlement were established and the immensity of the Amazon was now apparent.

After nearly three decades of extending its control in the Caribbean, punctuated by the occasional exploratory probe along the continental coastline (from Venezuela to Florida), Spain established its first viceroyalty in Nueva España (Mexico) in 1535. The second would be the Viceroyalty of Peru (1543), with its capital in Lima, that encompassed all Spanish holdings within South America. Subsequently, in the eighteenth century, two new viceroyalties were carved out of the Peruvian one—the Viceroyalty of New Granada (1717), based in Bogotá, and that of Rio de la Plata (1776), administered from Buenos Aires. The creation of the two eighteenth-century viceroyalties were in response to growing challenges to Spanish hegemony in the region. The Creole population of northern South America was ever more assertive and the French, British, and Dutch were all trading with them (technically contraband) in blatant contravention of Spanish law. A weakened and overly extended Spain was also unable to prevent all three European powers from claiming Caribbean islands and establishing enclaves on the northern South American coast. Meanwhile, in southern South America, both Great Britain and Portugal coveted Argentina and the future Uruguay. All three of the Spanish viceroyalties in South America shared boundaries with the expanding Luso-Brazilian continental empire.

Delimitation of the Lusitanian and Spanish spheres of influence was particularly confused in the Amazon basin, the continent's least developed frontier. In 1777, Spain and Portugal signed the treaty of San Ildefonso designed to draw the New World boundaries beween the two Iberian powers. However, this created its own disputes since Tabatinga

---

2. William A. Douglass, *Basque Explorers in the Pacific Ocean* (Reno: Center for Basque Studies, University of Nevada, Reno, 2015), 29–48.

now ostensibly fell within the Spanish orbit, yet it had been founded decades earlier by Portuguese.[3] Attempts by mapmakers to resolve such issues were failures, the ineffectuality exacerbated when Spain sided with Napoleon and Portugal remained allied to England, prompting a brief conflict (the War of the Oranges) beween the Iberian powers in 1801.

These alliances would be shattered when Napoleon attempted to impose French hegemony in Iberia, prompting the Peninsular Wars. By 1808, both Iberian monarchs had been deposed. Spanish king Carlos IV was forced to abdicate in favor of his son, Fernando VII, who was in turn replaced a few months later with Joseph Bonaparte—Napoleon's elder brother—who would reign until 1813 as King José I. By 1810, the Spanish parliament was functioning in the enclave of Cádiz. In 1812, it formulated a liberal Spanish constitution—albeit one that was repudiated by Fernando VII when, in 1814, he was restored as the Spanish monarch.

Meanwhile, the Portuguese royal family and court had moved to Rio de Janeiro in 1808. In 1815, the Prince Regent (his mother Queen Maria I was mentally ill) pronounced creation of the Kingdom of Portugal, Brazil, and the Algarves, with its capital in Rio. In 1816, he assumed the throne as King João VI after the death of the queen. Portugal itself was now a British protectorate, with a resident British ruler and a de facto colony of Brazil.

Then, in 1820, both Spain and Portugal experienced liberal revolutions that restored parliamentary process to Iberia. Fernando VII was forced to recognize the constitution, thereby paving the way for the ascent of a liberal monarchy when his daughter, Queen Isabella II, assumed the throne after her father's death in 1833.

In Portugal, the English were displaced and King João returned after having established a regency in Brazil for his son Prince Pedro. This was rejected by the parliament in Lisbon. In 1821, it abolished the Kingdom of Brazil and dispatched troops to bring it back under direct Portuguese authority. It also summoned Prince Pedro to Europe for possible grooming as the next monarch. This triggered hostilities in Brazil over the next two years.

In the event, whether ruled by absolute monarchs or liberal parliaments, both Spain and Portugal attempted to restore direct Iberian rule in the New World, and both initiatives triggered Creole resistance. In 1822, Prince Pedro pronounced creation of the Empire of Brazil with himself as its first emperor, Pedro I. It was to be a constitutional mon-

---

3. Juan Miguel Bákula Patiño, "Introducción" to Carlos Larrabure y Correa, *Colección de leyes, decretos, resoluciones y otros documentos referentes al Departamento de Loreto*, Vol. 1 (Iquitos: Monumenta Amazónica, CETA, 2006), 83.

archy with its own parliament (in 1889 a military coup converted Brazil into a republic with an elected president). While the Portuguese *cortes* disparaged this effort, and Portuguese troops in Brazil tried to destroy it, the hostilities were short-lived.

The independence movement in Hispanic America was particularly important for our story. In 1806, there was the first rumbling of rebellion in Venezuela against Spanish rule, and, by 1812, an exiled Venezuelan Creole (Basque-surnamed) Simón de Bolívar had assumed command. Compared to its North American counterpart, the South American Revolution was protracted and labyrinthine. In 1821, Bolívar became the first president of a newly created Gran Colombia (initially present-day Colombia and Venezuela). The following year, Ecuador gained its independence and formally joined, and it was then in Guayaquil that Bolívar met his counterpart, the southern South American liberator and fellow Basque descendant José de San Martín, to flesh out the continent's political future. Bolívar clearly had the upper hand, so San Martín had little choice but to defer and withdraw.[4] In 1825, Bolivia (named in Bolívar's honor) was established.

Shortly thereafter, Bolívar proposed holding the Congress of Panama to forge a united South America. He was the George Washington and Thomas Jefferson of South American independence rolled into one. He aspired to be not only a liberator, but also the unifier of a continent under an authoritative, albeit democratically elected, president for life (himself, to be sure). When it became apparent that such an ambitious scheme was unlikely to succeed, he proposed a more modest Federation of the Andes that would include the territories from Panama to Potosí that he had conquered. That initiative, too, proved abortive.

It had always been Spanish and Portuguese administrative policy to maintain direct control from Madrid and Lisbon over each of its New World settlements, while discouraging relations among them, as well as with foreign powers. Bolívar sought to create ex nihilo a common overarching "American" identity among the continent's Hispanic viceroyalties. He spent most of the decade of the 1820s doggedly pursuing this vision and therefore failed to define rigorously the internal boundaries among the future independent entities of Venezuela, Colombia, Ecuador, Peru, and Bolivia. The internal politics within Gran Colombia were stormy, and it ceased to exist in 1831, or the year after Bolívar's resignation as its president. Accordingly,

---

4. Marie Arana, *Bolívar: American Liberator* (New York: Simon & Schuster, 2013), 294–305.

No one knew more than Bolívar how imperfect the work had been. Independence had been achieved—enlightened forms of government considered—and yet the victors had emerged with no singleness of purpose, no spirit of collegiality. Warlords still wanted to rule their little fiefdoms, their undersized dreams a match for undersized abilities. It was as true in Bolivia as in Venezuela: notions of a larger union seemed pompous, foreign, vaguely threatening. The colonies were dead, but the colonial mentality was very much alive. The new republics were as insular and xenophobic as Spain had encouraged its American satellites to be. Venezuelans saw Peruvians as arrogant royalists. Coastal dwellers saw mountain dwellers as beknighted Indians. Southerners saw northerners as outlandish Negroes.[5]

So, with the collapse of Gran Colombia, the new South American nations of the continent's Hispanic realm faced an uncertain future without even the luxury of well-defined and mutually agreed territorial boundaries among themselves. Revolution had destroyed the viceroyalties that, in any event, were configured differently than their modern national successors, and the ecclesiastical jurisdictions of the old regime differed in some cases from the civil ones—thereby adding to the historical and geographical confusion.

Meanwhile, for centuries Portugal had expanded its influence in the basin from east to west, that is, from the Lower to the Upper Amazon (some Brazilian visionaries even believed that their country needed to have a Pacific port one day if it was to realize its foreordained destiny as a great world power), whereas, after the collapse of Gran Colombia, the nascent countries of Colombia, Ecuador, Peru, and Bolivia faced the challenges of defining not only their borders with Brazil to the east but also with one another on a north-south tangent.

Then, too, in 1823, an ascendent United States of America had issued the Monroe Doctrine that declared all of the Americas to be off-limits to aspiring European imperialists (while staking out the region for itself, of course). Nevertheless, in 1851, Great Britain signed a Treaty of Friendship, Commerce, and Navigation with Peru, one of the terms of which committed the Peruvians to cooperate with the British in the abolition of slave trafficking and the formulation of criminal laws prohibiting slavery within their territory.[6] Both France and Great Britain intervened at several times during the nineteenth century in the contest between His-

---

5. Ibid., 409.
6. Ibid., 96.

panic and Lusitanian interests in the Rio de la Plata region and were ultimately instrumental in brokering the creation of Uruguay as a buffer state. British, French, and German[7] migrants all settled in both the Hispanic and Lusitanian realms, and in particular Amazonia, where the foreigners carved out significant postions within shipping and commerce.

All of the above mentioned factors would influence the late nineteenth- and early twentieth-century developments in the Putumayo/Caquetá that are the subject of this study. Before proceeding, however, we should consider another historical influence of particular relevance. Our tale fundamentally regards the rationale behind the whites' treatment of Indians, itself a legacy of centuries of Iberian rule in the Americas. From the outset of Spain's Age of Exploration, or beginning with Columbus's first voyage in 1492, there had been a fundamental disagreement between some clergymen (particularly Bartolomé de las Casas) and the conquistadors over treatment of the Indians. Were they even human? If so, as Las Casas and many of his colleagues argued, then they should converted to Catholicism and civilized—an endeavor that would obviously influence the Europeans' treatment of them. The friars favored concentrating the indigenes into "reductions," or settlements, where they could be evangelized, educated in white ways and, above all, protected from the designs of European colonists. Conversely, if sub-human, then Indians could be enslaved and thereby come to constitute the critical labor force for the mining of precious metals and the establishment of commercial agriculture—plantations and ranches.

While the theologians might have prevailed technically when Pope Paul III issued his bull *Sublimus Dei* in 1537, declaring the Indians to be human beings with souls and excommunicating anyone guilty of enslaving them, in practical terms many secular whites in the Hispanic and Lusitanian New World spheres continued to regard the indigenes as sub-human and their territories targets of opportunity. The exploitation and occupation were justified in terms of a civilizing mission and patriotic commitment to the national destiny. Savages must be brought into the world capitalist system or perish; their "vacant" lands had to become a productive part of the national economy and polity. Consequently, the white colonists either ignored the papal bull or devised subterfuges to circumvent it. There developed a supposedly "humanitarian" exception that permitted raids against cannibalistic Indians, and their subsequent assignment to white overseers, in the interest of protection

---

7. Of the 161 Jesuits missionaries serving in the Mainas region (Amazonia), 19 percent were Germans. Ann L. Golob, "The Upper Amazon in Historical Perspective," PhD Diss. in Cultural Anthropology, City University of New York, 1982, 79

of the anthrophagi's potential victims.⁸ This, of course, was subject to interpretation and considerable abuse, not to mention the proliferation of supposed cannibals.

Throughout the expanding European empire in the New World, the elaboration of civilian and ecclesiastical authority went hand in hand. Catholicism encompassed the dual thrusts of the secular clerical hierarchy (often intimately entwined with the local civil administrators) and their designated diocesan jurisdictions, and the several religious orders—particularly Jesuits, Franciscans, Dominicans, and Augustinians—pursuing specific "missions" and reporting directly to the Vatican. As early as the sixteenth century, first the diocese of Cusco (1535–1546), then that of Quito (1546) and, finally, the bishopric of Popayán (1546) were charged with ministering to the vast area that included the Putumayo and Caquetá rivers and their hinterlands.⁹ The Audiencia of Quito continued to be concerned with creating missions in what would one day become Ecuador's Amazonian Oriente, primarily employing Jesuits in the endeavor.

Within three decades after the "discovery" of the New World, Christian unity within Europe was shattered by the Protestant Reformation. Spain's young monarch and Holy Roman emperor, Carlos V, became the champion of Catholicism. The Basque Iñigo de Loyola founded his new religious order, the Society of Jesus (1534—formally recognized in 1540 by Pope Pius III) to produce soldiers of God willing to serve anywhere in the propagation and defense of the faith after taking vows of poverty, chastity, and loyalty to the pope. The Society was committed to reforming the Catholic Church from within of its spiritual lassitude, venality, and corruption. Since much of the clergy of the day was poorly prepared, the Society insisted upon rigorous and lengthy formal training of its members. While not their stated purpose, the Jesuits quickly emerged as the phalanx of the Counter-Reformation, particularly since even in the sixteenth century they began founding universities throughout Christendom. In Catholic countries this, of course, gave the Jesuits considerable influence over the secular authorities (having educated many of them), while lending to the Society a certain elitist air of intellectual superiority.

From the outset, then, the Jesuits had their detractors and opponents both within and outside the Catholic Church. There was a certain cult-like

---

8. Ernani Silva Bruno, *História do Brasil, Geral e Regional*. Vol. I *Amazonia* (São Paulo: Ed. Cultrix, 1966), 48. Camilo A. Dominguez and Augusto Gómez, *Nación y etnias: Conflictos territoriales en la Amazonía Colombiana 1750–1933* (Bogotá: Disloque Editores, 1994), 42.
9. Augusto Javier Gómez López, *Putumayo: Indios, misiones, colonos, conflictos (1845-1970): Fragmentos para una historia de los procesos de incorporación de la frontera amazónica* (Bogotá: Imprenta Nacional, 1913), 98.

and secretive aura surrounding the order. Saint Ignatius insisted that the reform of the Church required the conversion of each individual's heart through participation in elaborate Spiritual Exercises (a four-week period of silence in which the individual meditates alone on the life of Christ and then discusses its meaning with a spiritual advisor). The experience—which can be repeated periodically—is also offered by the Society to interested laymen and has come to be referred to as a "retreat."[10] In modern terminology, then, each Jesuit and lay adherent was to be "born again."

While newcomers compared with the Catholic Church's other proselytizing religious orders, the Jesuits quickly became serious missionaries. Saint Ignatius of Loyola's fellow Basque and co-founder of the Society of Jesus, Saint Francis Xavier, became an active missionary in India and was engaged in an attempt to evangelize China when he died in 1552. A relative and contemporary of Saint Ignatius, (Basque-surnamed) José de Anchieta,[11] founded São Paolo and Rio de Janeiro in Brazil in the sixteenth century and came to be known as the "Xavier of the New World" and "Apostle of Brazil." He was a linguist and ethnographer as well, producing both descriptions and grammars of several indigenous languages and accounts of their cultures. In effect, he anticipated the Jesuit initiative in creating reductions—beginning in 1609 when (Basque-surnamed) Bishop Lizarraga requested missionaries for his diocese in Paraguay. In the Jesuit reduction the Indians were converted to Christianity but were allowed to preserve their languages and many of their customs. This, of course, required shielding them from the economic pressures of the European colonists and their governments. Over the next hundred and fifty years the Jesuits founded a continuous series of reductions in Paraguay, Argentina, and Brazil that, at their apogee, embraced 100,000 square kilometers and about 200,000 Indians—sometimes referred to as the Guaraní Republic.

In the Andes, while the indigenes had initially approached the whites out of both curiosity and a desire for metal tools, by the end of the sixteenth century there was considerable tension between the white *encomenderos* and various Indian tribes. The former sought to enslave the latter, employing roundups and an iron hand to enforce their will. The process had created factionalism amongst the indigenes, dividing them into those who submitted to Spanish authority and those who did not.[12]

---

10. Golob, "The Upper Amazon," 55–67. Heinrich Boehner, *The Jesuits: An Historical Study* (Philadelphia: The Castle Press, 1928).
11. The Basque Anchieta was actually born in the Canary Islands, son of a Gipuzkoan father who had moved there after being part of a failed coup in 1525 against King Charles.
12. Golob, "The Upper Amazon," 126–29.

The Jesuits prepared for their first missions in the Andes, based in the Audiencia de Quito, between 1602 and 1610. Most of the future missionaries there studied Quechua to be employed as the universal indigenous language for their prosyletizing in the lowlands. In 1610, the Reduction of the Loreto was created in present-day eastern Peru. Golob describes the process,

> The first step in founding a new village was to travel with some Christianized Indians until some infidels were discovered. A few infidels were to be brought back with the missionary who would teach them Quechua so that they could serve as interpreters for the missionary. He would give out a few gifts, and say a few words through the interpreter, and then leave. A few days later, it was expected that Indians would show up at the mission with presents for the Padre consisting of parrots, monkeys and other jungle goods. The Padre would be awaiting them with needles, knives and fishhooks. Through the interpreter the Padre would explain that if the Indians wanted to establish a Christian village no one would harm them. He would assure them that they would conserve their songs and dances, and that their liberty and material goods would be preserved. . . . Richter [superior of the Mainas missions] lamented that "if the Spaniards did not exploit the Indians so much they would all be Christian, therefore, in order to secure their confidence you must not take away their liberty nor expropriate their goods. With gifts you can obtain everything. . . ." Richter claimed that the Indians were so anxious to follow a Padre that all he must do is present the cacique with an axe, and the entire flock would follow him to the path of Christ. Richter's description omits the violence that was associated with taking the interpreters, moving Indians to a mission village, and converting them to Catholicism. His assumption was that the Indians would be impressed with the holiness and godliness of a Padre, and once they were convinced that they would not be exploited by the Spaniards, they would be willing to accept the authority of the missionary.[13]

---

13. Ibid., 132–33. Steel axes were in particular demand and, in indigenous worldview, equated to wealth and power. Furthermore, there was said to be the Spirit of the Axe that could be employed to cause cancerous tumors in an intended victim or even his death. Roberto Pineda Camacho, *Holocausto en el Amazonas: Una Historia social de la Casa Arana* (Bogotá: Planeta Colombiana Editorial, 2000), 65–66 [*Holocausto*]. Juan Álvaro Echeverri discusses the "philosophy of the axe" among the Andoque, Huitoto, and Muinane peoples in the Middle Putumayo/Caquetá. The Andoque even call themselves "the people of the axe." The myths concerning the axe go back in some cases to the pre-contact period of stone axes, in others to the acquisition of metal axes in the first contacts with whites.

In 1599, the Jívaros rebelled and drove the Spaniards out of several towns. In 1618, there was a military incursion into Amazonia led by Diego Vaca de Vega who established the settlement of San Francisco de Borja on the Marañón River in the territory of the Mainas Indians. It soon had more than 40 white-owned *encomiendas* and a weaving mill operated with the labor of coerced *indígenas* or indigenes. Treatment of them was so brutal and arbitrary that the mere mention of Spanish soldiers struck fear in Indian hearts. There were reports of suicide and infanticide to prevent themselves and their progeny from falling under Spanish governance.[14] Some fled farther into the forests; while others fought back. It became a tradition in Borja to assemble a well-armed contingent of whites and Christianized Indians to conduct an annual raid in search of heathens who had fled the settlement, as well as new ones for its work force—referred to as *correrías* (roundups). Borja and its hinterland became the first foothold of Jesuit proselytizing in Amazonia. Then, in 1635, one thousand Indians from several tribes struck Borja and killed at least nine *encomenderos* and their families. The following year a retaliatory expedition hunted down and hung many Indians for that rebellion.[15]

According to Golob,

> As settlement moved eastward from the Andes, Spaniards began to raid the lowlands for slaves to work on their encomiendas, in their fields, and mills. Raiding the lowlands began in the 16th century but was not restricted to that century; there are reports of Spanish raids in Mainas during the Jesuit tenure. . . . Laws were continually passed by the Viceroy in Lima to end these raids. In 1688, for example, an order was passed that no Spaniard could enter areas where the Jesuits were working. In 1701 it was decreed that highland settlers were prohibited from taking Indians from the Jesuit mission villages; violation of this order carried a fine of 2,000 pesos. In 1703 another order was passed that all Indians

---

Some of the stories are cautionary and regard the pathological properties and adverse consequences of axes for traditional power relations in indigenous societies; others celebrate the abundance offered through the far greater efficiency (with respect to the stone axe) introduced by the new technology in the felling of the forest. Juan Álvaro Echeverri, "The People of the Center of the World: A Study in Culture, History, and Orality in the Colombian Amazon" (PhD diss. New School for Social Research, 1997), 90–100. There is no question that the introduction of the metal axe (and other metal implements such as the machete, knives, fishhooks and sewing needles) proved nearly irresistible to many indigenes and would constitute the thin edge of the wedge of their conquest and subjugation by outsiders.

14. Ibid., 142.
15. Ibid., 140–46.

taken from Mainas were to be restored to their lands. These orders seem to have been to no avail, slave raiding continued.[16]

For their part, the Jesuits targeted young boys (often in sneak attacks against unconverted Indians) to be trained in their seminary or college in Borja both as interpreters and as impressionable future advocates of the new religion with their parental generation. Consequently, the interpreters themselves came to be viewed by many Indians as both lackies of the external authorities and subverters of indigenous ways.[17] Furthermore, despite Richter's reporting of a rather benign relationship between the Indians and the padres, in fact the Jesuits were concerned about their personal security (a few died at the hands of Indians). Golob states,

> The missionaries lived in fear of the Indians successfully overrunning them. They developed strong militias in the mission villages, and retained the use of the Spanish militia in Borja in order to keep the Indians in a state of fear.[18] Jesuit policy then was that any time a missionary was killed or Jesuit authority questioned, massive retaliatory attacks were organized by soldiers and fellow Jesuits to capture and hang the organizers of the uprising and to forcibly take captives to found new villages.[19]

While the missionaries collaborated with civil authorities and the Spanish military, they had their differences. The mission reduction was designed to be a self-contained, centralized settlement with its own degree of indigenous self-government, albeit ultimately under the authority of the padre. It was likely situated on the banks of a navigable river for ease of access and transport. From there the missionaries sent expeditionary parties into the interfluvial interior in search of new converts. The *encomenderos* and other Spanish interests were interested in an enslaved, or at least indentured, work force, disseminated among the *encomiendas*. While the Jesuits were far from tolerant at times in their treatment of their Indian wards, they were considerably less punitive and arbitrary than were their white civilian counterparts. Thus, some indigenes viewed the missionaries as protectors from the harsher alternative, and there emerged a distinction between Christian Indians and the remaining hostile or indifferent heathen ones.

The Portuguese raiders remained a threat throughout. Regarding the year 1709, (Basque-surnamed) Governor Iturbide reported that the Por-

---

16. Ibid., 133–34.
17. Ibid., 154–55, 189–90, 233–37.
18. Ibid., 164.
19. Ibid., 163.

tuguese had captured 5,040 Indians in the Mainas. In 1710, there was a new attack entailing no fewer than 1,500 Portuguese and 4,000 Indians. In their aftermath, the 38 reductions along the Amazonas and Napo rivers had been reduced to 5.[20]

In 1741, the governor of Mainas ordered the Jesuits to conduct new raids to bring in more converts for missions depleted by the substantial numbers of runaways. So it is clear that many of the Indians who came to the missions in search of utensils and temporary security had no intention of remaining permanently. Others no doubt changed their minds after experiencing white authority. In any event, chasing down runaways and searching for new converts sapped the energies and resources of the missions.[21]

We do not know with certainty how many mission villages the Jesuits founded in the Mainas over the period between 1636 and 1767. Golob places the total at about 60 but notes that there are definitional issues. Many were quite ephemeral and lasted but a few years. She also states that, during the sixteenth century, at any one time there were but 8 to 10 missionaries in Mainas, and throughout the entire period there were only 161 missionaries there in all. They were assisted by anonymous (in the documents) lay brothers, but still the number of Christian religious working in such a vast area was miniscule.

The Jesuits were entrenched mainly in the Borja hinterland, or along the Marañon and Amazonas rivers, including such tributaries as the Huallaga, Pastaza, Morona, and Ucayali. They also subsequently founded some missions along the Napo River.[22] However, there was also evidence that,

> ... in the 1630's a group of Encabellados Indians of the Napo River murdered a number of Spanish soldiers and their captain, Captain Palacio, who were accompanying two Franciscan friars through the region. The Franciscans escaped down the Napo and on to the Marañon. They made it all the way to Pará where they

---

20. Ibid., 205–206.
21. Ibid., 169–71.
22. Ibid., 148, 216. The Napo served as the link between the Andes and Amazon for Francisco de Orellana in 1540. He would become the first European to descend the entire Amazon River to its mouth. Twenty years later, two Basques, the expedition's commander, Pedro de Ursua, and his lieutenant, Lope de Aguirre, descended the the Upper Marañon in search of El Dorado. The latter mutinied and executed the former. Continuing down the Amazon and then along the northern South American coastline to present-day Venezuela. Aguirre became the first "American" to openly challenge the authority of an Iberian absolute monarch (Felipe II).

secured help from the Portuguese. The Portuguese returned to the Napo in 1638, enroute to Quito, and spent 11 months in the region, killing Indians and destroying stores of food in their attempt to find and punish the murderers of the Spanish soldiers.[23]

It should be remembered that Spain and Portugal were unified under the Spanish Crown for the sixty-year period between 1580 and 1640. So the Portuguese expeditionary force to the Napo would have enjoyed official sanction. It is interesting that the Jesuit chroniclers were prone to blame slave raiding in Mainas on the Portuguese, while remaining largely silent regarding the raids by Spanish *encomenderos* and the missionaries' close collaboration with the Spanish military.[24]

The Napo River parallels and is adjacent to the Putumayo River (or the central setting of our study). However, Golob provides no evidence that the Jesuits ever worked anywhere along it (her map displays no Jesuit missions there).[25] It should be noted, however, that missionary activity on the Napo would have impacted the interfluvial area between it and the Putumayo, including the "destruction of any overland routes of communication or trade that existed pre-colonially."[26]

Iquitos began as a Jesuit mission village, but was abandoned after the departure of its missionaries. Consequently, "With the expulsion of the Jesuits in 1767, the reductions collapsed and the Indians were forced to return to the forests."[27] On balance, the Jesuit reductions were largely unsuccessful. The missionaries struggled with the plethora of indigenous languages throughout Mainas and they "spent the largest part of their energies collecting and recollecting Indians and trying to force them to stay."[28]

It should also be noted that during the period the Indian population may have declined by as much as 90 percent through a combination of the unintentional introduction of European illnesses among the vulnerable Indians (exacerbated by concentrating them in the reductions), purposeful violent raids into their territories, and the abuse and enslavement of subjected tribes. There were also the adverse health consequences of dislocation as some tribes fled deeper into the forest fastnesses.[29] Bákula Patiño

---

23. Ibid., 175.
24. Ibid., 133–34, 147.
25. Ibid., 217.
26. Ibid.
27. Francesco Turvasi, *Giovanni Genocchi and the Indians of South America (1911–1913)* (Roma: Editrice Pontifica Università Gregoriana, 1988), xix.
28. Golob, "The Upper Amazon," 169.
29. Bákula Patiño, "Introducción," 29–30; 32; 35; 37; 38.

places the total of Jesuit reductions established in the Peruvian Amazonia at 130, whose indigene inhabitants numbered approximately 160,000 in the year 1700; by 1764 there remained but 30 with a combined population of 18,000. By 1848, these had declined further to 15 functioning villages with a total population of but 4,414 inhabitants.[30]

The Franciscans were the largest and longest-standing missionary corps in Northwest Amazon. As early as 1557, Friar Antonio Jurado had served as interpreter to an exploratory expedition into the Peruvian Huánuco region. He knew the language of the Chupachos and prepared a lexicon of it. By 1572, there were Franciscan missionaries established half a league from Sibundoy, gateway to the Upper Putumayo.[31] Then, in 1580, Franciscans branched out from Moyobamba and the Franciscan College in Huánuco to evangelize more than a dozen Indian tribes in the Upper Putumayo/Caquetá. This activity intensified throughout the seventeenth and early eighteenth centuries.[32] Indeed, in 1725, the Franciscans established the Colegio de Ocopa, whose sole purpose was to prepare missionaries for work among the pagan Indians of the Peruvian Amazon. It would be staffed almost exclusively throughout its existence by Franciscan priests and laymen from Spain. During the nearly two centuries from the founding until it ceased being a mission college in 1907,[33] there were 313 priests and 153 Franciscan lay brothers at the Colegio de Ocopa. Seventy-six of these would die while assigned to forest missions—most killed by hostile Indians.[34]

During the second half of the sixteenth century, and then the early seventeenth, several Spanish captains general and missionaries explored the Putumayo and even settled its upper and lower regions.[35] We know that by the late seventeenth century there was an *encomendero* in the Sibundoy Valley, the headwaters of the Putumayo River, who

---

30. Ibid., 37; 43.
31. P. José Restrepo López, *El Putumayo en el tiempo y el espacio* (Bogotá: Centro Editorial Bochica, 1985), 13.
32. Fr. Félix Sáiz Díez OFM, *Centenario de la Creación de las Prefecturas Apostólicas en el Perú. Después de tres siglos de Acción Misional en el Oriente Peruano* (Lima: Editorial Sin Fronteras, 2000), 17–21.
33. In 1824 Ocopa was converted into a college of education by virtue of a decree from Simón Bolívar. However, in 1836, it was restored to its original evangelical mission and returned to the Franciscans (Bákula Patiño, "Introducción," 25).
34. Ibid., 21, 26, 34.
35. Hector Llanos Vargas and Roberto Pineda Camacho, *Etnohistoria del Gran Caquetá* (Bogotá: Fundacion de Investigaciones Arqueológicas Nacionales, Banco de la República, 1982), 19–21. Cornelio Hispano, *De París al Amazonas. Las fieras del Putumayo* (Paris: Librería Paul Ollendorff, n.d. [1913]), 246–47.

received tribute from the Sibundoy Indian tribe. Many of the latter, by the beginning of the eighteenth century, secured title to their lands that were recognized by the government, but lost them through fraud to intending white settlers by century's end. The whites believed, erroneously as it turned out, that they had destroyed all of the documents supporting the Indians' rights to the rich valley.[36]

In 1739, the Franciscans had seven mission reductions in the Upper Putumayo. Nevertheless, they were under attack from rebellious Indians,[37] and, by 1791, the missionaries were forced to abandon their settlements.[38] In the early nineteenth century, there were a couple of failed attempts by a few hardy Franciscans to refound the missions in the Upper Putumayo;[39] but there were simply none in the middle stretches of the river. It would seem possible, however, that Franciscan missionaries passed through the Middle Putumayo at one time or another. In 1887, a Concordat was signed between the Holy See and Colombia that again made Catholicism the official religion and restored to it state subsidies. On December 16, 1890, Congress passed a law authorizing, "the government to, in accord with the ecclesiastical authority, proceed to organize missions to reduce to civilized life the savage tribes that inhabit Colombian territory bathed by the Caquetá and Amazon rivers and their tributaries."[40]

By the late nineteenth century, Capuchin missionaries were granted sway over the entire Colombian Putumayo/Caquetá. There were several exploratory trips by Capuchin friars to the Middle Putumayo, including the future Aranalandia, in the first decade of the twentieth century, but they failed to establish any kind of residence (mission) in the area. De facto, their activities were based out of Sibundoy on the Alto Putumayo and extended but a relatively short distance downstream.[41] In short, from an evangelistic point of view, by the nineteenth century the Middle Putumayo was truly the river that God had forgotten.

For the Portuguese, the Putumayo/Caquetá/Napo region was extremely distant from any settlement of importance and was therefore

---

36. Victor Daniel Bonilla, *Servants of God or Masters of Men? The Story of the Capuchin Mission in Amazonia* (Middlesex: Penguin Books, 1972), 22–32.
37. Michael Edward Stanfield, *Red Rubber, Bleeding Trees: Violence, Slavery, and Empire in Northwest Amazonia, 1850–1933* (Albuquerque: University of New Mexico Press, 1998), 100.
38. Ibid., 102.
39. *Misiones católicas del Putumayo: Documentos oficiales relativos a esta Comisaria* (Bogotá: Imprenta Nacional, 1913), 57.
40. Misrael Kuan Bahamón, S. J., "La misión capuchina en el Caquetá y el Putumayo 1893–1929." MA thesis in history, Pontífica Universidad Javeriana, 2013, 75.
41. Ibid., 46–48.

solely the object of occasional incursions by Brazilian slavers and/or raiders seeking Indians or in search of booty. However, this creation of the Captaincy General São Jose del Río Negro, in 1755, intensified contacts between the Portuguese and Indians in the region—probably introducing trade goods (particularly steel axes) among the tribes while causing some of them to flee further into the forest to escape marauders.[42]

Despite the fact that Spain and Portugal had signed the Treaty of San Ildefonso in 1777, whereby Brazil recognized that the Putumayo was Spanish, a 1783 account suggests that Brazilian slavers continued to enter the area to ravage the countryside. Clearly, there were some established white frontiersmen present there as well who were also enslaving Indians—with the collusion of Spanish officials.[43] The Bavarian naturalist, Carl Friedrich Philipp von Martius, reported, during his scientific expedition in the Amazon (1817–1820), visiting a small Yuri Indian settlement above the Cupatí rapids on the Lower Caquetá that traded slaves for goods with Brazilian *mamelucos*,[44] "serving as the point of the lance for the capture and deportation of Indians toward Brazil, coming from the Upper Caquetá."[45]

Colombia was more active than Peru in the Putumayo/Caquetá. Beginning about 1835, merchants from Pasto had begun transporting shoes, cigars, and varnishes down the Putumayo, to be sold eventually in Manaos and Pará before returning to Colombia with Brazilian and European goods.[46] There is also ample documentation that Pasto was the destination of considerable Brazilian contraband traffic.[47] Indeed, Pasto, the Department of Cauca and its capital city of Popayán will all constitute critical geographical points of reference in our account as the key source of Colombian influence and settlement in the Putumayo/Caquetá. They also occupy a unique position within Colombian history. During the War of Independence, the Cauca was a contested liminal zone between the future Ecuador and Colombia, and hence between Bolívar's and San Martín's spheres of influence. Pasto remained a royalist city described and bemoaned by General Santander in a missive sent to Bolívar, his commander-in-chief,

---

42. Pineda Camacho, *Holocausto*, 26.
43. Gómez López, *Putumayo*, 168.
44. Brazilian term for persons of mixed racial descent equivalent to *mestizos* in the Hispanic world. There is also the Brazilian term *caboclo* that refers to the Amazonian person of mixed Indian and some other ethnic descent.
45. Jon Landaburu and Roberto Pineda Camacho, *Tradiciones de la gente del hacha: Mitología de los indios andoques del Amazonas* (Bogotá: UNESCO, 1984), 29 [*Tradiciones*].
46. Gómez López, *Putumayo*, 157.
47. Landaburu and Pineda, *Tradiciones*, 13.

I hardly know what policy to follow; they are terrible peoples, the wall against which all our efforts since 1811 have always, always come to grief. They know admirably how to wage a partisan war. I am going to give orders for the chief leaders, rich, noble or plebeian, to be hanged in Pasto, and the rest of the population to be transported away to Venezuela, so that no one is left but the women and children, who can do no harm to us for the present and may change their minds.[48]

Even so, according to one chronicler, Pasto held out for more than two years against overwhelming odds,

Pasto refused to surrender and fought on, but at last was stormed, and writes O'Leary, "in the horrible massacre that ensued, soldiers and civilians, men and women, were sacrificed promiscuously".... Bolívar himself arrived in Pasto on January 2nd, 1823. He was inexorable and cruel. He confiscated all estates and gave them to his officers, drafted all the men into the army and sent them away, and left Salom in charge with orders to imprison and send to Quito and Guayaquil all who had not been militarized. Many committed suicide.[49]

During the 1830s and 1840s, the Colombian prefect of the Caquetá was appointing *corregidores* (overseers) to Indian villages in the Upper Putumayo/Caquetá. Nevertheless, there were also reports that Brazilians were still ascending the Putumayo in the mid-nineteenth century to pit enemy Indian tribes against one another in order to purchase war captives. Other Brazilian slavers were ascending the Caquetá to the rapids at Araracuara to buy Indians.[50]

By one account, in 1860, hundreds of *indiecitos*, or Indian children, were being shipped down river from the mouth of the Cahuinarí (where it met the Caquetá) for sale in Brazil. The region up that river was being governed by a Brazilian who claimed to have been commissioned "by the authorities."[51] Colombia's administrative initiative in the Putumayo/

---

48. Salvador de Madariaga, *Bolívar* (New York: Pellegrini & Cudahy, 1952), 417.
49. Ibid., 446. In 1860, the Federal State of Cauca declared its independence from Colombia with Popayán as seat of the triumphant phalanx in a civil war against the government in Bogotá. Colombian journalist and essayist Juan Lozano y Lozano would subsequently describe Popayán as, "a democracy of the Athenian type of Pericles. That is to say, a democracy of aristocrats who exist and philosophize at the expense of plebeians without rights." Juan Lozano y Lozano, *Ensayos críticos y mis contemporáneos* (Bogotá: Biblioteca Colombiana de Cultura, 1978), 137 [*Ensayos*].
50. Ibid., 170–72.
51. Lozano y Lozano, *Ensayos*, 170–72.

Caquetá was interrupted by its Federal War of 1860–1863, a conflict that afforded the opportunity for Brazilian slavers to again penetrate the region.[52] In 1870, thirty to forty Brazilian canoes were still coming to the Cahuinarí annually to transport Indian slaves to Brazil. As late as 1884, the narrator witnessed a Brazilian vessel in the Iparaguaraná on a slaving mission.[53]

Indeed, the practice seems to have persisted until the 1890s. Dominguez and Gómez cite a protest in 1896 by the Colombian Ministry of Foreign Affairs to the Brazilian government regarding the activities of Brazilian nationals in the Caquetá borderlands between the two countries. To wit,

> They enslave those unfortunates [Indians] who, whether through deception or manhunts, with shotgun and dogs, they manage to imprison. They debase the women, separate them from their children, in order to later sell them as slaves; cases of the shameful mutilation as well of grown men and boys younger than ten not being rare. Faced with such calumnies, committed on Colombian territory, against Colombians that the government has the duty to defend and protect, it is nothing less than plundering that those same traffickers commit in the forests of our Putumayo and Caquetá, producing great devastation of the principal wealth that we have there.[54]

Nevertheless, Colombian *caucheros* (i.e., rubber entrepreneurs) were equally guilty regarding their treatment of Peruvian Indians and would surely do the same where they were now settling in the Putumayo if the Colombian government failed to protect its Indians there. That, however, would be quite difficult as the Putumayo flows into Brazil and the border between the two countries had yet to be agreed. Then, too, the Putumayo was too far from Manaos for the Colombian consul there to impede the crimes that might be committed.[55]

Meanwhile, in 1853, a prominent North American explorer, Matthew Fontaine Maury, published his book *The Amazon, and the Atlantic Slopes of South America* in which he described the boundless wealth in Northwest Amazonia, while noting that the region's sovereign nations were incapable of developing its economy properly.[56] The work, which was

---

52. Stanfield, *Red Rubber*, 71.
53. Cited in Gómez López, *Putumayo*, 39–42.
54. Quoted in Camilo A. Dominguez and Augusto Gómez, *La economía extractiva en la amazonía colombiana* (Bogotá: Tropenbos-Colombia, 1990), 174.
55. Ibid., 174–75.
56. Matthew Fontaine Maury, *The Amazon, and the Atlantic Slopes of South America: A*

highly influential in the United States, called for the area to be opened up to the world's industrialized economies. Brazilians were disturbed and urged their neighbors to exert greater control over their Amazonian lowlands.[57]

Henceforth, political hegemony of parts of Northwest Amazonia, their respective Amazonian lowlands, or "Orientes" as they were called, would be disputed by Bolivia, Colombia, Ecuador, and Peru. During the initial phases of this territorial struggle the stakes were low. Indeed, the region seemed a world away from all of the national capitals and supposed a dubious drain upon national treasuries. It lacked obvious mineral wealth and, while the agricultural frontier was expanding, the Putumayo/Caquetá remained too distant from national (let alone international) markets for its agriculture to be anything more than of the subsistence variety.

The year 1850 is frequently cited as the starting point of two partly overlapping tropical products' booms that would transform the economy of the Amazon Basin. Reference is to a pronounced world demand for quinine, on the one hand, and rubber or latex, on the other.

The curative powers for malaria of quinine, the extract from the several varieties of the cinchona tree, were known to the Quechuas and transferred by Spaniards to Europe as early as the seventeenth century. However, it was nineteenth-century European imperial expansion across the planet, impeded in part by the whites' susceptibility to tropical diseases—particularly yellow fever and malaria—that produced a huge demand for the Amazon's cinchona. By 1849, it was being gathered and exported in quantity, particularly in the eastern mountaine regions of Colombia, Peru, Ecuador, and Bolivia.

In 1860, a British expedition led by Clements Markham smuggled cinchona seeds and plants out of South America that were then propagated at Kew Gardens in London and seeded in Burma, India, and Sri Lanka. By 1883, there were 64,000 acres of cinchona under cultivation in Sri Lanka alone, and its 1886 export of the product reached 15 million pounds. The gathering of wild cinchona bark, which often entailed destruction of the trees, proved incapable of competing with its cultivation under controlled plantation conditions.

"Rubber" is a generic term for a number of tropical trees and vines growing primarily in the Amazon and Congo river basins, capable of producting latex. In South America, while some of its properties were known to the Indians in pre-Colombian times—latex was used to make the balls

---

Series of Letters Published in the National Intelligencer and Union Newspapers, under the Signature of "Inca" (Washington, D.C.: Franck Taylor, 1853).
57. Dominguez and Gómez, *La economía*, 115.

for certain games, as water-proofing material in their (scant) clothing and footgear, and as a slow-burning material for torches—it was not a key contributor to the quality of native life. One of its limiting characteristics was rubber's sensitivity to air and cold which made it hard and brittle. This obviously reduced its attraction to Europeans as well.

However, in 1839, Charles Goodyear discovered the process whereby adding sulphur to raw latex and heating it to a high temperature made rubber permanently flexible and stable. "Vulcanized" rubber therefore entered the Industrial Revolution as a critical raw material, its consumption in industrial economies exploding with the invention of the bicycle and automobile (tires) and as coating for electrical wiring. It was employed as well in gaskets, in tubing, and as the buffer between railway cars. It also became a key component in the manufacture of waterproof garments and footgear. Demand quickly outstripped supply.

Collection of rubber in South America began near Belem (where there were as many as 25,000 gatherers by the 1850s)[58] and then spread throughout the Brazilian Amazon, and ultimately into the tropical lowlands of the adjacent countries of Bolivia, Peru, Ecuador, and Colombia. There was also an industry in the drainages of the Orinoco (Venezuela) and the Esequibo (Guyana).[59]

The shortage of labor for the rubber industry was a perennial issue. Both the Amazon and Congo basins were host to some of the planet's worst diseases. Caucasians, in particular, had low resistance and experienced high mortality rates there. The conventional wisdom of the day

---

58. Warren Dean, *Brazil and the Struggle for Rubber: A Study in Environmental History* (Cambridge, New York, New Rochelle, Melbourne, Sydney: Cambridge University Press, 1987), 11.

59. While we shall consider its relevant highlights, a comprehensive history of the rubber industry is beyond our scope. See John Tully, *The Devil's Milk: A Social History of Rubber* (New York: Monthly Review Press, 2011). For an overview of the South American case, there is the iconic work by Barbara Weinstein, *The Amazon Rubber Boom 1850–1920* (Stanford, CA: Stanford University Press, 1983). The book by Warren Dean, cited in the previous note, tells the Brazilian story. Guido Pennano's *La economía del caucho* (Iquitos: Centro de Estudios Teológicos de la Amazonía, 1988) [*La economía*] and José A. Flores Marín, *La explotación del caucho en el Perú* (Lima: Concejo Nacional de Ciencia y Tecnología—CONCYTEC, 1987) do the same for Peruvian rubber history. For similar treatment of the Colombian story, see Camilo Domínguez and Augusto Gómez, *La economía extractive en la Amazonia colombiana, 1850–1930* (Bogotá: Tropenbos-Colombia/Araracuara: Corporación colombiana para la Amazonia, 1990), 79–109. Finally, for extensive discussion of rubber collection in the Loreto against the backdrop of its history throughout the Amazon, see Fernando Santos-Granero and Frederica Barclay, *Tamed Frontiers: Economy, Society and Civil Rights in Upper Amazonia* (Boulder, Colorado: Westview Press, 2000), 22–33.

questioned whether whites were even capable of manual labor under tropical conditions. In South America, there was a mixed-race population (whites and Indians, whites and blacks, blacks and Indians, etc.) constituting the rural *peón* laboring class. Even when recruitment of them was extended beyond the Amazon basin, the available labor was far from adequate. The collective terms of reference in Spanish for these people were *blancos* (despite their many hues) and civilized *racionales* ("reasoners") as opposed to the barbarous and sinister sub-human *indios* or *salvajes* ("savages"). In the Congo there was no such *mestizo* population of any note.

Each of the South American rubber-producing countries would experience its own boom until World War I, at which time wild rubber could no longer compete with the product emanating from plantations in Southeast Asia.[60] It was a *déjà vu*–like repetition of the quinine story. In 1876, a British explorer named Henry Wickham, had transported seventy thousand seeds of *Hevea brasiliensis*, Brazil's best rubber-producing tree, to England, where they were germinated at Kew Gardens to become the seedlings dispersed to places like Sri Lanka, Malaysia, and Indonesia.[61]

Throughout this vast Amazonian expanse there was a complete lack of roads and railways. While the possibility of building both would be vetted regularly by the national governments in question during the second half of the nineteenth century, in point of fact little was constructed until the early twentieth century.[62] That left the extensive fluvial network provided by the Amazon River and its many navigable tributaries, all of which flowed to the Atlantic. In the cases of the four Andean countries, their Amazonian regions were closer to Europe and North America than to the national capital. Travelers to the latter from their respective Orientes were better advised to descend to Belem/Pará, and then voyage either around Cape Horn or to Panama (for a land crossing of the Isthmus before proceeding by sea), rather than to undertake the month-long arduous trek across the Andes.

---

60. Wade Davis, *One River: Explorations and Discoveries in the Amazon Rain Forest* (New York: Simon and Schuster, 1996), 302–307.

61. Wild rubber "groves" averaged about two trees per hectare, whereas plantations supported 350. The cost of rubber production on the latter was less than half than on the former. Roberto Santos, *História economica da Amazonia (1800–1920)* (São Paulo: T. A. Queiroz, 1980), 234. See Dean, *Brazil*, 7–35 for the best English-language account of this story.

62. The first significant initiative would be Brazil's Madeira-Mamoré railroad line (1907–1912) that came to be known as "the Devil's Railway," given the thousands of fatalities during its construction. Indeed, implementation of the first land links in many parts of Amazonia is still a work in progress. And, of course, air travel, which is critical today, was science fiction at the time.

By the mid-nineteenth century, or even before the rubber boom, Peru recognized that river transportation was fundamental to the economic development of its Oriente. In 1841, the Peruvian government negotiated with Brazil regarding freedom of navigation on each other's rivers, an arrangement that was further refined in 1851[63] when Brazil effectively relinquished any political claim to the Putumayo/Caquetá by agreeing in principle upon its modern border with Peru. Brazil retained control over the Lower Putumayo, renamed the Iça River upon entering its territory. That meant that any river traffic between Iquitos and the Putumayo had to pass through Brazilian territory.[64]

Ecuador was initially energetic in attempting to exert direct political hegemony over its Oriente. Indeed, in 1853, when Peru established both political and military administrations in a newly created Department of Loreto, Ecuador protested that this constituted infringement upon its national sovereignty. Ecuador then declared navigation on the entire length of the Putumayo and Napo rivers to be open to anyone, while at the same time devising plans to colonize the region.

In the same year, the Peruvians conceded a monopoly on river traffic to Nauta to the Brazilian *Compañía de Navegación del Amazonas*. That firm was to receive for five years an annual subsidy of 20,000 pesos and have a first right of refusal regarding initiatives on other river systems in the Peruvian Oriente. In return the *Compañia* would support exploration of it, found ten colonies of foreigners (whose nationality would be designated by Peru), publicize the attractions of the region's opportunities for intending immigrants, and carry the mails and certain government personnel gratis.[65]

In 1857, Ecuador tried to cancel part of its national debt by ceding control of a vast tract of its Oriente to its English creditors—a move that prompted Peru to break off diplomatic relations with its neighbor. However, between 1854 and 1861, Ecuador experienced civil war during which an opportunistic Peruvian Army occupied Guayaquil. A condition of its withdrawal was Ecuador's recognition of Peruvian sov-

---

63. Carlos Larrabure y Correa, *Colección de leyes, decretos, resoluciones y otros documentos oficiales referentes al Departamento de Loreto* (Lima: Imprenta de "La Opinión Nacional," 1905), vol. 2, 18–23, Colección II.
64. If, over time, the effects of that arrangement would lead on occasion to some misunderstanding and tension beween the two countries, genuine confrontation between them was over. In fact, there was only one last military conflict between Northwest Amazonia's Lusitanian and Hispanic worlds, the so-called Acre War (1899–1903) between Brazil and Bolivia that resulted in the latter's defeat and concession to the former of sovereignty over the contested Acre River drainage.
65. Ibid., 29–40.

ereignty over all navigable rivers (including the Putumayo) flowing to the Amazon.[66]

Then, in 1861, Peru authorized a "maritime military department" (naval base) near the confluence of the Ucayali and Marañon rivers[67] and commissioned construction of two vessels in England to provide the country with a fluvial link with Pará. At the same time, it mandated construction of a ship building facility in Iquitos.[68] The contract with the Brazilian company was annulled as of 1861, and, in 1862, six Peruvian vessels were under construction in England named the *Ucayali, Mocotra, Napo, Putumayo, Yapurá,* and *Yavarí,* after the major river systems draining the areas claimed by Peru in its Oriente. The first two vessels were designed for commercial river traffic, whereas the smaller *Napo* and *Putumayo* were intended for exploration in the Oriente (the last two were for service on Lake Titicaca).[69] In 1863, it was declared that all of the new boats in the Oriente would link the Loreto directly with Pará as well.[70]

If Iquitos had but eighteen inhabitants in 1814, and still remained a sleepy Indian *ranchería* in 1861, by 1864 it berthed several steamships engaged in river trade throughout Upper Amazonia. Thenceforth, the growth of Iquitos would be spectacular, converting it into the hub of not only the Loreto but also of all of northwestern Upper Amazonia's commerce. It also became the Peruvian counterweight to Pasto/Popayán in the Department of Cauca with respect to control of the Putumayo/Caquetá.

The second half of the nineteenth century was a period of river exploration in Peru's Oriente carried out by Peruvian authorities and missionaries, as well as foreigners. In 1867, Peru established for this purpose, the Hydrographic Commission of the Amazon under the aegis of North American admiral John Tucker. It would organize and patronize systematic exploration of this vast fluvial domain.[71]

In 1874, there is documentation in Bogotá archives expressing a concern. Colombian citizens were spreading throughout the area, and indeed a number of them were collecting both quinine and rubber on the Napo River on "concessions" issued by Ecuador. Meanwhile, Peruvians were

---

66. Stanfield, *Red Rubber*, 75.
67. Carlos Larrabure y Correa, *Colección de leyes, decretos, resoluciones y otros documentos oficiales referentes al Departamento de Loreto* (Lima: Imprenta de "La Opinión Nacional," 1905), vol. 1, 23–24 [*Colección I*].
68. Larrabure y Correa, *Colección II*, 69.
69. Ibid., 70.
70. Ibid., 73-74.
71. Pilar García Jordán, "El oriente peruano territorio de confrontación social, económica, ideológica y política, 1821–1930," in Clara Inés García (ed.), *Fronteras, Territorios y Metáforas* (Medellín: Hombre Nuevo Editores, 2003), 232.

penetrating everywhere, and so were Europeans in search of rubber and other business opportunities. It would be only a matter of time before one or more of the foreign powers themselves began coveting outright political hegemony in the region. It was urgent that Colombia defend actively its territorial claims.[72] That same year, Peru and Brazil signed an agreement in Lima fixing their common border along the Yapurá River.[73] In 1876, Peru and Brazil negotiated yet another treaty regarding free navigation of both countries on the Putumayo/Iça.[74] In 1882, Peru established the customs house for its Oriente in Iquitos,[75] with a branch on the Cotuhé River to administer traffic on the Putumayo.

Peru suffered a disastrous defeat in its Pacific War (1879–1883) with Chile. This loss of a piece of the southern part of the country further focused national attention upon the Loreto to the east. Bolivia had sided with Peru in the conflict and lost its access to the Pacific over it. This placed an even higher premium upon development of its own Oriente as the now-landlocked country's critical window upon the Amazonian outlet to the Atlantic and beyond. In sum, by this time the Orientes of all four Andean countries were more oriented towards the Amazonian outlet to the Atlantic as their window upon the wider world.

By 1888, Peru was constructing two gunboats in England designed to protect its territorial claims in the Oriente.[76] In 1890, the Peruvian parliament was asked to approve a monopoly and subsidy for the British navigation company that would inagurate direct service between Liverpool and Iquitos. It would be paid an annual fee of 12,000 soles and an additional 2,000 soles for each voyage.[77] Then, in 1891, Peru and Brazil renegotiated their treaty ensuring free navigation on the Yavarí, a major river that served as a lengthy stretch of the frontier between Peru and Brazil (whose watershed includes the Putumayo). It established a single fixed tariff of 10 percent of the value of rubber exported from the region and 7 percent of all other imports and exports to be collected by the Brazilian customs house at Tabatinga.[78]

The first rubber shipment from Iquitos was in 1887. Political power would accompany economic clout. Until 1897, salubrious Moyobamba in the Peruvian montane and dynamic Iquitos were contesting political con-

---

72. Dominguez and Gómez, *La economía*, 63.
73. Larrabure y Correa, *Colección I*, 100–102.
74. Larrabure y Correa, *Colección II*, 87–90.
75. Larrabure y Correa, *Colección I*, 255–57.
76. Larrabure y Correa, *Colección II*, 98–99.
77. Ibid., 104–105.
78. Ibid., 105–15.

trol of the Loreto. However, in November of that year, the balance tipped in favor of the port town. While Moyobamba remained the nominal capital, the governing council (*Junta Departamental*) and its fiscal agents were all moved to Iquitos—making it the de facto operational nerve center of the entire region.[79] Rambunctious Iquitos had emerged as the second most important port city in Upper Amazonia after Manaos.[80] It was also the focus of Loretan regionalist sentiment, fanned by the belief that Lima ill-served this remote part of the country.[81]

The first years of the 1900s were the zenith of Loretan maritime and fluvial traffic. By 1900, the Booth Steamship & Co. line was providing direct service between Iquitos and Liverpool via Barbados. Then, that same year, the Red Cross Steamship & Co. announced in New York direct service to Iquitos via Panama, and a French company initiated service from Le Havre. In 1901, the Booth Steamship people contracted to construct a wharf and customs warehouse in Iquitos. Meanwhile, local shipping companies inaugurated service on several of the Oriente's river systems. More than one plied the routes between Iquitos and Yurimaguas, and from the headwaters of the Napo River to Leticia.[82]

While both Colombia and Peru would remain wary of Brazil, the real dispute over political sovereignty of the Middle Putumayo would eventually come down to the two Hispanic countries. If they had signed a treaty in 1829 recognizing Colombia's sovereignty over the Putumayo, by mid-century the Peruvians were ignoring it (even arguing that its terms had

---

79. Stanfield, *Red Rubber*, 94.
80. Roger Rumrrill and Pierre de Zutter, *Amazonia y capitalismo: Los condenados de la selva* (Lima: Editorial Horizonte, 1976), 19–21. Iquitos currently has a population of about 450,000 inhabitants.
81. In 1888, Peru appointed a commission to study the Loreto and organize the administration of it. The instructions provided to the commissioners were both thorough and enlightened. For example, they stated,

> ... there will be places and circumstances in which the isolation of the place or the nature of its exigencies or the matters that have to be resolved make it more necessary to depend upon the chief of a tribe rather than the mayor of a municipality, or a priest rather than a department prefect; but in the cardinal propositions like the establishment of a border guard, customs, the collection of taxes, the use or repository of funds, the means of communication, the designation of places for road construction, and that which regards the military defense of the territory with its police in communication with those of bordering departments; and in sum with all that concerns the system of administration that one wishes to reform or create, agreement with the departmental authority is necessary, subject to governmental appeal or revision in the cases of resistance, doubt or the importance of the matter under consideration. Larrabure y Correa, *Colección, II,* 269.

Unfortunately, these remained pretty words rather than the formula for real action.
82. Bákula Patiño, "Introducción," 106.

"expired") while seeking aggrandizement at Colombian expense within Northwest Amazonia by negotiating recognition of its new territorial claims by Brazil and Ecuador. On occasion, Peru even sent military forces into the Putumayo/Caquetá and, in 1867, its negotiator of the resulting disputes with Colombia, (Basque-surnamed) Barrenechea, could be quoted as saying, "I desire that there transpires in Peru an epoch in which it is believed that the art of diplomacy consists in obtaining for itself through trickery the greatest advantages at the expense of the other party [Colombia] to the agreement."[83] Nevertheless, the real contest between the two countries would be a *fin-de-siècle* phenomenon. As late as 1887, influential Peruvian author and academic Carlos Lissón listed and discussed all of Peru's border disputes (Ecuador, Brazil, Bolivia, and Chile), while failing to even mention Colombia.[84]

According to Stanfield,

> ... governments used caucheros and local functionaries to further territorial ambitions, and, symbiotically, caucheros cloaked their economic motives in the guise of nationalism. Much of the violence in the Putumayo then was not the result of a lack of government regulation—for Colombia, Ecuador, and Peru maintained Amazonian administrations after 1850—but rather of governments consciously employing violence as a mechanism for territorial aggrandizement and economic gain.[85]

In sum, during four centuries of Luso-Hispanic colonization, the European civilian and ecclesiastical presence in the Middle Putumayo/Caquetá, while not entirely absent, had been episodic and ephemeral. Indeed, by the late nineteenth century, the area languished beyond the pale of effective state and Church control. Brazilian traders and slavers were now gone. Peru, Ecuador, and Colombia were locked in an acrimonious

---

83. Hispano, *De París*, 224.

84. Carlos Lissón, *Breves apuntes sobre la sociología del Perú en 1886* (Lima: Imp. y Librerías de Benito Gil, 1887), 80–88 [*Breves apuntes*].

85. Stanfield, *Red Rubber*, 64. It has recently been argued that it was in the interests of the national governments to allow such "no man lands" to remain largely lawless and under the sway of "gangs" like the Casa Arana ruffians. The combination of exerting through them a degree of border security at practically no cost while having a refuge in which to warehouse prisoners and malcontents, not to mention the income from the extractive industries, illicit substances, contraband, and bribes, were all incentives for both Lima and Bogotá to keep hands off their borderlands—let alone take a hard look at the atrocities and enslavement of barbarous indigenes. Margarita Serge, "Fronteras carcelarias: Violencia y civilización en los territories salvajes y tierras de nadie en Colombia," in Clara Inés García (ed.), *Fronteras, Territorio y Metáforas* (Medellín: Hombre Nuevo Editores, 2003), 189–97.

dispute regarding political sovereignty of the region, and there was no longer a presence there of any country's secular Catholic clergymen or the Vatican's missionary orders. Yet the region, endowed with both rubber and tens of thousands of potential indigenous gatherers, remained a near-perfect political vacuum awaiting its entrepreneurial exploiters.

Regarding such, I would argue that, its physical isolation notwithstanding, the Middle Putumayo/Caquetá stood center stage regarding a plethora of historical contestations of a conceptual nature. It was one of the last places in which the frontiers between Hispanic and Portuguese hegemony were finally determined.[86] In that regard, Brazil's relinquishment of any claim to the Middle Putumayo/Caquetá largely represented the termination of the global tug of war[87] between the two Iberian powers of Spain and Portugal, which had been set in motion by the papal division of the world between them through the Treaty of Tordesillas of 1494.

Although there were local variations of the system, the manner of recruiting *peón* labor for the rubber industry was similar throughout the Upper Amazon. *Mestizos* became the subjects of *enganche* whereby river traders (called *aviadores* in Brazil, Ecuador, Colombia, and Peru[88] and *habilitados* in Bolivia)[89] recruited, settled, and provisioned the impoverished *peón* manual laborers, or *caucheros*, on river banks throughout the rubber districts. The merchants provided them everything from tools, firearms, ammunition, and clothing to foodstuffs and alcohol (particularly alcohol).

The *caucheros* (some as loners and others with families) established routes (*estradas*) into the forest for the collection and processing of the latex. The trees could be felled or bled so severely as to kill them—a practice employed extensively in the foothills of the Andes. However, in the lower elevations there was considerable permanency to the arrangement. The

---

86. This is not to say that they are entirely uncontested at present or will necessarily be permanent in the future. Border disputes among South American nations continue to flare periodically two centuries after their independence movement in the early nineteenth century.
87. Douglass, *Basque Explorers*.
88. The word *aviador* comes from Brazilian Portuguese. Given that the South American rubber boom began in Brazil and reached its zenith there, not to mention the facts that crossed borders throughout the Upper Amazon to requisition labor (whether by *enganche* of *peones* or the enslavement of Indians) and Manaos became the focal point for exportation of the product to European and North American markets, it is scarcely surprising that Brazilian Portuguese provided much of the terminology of the rubber industry. The word *cauchero* was applied interchangeably to anyone engaged in the exploitation of latex—from the *peón* to the baron.
89. Frederic Vallvé, "The Impact of the Rubber Boom on the Indigenous Peoples of the Bolivian Lowlands (1850–1920)," PhD diss., Georgetown University, 2010, 129, 132.

trees were tapped throughout the dry season rather than felled. Incisions were made into the trunks and air roots, and the sap was collected in little tin cups. It was then processed with an admixture of common soap and/or natural exposure to the air, or by smoking. The resulting balls of crude rubber were then ready for sale to the *aviadores*. If bled prudently, usually for just six months out of the year, the tree recovered and its production was sustainable for as long as fifty years. Resting and rotating of the trees meant that the *peón* was likely engaged in other activities as well, such as subsistence agriculture and livestock-raising on a modest scale in clearings cut out of the forest.

A typical *estrada* consisted of 100 to 140 rubber trees, and becoming a *cauchero* required little initial investment. One needed a machete, an axe, maybe a thousand metal collecting cups, a vessel for curing and concentrating the latex sap, a firearm used for hunting but also protection, personal effects, and about six months' rations.[90] In all, the outfit likely cost in the range of 200 to 250 *soles*—not an enormous amount, but one that was beyond the means of the majority of intending *caucheros*.

Consequently, most *peones* were indebted from the outset to their provisioner and highly dependent on credit from him. Through a combination of high transportation costs and monopoly, the prices of the *aviador*'s trade goods were usually high, when not excessive. The *peón*'s only means of paying off his debt was with rubber, and he was obviously in a poor position to influence its price. *Aviadores* each had their territories and rarely impinged upon one another. They controlled most shipping on the rivers, so a *peón* had little choice but to sell his rubber to his *aviador*.

Most *caucheros* started out in debt and never managed to get out of it. If they absconded, under the law, they could be pursued, imprisoned, and subjected to physical punishment. Their debts were inherited by their heirs, thereby binding successive generations to the system.[91] Consequently, the system has been deemed debt servitude, and even a form of slavery. The *peones* were little educated and consequently far from sophisticated. It is worth pondering the words in a February 20, 1902, letter written by Augustinian Friar Plácido Mallo to his superior,

> If I remember well, I promised to inform you at length and in detail of the exploitation of rubber, the principal source of this Department's [the Loreto's] wealth, and of the abuses that the *caucheros* commit against the *peones*. The whites or *mestizos* who

---

90. Hildebrando Fuentes, *Loreto: Apuntes Geográficos, Históricos, Estadísticos, Políticos y Sociales* (Lima: Imprenta de la Revista, 1908), vol. 1, 208–10.
91. Santos-Granero and Barclay, *Tamed Frontiers*, 39.

have capital of at least a thousand *soles*, do not need great ability or astuteness in order to become potentates, more feared than the ancient *Señores* of the noose and knife; as long as they belie God and conscience they have an expedited path. When they have found some lands to exploit, they begin by intruding themselves into the houses of the poor and the foolish, and, pretending to be sympathetic to their misery, they make offers and exaggerate their resources, tricking them with gifts, usually liquor, and they invite them to meetings in taverns, where all tie one on and, in the end, the proposal of a hefty salary is blurted out if only the *peones* would like to place themselves in the service of the exploitation of rubber. And, in this fashion, whole families swallow the hook and abandon their homes thinking they will gain all this and heaven too. The patron promises to set up the temporary living quarters of the workers and provide them with clothing, provisions, etc., on credit. And here is the trap, since within the arrangement the price of the goods is always arbitrary. If something costs the patron one *sol*, he charges ten to the *peones*, and since their salary never covers the consumption, no matter how they economize, instead of getting ahead they become further entrapped, and in fact never manage to escape the network in which the *cauchero* has them enmeshed. In order to underscore their enslavement more, the patrons present them an accounting at the end of the year, intimidating them with the demand that they liquidate their debts by working even harder than normal. If anyone resists he is not denounced to the justice courts, let alone fired; rather the whip is brought out and the backs are lashed until the profligate is bedridden to contemplate for a few days the advisability of submission to the despotism of his Señor, and on the risks to life if one does not conform.[92]

There was another player in the rubber mix—the would-be entrepreneurial *cauchero* in search of virgin forests and untamed tribes with which to gain a foothold in the industry. These men, too, would likely become dependent upon an *aviador* for supplies and the export of their rubber; they themselves might come to underwrite a few to several dozen *peones* and control several hundred Indians. Unlike the *peón*'s family operation, this production was now on a modest industrial scale.

---

92. Isacio Rodríguez Rodríguez, OSA, and Jesús Álvarez Fernández, OSA (editors), *Monumenta historico-augustiniana de Iquitos, Volumen Primero 1894–1902* (Valladolid: Centro de Estudios Teológicos de la Amazonía [CETA], 2001), 386–87.

While the *aviadores* were generally better off than their *peones* and the entrepreneurial *caucheros*, the life of the river trader was far from posh. In fact it entailed lengthy periods (weeks at a time) on the water exposed to danger and in a tropical environment that posed considerable health risks. Boating accidents (fires and shipwrecks) were common and could easily prove fatal. Attacks by the indigenes were at least a threat, if seldom a reality. Then, too, there were the many diseases—including such nasty ones as malaria, yellow fever, and beri beri. In some ways the supreme enemy was the boredom that easily induced alcoholism, gambling addiction, and whoring, with the attendent venereal diseases.

The *aviador* was dependent upon one of the major merchants, also called *caucheros*, in such shipping centers as Pará, Manaos, and Iquitos. They, too, extended goods on credit to their clients (*aviadores*) and purchased rubber from them for shipment abroad. There was a premium on loyalty that made it difficult for an *aviador* to change his "rubber baron" *cauchero*. Like his *peones*, once indebted, an *aviador* was hard-pressed to ever satisfy all of his obligations and he remained dependent upon his urban supplier (and rubber buyer) for further credit.

The debts of the modest *peón caucheros* became a form of capital that was bought and sold among the river traders and the industrial *caucheros*. A debtor might be ordered by his new creditor to relocate to another river (or even country). Flight was always a possibility, but carried its own risks. The debt peonage system was legal in some form in all the countries in question and enforceable under their laws. A man could be indicted, pursued, and imprisoned for shirking his debts; anyone who harbored him was liable for fines or more severe legal consequences.[93]

The obvious alternative labor solution in both the Congo and the Amazon was to involve the natives. Not only were they knowledgeable concerning the lay of the land (including the locations of the widely distributed trees and vines in the dense forest), but they were also potentially the consummate gatherers (adapted as they were physically to local conditions and wise in the ways of the jungle). However, there was a serious problem. The indigenes were only moderately interested in the civilized world's capitalist economy. South American Indians and Congolese blacks would both prove remarkably impervious to the "normal" or "civilized" material inducements to abandon their traditional lifestyles in order to gather rubber.

In both the Congo and South America there was a remarkably similar pattern in the treatment of the indigenes. The Belgian King Leopold

---

93. Stanfield, *Red Rubber*, 50–51.

II,[94] as well as the South American rubber barons, either simply usurped through force or tricked unsophisticated natives into signing agreements that under "civilized" legal systems gave the whites virtual sovereignty over native territories and their inhabitants, while requiring them to furnish fixed quotas of rubber on a periodic basis and without much (if any) compensation—or suffer severe consequences.

When the indigenes balked, they were tortured, maimed, and, in some cases, murdered. Women and children were kidnapped and held to ensure that their menfolk would neither run away nor fail to collect rubber. Some of these hostages were sold into slavery. Enforcement was realized by bands of armed men constituted by a tiny percentage of the natives (who turned upon their fellow indigenes in order to enjoy certain privileges) and commanded by mercenaries recruited from the outside (Zanzibar in the Congo and Barbados in Arana's case). The mercenaries and their immediate European overseers alike were often little more than sadistic thugs. They regularly conducted raids both to capture more tribesmen and to punish rebels and absconders. They tortured and murdered for amusement, while regularly raping the indigenous women (and even young girls) for pleasure and as another form of intimidation.

In both venues, the atrocities were most pronounced at the furthest flung peripheries of white control. While there are no precise counts, indigene deaths clearly numbered in the tens or hundreds of thousands in the Amazon and have been estimated in the millions in the Congo.[95] The much smaller absolute number of fatalities in the Amazon was not the

---

94. King Leopold (whose reign began in 1865) was convinced that his country needed colonies if it was to be taken seriously by Europe's imperial powers. He was well connected through kinship ties with several European monarchs. His sister, Carlotta, was empress of Mexico, and he was a first cousin of both England's Queen Victoria and Prince Albert as well as of the king of Portugal. His colonial pursuit eventually settled upon Africa, and, in 1885, he was given control (by an international congress of all the European powers and the United States) of the "Congo Free State"—a territory with an estimated 20 million inhabitants that was 76 times larger than Belgium itself. Leopold appointed all of its administrators and was in charge of the Force Publique. At first its economy turned upon the ivory trade, but by the 1890s the real money was in rubber. See Adam Hochschild, *King Leopold's Ghost* (Boston and New York: Houghton Mifflin Company, 1999).
95. Tully reviews the mortality estimates and notes that, between 1880 and 1920, the population of the Congo was at least halved from an estimated 20 to 30 million in the 1880s. While it is also true that sleeping sickness reached the region during this period and accounted for many deaths, it is still likely that Leopold's policies were directly and indirectly responsible for about ten million fatalities (*The Devil's Milk*, 116–17).

result of more lenient policies. Rather, the indigenous population of the entire Amazon was likely fewer than one million by the late nineteenth century.[96] In both venues, the slaughter eliminated whole tribes, and many others were brought to the brink of extinction.[97]

Despite the incredible number of victims in the Congo, in that venue there were actually certain countervailing forces limiting the slaughter. These included theoretical international oversight by the nations that had awarded Leopold his concession. Christian missionaries were present throughout the Congo basin, and there was an administrative and judicial structure of sorts in place there. In theory, at least, a scandalized missionary, or even the aggrieved native, could complain to the authorities and receive some due process (notwithstanding that its outcome was usually foreordained). Nor was it possible for Leopold to escape entirely international scrutiny by religious and humanitarian groups, not to mention a novelist named Joseph Conrad, who set his most famous work, *Heart of Darkness*, in Leopold's Congo. That book alone prompted immense public interest and subsequent outrage on an international scale.

It is also important to recall that the Indians of Upper Amazonia were in varying degrees of contact with whites from at least the middle of the sixteenth century. They were the objects of missionizing efforts. Indeed, from the sixteenth century on there were two classes of Indians— the "semi-civilized" who had been Christianized to varying degrees and

---

96. Historical demographer William M. Denevan estimates the pre-Colombian population of all of Amazonia to have been about 6,800,000. William M. Denevan, "The Aboriginal Population of Amazonia," in William M. Denevan (ed.), *The Native Population of the Americas in 1492* (Madison: The University of Wisconsin Press, 1976), 229. The numbers for Peru's entire Oriente, of which our study area in the Putumayo forms a small part, are slightly under 500,000 in pre-Colombian times, compared with only 126,000 remaining indigenes in 1965. (Ibid., 227).

97. Clearly, the natives possessed little or no immunity to the Europeans' common ailments and died from those diseases in the hundreds of thousands, if not millions. Until recently, anthropologists and historians subscribed to the view that the so-called "Colombian exchange" (Alfred W. Crosby, Jr., *The Columbian Exchange: Biological and Cultural Consequences of 1492* [Westport, CT: Greenwood Pub. Co., 1982]) exposed New World populations to European diseases that triggered devastating epidemics that raced across the New World continents ahead of actual contact between whites and indigenes (the diseases spreading from tribe to tribe more rapidly than the movement of European explorers and colonists). There is now revision of that unicausal thesis that sees epidemics as but one facet among several leading to the mortality rates of indigenes. Violence and various forms of maltreatment of subjected peoples contributed considerably to their capacity to resist disease. Catherine M. Cameron, Paul Kelton, and Alan C. Swedlund (eds.), *Beyond Germs: Native Depopulation in North America* (Tucson: The University of Arizona Press, 2015), particularly David S. Jones, "Death, Uncertainty, and Rhetoric," 16–49.

the "savages" in the forest fastness who were raided for slaves and enticed by trade. The latter were sometimes victims who fled to the interior from outside influences along the river arteries, and, at others, curiosity seekers who presented themselves in pursuit of desired trade goods—particularly iron ones such as machetes, axes, and firearms. The French explorer Jules Crevaux, for example, reports being perplexed after visiting an Indian chief in the Upper Amazon in the mid-1870s who had:

> No fewer than ten rifles and a similar number of cavalry swords, and some veritable cuirasses. Although he lived at a distance of 200 leagues from the Amazon River, he possesses four trunks filled with all the objects that serve civilized life. Why then are the savages of the interior better provisioned than the inhabitants along the Amazon?[98]

There were certain fundamental longstanding issues in play regarding the interactions between whites and Indians throughout the Upper Amazon (Colombia, Ecuador, Peru, Bolivia, and parts of Brazil). The sixteenth- and seventeenth-century dynamics that we considered earlier when reviewing Jesuit missionary activity in Mainas were not all that unlike those obtaining in the nineteenth century. Whether rounded up for Christ's reductions or as forced labor for Mammon's agricultural *encomiendas* and/or forest enterprises, Indians were sub-human barbarians to be civilized for their own good. From a white standpoint (of government officials, military officers, missionaries, and entrepreneurs), indigene lands were uninhabited voids in the national territory and impediments to the nation's economic and political development. While the missionaries were somewhat tolerant of Indian cultural differences, and particularly maintenance of the local language(s), they saw themselves as civilizers of savages. Christian morality demanded covering the bodies of semi-naked pagans. The cannibalism of some of the tribes was simply intolerable to all whites—and was prohibited and punished accordingly.[99]

While the linguistic diversity among the region's dozens of tribes was considerable, they actually shared many cultural similarities. Most engaged in slash-and-burn agriculture with the fields cared for primarily by women, while the men hunted and fished. The diet was also supplemented by the gathering of forest fruits, roots, and insects by all, including children.

---

98. Docteur J. Crevaux, *Voyages dans L'Amerique de Sud* (Paris: Librairie Hachette et Cie, 1883), 372.
99. Pineda Camacho, *Holocausto*, 107–16.

The *maloca*, or longhouse, was generic throughout the region. As many as a hundred or more people, generally extended clansmen, resided beneath the same roof in an enormous palm-thatched structure.[100] They were exogamous, patrilineal, and patrilocal, and were presided over by a strongman, or *cacique*, and possibly a council of advisors. In larger tribes there were several outlying *malocas* that recognized the authority of the supreme *cacique*. His *maloca* was the tribe's ceremonial center where dances of both a spiritual and recreational nature were held. It was also the locale where the *cacique* and his advisors discussed political affairs and formulated policy, including whether or not to go to war.[101]

Use of the *manguaré*, or large drum, was central to both the ceremonial (dancing and the consumption of hallucenogenics, including a tobacco-chewing ritual) and religious life of the community. Many tribes used it both to summon the inhabitants of the outlying *malocas* to the main one and to communicate all kinds of other information over considerable distances. Whites were made uneasy by the sound of the *manguaré*, since it was sometimes the clarion of violent rebellion and/or passive resistance against their authority.[102] Pineda Camacho described the sources of white fear of the Indians in insightful terms:

> One must understand that the *caucheros*, incrusted in their barracks and surrounded by the exuberant forest, scarcely counted upon information that would permit them to "analyze" and filter out the rumors, oral histories, and narrations of their immediate assistants, or the far-off sounds of the *manguarés* or *yadikos* [empowering rituals] that resounded in the forest. The sense of isolation was accentuated, besides, because with frequency the outsiders did not know the native languages, even if some of them predominated to a considerable degree. Notwithstanding that Huitoto was employed as a kind of universal language, the native communities continued communicating among themselves in their own languages. Some of them were unintelligible for the Huitotos, although there was multilingualism in certain areas. If in fact the *caucheros* arranged the population according to certain local cultural canons (language groups, "ethnic identity," etc.), the

---

100. See report by missionary Fray Basilio de Pupiales (1903) in Augusto Javier Gómez López (ed.), *Putumayo: La vorágine de las caucherías. Memoria y testimonio* (Bogotá: Centro Nacional de Memoria Histórica, 2014–2016); *Segunda parte, Documentos relativos a las violaciones del territorio colombiano en el Putumayo (1903–1910)*, 28 [*Putumayo*].
101. Pineda Camacho, *Holocausto*, 50–51.
102. Ibid., 53; 109; 113.

frequent displacements, the flight of diverse groups, the massive gatherings to turn over the rubber, created a "Tower of Babel," even for the natives themselves. All of this contributed to the isolation of the *cauchero* and the configuration of his psychology. The cultural distance between the forms of social and cultural organization of the Indian societies and the cultural patterns of the *cauchero* himself increased the social abyss.[103]

The use of tobacco was associated in the whites' view with rituals involving cannibalism and was therefore discouraged. Indeed, the belief that the natives were cannibals fueled both fear of them and ratification of the "Christian" civilizing mission that justified the whole exploitative system.[104] When Henry Pearson toured the rubber districts of the Amazon in the first decade of the twentieth century, he reported that the caucheros were sometimes harassed by "wild Indian tribes," citing as examples the Tauapery, who killed and ate the whites, saving their right leg as a trophy, and the Acarinus, famed for "carrying away their heads as trophies."[105]

Pineda Camacho questions how a handful of whites were able to so quickly dominate thousands of Indians. He concludes that there were several factors involved. There was the discrepancy in "firepower" once the whites disarmed the indigenes by supplanting blowguns, spears, and bows with shotguns. The white overseers armed themselves, their *peones*, and some of their Indian subordinates with the Winchester rifles—thunderbolts that killed from afar. In the case of the Putumayo, the armed Indians were called the *muchachos de confianza*, or "trustworthy boys," and were thoroughly despised by the others (shades of the Jesuits' interpreters).[106] They were usually orphaned or taken from their parents and raised like Spartan helots to be the loyal minions of their white overlords. They were often the first phalanx in the pursuit and punishment of runaway Indians. There would be numerous occasions in which a handful of riflemen were able to kill, with near impunity, scores of Indians. In addition to firearms, the weapons of intimidation included fierce dogs (greatly feared by the Indians), whipping posts, and stocks.[107] The prospect of making strategic alliances with such a pow-

---

103. Ibid., 109–10.
104. Ibid., 110–13. Despite this, there was some evidence that the Indians were repulsed by the very odor of whites. That would certainly have reduced their attraction as a culinary treat! Ibid., 115–16.
105. Henry C. Pearson, *The Rubber Country of the Amazon* (New York: The India Rubber World, 1911), 136.
106. According to a Colombian account, the *muchachos* "were good shots, swimmers and paddlers and without rival in their knowledge of the jungles and rivers." Cited in Gómez López, *Putumayo*, 135.
107. Pineda Camacho, *Holocausto*, 91–95; 107.

erful force against one's traditional enemies was compelling.[108] "Friendly" Indians thereby became the shock troops in the conquest of holdouts.

Also, there was the attraction for stone-age peoples of the whites' trade goods, and particularly of metal implements such as axes and machetes.[109] Undoubtedly, it was for this reason that they allowed whites in their midst and acquiesced in the initial demand for rubber; but certainly without appreciating the full magnitude of white intentions and their consequences. The introduction of steel axes had a profound impact upon indigenous social structure and gender relations. It now required a fraction of the time to clear forest for fields, freeing up the men for other activities. It also meant that less communal labor was required, making it possible for disseminated families to establish themselves as autonomous agricultural households. Then there was the disruption in the long-standing indigenes' trade relations. For example, the Andoques had been the source of stone axes that moved among tribes; a trading network that was threatened by the steel axe.[110] Subsequently, by the mid-eighteenth century, Andoque *caciques* within Arana's future realm attempted to maintain their status by purchasing steel axes from Brazilian traders in exchange for captured Indian adversaries who were now destined to be slaves in Brazil.[111]

Then, too, there were the diseases introduced by the whites that decimated Indian populations, shattering existing social structures.[112] Furthermore, the indigenes processed the rampant diseases in terms of their beliefs in witchcraft:

> The native interpretation of the epidemics as a form of witchcraft had not only consequences for their therapeutic practices, it also affected decisively everyday life in terms of the relations among persons of one community and between communities. The proliferation of the epidemics surely exacerbated the internal social conflicts; creating an atmosphere of generalized fear and mutual accusations. Perhaps this had much to do with the collaboration of certain groups in denouncing or pursuing other natives, or the "febrile" imagination of the "muchachos."[113]

---

108. Ibid., 63; 65; 149.
109. Ibid., 63–64; 84. One Huitoto oral history reported that the first contacts with the whites at La Chorrera were very amiable and peaceful. Ibid., 84–85.
110. Landaburu and Pineda, *Tradiciones*, 26–27.
111. Ibid., 27.
112. Pineda Camacho, *Holocausto*, 116. One friar reported devastating epidemics of both smallpox and influenza in the Putumayo in 1905, with high mortality among the indigenes.
113. Ibid., 119.

While the "clash of cultures" was far from a contest between equal forces, the *caucheros* viewed themselves as surrounded by hostile indigenes engaged in sinister black magic and cannibalism when not assaulting the intruders outright.[114] There were sufficient fatal attacks of Indians upon whites to reinforce the fears of the latter. Fray Basilio Pupiales reported that, in 1901, no fewer than seventy-one whites were slaughtered one night in an Indian attack in Aranalandia—no doubt for their meat.[115] Then, in 1903, some indigenes on the Caquetá assaulted a party of white raiders, killing them all and displaying their heads on their drums, while conserving their arms and legs in water. That same year, an uprising by the Boras killed several *caucheros* and brought rubber collection to a halt. Both whites and "friendly" Indians were afraid to go into the forest. According to Huitoto oral history, around this time the head Colombian *cauchero* on the Caraparaná was killed when he fell into a pit and received a mortal wound in his chest from a poisoned spike.[116] There was a legendary Bora leader named Makapaamine. He obtained rifles and continually attacked Casa Arana river traffic on the Cahuinarí. He had once been a *muchacho de confianza* of the whites and had even spent time in Iquitos before returning to Aranalandia to try and expel the *caucheros*.[117]

The objectives and strategies of white missionaries, on the one hand, and the secular exploiters of natural resources and the Indian work force, on the other, differed fundamentally. The friars (often of foreign, particularly Spanish, origin) had an intrinsic interest in civilizing reductions in which the church was the dominant architectural feature. The missionaries would live in an adjacent house that was likely the community's finest residence, possibly with an appended school. The Indian faithful lived in more modest dwellings as individual nuclear families—affording the privacy implicit in monogamy. Communal life within the *maloca*, with the threat of casual sexual unions and/or polygamy, was deemed unacceptable by the Christianizing mission. The centralized settlement was surrounded by the fields that sustained its inhabitants. There was likely present one or more merchants selling essential trade goods—such as

---

114. Ibid., 110–13.
115. Fray Basilio Pupiales stated in a report dated January 13, 1903, that, "The Huitotos are cannibals and therefore have not relented even amongst themselves; given that the strongest tribes capture the weakest and at times destroy them altogether, leaving only their names behind. It is a certain truth that the tribes hunt one another; they always do so beause, according to them, they desire meat, of which human is the best, that which appeals most to them." (Cited in Gómez López, *Putumayo*, 27.)
116. Pineda Camacho, *Holocausto*, 140–41.
117. Ibid., 150.

steel machetes, axes, clothing, canned and dried foodstuffs, medicines, and so on.

The mission settlement was conceived as permanent, and the rhythm of life within it required one's regular presence at religious services and the daily schooling of both adults and children in western ways—as well as the Spanish language, to be sure. From the national government's viewpoint, the mission villages, along with the settlements in the tropical forest fastnesses of both domestic and foreign migrants, were the building blocks of national destiny. Thus, the missionaries were engaged as much in producing "Peruvians" or "Colombians" as "Christians."

In contrast, the white *caucheros* favored a disseminated settlement pattern in which existing Indian *malocas* were under their nominal control and capable of being relocated as the rubber trees of a particular area were exhausted, or wherever the purview of the *cauchero* was expanded into a previously unexploited area. The white overseers were likely to construct a large, fortified headquarters dwelling, normally on the shore of a major river capable of supporting fluvial maritime traffic. The structure was typically two-stories tall (to facilitate surveillance) and housed the station's white supervisors and possibly its general store. Imported foodstuffs (canned and dried) were dispensed from there to the Indians and *peones* alike. *Peones* and entrusted Indians lived in simpler dormitory structures at some distance from the main structure. A considerable area around the whites' dwelling was cleared, both as a security measure and for cultivation of vegetables, fruits, and some livestock pasturage.[118]

Indian men were expected to concentrate their efforts upon rubber gathering and processing at the expense of their former hunting and fishing activities. Control of their women and children was an effective means of securing if not loyalty then at least obedience. The Indians were recruited (often intimidated) into transporting heavy burdens (at times weighing over a hundred pounds) from their station to the headquarters. The journey might take a few days, and there are many accounts of Indians, including women and children, arriving in an exhausted and semi-famished condition.

While the missionaries and white entrepreneurs both sought to settle Indians permanently, flight of *their* respective indigenes (in some cases after themselves burning down their houses and crops)[119] was a pervasive possibility in Indian–white relations. There emerged a clear difference throughout the Oriente between the riverine population (immigrant colonies, the *peón* families and their *cauchero* overseers, and the missions

---

118. Pineda Camacho, *Holocausto*, 103–104.
119. Landaburu and Pineda, *Tradiciones*, 144.

with their "domesticated" Indians), on the one hand, and the interfluvial indigenes who had fled from the civilizing juggernaut into the forest fastness, on the other.[120]

Finally, some *aviadores* (like Arana) themselves rose to the rank of rubber baron, based out of Iquitos and/or Manaos and in control of the exportation of South American rubber to European and North American markets.[121] In point of fact, while few in absolute numbers, the rubber (and quinine) barons came in various guises (that even included consortia of local and/or foreign capitalists).

To appreciate this variety, we can consider some examples. In 1865, the Colombian government conceded a vast tract (40,000 hectares) of prime quinine-rich forest in the state of Tolima to a privately owned Compañía de Colombia, treated as "empty lands," that is, without any regard to the indigenous inhabitants. The petitioners were well connected politically and probably paid kick-backs to their benefactors. There was also considerable manipulation of Colombian land laws to maximize their control over this private empire on public lands. The enterprise would ultimately employ somewhere between 1,500 and 2,000 *peón* gatherers. It also defended with arms its interests against the army of entrepreneurial individuals who combed the country's forests in search of cinchona trees.[122]

Then, in 1887, a public Compañía de Caquetá was formed in the Cauca region, Tolima Department, to seek a twenty-year concession from that state government to settle and farm along the shores of the Caguán and Orteguaza tributaries of the Caquetá River. Nevertheless, its primary activity would be the collection of rubber; the Caguán was particularly endowed with it.[123] The enterprise agreed to build a key road

---

120. It might be noted that even by the late twentieth and early twenty-first centuries the notion persists that there may still be Amazonian tribes living in the Stone Age. Both western missionaries and anthropologists have been (and even continue to be) motivated to search them out—albeit for very different purposes. In the process, they have contributed to the myth. In point of fact, after half a millennium of European activity on the South American continent, including thorough penetration of it by explorers, traders, evangelists, geographic, and military missions, and so forth, it is extremely unlikely that there remain today (or have even for some time) "pristine" Amazonian naked savages *entirely* oblivious to the outside world. This is not to say that there are not a few remaining extremely isolated groups—whether through choice or circumstance.
121. Weinstein, *The Amazon*, 75–76.
122. Dominguez and Gómez, *La economía*, 45–52.
123. Ibid., 142–43. By the late nineteenth century, due both to the destruction of the resource (the felling of trees to strip them of their bark) and competition from Asian plantations, the Compañía de Colombia, regarded at one time to be Colombia's most important firm, ceased operations—leaving in its wake a devastated and depopulated countryside.

and in return received exclusive usufruct of 10,000 hectares and a subsidy of 8,000 pesos.[124]

In Peru, there is the remarkable story of Carlos Fermín Fitzcarrald and his subsequent Bolivian rubber magnate associates. Son of a North American of Scottish (or possibly Irish) descent, William Fitzgerald (or "Fitzcarrald" as the surname became), had a nebulous past. After fighting as a teenager in the War of the Pacific,[125] there were accusations that he had been an agent of the Chilean government. There may have been criminal charges against him as well, and at one point he was stabbed in the stomach by the region's most notorious bandit, (Basque-surnamed) Benigno Izaguirre. In any event, Fitzcarrald left his natal town of San Luís de Áncash in the highlands and disappeared for eight or ten years into the Amazon forests. There, he became an accomplished explorer and then resurfaced as the dominant pioneer *cauchero* in the tributaries of the Ucayali River. By his late twenties, his accomplishments were legendary. He proved ruthless in his conquest of several Indian tribes, assembling formidable forces of white and friendly Indian allies to conquer the recalcitrants. The fatalities on both sides of these campaigns numbered in the hundreds. The goal was to establish *caucheros* dependent upon Fitzcarrald for their supplies and the export of their rubber.[126] His feat in dismantling and dragging the parts of a steamship over a mountain ridge from the Serjali River in the Ucayali drainage to the Cashpajali River (a tributary of the Manu/Madre de Dios river system) has been immortalized by Werner Herzog in the film *Fitzcarraldo*.[127]

Fitzcarrald received from the Peruvian government the exclusive navigation rights on the Ucayali, Urubamba, and Manu/Madre de Dios rivers.[128] His activities on the latter were particularly noteworthy. The fierce Indians there, particularly the Mashcos and Huarayos, had always resisted white penetration of their territory. Consequently, from a white viewpoint, it was not even understood that the Manu and Madre de Dios were part of the same drainage. However, well downstream, or generally where Brazil, Peru, and Bolivia met in another ill-defined border region,

---

124. Ibid., 147.
125. Fought between Colombia and Bolivia against Chile between 1869 and 1872. The Chilean victory cost Bolivia access to the Pacific Ocean and Peru a significant amount of its southern coastal territory.
126. Zacarías Valdez Lozano, *El verdadero Fitzcarrald ante la historia* (Iquitos: Imprenta El Oriente, 1944), 12–28.
127. See Werner Herzog, *Fitzcarraldo: The Original Story* (San Francisco: Fjord Press, 1982).
128. Pilar García Jordán, *Cruz y arado, fusiles y discursos. La construcción de los Orientes en el Perú y Bolivia, 1820–1940* (Lima: Instituto Francés de Estudios Peruanos, 2001), 167.

there were established Bolivian, Peruvian, and Brazilian *caucheros*.

The situation was exacerbated by the steady penetration of Brazilian rubber men into the adjacent Acre region of Bolivia. Indeed, a few years later, after the brief conflict known as the Acre War (1899–1903), it would be conceded by Bolivia to Brazil in return for the latter's commitment to build the Manaos–Mamoré railway. That particular project was deemed essential by Bolivia to the development of its rubber industry in the vast Amazonian forests of the Beni region. While the Madeira River drained the area and provided a direct fluvial link with the Brazilian rubber capital of Manaos, in fact a series of rapids and waterfalls near the Bolivian border, not to mention flats that made the river impassable in the dry season, meant that the journey up or down the Madeira was extremely exhausting and dangerous. Lost lives and cargoes were common, and the trip to or from Manaos tended to be measured in weeks. The added costs to the value of trade goods and the exportation of rubber were prohibitive. A Madeira-Mamoré railway would circumvent the natural obstacles and presumably trigger a new rubber boom.

It was now that Fitzcarrald's bold move into the upper reaches of the Manu portended to be a game-changer. His initial downstream exploration with an armed expeditionary force demonstrated that the Manu and Madre de Dios were a single system. His arrival in their town astounded residents of El Carmen in the Bolivian-Peruvian borderland. Even though they possessed steamships, its largely Bolivian population never explored upstream out of fear of the Indians. According to one of Fitzcarrald's companions, "they regarded us as if we had appeared out of nowhere or from another world." When the visitors departed for home several weeks later, most of the residents saw them off dressed in mourning, convinced the Peruvians would simply perish from disease or the attacks of wild animals and savages.[129] They did not.

Fitzcarrald organized two subsequent journeys down the Manu/Madre de Dios with valuable cargos that he sold to the El Carmen agent of Nicolás Suaréz, Bolivia's wealthiest *cauchero*. Both shipments were paid with a letter of credit on Suaréz's London bank. The Bolivian's rubber company (Suaréz Hermanos) was incorporated in that city. On his return to Iquitos, Fitzcarrald sold them both to the German merchant firm of Wesche and Co. When Fitzcarrald and Suaréz eventually met in 1897 they formed an Iquitos-based partnership (Suaréz–Fitzcarrald) with a trading house in the town. They intended to improve the short land crossing (eventually with even a rail link) between the Manu/Madre de Dios

---

129. Valdez Lozano, *El verdadero*, 26–27.

and Ucayali drainages, while placing a fleet of steamships in both in order to provision the Beni while exporting its rubber through Iquitos.[130] Suaréz would become the godfather of Fitzcarrald's children.[131]

Fitzcarrald's route attracted the attention of another wealthy Bolivian, Antonio Vaca Diez, who had extensive rubber holdings near Riberalta. He likewise was incorporated in London as the Orton Rubber Company.[132] He met with Fitzcarrald and the two new acquaintances sailed together on July 9, 1897, on the steamship *Bermudez* down the Alto Ucayali River, discussing details of forming their own partnership. Fitzcarrald was now in his mid-thirties and clearly the most powerful man in the Loreto. In the end, the captain lost control of the vessel in some rapids and it sank. While the skipper and most of his crew survived, four men drowned—including Fitzcarrald and Vaca Diez.

Fitzcarrald's empire unraveled quickly. His widow and his brother, Delfín, tried to hang onto the Ucayali business. However, it was now that the Purús and Acre rivers had become the new rubber frontier, and Delfín transferred his initiative there (he would soon be killed by hostile Indians). She sold her interests and moved to Paris, France, with her children.[133]

There is considerable evidence to suggest that Fitzcarrald was scarcely the only Peruvian rubber baron to exploit and abuse indigenes. We might consider the case of the Spaniard (Asturian) Máximo Rodríguez. He arrived in Iquitos in 1888 and then worked as an employee for a *cauchería* on the Ucayali and Sepahua Rivers. In 1905, he formed a partnership with Rafael de Souza to create a rubber station, christened Iberia, on the Tahuamanu River in a no-man's land (shades of the Putumayo/Caquetá) of the Madre de Dios region.[134] It controlled about 4,000 square kilometers. Given the lack of state authority in the area, it was run like a feudal fiefdom. Rodríguez's biggest adversary was Bolivian rubber baron Nicolás Suárez—who came to have 16,000,000 acres and 10,000 workers in 1910 after three decades of expansion.[135]

Both men launched raids to capture Indians that were brought back as slaves to work their holdings. Between 1905 and 1910, Rodríguez also

---

130. Ibid., 36–37.
131. Ibid., 44.
132. Vallvé, "The Impact," 226.
133. Valdez Lozano, *El verdadero*, 40–47.
134. Klaus Rummenhoeller, "Shipibos en Madre de Dios: La historia no escrita," in Beatriz Huertas Castillo and Alfredo García Altmirano, *Los pueblos indígenas de Madre de Dios: historia, etnografía y conyuntura* (Lima: Grupo Internacional de Trabajo sobre Asuntos Indígenas—IWGIA, 2003), 166–67.
135. Fiebre del Caucho, *Wikipedia*, http://es.wikipedia.org/wiki/Fiebre_del_Caucho.

purchased many Indian families, including Huitotos from Aranalandia, for installation in Iberia—his largest contingent was Shipibo.[136] In 1914, a Peruvian official accused the Bolivian Suárez of conducting cross-border raids that had enslaved 600 Peruvian indigenous families.[137] The two rivals also engaged in armed conflict against one another. The situation was so tense that Peru and Bolivia asked a British commission to settle their common boundary between the Acre and Madre de Dios rivers, and in 1912, Rodríguez was ordered to stop engaging in a private war and remove his men from the new frontier.[138] The Iberia enterprise lasted for nearly four decades and employed Asturians exclusively as its foremen and administrators. It the 1930s, there were about one hundred Loretan, Bolivian, and Peruvian *peones* working Iberia, down considerably from the peak years of the earlier boom.[139] The "social order" was maintained with severe punishment that included imprisonment, beatings, public ridicule, and death (the last for disobedient or thieving *peones* more than for Indians).[140]

All of the foregoing fails to treat the life histories of Brazil's rubber magnates and the many international rubber firms based out of Manaos. Bolivia,[141] Peru, and Colombia[142] all had many foreign firms—including North American and British ones—searching for rubber territories. In short, by the time that Arana was establishing his empire in the Putumayo/Caquetá, he was aspiring to the elite status of a select few fabulously wealthy and powerful individuals, partnerships and foreign concerns that overarched the collection and exportation of Amazonian rubber.

Arana joined their ranks just in time to experience the South American wild latex's bust. The world's consumption of rubber had gone from 2,000 tons in 1850 to 381,497 tons in 1919.[143] By the latter date, harvesting in the wild had all but been eclipsed by plantation production, particularly in southeast Asia and, subsequently, Liberia.[144]

It is interesting to compare Harvey Firestone's concession in 1926 from the Liberian government to establish the vast rubber plantations that would

---

136. Rummenhoeller, "Shipibos," 170.
137. Ibid., 168.
138. Ibid., 170.
139. Ibid., 172.
140. Ibid., 175–76.
141. Vallvé, "The Impact," 227.
142. Dominguez and Gómez, *La economía*, 155–64.
143. Santos, *História*, 236. It should be noted that these figures likely refer to British "long tons," or 2,240 pounds—a measurement usually employed internationally in calculating rubber production.
144. Tully, *The Devil's*, 195–200; Davis, *One River*, 338–40.

come to dominate the country's economy to the present day with Henry Ford's failed Fordlandia and Belterra projects, beginning in 1928 and ceasing in 1945, along the Tapajós River near Santarem in Brazil. Firestone's initiative provided salaried employment to the masses in a populous society with few alternatives for its unemployed workers. Ford purchased a vast tract of nearly vacant land and then had to recruit and move his work force there. Not only did Fordlandia's (and his subsequent Belterra plantation's) rubber trees have the bad luck of being struck by a particularly fatal leaf blight, the equally intractable Henry created his own labor problems by seeking to impose upon his Brazilian workers a Yankee work ethic and Puritan mores (he prohibited them to smoke or drink alcohol even in their own homes!).[145]

In 1940, President Franklin Roosevelt, designating rubber to be a "strategic and critical material," created the Rubber Reserve Company, which was designed to stockpile natural rubber and regulate the production of the synthetic product. Japan's rapid control of southeast Asia after Pearl Harbor deprived the Occident of 90 percent of Asia's rubber supply. Firestone, B. F. Goodrich, Goodyear, and U.S. Rubber agreed to work together with the U.S. government in the production of synthetic rubber to resolve the country's wartime shortage.[146] There quickly emerged among them the public/private partnership known as the Rubber Development Corporation.

The thrust regarding the supply of natural rubber was two-pronged—expanding plantations and exploration for new sources in the wilderness. The attempts to establish plantations in South America, most notably the initiatives of Henry Ford, but also the rather peripatetic ones of each of the South American rubber-producing countries, largely failed due to the presence of the leaf blight endemic throughout Amazonia. The outbreaks were unpredictable but inevitable. Southeast Asia and Africa were free of the blight but were obviously hugely vulnerable to "eco-terrorism" should anyone choose to introduce the spores there. Fungicides were capable of controlling the disease but far too costly.[147] A process of grafting crown clones onto mature trunks was developed, beginning in 1936, at Ford's Belterra Brazilian plantation, but it was a both cumbersome and costly process.[148]

In 1940, the U.S. Congress authorized funding for a broad cooperative program with fifteen Latin American nations, including Colombia.

---

145. Tully, *The Devil's*, 194–99.
146. "Rubber in World War II at a Glance," http://www.nationalww2museum.org/learn/education/for-students/ww2-history/ataglance/rubber.html, np, nd ["Rubber"].
147. Davis, *One River*, 336; 371.
148. Ibid., 340.

The main research facility regarding development of blight-resistant varieties of rubber trees was established at Turrialba, Costa Rica. In 1942, Hans Sorensen, a Danish botanist, was hired by Colombia to establish three experimental plantations on the northern coast near Urabá, the key one being at Villarteaga.[149]

The obvious real answer was to find trees in the wild that were impervious to the blight that could then provide the nursery stock for establishing reliable future rubber plantations. So it was that, between 1942 and 1945, the United States and the four rubber firms, through their newly established Rubber Development Corporation, engaged Colombia, Peru, and Brazil in what was labeled "the battle for rubber." There was by then a Rubber Investigations Division of the Bureau of Plant Industy in the U.S. Department of Agriculture.[150]

During the Second World War, Peru expropriated the Iberia station in order to turn it over to the North American Rubber Development Corporation. Máximo Rodríguez had retired to Spain by then. He returned to conclude his ultimate transaction and then died a few months later (1943). When the Peruvian and North American administrators inventoried the property, they found (to their chagrin) that it had a number of Shipibo Indian slaves. In 1944, they constructed an airstrip at Iberia and offered to fly Shipibos gratis to Pucallpa—their homeland. Many accepted, but others refused. They had had no contact with their native area and relatives there for forty years.[151]

The South American wild rubber industry was, by then, moribund, albeit not entirely dead. There was now the possibility that the rising prices of an increasing demand could resuscitate that gathering, at least in part. It was clear, however, that the known rubber areas had been overworked—in most cases to the point of exhaustion. Rubber collection in Brazil had declined to the point that there were only 35,000 rubber workers in the entire Brazilian Amazon. The goal now was to increase production there from 18,000 to 45,000 tons annually.[152] This required recruitment of gatherers from throughout the nation. It became a tale of misery (like that of the construction of the Devil's Railway). The ill-adapted men succumbed from tropical diseases, heat prostration, snake bite, and so on. An estimated 15,000 of them died.[153] Peru cooperated, but, with a few

---

149. Ibid., 334.
150. Ibid., 296.
151. Rummenhoeller, "Shipibos," 180.
152. Dean, *Brazil*, 170.
153. Xenia Vuconic Wilkinson, "Tapping the Amazon for Victory: Brazil's "Battle for Rubber" of World War II," PhD Diss. Georgetown University, Washington, D.C, 2009, 186.

exceptions (the notable one being at Iberia on the Tahuamanu River in the Madre de Dios drainage), there proved to be few Peruvian prospects. Colombia was also an obvious area for prospecting.

One of the most accomplished plant botanists of all time was Richard Schultes, based at Harvard University. Schultes had literally spent years, while undergoing unimaginable hardships, collecting plants throughout Amazonia. He had discovered, named, and described literally thousands of new species. In the fall of 1942, three wealthy Colombian absentee landowners with extensive rubber holdings on the Caquetá approached the Rubber Reserve Company offering to finance a survey of their distant holdings to determine if they would be reliable rubber producers. A somewhat reluctant Schultes was essentially ordered by the U.S. government to do that work. He undertook the harrowing and difficult overland and riverine journey and finally approached Aranalandia. It was near there that an aged former employee at El Encanto told the American that the rubber resources of the Caquetá had long-since been exhausted. Schultes did discover a new sub-species of *Hevea* growing on a mountain that seemed blight-resistant, as well as a previously untapped, fairly extensive stretch of rubber trees along the Apaporis River that he believed would qualify for commercial gathering.[154]

Meanwhile, in 1942, Sorensen was undertaking his own field explorations. Near Leticia he found a variety of wild rubber tree that also seemed impervious to the leaf blight, particularly stock descended from one "mother tree." Subsequent experiments with it at the Villarteaga experimental station bore this out.[155]

In July of 1944, Schultes was recruited into the campaign to establish blight-resistant rubber plantations in the Americas. They might not come on line in time to address the war shortage but could hold the key to the rubber industry's future.[156] He therefore traveled to the Leticia area to continue his explorations for possible species that were both blight resistant and heavy latex producers. In particular, he was seeking Sorensen's mother tree.[157]

---

154. Davis, *One River*, 307–309.
155. Ibid., 342.
156. At this time, the future of synthetic rubber was unclear. Davis provides an excellent synopsis of this history (Ibid., 360–68). From production of but 3,721 tons in the United States in 1942, by 1945 it provided 756,042 tons. Today, more than 70 percent of the world's rubber is synthetic and its annual consumption of natural rubber is approximately only 12,000 tons ("Rubber").
157. Ibid., 342–43.

Initially, he stayed as a guest at the Capuchin mission in Leticia, but soon moved to the rubber station at the mouth of the Loretoyacu River owned by a Colombian-Basque named Wandurraga. Davis reports,

> Wandurraga was a merchant, but his abiding interest was the welfare of the Indians and the protection of their forests. An honest and decent man, he operated with a fixed markup of 10 percent, refused to barter in liquor, and provided for the wives and children of all the rubber tappers who worked for him. In return, the Witoto and Bora, Mirañas and Tikunas, did what they had vowed never again to do: gather latex from trees that a generation before had been the reason for the misery and torture of their parents.[158]

Schultes collected budwood for cloning, and the following year offered to buy seeds gathered in the forest by locals that he then processed for shipment in an abandoned church leant to him by his friend, the Capuchin missionary. Both his cuttings and seeds were dispatched to the experimental stations in Colombia, Peru, and Costa Rica.[159]

In 1945, Schultes visited the Madre de Dios region in the company of fellow botanist, Russell Seibert, who had been dispatched by the Rubber Development Company to explore for rubber in Peru. Seibert had confirmed that there were extensive mature groves of the species known as *Acre fino*, the "finest rubber in the Amazon," on the Iberia concession, and Schultes wanted to see them. The two botanists agreed that the area was exceptional, and Seibert would remain there for the next three years searching for new species and collecting 350 budroots that he sent to Turrialba.[160] Unfortunately, in the immediate post-war period, shortsighted bureaucrats in Washington, DC, terminated the U.S. government's commitment to Latin American rubber exploration and experimentation. Even the facilities at Turrialba were dismantled and its extensive test plantations cut down.[161]

Beginning in 1950, Schultes would initiate another stint in the Colombian Oriente, this time combining his scientific quests with entrepreneurial enterprise. Miguel Dumit, an Antioquia businessman, had purchased an aircraft and established commercial flights to the Vaupés. He and Scultes had a chance meeting in Bogotá and Dumit convinced the American to help him establish a trading post and rubber station some-

---
158. Ibid., 346.
159. Ibid., 350–52.
160. Ibid., 353–54.
161. Ibid., 363–66.

where in the region. He promised to treat the indigenes fairly. Schultes agreed and knew just the place: the stretch on the Apaporis where he had earlier discovered the promising stands that had never been exploited during the war. He and a young field assistant went to the area and established the Saratama enterprise, named after Dumit's birthplace. Over the next three years Schultes and Dumit collected thousands of plants in the Apaporis area.[162]

Since the area was largely unpopulated, the Soratama enterprise recruited its Indian workers from elsewhere. This created a problem when some of them were seized by Alejandro Betancourt, "a rubber trader of ill-repute," for non-payment of debt. They were obviously under *enganche* contracts elsewhere when they agreed to come to Soratama. Their women were seized as well, despite Schultes's efforts to prevent it.[163] Nevertheless, the Soratama enterprise was so successful that Dumit established a second Apaporis River operation some miles down from it. For his part, Schultes was little interested in money and concentrated upon his plant collecting.[164] In 1953, Schultes left Amazonia never again to return.[165]

Enter Reyes

By the late 1860s, Rafael Reyes (1849–1921), the future president of Colombia (1904–1909), and his brothers had a family firm exploiting quinine in the Cauca Department between Popayán and Pasto.[166] In 1868, the Reyes brothers were successful in having their partner in

---

162. Ibid., 465–66.
163. We have an account of a rubber operation in the early 1970s on the Apaporis River, owned by Efraín and Jorge Sánchez. They recruited *peon* collectors in the highlands with exaggerated promises of wealth and then charged all of their expenses against their rubber production. While they did not have formal contracts with the men, the arrangement was otherwise similar to the *enganche* system. They also had dozens of Indians under their control who they seldom paid and sometimes beat. The Indians feared being chased to ground should they flee. The Sánchez brothers were also said to own the local judiciary, so any attempt by *peones* or indigenes to seek legal redress for maltreatment were doomed to fail. Germán Castro Caycedo, *Mi alma se la dejo al diablo* (Bogotá: Plaza & Janes, 1982), 100; 130; 133 and elsewhere [*Mi alma*].
164. Ibid., 477–78.
165. Ibid., 490–91.
166. Rafael Reyes, *Memorias. 1850–1885* (Bogotá: Fondo Cutural Cafetero, 1986), 70, 105 [*Memorias*]. This work was produced from eleven notebooks or diaries in the possession of Reyes's grandson Ernesto Reyes Nieto. The precise dates of each of them remain unclear and the material seems in some cases to have been rearranged later for publication. There is internal evidence that Reyes (who died in 1921) was still writing some or all of the notebooks in 1914 (Ibid., 151, 159, 190).

the Compañía de Caquetá, (Basque-surnamed) Pedro F. Urrutia, named prefect of the Caquetá Department.[167] He appointed new *corregidores* (administrators) up and down both the Putumayo and Caquetá rivers and requested twenty-five soldiers and a cohort of missionaries to thwart Peruvian initiatives from the Loreto.[168]

Rafael traveled to both Europe and the United States, establishing important business contacts for the family firm on both continents. After an absence from office, in 1874, Urrutia, now an employee of the Reyes, was reappointed prefect of Caquetá—providing key political underpinning to the expansion of the family's activities. By this year, the newly denominated firm of Elías Reyes y Hermanos, operated out of Popayán by Rafael's elder half-brother, was fully engaged in the collection of quinine in the mountains around Mocoa (or the headwaters of the Putumayo). Rafael established a base downstream, at the point where the Putumayo becomes navigable, that he christened La Sofía in honor of his beloved fiancée.[169] Younger Reyes brothers were involved as well. Andrés ran an office in Pasto[170] (astride the land route for exporting their product out of the port of Tumaco on the Pacific coast) and Ernesto (who was in charge of the quinine collection station at La Sofía). Most of the latter's gatherers were local Indians.

Rafael believed that it would be far cheaper to take their quinine by steamship down the Amazon. So, in February of 1874, he and his companion, the *Pastuzo*[171] (and Basque-surnamed) Benjamín Larrañaga,[172] explored the entire Putumayo to its convergence (as the Içá River in Brazil) with the Solimões. Rafael was outfitted with the documents and letters of recommendation that would assist him in his negotiations with the Brazilian authorities for navigation rights.[173] In Mocoa, they acquired canoes and several Indian oarsmen from the Upper Putumayo. These guides were unwilling to proceed beyond a certain point (the outer limit of their knowledge of the river). Below that all of the indigenes were said to be

---

167. Stanfield, *Red Rubber*, 71.
168. Ibid. Today this river port is named Puerto Asís.
169. Sofía Angulo Lamis (1860–1890) died prematurely. When Reyes penned his autobiography, he idealized and idolized his bride at considerable length.
170. Pasto became a strong point of reference for Rafael since his mother and sister were now living there as well. In describing the town he underscored its Basque legacy, "Pelota is the most popular game in Pasto, no doubt established there by Basque settlers of whom there are surnames such as Zarama and Astorquiza, who are of [the town's] most prominent families" (Reyes, *Memorias*, 233).
171. *Pastuzo* is the Spanish term for a native of Pasto.
172. This surname is sometimes rendered as "Larraniaga" by Anglo and even Colombian writers. I will employ the common Basque rendering throughout this narrative.
173. Ibid., 107–37.

ferocious cannibals. In short, like a contrarian Charles Marlow, Reyes was about to descend (rather than ascend) into the Heart of Darkness.

At one juncture, their canoe was swamped by a whirlpool, and they lost most of their ammunition, supplies, and documents. By the time Reyes and his companions reached the realm of powerful[174] Chief Chua of the Mirañas tribe, the explorers were in extremely bad condition. They received gracious hospitality for the next two weeks while they recovered. The two leaders bonded and the Indian chief asked for permission to adopt the name "Rafael Chua." He then provisioned his guests with canoes and supplies. Reyes describes the indigenes as totally naked savages. When he told Chua of his plan to establish a steamship on the river within the next few months, the chief requested that Reyes bring back some western clothing for himself and five of his wives.[175]

Rafael continued his journey and, about one hundred leagues upstream from the Solimões, came upon the boundary markers placed on the riverbank by Brazil and Peru. In his memoir Reyes comments that they had been erected ". . . in accord with treaties celebrated between these two countries and without the intervention of Colombia, which is the true owner of those territories."[176] He uprooted the stakes and threw them into the Putumayo.

In Brazil, and now devoid of his documentation and letters of credit, Rafael was unable to secure passage to Pará and was forced instead to sign on as a deckhand on a steamship. In Pará,[177] he reestablished his credentials and retained as his agent an influential rubber trader, Manuel Pinheiro. Reyes obtained two small steamships in Pará and then traveled to Iquitos, where he purchased a larger one (that he christened the *Tundama*) for service on the Putumayo. The Brazilian government assigned a launch to accompany (and protect) him. Rafael, his Colombian captain, and his twenty-man crew then sailed the *Tundama* upriver to La Sofía (dropping off Chua's garments along the way).

In La Sofía, Rafael loaded a cargo of Reyes brothers' quinine for the return descent of the Putumayo/Amazonas to Pará. The Brazilian launch left earlier as it was to stop along the way to cut wood to fuel the *Tun-*

---

174. It seems that Chua's sway spread over hundreds of square kilometers and encompassed about sixty thousand Indians, including the recently conquered Orejones, a subtribe of the Huitotos (Ibid., 132).
175. Ibid.
176. Ibid., 135.
177. Thanks to the rubber boom, at one point Pará actually came to have the largest merchant fleet in the world. Roger Rumrrill, "El sueño del celta y el paraíso del diablo" http://marioelescribidor.blogspot.com/2011.01/el-sueno-del-celta-y-el-paraiso-del.html, 5.

dama's steam engine. It was the dry season and navigation on the shrunken and braided stream was tricky. At one point, Reyes made a critical mistake by selecting a wrong branch and the steamship ran aground. The twelve-man contingent now faced the possibility of being stranded for months until the next wet season.

After several weeks, many of the men had died from yellow fever, and the rest were in a mutinous mood. They wanted to take the small life craft and attempt to descend to civilization on the Amazon. Reyes warned them of the mortal danger of passing through the territories of the many cannibal tribes along the way. Then, just when things seemed most desperate, an upstream thunderstorm provided the freshet that raised the Putumayo's level enough to free the *Tundama*.

This first voyage down the Amazonas became a triumphal journey that was feted in all of the Brazilian settlements along the route, and particularly in Manaos. Newspapers throughout the Amazon, as well as the international press, acclaimed the accomplishment, and Reyes was received in Pará as a celebrity.[178] He would lay claim to having been the first to pioneer steamship travel in Northwest Amazonia.[179] He had also just delivered the first cargo of sub-Andean quinine to ever be exported from Brazil (previously the product had always gone overland from the eastern Andes to Pacific ports in Ecuador, Peru, and Colombia). The sale of the shipment fetched the fabulous sum of more than one hundred thousand dollars.[180] Thus, according to Stanfield,

> Rafael accomplished much for the Compañía de Caquetá in two years: he had contacted Indian tribes who would collect quina; he had found an important import/export partner in Pinheiro; he had negotiated an agreement with the Brazilian government (although he was not a representative of his government); and he had acquired steamboats for the company and commenced importing and exporting on a grand scale. Reyes was thus hailed the "Amazonian hero," responsible for opening the Putumayo to progress.[181]

In 1876, the Reyes brothers petitioned the Colombian government for funds to build a road from Pasto to Mocoa and an annual subsidy to create a fleet of river boats that would ply the waters of all of Northwest

---

178. Reyes, *Memorias*, 183–84.
179. Ibid., 147.
180. Ibid., 181–85.
181. Stanfield, *Red Rubber*, 16.

Amazonia, linking them with the Atlantic, and thereby turn the region into one of the world's primary centers of fluvial navigation. Rafael obtained from the Brazilians permission for fifteen years to navigate their rivers, as well as exemption from Brazil's taxes and import duties.[182] However, the project was ill-fated from the outset. On its second voyage, the *Tundama* was shipwrecked in the Upper Putumayo. Reyes leased another vessel and managed to reach La Sofía with a large cargo of freight with which he introduced into the Pasto market heretofore unknown foreign luxury goods.[183] The project came to naught when Colombia plunged into another of its many internecine civil wars.

Like most whites who spent considerable time in the Amazon, Rafael Reyes would endure many bouts of tropical disease, some of which were quite life-threatening. During that first voyage from La Sofía to Pará, his health suffered considerably, and he was advised to continue on to the United States and Europe for medical treatment. While in the former, he contracted (in Wilmington, Delaware) for construction of a steamship, to be called the *Colombia*, with which he intended to explore and exploit the navigable rivers of Northwest Amazonia. He returned to Pará some months later to await arrival of the recently launched steamship, but it failed to show up. Later it was learned that the *Colombia* was shipwrecked in a storm off of Trinidad.[184]

Consequently, Rafael contracted for a Brazilian steamship, the *Fortaleza*, and returned to La Sofía for his next cargo of quinine. And thus began a curious adventure that underscored the growing tensions between Colombia and Peru regarding sovereignty in Northwest Amazonia. In 1876, in part to counter the machinations of the Reyes brothers in the Putumayo/Caquetá, Lima ordered its naval commander in the Loreto to station military, maritime, and police officials in the Putumayo.[185]

We are now in 1877,[186] and, when the *Fortaleza* reached the point on the Putumayo that Brazil and Peru had earlier accorded to be their border, a Peruvian named Zacarías accosted them and said that he was the frontier commander. Reyes demanded to see documentation of his commission and the man could produce nothing. So, Rafael took him prisoner and threatened to hand him over to authorities in Iquitos as an imposter. Zacarías confessed that his appointment was "informal" and

---

182. Dominguez and Gómez, *La economía*, 62, 70.
183. Ibid., 73.
184. Reyes, *Memorias*, 184–85.
185. Stanfield, *Red Rubber*, 16.
186. During this voyage Reyes and Francisco A. Bisao, the Portuguese captain of the *Fortaleza*, co-authored the first comprehensive map of the entire Putumayo River.

begged not to be punished. He swore loyalty to Reyes and agreed to be put ashore with tools and supplies. He promised to cut firewood for use in the steamer's engine on its return trip upriver.

Rather than fulfill his bargain, Zacarías abandoned the post and absconded to Iquitos with the supplies. There he spread calumny against Reyes—accusing him of trampling on the Peruvian flag and threatening to imprison any Peruvian who entered the Putumayo. Rather than a loyal woodcutter, then, what awaited Reyes on his return upriver was a Peruvian gunship, the *Morona*, under orders to seize the Colombian's cargo boat. However, the Brazilians had gotten wind of that plan and sent a gunship, the *Mearín*, to protect both Reyes's person and vessel under the terms of their freedom of navigation agreement with him.

The commander of the *Mearín* agreed to accompany Reyes aboard the *Morona* to give his side of the story. Rafael was able to convince both the Peruvian commander and his crew of his innocence and was allowed to proceed upriver to La Sofía. After delivering his next cargo in Pará, Reyes boarded the *Augusto* bound for Iquitos. However, the months of calumny spread there by Zacarias had earned the Colombian the sobriquet of "the pirate of the Amazon" in certain Iquitos circles, making it too dangerous for him to disembark publicly. Rather, Rafael snuck off the *Augusto* incognito and made his way to the governor's residence. There he was able to convince his interlocutor of the true facts of the case. Within days Reyes was being hailed in Iquitos as an important explorer who was opening up Northwest Amazonia for development.[187]

By 1879, the Casa Elías Reyes y Hermanos firm was employing hundreds of quinine gatherers, including Indians.[188] In his memoirs, Rafael Reyes noted that many of the latter now used western clothing and household furnishings and were no longer "semi-savages." He also claimed to have stopped (with the cooperation of the Brazilian government) the maltreatment and enslavement of the Indians in his Colombian jurisdiction.[189] Mocoa had become a boomtown through the export of cinchona collected in the Upper Putumayo and adjacent river drainages.[190] It came into its own just about the time that the collection of wild quinine in South America would be entirely eclipsed by the plantation production in southeast Asia. If, between 1881 and 1883, quinine represented fully 31 percent of Colombia's exports, by the following year it had disappeared

---

187. Reyes, *Memorias*, 186–88.
188. Gómez López, *Putumayo*, 160.
189. Reyes, *Memorias*, 171–72.
190. Gómez López, *Putumayo*, 158–61. Its population peaked at around three thousand during this period (Restrepo López, *El Putumayo*, 1).

from sight. Much of the harvest was left to rot, since the precipitous drop in its price would not even cover the cost of transporting it from the forest fastnesses.[191]

In 1881, or the initial stages of the rubber boom, Rafael and his youngest brother, Nestor, recruited a large contingent of *mestizos* and blacks as collectors and descended 500 kilometers down the Putumayo with them. They carved a station (named La Concepción) out of the wilderness. However, the initiative proved to be pretty much a disaster as three quarters of the men died of yellow fever. Nestor led a small party further down river to explore Huitoto territory—believed to be rich in rubber. These Indians were thought to be cannibals. When the explorers failed to return after a month, Rafael went in search of them,

> My anxiety and my pain were immense thinking that my poor brother had been sacrificed by those savages or that he had died of hunger in the jungles.... Torrential rains fell constantly. We sailed day and night, given that my anxiety was so great that I imagined that my unfortunate brother might be agonizing from hunger along the river banks and that every hour that I lost in reaching his side could cause his death. I fired my rifle constantly and shouted loudly, hoping that my brother would hear me.... Around every curve of the river I anxiously scanned both banks whenever the dense forest surrounding us permitted, in the hope of discovering him.... [T]he seventh day I saw a canoe descending ... [and] we recognized the Indians who had accompanied Nestor. One of them said: "Your brother turned up." My heart leapt and I made a sign of thanks to God. "Where is my brother?" I replied. And the Indian answered: "We only found on a beach his bones and part of his clothing; the Huitotos ate him."[192]

They retrieved Nestor's singed remains, including his skull. Reyes claimed that a few months later the Huitotos killed and ate another of his employees, (Basque-surnamed) Ochoa.[193]

Reyes left the Putumayo at this point, although his brother Enrique, Benjamín Larrañaga, and a few other stout Colombians set up more suc-

---

191. Dominguez and Gómez, *La economía*, 37, 43–44.
192. Reyes, *Memorias*, 174–76.
193. Ibid., 176. A contemporary of Reyes [Demetrio Salamanca] claimed that Nestor had actually killed himself (Dominguez and Gómez, *La economía*, 140). This raises the interesting question of whether Reyes concocted the cannibalism as a means of protecting Nestor's memory from the opprobrium of suicide. While possible, it seems a stretch. Reyes could have simply left his brother's death out of his *Memorias*, rather than providing an extremely personal and emotional fabricated account of it.

cessful rubber operations in the Middle Putumayo. In 1884, the market for wild cinchona bark collapsed. By then, however, the entire Putumayo/Caquetá region was beginning to feel the rubber boom. Enrique Reyes remained on the Putumayo in a rubber partnership with a Frenchman, Charles Mourrail, a resident of Paris and Iquitos. By this time, both the Ecuadorian and Peruvian governments were trying to tax the *caucheros* of the Putumayo.[194]

As early as 1887, the Colombian consul in Iquitos was sounding an alarm regarding destruction of rubber trees in the Putumayo/Caquetá. It seems that the *Castilloa* variety in question, the exclusive one in the upper rivers (while occurring along with *Hevea* in the middle and lower reaches), did not lend itself to "bleeding" because its sap coagulated in the incisions. So the trees were being felled for a one-time bonanza. Most of those close to the river systems were gone, and it was now necessary to go farther into the forest in search of new trees.[195] Yields plummeted sharply, and, indeed, by the turn of the century, Mocoa was in crisis. Many houses had simply been abandoned. Furthermore, the Indians were leaving the mission settlements and returning to the forest, prompting the fear that they might revert to their former savage ways.[196]

Actually, given its own internal strife (eight major civil wars and fourteen local ones in the nineteenth century alone), it is surprising that Colombia managed to maintain any sort of presence in the Putumayo/Caquetá. It was the most remote region of the country and the overland journey to there from Bogotá required weeks. Then, too, the Upper Putumayo often served as a dumping ground for exiled prisoners and political dissidents. Colombia's only real means of reinforcing its claim to the region was by establishing settlers on both the Putumayo and Caquetá rivers.

Peru possessed the geopolitical advantage that would prove decisive, at least over the short term. In trackless Northwest Amazonia, the rivers were the only "highways," and they all converged on the Loreto before accessing the Amazon and the wider world beyond. River transport was by far the cheapest and quickest means. Consequently, not only Peruvians but also Ecuadorians and Colombians living in Amazonia, came to depend upon the Loreto for trade, supplies and export. Unsurprisingly, many cross-national business partnerships emerged, as did the predominance of Iquitos within the regional economy.

---

194. Gómez-López, *Putumayo*, 162–65.
195. Ibid., 167–68; Stanfield, *Red Rubber*, 23–25.
196. Gómez López, *Putumayo*, 164.

Enter Arana

Julio César Arana del Águila Hidalgo was born on April 12, 1864, in the tiny town of Rioja[197] in the province of Moyobamba[198] of the Department of Loreto in the montane district of the Peruvian Amazon. He was a "[s]on of the matrimony of Don Martín Arana Hidalgo with the Señora María del Águila, both descendants of the Spanish race (Basque) and dedicated to agriculture and the manufacture of hats in said District."[199] The powerful Arana clan ("a network of pioneers, capitalists, and politicians"[200]), had roots in the highland City of Cajamarca—"where Pizarro and the Incas first came face to face."[201] The original ancestor had four sons, one of whom remained in Cajamarca and traded in precious metals. Another, Benito Arana, settled in Loreto and become that Amazon state's *prefecto* or governor. Gregorio went to Ayacucho and Huancavelica to the silver and mercury mines. Meanwhile, Martín, Julio César's father, established his hat business in Rioja.[202]

---

197. There is a suggestive Basque role in the founding of Arana's birthplace. Legend has it that it was settled in 1722 by three founding figures out of the remnants of several communities that were destroyed in an epidemic. Ovidio Lagos, *Arana, rey del caucho* (Buenos Aires: Emecé Editores, 2005), 14 [*Arana*]. One of the founders was the (Basque-surnamed) Félix de la Rosa Reátegui Gaviria. The full name of the new settlement was San Toribio de la Nueva Rioja. Calling the settlement "New Rioja" may have been invocation of the Rioja region of Araba in the southern Basque Country. San Toribio was from adjacent Santander and a saint with considerable devotion throughout northern Iberia.

198. Historically, the Peruvian reaction to the rubber boom in Middle and Lower Putumayo came from Moyobamba and its Huallaga hinterland. There was a collective response among Huallaga's *mestizos* and Christianized Indians as well. The rubber boom triggered an exodus that reduced Moyobamba's population from 15,000 in 1880 to 7,000 in 1904. The Christianized Indian communities of the Huallaga were all but abandoned. Indeed, these Indians, accustomed as they were to work under white domination, proved to be the most reliable initial source of labor and were contracted as *enganche* gatherers by the aspiring *caucheros*. The *mestizo peones* were also candidates for *enganche*, but were less given to hard work and more prone to strike out on their own into virgin forest in search of rubber. Fernando Santos Granero and Frederica Barclay, *La frontera domesticada: Historia económica y social de Loreto 1850–2000* (Lima: Pontificia Universidad Católica del Perú, Fondo Editorial, 2002), 65 [*La frontera*].

199. Quoted in Fray Gaspar de Pinell, *Excursión apostólica por los ríos Putumayo, San Miguel de Sucumbios, Cuyabeno, Caquetá y Caguán* (Bogotá: Imprenta Nacional, 1928), 196 [*Excursión*]. This quote was taken from a mid-1920s issue of the illustrated magazine *Cultura Peruana* (Lima). It published a series of laudatory articles on Arana that are quoted verbatim and at length by Pinell. I have been unable to consult the originals and will rely upon Pinell's accuracy and veracity.

200. Marie Arana, *American Chica: Two Worlds, One Childhood* (New York: The Dial Press, 2001), 42.

201. Ibid.

202. Ibid.

When Julio César was but three, his uncle Benito was navigating Peru's Ucayali, Pachitea, and Palcazu rivers to demonstrate that they could be opened to commerce. Potential entrepreneurs were put off by the disappearance of two young sailors, alleged to have been killed and eaten by cannibalistic Cashibo Indians. In 1867, Benito went to their main village and demanded to see Chief Yanacuna. His furious wife came out of the main dwelling instead,

> The chief's wife flung down two jawbones—two sets of teeth—at Governor Arana's feet. *There are your "boys,"* she snarled. *Take a look. The same could happen to you.* When Benito Arana returned to Iquitos, it was as Moses descending from Mount Sinai with commandments: The rain forest Indians were beasts, not people. They were less than simian, incapable of real, human feeling. Henceforward they would be dealt with as animals. And with that, a road was paved: two decades later, my ancestor Julio César would travel it.[203]

When he was fifteen, Julio César was sent by his father to Yurimaguas to study bookkeeping and business for a couple of years. The move was designed to keep him out of the War of the Pacific with Chile.[204] In his late teens he traveled many of the rivers of Northwest Amazonia selling his father's hats, and probably other goods as well. This early background as an *aviador*, or river trader, provided Arana with the intimate knowledge and critical contacts to take full advantage of the three-decade rubber boom that would engulf the area in the late nineteenth and early twentieth centuries.

There was a sensitive and caring side to Julio César. When he was but twelve years old he was writing romantic poetry to his next-door neighbor and the future love of his life—fifteen-year-old (Basque-surnamed) Eleonora Zumaeta. She ignored her young suitor, and it wasn't until several years later, and only when returned to the Amazon after spending two years in Lima studying to become a schoolteacher in the Yurimaguas district, that she was impressed enough by the persistent young entrepreneur to marry him in 1887.[205]

Shortly after his marriage, Julio César opened a trading business in nearby Tarapoto in partnership with his new brother-in-law, Pablo Zu-

---

203. Ibid., 43. Shades of Rafael Reyes's experience with cannibals on the Putumayo fourteen years later!
204. Jordan Goodman, *The Devil and Mr. Casement* (New York: Farrar, Straus and Giroux, 2010), 36 [*The Devil*].
205. Lagos, *Arana*, 16–20.

maeta.²⁰⁶ Pablo would become his most trusted associate throughout Arana's business career. Henceforth, the jungles and many rivers of the Upper Amazon were Arana's playing field. Initially, Arana left his wife and growing family behind in Yurimaguas, visiting them only periodically. He often spent months at a time plying the rivers as a hands-on *aviador*. Indeed, his domestic life for the next two decades would be punctuated by such lengthy absences. At one point, he returned to Eleonora suffering from beriberi and on the verge of death. It took him six months before he could resume his hectic schedule.²⁰⁷

The life of a river trader had its many challenges. In addition to the constant threat of tropical disease, there was the almost endless boredom during the weeks of travel throughout the vast watery network. It was not unlike serving a prison sentence. This could lead to a profligate lifestyle. Many married men brought along their concubines and/or availed themselves of brothels along the way. Most struggled with alcohol and the temptation to gamble compulsively in order to pass the time. Julio César Arana appears to have been one of the exceptions²⁰⁸—an extraordinary man who configured both self-discipline and self-improvement out of the enforced solitude and patience. While never endowed with much formal education, Arana was an autodidact and avid reader who would eventually own the largest personal library in the Peruvian Amazon.²⁰⁹

In 1890, the soaring prices of rubber prompted Julio César Arana to recruit twenty workers from distant and drought-stricken Ceará in northeastern Brazil, now tied to their patron through inescapable debt peonage.²¹⁰ Arana established them as gatherers on some rubber concessions that he had obtained near Yurimaguas. His need to go so far afield for his workers itself underscored the industry's acute labor shortage.

By then, there was a tight network of interrelated rubber extrac-

---

206. Ibid., 29.
207. Collier, *The River*, 51–52.
208. There is no evidence that Julio César was ever unfaithful to Eleonora—perhaps extraordinary given their lengthy separations and the cultural acceptance among the South American elite of the extra-marital affairs of married men. Regarding his sobriety, Arana's words when testifying before the Select Committee were, "I am not in the habit of drinking." British Parliament, *Special Report*, 482.
209. Lagos, *Arana*, 27. One might surmise that the boring, weeks-long river trips leant themselves to the escapism afforded by reading.
210. Each owed Julio César 30 pounds for his fare and was then charged 70 more for his provisions and gear (that had an actual value of about 4 pounds) before setting out for a three-month stay in the forest. Dominguez and Gómez, *La economía*, 179.

tors and shippers based in the area.[211] The Morey Arias brothers (Luis Felipe and Adolfo), sons of a Spanish naval officer and an Ecuadorian mother, were both born in Tarapoto and would eventually become the predominant traders and shipping magnates of the Loreto. Luis Felipe was educated in Pará, providing him with Brazilian connections and fluency in Portuguese that would later serve him well. The Morey brothers got their start, as did Julio César, in the Panama-hat business. Along with a third brother, Juan Abelardo, they became rubber barons in the firm Morey Brothers, incorporated in Yurimaguas in 1885. It was dissolved three years later and, in 1888, Luis Felipe formed the Morey and Dávila shipping company in Yurimaguas and the next year another called Morey, Dublé, and Company operating on the upper Ucayali River. In 1890, he constituted Vega, Morey, and Company in Iquitos in association with his brother-in-law Juan C. del Aguila and Colombian rubber baron and rubber trader Juan B. Vega. Their new firm quickly became the leading trading house in the city. The *Pastuzo* Vega was the Colombian consul in Manaos. In 1892, Morey and Del Aguila formed another shipping company in Iquitos with a branch on the Yavari River and that same year purchased the Marcial A. Piñón Company, which owned several steamships. In 1896, Luis Felipe entered briefly into association with the Frenchman Charles Mourraile and Cecilio Hernández (uncle of Julio César Arana's wife Eleanora Zumaeta) in the purchase of a steamship to service a direct route between Iquitos and Liverpool. Again, the importance of family ties is evident in that two of Hernández's sons were married to two of Morey's daughters. Then, in 1897, Mouraille, Hernández, and Magne sold their largest steamship to Fermín Fitzcarraldo and his partner Adolfo Suárez. Mouraille left the firm and, in 1900, Hernández, Magne and Company controlled 9 percent of Iquitos's rubber exports. In 1904, Cecilio sold out and set up his own new firm with his sons.

Adolfo Morey Arias, born in Tarapoto in 1862, moved to Yurimaguas in the early 1880s to engage in commerce and the hat business with his sister. In 1885 he was in Morey and Brothers in rubber collection and in 1888 moved to Iquitos and founded the merchant house A. Morey and Company, with a branch in Yurimaguas. Two years later it was among the city's top twenty-five trading firms and controlled 6 percent of the Loreto's rubber exports. Along with Anselmo del Aguila he purchased river boats that sailed under both the Peruvian and Brazilian flags. By 1904 he owned four steamships that he had built in Great Britain, France, and Germany. They represented 22 percent of Iquitos's

---

211. See Santos-Granero and Barclay, *Tamed Frontiers*, 59–63, from which the following synopsis is derived.

merchant tonnage. He ran one of two shipping links between Iquitos and Yurimaguas. Beginning in 1914, after the rubber crash, he engaged in intercontinental navigation with lines between Iquitos and Europe and the United States.

It was in 1889 that Julio César moved to Iquitos whence to trade along the Yavarí in partnership, beginning in 1891, with the *Pastuzo* Juan B. Vega.[212] The firm of Vega and Arana soon (1893) associated with Mourraille,[213] Hernández, Magne, and Co., an operation headquartered in Nazaret on the Yavarí River.[214] The Hernández was Cecilio, Julio César's affinal uncle.

By 1896, Julio César was able to buy out his partners and became the primary owner of J. C. Arana y Hermanos—also known as the Casa Arana. His partners were his brother Lizardo and two of his brothers-in-law, Pablo Zumaeta and Abel Alarco. Julio César purchased a "fine" residence in Iquitos and moved his family there. According to Collier,

> His prodigious industry, his contempt for Manaos' free-spending "colonels," led journalists to dub him "The Abel of the Amazon." From 1896, the year he moved his family to a fine ten-room house at Iquitos, on the Peruvian Amazon, he had made his credo plain. Though a family group of twenty-four often sat down to dinner, no business contact ever got past his threshold. The nearest thing to luxury was a cow grazing on a waste plot to ensure milk for his children. And in place of glittering *azulejo* tiles, Arana faced his house with the motto, carved in stone, that was now his watchword: "Activity, Constancy, Work."[215]

Collier recounts how Julio César and Pablo Zumaeta were attacked by Indians from ambush. Arana was in grave difficulty when Zumaeta weighed in and pulled off an attacker. The two men then stood shoulder to shoulder and bested the Indians in fisticuffs. According to Collier,

> From this encounter, Julio bore away two resolutions. He wanted men around him united by blood ties, men who, whatever their shortcomings, he could literally trust with his life, and in the years that followed he was to find work not only for Pablo but for five more of Eleonora's relatives and for his own brothers Francisco and Lizardo. For nothing, neither Indians nor white men nor dis-

---

212. Pineda Camacho, *Holocausto*, 72.
213. The same Mourraille who was involved earlier in the Putumayo with the Reyes Brothers.
214. Pinell, *Excursión*, 197.
215. Collier, *The River*, 50.

ease nor the latent horrors of the jungle, would deflect him from what he had set out to do.[216]

During that same year, 1896, there was a separatist insurrection in Iquitos that fizzled out before the government's expeditionary leader, (Basque-surnamed) Colonel Ibarra, arrived on the scene to become the Loreto's military commander.[217] Ironically, the uprising coincided with publication of former governor (and Julio César's uncle) Benito Arana's book of correspondence (including his own) regarding his 1867 exploration. The book underpinned the argument favoring completion of a road from the highlands to Huánuco and then on to Mayro to make it the terminus of a fluvial transport route down the Ucayali River to Iquitos—thereby bonding the Loreto to its national capital, while opening the Oriente to enterprise, immigration of new settlers, and ultimately linking of South America's Pacific and Atlantic coasts.[218]

The Casa Arana soon had business ties in Lisbon, Paris, Manchester, and particularly London.[219]

Arana played his political cards well in Iquitos throughout this period. He was appointed first president of the board of the Junta Departamental (1897). He championed a school tax and became the school superintendent.[220] He also founded a bank in town. Arana later presided over the Iquitos Chamber of Commerce and, in 1902, he would be elected mayor. In short, Arana's economic clout was now firmly wedded to growing political influence.

Meanwhile, by the last decade of the nineteenth century, the rubber boom had created its own multinational rush into the Putumayo/Caquetá. First, we might consider the Colombian narratives of the events. In 1896, the Colombian consul in Iquitos contended that many intending Colombian rubber entrepreneurs were entering that region and that there was already a substantial Peruvian presence there. He surmised that the latest influx might very well decimate the Indians, "as has been done in the Peruvian forests where the grimmest crimes have been perpetrated upon the unfortunate savages." The Colombians were usurping lands that had been cleared for cultivation by the Indians—"an atrocious crime because it condemned their owners to death from starvation."[221] In an 1899

---

216. Ibid., 44.
217. Stanfield, *Red Rubber*, 93.
218. Benito Arana, *Lima al Amazonas* (Lima: Imprenta y Librería de San Pedro, 1896). Ex-Prefect Arana's book makes no mention of his nephew, Julio César.
219. Collier, *The River*, 52.
220. Stanfield, *Red Rubber*, 94.
221. Gómez López, *Putumayo*, 167.

dispatch to Bogotá, he claimed that there were "about 300 Peruvian men exploiting Putumayo rubber and some of them were dedicated to committing acts of piracy, inflicting armed assaults upon the unfortunate Indians and exterminating them or taking them prisoner to be sold later."[222]

For our purposes, one of the critical questions becomes: How many Indians were there in Aranalandia? Joaquín Rocha, writing about his visit to the area in 1903, referred to the broader region that included much of the Middle Napo, Putumayo, and Caquetá as Huitocia. He placed its overall population at as many as 250,000 Indians, although there were clearly many other tribes within it that were not Huitotos.[223] Pineda Camacho calculates the population of more or less the same area as about 72,000 in the year 1900.[224] Calderón believes that by this time there were 15,000 indigenes on the Igaraparaná River and 10,000 on the Caraparaná (the core of the future Aranalandia).[225] According to ship's captain Enrique Espinal, in his report filed in Iquitos on October 25, 1902, there were 50,000 Indians in Aranalandia.[226] In 1903, the Capuchin missionary Basilio de Pupiales, M. C. corroborated this estimate by stating that there were 20,000 Indian workers in the Igaraparaná region (the indigenous numbers would rise to 50,000 when women and children were included), as well as 120 whites by then.[227] Under continuing Colombian and Peruvian domination, the estimated Indian population of the area had been reduced to somewhere between 7,000 and 10,000 just a few years later.[228]

---

222. Ibid., 172. However, reference must be to the entire Putumayo drainage, since Pinell provides a list of all such operators on the Caraparaná and Igaraparaná and not a single one was Peruvian (Ibid., 210–11). Indeed, that pretty much remained the case as late as 1905 (Ibid., 211–12).
223. Joaquín Rocha, *Memorándum de viaje: Regiones amazónicas* (Bogota: Casa Editorial de 'El Mercurio', 1905), 105–108, 111, 124.
224. Pineda Camacho, *Holocausto*, 48.
225. José Gregorio Calderón, "Importantes documentos históricos. Contestación del señor José Gregorio Calderón," *Huila histórico* 11 (September 1933), 272.
226. Enrique Espinar, "Viaje al Igara-Paraná, afluente izquierdo del río Putumayo, por el Capitán de Navío Enrique Espinar," in Carlos Larrabure y Correa, *Colección de leyes, decretos, resoluciones y otros documentos oficiales referentes al Departamento de Loreto* (Lima: Imprenta de "La Opinión Nacional," 1905), vol. 4, 220.
227. Cited in Gómez López, *Putumayo*, 26.
228. Casement Report reproduced in W. E. Hardenburg and Introduction by C. Reginald Enock (ed.), *The Putumayo, the Devil's Paradise; Travels in the Peruvian Amazon and an Account of the Atrocities Committed upon the Indians Therein*, 2nd edition (London: T. Fisher Unwin, 1912), 336. Prefect Hildebrando Fuentes placed the number of indigenes working for the Arana y Vega company (formed in 1904 and dissolved about two years later) in Arana's domain at 13,603, but the figure includes tribes that most regarded as distinct, such as the three thousand Boras. Hildebrando Fuentes, *Loreto: apuntes geográficos históricos, estadísticos, políticos*

There is lack of precision regarding first white settlement of the remote region. We do have the Rocha account, which attributes the Colombian mulatto Crisóstomo Hernández with having been its first non-Indian settler. Hernández moved to the Garaparaná [sic] with a few followers in 1896 or 1897, fleeing punishment under warrants issued to arrest him for crimes (including homicide) committed in the Caquetá. By the time of Rocha's visit, Hernández had been shot to death by one of his associates. However, the Colombian mulatto was certainly a legend—brave, intelligent, bold, and calculating, yet passionate and quick-tempered. He was extemely fluent in Huitoto and was even a skilled orator in the language, capable of winning over its leaders with a combination of persuasion and ruthlessness—while treating whites and Indians equally in this regard. In the spirit of "work and peace or perpetual war," he conquered all of the Caraparaná and much of the Igaraparaná[229] by employing tribes that had already acquiesced against the recalcitrant holdouts. Within a short time he had become the Huitotos' "King and God." He divided the "nations" (defined as the occupants of two or three *malocas* who recognized the same supreme *cacique*) among his followers. He had his own vessel that brought trade goods from Manaos and took to there the region's rubber.[230] This was a faster and far more economical journey than the one to Pasto or any other Colombian town of consequence. Indeed, the Putumayo was closer to Manaos than to Iquitos.

It had been shortly after 1896 that Julio César Arana began selling goods to the Colombians in the Putumayo when,

> Pablo Zumaeta, Lizardo Arana, and Abel Cardenas sent their steam-launch the *Galvez* to the Caraparaná, where [ship's captain] Lizardo Arana found many Colombians established and brought from them 6 tons of rubber.[231]

When Rocha arrived in the Putumayo, there were still many Colombians, although Arana was by then its most powerful person. Rocha noted that the year before his arrival there, the Indians of Huito-

---

y sociales (Lima: Imprenta La Revista, 1908), vol. 2, 114. Casement, for example, notes that while the Boras and Huitotos were probably distantly related in some remote past, they currently exhibited marked cultural and linguistic differences. Casement Report reproduced in Hardenburg, *The Putumayo*, 269. Judge Rómulo Paredes, at the time of his visit there, calculated the indigenous population in Aranalandia to be seven to eight thousand persons, divided into seven distinct tribes. Alberto Chirif (editor and annotator), "Los informes del Juez Paredes," in Alberto Chirif (ed.), *Cien años después del caucho: Cambios y permanencias en las relaciones con los pueblos indígenas* (Lima: Tarea Asociación Gráfica Educativa, 2009), 111.
229. Rocha, *Memorándum*, 105.
230. Ibid., 138.
231. Mitchell, *Sir Roger*, 688.

cia had risen up and killed thirty whites. The killings seem to have been widely dispersed yet simultaneous and hence likely coordinated. He noted that [José] Gregorio Calderón subsequently convened several Indian *caciques* and asked them why they had rebelled. The answer was that the whites had usurped their lands, and the Indians wanted them back. Calderón replied that the Colombians (there were about eighty at the time on the Caraparaná and Igaraparaná)[232] were cultivating land that the Indians previously ignored, and that once the whites were done with the area (presumably after the rubber ran out), the fields would revert to the Indians. There was a truce in effect, and, in an aside, Rocha suggested that only time would tell if the white man's reasoning would prevail.[233]

Hernández may actually have settled the area earlier than 1896, given the account of one of its most prominent Colombians—José Gregorio Calderón. Calderón first settled in the Caquetá region in 1886 and then, in 1896, joined his brother Teófilo on the Caraparaná River. By then, Teófilo had been there but two months and was a partner in rubber gathering with Benjamín Larrañaga,[234] the earlier Putumayo explorer and companion of Rafael Reyes.

This Colombian partnership then set about dominating three Huitoto sub-tribes, as well as Ocainas, Andokes, and Boras. According to Calderón, "It was pretty risky for us to conquer those barbarians," because the first whites to enter their area as traders (two men and the sister of one of them) learned that the Indians were fomenting a plot to kill and eat them.[235] Calderón noted that one Indian chief alone, Cheigamuy, had killed over one hundred whites.[236]

Calderón and his brothers, too, became the objects of Indian plots, but managed to head them off with marvels. At one point, he removed and then replaced his false teeth, to the amazement of the Indians. On another, when there was a plot to kill the outsiders, one of the Colombians hid and then set off a charge of explosives that terrified the Indians. They then convinced the indigenes that it had been the work of Calderón, who, as a witch doctor (*brujo*), knew about their plot and punished them with the lightning-like blast. According to José Gregorio, "In this fashion and by gentler means we

---

232. Rocha, *Memorándum*, 128.
233. Ibid., 125.
234. Demetrio Salamanca knew Larrañaga in Mocoa in 1876 and described him as having "pronounced signs of lower-class upbringing, ignorant and extremely uncultured; he was then a worker in the extraction of quinine in the agencies of Campucana; he had a bad reputation; and, in effect, he was a moral low-life; a regular drunkard, handsome, a bully." Quoted in Dominguez and Gomez, *La economía*, 141.
235. Calderón, "Importantes," 270.
236. Ibid., 273.

made them love and obey us." So the Calderóns and Larrañaga were able to convince the Indians to gather rubber in return for trade goods.[237]

Then, in July 1898, José Gregorio and Teófilo Calderón, accompanied by Larrañaga, descended the Igaraparaná by canoe to its confluence with the Putumayo where they "discovered" rapids (La Chorrera) and a large Indian settlement. They continued down river with a large cargo of rubber all the way to Pará in Brazil where Larrañaga sold it.[238] However, according to Calderón, Larrañaga cheated the brothers, and they had a falling out. The Calderóns settled on the Caraparaná, where they subjected and incorporated several sub-tribes into their operation.[239]

There is a competing account, varying considerably from the foregoing in its details, published by a different (and apparently unrelated) Calderón—Abel. It recounts his two-year voyage through Huitocia, beginning on January 18, 1900.[240] Abel reports visiting several Colombians in the Middle Caquetá, including Paulino Solís, a *cauchero* who had two hundred *arrobas* of rubber stolen by river "pirates," and the trading agency of the merchant Mauricio Cuellar. He also stopped at the property of the Colombians Emilio and Urbano Gutierrez, "a site famed

---

237. Ibid., 270-71. Rocha observed that after the death of Hernández the Indians were never subjected to violence by the Colombians, in contrast to the treatment meted out by the more stern Peruvians. Rocha, *Memorándum*, 124. As we shall see, this was not exactly accurate regarding the Larrañagas.

238. There is a competing account in a 1904 report sent to the Colombian minister of foreign affairs by Gerardo de la Espriella contending that the there were three Calderón brothers involved, as well as some Gutiérrez ones. The latter took their share of the rubber up the Caquetá for sale in Tolima, whereas the Calderóns and Larrañaga proceeded to Iquitos with theirs. Cited in Gómez López (ed.), *Putumayo: La vorágine*, 80. It actually makes more sense that Iquitos, rather than distant Pará, was their destination.

239. Calderón, "Importantes," 271.

240. Abel Calderón, *Viajes de Caquetá y Putumayo* (Bogotá: Imprenta Hernando Santos, 1904). Calderón, a political conservative, was banished as a political prisoner to Paiguarra in the headwaters of a tributary of the Caquetá. He escaped from there and fled downriver into the jungle with his brother-in-law, Aquileo Torres, a future *cauchero* in Aranalandia. When he published his booklet in 1904, he dedicated it to Rafael Reyes, who had just assumed the presidency of the nation. The dedication reads,

> Upon launching this account of my travels through the deserted regions of the Caquetá and Putumayo to the Amazon, I have believed it to be my duty to single out your name, as one of the most audacious and bravest explorers. The blood of your brothers bathed these expansive regions that you brought to light with your actions, and the Colombian who traverses them can discern everywhere the signs of your passage.

Calderón states that it is his purpose in writing to warn of the dangers that Peru and Brazil pose to Colombian sovereignty in this rich territory, while believing that saving it for his nation was one means of assuaging bruised national honor after the Panama debacle (Ibid., 7).

for the series of outrages and murders that have been perpetrated there."[241]

Regarding the first white settlement of Aranalandia, Calderón attributes one Sr. Mejía, who conquered the Huitotos of the Upper Caraparaná and was later attacked and besieged for three days by the Indians, with being the pioneer. Once the Huitotos were defeated, they resettled in the Lower Caraparaná and on the south bank of the Putumayo. Meanwhile, Mejía abandoned his trading post and went to Brazil, his place being taken by the Colombian Pizarro brothers who were still there when Abel Calderón arrived in the area. Abel notes that the nearby area known as Cuirá was first discovered "by the *Pastuzos* Crisóstomo Hernández, Domingo David, Antonino Ordóñez, Antoni Martínez, Carlos Lemus and others."[242] However, Hernández scarcely stands out in this account. Indeed, according to Abel Calderón, many Colombians predominated throughout Huitocia, particularly persons from Pasto, including Benjamín Larrañaga.[243]

Calderón notes that Larrañaga entered the Putumayo region with trade goods (date unspecified—probably mid-1890s) and discovered a large Indian population in the Upper Iguaraparaná. He exchanged his wares for their beeswax and venom, both of which had value in Pasto. He was able to place the Indians under his control. Larrañaga was the first to discover the latex called *siringa* that he claimed to be superior to the familiar "black rubber," or *hevea*, that was the main target of the industry throughout the Amazon. He was deemed "crazy" until he managed to bring eighty *arrobas* of *siringa* from the Igaraparaná back to Pasto.[244]

On his return trip to the Igaraparaná with more trade goods, Larrañaga purchased three hundred *arrobas* of *siringa* along the way, but was then intercepted by the same twenty-five pirates who had assaulted Solís earlier. Greatly outnumbered and accompanied by his adolescent son, Rafael, Benjamín had little choice but to surrender his goods to the usurpers. He therefore resolved to avoid that danger in the future. He next accumulated three thousand *arrobas* of rubber and set out down the Putumayo (much of which was unknown to him) to its convergence with the Ama-

---

241. Ibid., 12.
242. Ibid., 13.
243. Dominguez and Gómez (*La economía*, 167–69) cite a list of Colombian caucheros who settled in Aranalandia between 1901 and 1903 on a total of fifty-one properties. These included no fewer than nineteen owned by *Pastuzos*, and twelve each by persons from Tolima and Huila respectively. When an additional three by owners from "Cauca" are added to the list, there is almost unanimous provenience of ownership from that department (the only other two properties being in Portuguese hands).
244. Calderón, *Viaje*, 15.

zon. He then transferred his cargo to a river boat and sailed all the way to Pará-Belem. It was there that he sold his rubber for the equivalent of 200,000 soles.[245]

Meanwhile, according to Abel Calderón, in 1900 Larrañaga traduced his fellow Colombians by putting himself under the protection of the Peruvian military in return for the concession to exploit all of the rubber and Indian tribes in the region. Larrañaga thereby became "the most efficacious instrument that Peru had in order to penetrate the Putumayo drainage, because of the knowledge that this gentleman had of the river and the zone drained by it and its affluents."[246] Larrañaga was also capturing Indians and sending them to be sold in Iquitos.[247]

By late 1900, Benjamín Larrañaga was being transported throughout the Igaraparaná by the Peruvian warship *Cahuopana*. Several Colombians were accused of having collected rubber illegally in Larrañaga's (Peruvian) concession and were ordered to deposit their harvest with him for safekeeping at La Chorrera. These included Abel Calderón and his brother-in-law Aquileo Torres. Then, the following year, an armed ship named the *Putumayo* (now owned by Arana) ascended the Igaraparaná, but, instead of returning their rubber to them as promised, several Colombians were detained for trespass. The Peruvians sailed for Iquitos with their captives and turned them over to the authorities. Thanks to the intervention of the city's Colombian consul, the prisoners were soon freed. Meanwhile, Abel Calderón had not been arrested (he does not state why), and he sounded the alarm among the remaining Colombians.[248]

He now traveled down river on the *Putumayo*, accompanied by an Indian infant boy given him by the parents to be raised and educated along with Abel's own children. The ship's captain offered 500 soles for the Indian child. When told that "in Colombia there was no slavery and that here we are not accustomed to sell [our fellow humans] like animals or things," he was informed that the money was to recom-

---

245. Ibid., 15–16.
246. Ibid., 17.
247. Lagos, *Arana*, 78. A 1903 report by Pedro Antonio Pizarro to the Colombian minister of foreign affairs noted that Benjamín and Rafael, father and son, were guilty of "numerous assassinations, arsons and thefts," and they therefore sought Peruvian protection by incorporating their affairs in Iquitos. The Larrañagas and their partners came to control the richest and best part of the Putumayo/Caquetá by "passing over the cadavers of many Colombians and the remains of various establishments that pertained to Colombians, including mine." Cited in Gómez López, *Putumayo*, 31–32. Pizarro was subsequently named Colombia's honorary consul in the Putumayo/Caquetá.
248. Calderón, *Viaje*, 17–19.

pense him for the time and expenses expended thus far on the child's behalf. The Colombian concluded, "They [the Peruvians] disguise with subterfuges this hateful trade in new slaves; but it does not remove the evident truth of this infamous commerce."

The Peruvian commander then produced a map that had been commissioned in England by the Peruvian government. It showed all of the Putumayo drainage to be Peruvian, not to mention an "enormous part of Ecuador" and much of the Caquetá as well. He added that it was the Peruvian flag that was displayed throughout the area. It was Peru, not Colombia, that maintained the only customs house in the region (where the Cotuhué joined the Putumayo). Calderón's response was that a map published by one of the interested parties was scarcely proof of anything, and that the rest was evidence of neglect by the Colombian government of its own citizens rather than recognition of Peruvian sovereignty over the disputed area. As further evidence of his government's failure to exert effective governance in the region, Abel noted that Brazilians were pretty much in control of the lower Caquetá and would have ascended further if not for the formidable cascades at Araracuara.[249]

The following year, the *Putumayo* ascended the Caraparaná in search of more trespassers, but was stymied by the forewarned Colombians. The intent to take them as prisoners to Peru failed, but the Peruvians carried off many Indians. This is the confrontation reported in 1902 by José Gregorio Calderón, when the "ignoble" Larrañaga brought Peruvian authorities to La Chorrera and dispatched others up the Caraparaná in an attempt to impose his will on the whole region. The Calderóns had resisted and, despite two indiscreet shots directed at his gunboat from ambush by the defenders, the Peruvian commander proved understanding. He also got the clear message that the patriotic and fortified Colombians were not about to tolerate without a fight the incursion of Peruvian authority into their territory.[250]

Abel Calderón recounted the sad case of two Colombians named Muñoz and Silva who were settled on the banks of the Campuya River (an affluent of the Putumayo). One day while they were absent, a Peruvian boat arrived, and its crew began boarding their rubber. When their Indians resisted, many were slaughtered, and the survivors were placed in bondage for sale in Peru. The marauders then burned down the house. When Muñoz and Silva suddenly arrived on the scene, the Peruvians hesitated to kill them and instead enticed them on board with the promise that they would all sail to Manaos where the rubber would be sold and

---

249. Ibid., 18–19.
250. Calderón, "Importantes," 272, 274.

the Colombians paid for it. Instead, a few days into the journey, they were ordered off the boat and abandoned on a forelorn sandbar. They would have perished had not the Calderón brothers happened to be passing by a few days later.[251]

From their El Encanto base, the Calderóns had established several sub-stations along the left bank of the Igaraparaná, but their troubles with the Larrañagas were ongoing. At one point, they dispatched their employee, Alberto Valderrama, to Bora country across the Igaraparaná, where he founded a station called Providencia and conquered thirty-seven sub-tribes. But Valderrama was killed by Peruvians and the Calderóns then sold Providencia to Julio César Arana to avoid more conflict with the "calamitous" Larrañaga. They also founded La Unión station on the banks of the Caraparaná, but abandoned it more out of fear of trouble with the whites (particularly Larrañaga) than with the Indians.[252] Even so, as late as 1903, Larrañaga maintained his ties with Sibundoy and Pasto, enticing people from there to come and settle in the orbit of La Chorrera by advancing them expense money.[253]

In 1899, Larrañaga sold his rubber for the first time to the firm of Arana y Hermanos of Iquitos. Always the consummate *aviador*, Julio César's strategy was to extend credit to the established Colombians—entering into partnership with some of them. Meanwhile, Larrañaga formed a partnership with F. Barchillón, a Jew from Tangiers who was living in Brazil, and J. M. Moris Ramírez, an influential Peruvian and Arana partner. The Peruvians who preceded Arana in the Putumayo were notorious for having captured its Indians to be sold as slaves throughout Upper Amazonia.[254] As late as 1904 it could be stated,

> In Cotuhé there is a big Peruvian settlement with the character of a customs house, and that serves as the point of departure for the innumerable invasions that are organized and equipped for the hunting of Indians. These bands of executioners who are so inappropriate called rationals, are those that are used by Peruvian high officials to dominate our territory. There are always 20 or 30 of these soulless men who go armed with Winchester carbines to each tribe, and, it is embarrassing to have to say, accompanied by Colombians who have embraced the Peruvian flag, some out of bad instincts and some others due to

---

251. Calderón, *Viaje*, 20.
252. Calderón, "Importantes," 273.
253. Dominguez and Gómez, *La economía*, 142.
254. Demetrio Salamanca T., *La Amazonia Colombiana: Estudio geográfico, histórico y jurídico del derecho territorial de Colombia* (Bogotá: La Imprenta Nacional, 1916), 555 [*La Amazonia*].

the neglect with which our governments have treated that marginated region.²⁵⁵

Regarding the presence of latex in the Putumayo/Caquetá, according to Stanfield,

> A number of different rubber trees grew throughout the Putumayo. Various species of *Sapium* flourished in the temperate climes of the Upper Putumayo and yielded a latex called *caucho blanco*. Two *Castilloa* (*Castilla*) species, *Castilloa ulei* and *Castilloa elastica*, grew alongside *Sapium* but also adapted well to hotter regions of the Middle and Lower Putumayo; the plentiful *Castilloas* produced a rubber called *caucho negro*. *Hevea brasiliensis*, which produced the finest rubber in all Amazonia, called *jebe fino* or *siringa fina*, existed in great quantity in Brazil but was confined to the far southwestern corner of the Putumayo. Its relative *Hevea guianensis*, however, could be found in the Middle and Lower Putumayo and relinquished a lower-grade jebe, called "jebe débil." Caucho and jebe were the two most important varieties of Amazonian rubber exports, but they were often found in different soils, and their latex was extracted in distinctive manners. *Hevea* and *Castilloa* trees could be found close to one another, but it was unusual to find *Castilloas* in low-lying flooded areas, conditions *Heveas* could stand. Therefore, *Castilloas* could be exploited almost all year, for they grew above flooded areas; on the other hand, *Heveas* could only be tapped about six months of the year because the rainy season and flooding disrupted operations. Another crucial difference between them centered on extraction methods; jebe latex flowed down a series of shallow incisions cut into the *Hevea* bark, and a tapper collected this rubber over a period of years without severely damaging the tree; conversely, caucho latex could not be tapped because the *Castilloas* yielded little latex and incisions often invited lethal fungi, so the tree was drained of all its latex at one time. Deep cuts in the trunk, branches, and roots of the *Castilloas* exposed a great quantity of rubber, but also killed the tree in the process.²⁵⁶

By the time that the Casa Arana began expropriating the future Aranalandia (1901–1907), most of its *Castilloas* were gone, but the region

---

255. Cited in Gómez López, *Putumayo*, 81.
256. Stanfield, *Red Rubber*, 23.

was rich in *jebe débil* (*Hevea guianensis*) and *balata*,²⁵⁷ and the international rubber market had begun to accept them, albeit at a lower price than *jebe fino* (from the *Castilloas*). Collection of *jebe débil* and *balata* entailed the periodic bleeding of trees along established *estradas*. Consequently, the former system, whereby indigenes periodically felled *Castilloas* to satisfy their debts with the *caucheros*, was replaced by one in which the adult male Indians were expected to become full-time gatherers—to the detriment of their former hunting, fishing, and subsistence agriculture. This implied a radical reorganization of the indigenous work force and was antithetical to its cultural values. The fact that their product fetched a lower price increased the pressure on the indigenes to produce greater volume. That, in turn, provided them with incentive to fell trees in order to meet the immediate demands of their white overseers. But, of course, this limited the future supply and forced the Indians to search further afield for new groves to fell. The alternative was to simply abscond. In any event, the dynamics for ever increasing abuse of the indigenes had clearly been set in motion.²⁵⁸

Julio César claimed to have made his first personal trip to the Putumayo in 1901, ostensibly to work out with Larrañaga repayment of an outstanding debt, but actually—according to Lagos—to reconnoiter the area as part of the planned take over.²⁵⁹ When Arana first entered the Putumayo, Colombia was embroiled in civil war—the punishing War of a Thousand Days (1899–1902) that had all but destroyed connections between Bogotá and the region. Ecuador was also on the verge of its own civil war and was in no position to exercise effective control over its Oriente. Indeed, a powerful Ecuadorian rubber magnate and former official in Tiputini, David Elías Andrade, became a naturalized Peruvian in 1900. He agreed to be Peru's *comisario* (commissary) on the adjacent, and also contested, Napo River.²⁶⁰ As late as 1902, Colombia was still claiming jurisdiction over the Napo River,

> On the shores of the Napo River in the territory of Colombia there are many Ecuadorian and Peruvian individuals, established in commercial businesses and agricultural holdings and they are governed, the same as those Colombian indigenes, by recently-

---

257. The word *balata* usually refers to the product of a latex tree species found primarily in Guyana and the West Indies. Stanfield notes that, in 1928, Arana was exporting considerable quantities of both jebe debil, or "Putumayo tails," and *balata* from Aranalandia. Stanfield, *Red Rubber*, 198. He further notes that balata was being collected in the hinterland of nearby Güepi as well. Ibid., 202.
258. Santos Granero and Barclay, *La frontera*, 79–83.
259. Lagos, *Arana*, 78.
260. Stanfield, *Red Rubber*, 100.

created Peruvian and Ecuadorian authorities, who then reclaim the territory as Ecuadorian or Peruvian and charge tariffs on the introduction there of goods.[261]

In 1901, Julio César purchased the shares of Moris and formed Larrañaga, Arana and Company as a branch of J. C. Arana and Company of Iquitos. That same year, on the nearby Caraparaná River, he formed the Calderón, Arana, and Company with the Calderón brothers, who were based at El Encanto. From these initial footholds, Arana extended his influence throughout the region, often entering into partnerships that were sub-branches of either the La Chorrera or El Encanto operations.[262]

In 1902, Enrique ship's captain Enrique Espinar visited and mapped the Igaraparaná. He reported that the Larrañaga and Arana firm had conquered five tribes ("inappropriately called 'nations'") with eighteen subgroupings. Each had a minimum of three hundred workers and somewhere between five hundred and one thousand adult males in all. The firm had no fewer than twelve thousand Indians on its books as indentured rubber gatherers. The Indians were quite willing and anxious to provide the product in return for the trade goods that Espinar characterized as cheap and trashy—but flashy. Two to four Arana employees could supervise a grouping, proof that the indigenes were docile. Larrañaga and Arana were continuing to probe unconquered tribes in adjacent areas.[263]

Arana's economic penetration of the Middle Putumayo was furthered by the area's political movers. Pedro Portillo—Loreto's prefect from 1901–1904—was a firm believer in imposing indisputable Peruvian hegemony over the rivers within Iquitos's orbit, including the Putumayo. He sent many military expeditions into them (even commanding some himself). At one time or another, he clashed openly with Ecuadorians, Colombians, and Brazilians. Portillo appointed minor officials to a nascent Peruvian administrative structure throughout the region.[264] He was openly supportive of Arana's expansion into the Putumayo, despite being fully cognizant that the *caucheros* of the Oriente were enslaving and abusing its Indians.[265] He admonished that it was better to remain silent until the full weight of the Peruvian government could be brought to bear upon, "those despicable men who know no other law than that of profit and pleasure." Indeed, Portillo himself reported,

---

261. Quoted in Dominguez and Gómez, *La economía*, 177.
262. Pinell, *Excursión*, 197.
263. Espinar, "Viaje," 221–22.
264. Lagos, *Arana*, 98–100.
265. Turvasi, *Giovanni Genocchi*, 28.

Our natives were few and they are disappearing. The tribes attack and destroy each other. Infanticide reigns. Children and women, the spoils of war, are sold. The present rubber traders of the Ucayali and other rivers go there, not to build families and prosperity, but to obtain riches and Peruvian slaves, to exchange them for pounds sterling. . . . For every Chinese man who enters,[266] ten, twenty, or thirty Peruvian Indians are sold abroad.[267]

In 1903, Peruvian officials summoned Larrañaga and his son to Iquitos for questioning regarding their alleged massacre of Indians. According to the accusation, when the Indians killed two of his employees, Benjamín and his son, Rafael, tricked about twenty-five Huitotos and Ocainas into coming to La Chorrera. The Indians were captured and then whipped, tortured, and shot. By some accounts, kerosene was poured upon wounded survivors who were then burned to death.[268] According to Lagos, with his son languishing in the local jail, known euphemistically as the "Office of the Casa Arana," Benjamín was forced to sell his interest in La Chorrera to Julio César for 25,000 pounds. The Colombian owed Arana 70,000 pounds, a debt that was forgiven. There is the implication that the transaction likely involved considerable coercion. In any event, Julio César's new acquisition was consummated at well below the property's real worth.[269]

By his own account, that same year of 1903 Julio César made a second (brief) trip to the Putumayo to negotiate with his Colombian debtors there.[270] By then, Julio César had opened an office in Manaos and forged a partnership in New York (Arana & Bergman Co.) that regarded own-

---

266. By this time, Peru had received thousands of indentured Chinese and some voluntary Chinese immigrants (including railway construction crews recruited from the colony remaining in California after having participated in the United States' transcontinental railway project). While concentrated at first on the *haciendas* in coastal Peru, the Chinese spread throughout the country. In the Oriente they became traders and merchants of note, and some even worked as *aviadores* who bartered for rubber in exchange for provisions and utensils. By the turn of the twentieth century, there was an established Chinese community in Iquitos with its own beneficent society. Isabelle Lausent Herrera, "Los inmigrantes chinos en la Amazonía peruana," *Bull. Inst. Fr. Et. And.* 15, nos. 3–4 (1986), 49–60.
267. Pedro Portillo, *Las Montañas de Ayacucho y los Ríos Apurímac, Mantaro, Ene, Perené, Tambo y Alto Ucayali* (London: Forgotten Books, 2013 – original work published in 1901), 53.
268. Lagos, *Arana*, 78.
269. Ibid., 83.
270. Julio C. Arana, *Las cuestiones del Putumayo* (Barcelona: Imprenta Viuda de Luis Tasso, 1913), 9.

ership of river steamers in both Brazil and Peru. Arana's expansion into shipping tightened his growing stranglehold on the Middle Putumayo. The area came to depend entirely upon his services, both for the importing of supplies and for the exportation of its rubber to Brazil and beyond. Arana's flagship on the Putumayo was the fairly luxurious *Liberal*, captained by (Basque-surnamed) Carlos Zubiaur. Accessing the Middle Putumayo from Iquitos meant descending the Peruvian Amazonas River to the Yavarí, and then the Solimões in Brazil, before then ascending the Içá River to where it becomes the Putumayo. The boat journey from Iquitos took about three weeks.

The Iquitos headquarters of the Casa Arana was under the administration of Pablo Zumaeta, and a future key player in the Arana firm, Abel Alarco, Pablo's brother-in-law, continued to reside in La Rioja.[271] Since Arana would now spend considerable time in Brazil, he and Eleonora decided that she should move with the family to Europe. By then, it was common for wealthy South Americans to do so in order to better educate their children. The Aranas opted for Biarritz, where they rented a fine house and hired staff, including tutors for their four children. With his family settled abroad, for most of the next three years Arana would direct his expanding empire from Manaos. His key employee there was his brother Lizardo.

By late 1903, Lima had accorded Arana at least implicit recognition as its spearhead in the Putumayo/Caquetá. A November 4, 1903, letter from Minister of Foreign Affairs José Pardo asked Julio César to accommodate the field research in Aranalandia of a French explorer and would-be ethnographer, Eugène Robuchon. If the Casa Arana would kindly pay the Frenchman's salary of 35 pounds sterling monthly, and also advance his field expenses, the ministry would later reimburse it. Never one to fail to exploit an opening, Julio César received Robuchon in Iquitos graciously and presented him with a contract (September 2, 1904) that he forwarded to Lima, along with a cover letter stating:

> It is equally our pleasure to inform you that our firm has decided to assume all of the expenses that the entrusted mission of Mr. Robuchon should incur, it being our desire, if only modestly, to contribute to the patriotic ends that our government is pursuing.[272]

---

271. Joaquín Rocha during his travels in late 1903 met Alarco there and (ironically as it would turn out) praised him as a literate man enthralled with Colombia's best writers and favorably inclined toward the country. Rocha, *Memorándum*, 164–65.

272. This letter is reproduced in the book Eugenio Robuchon, *En el Putumayo y sus afluentes: Edición oficial* (Lima: Imprenta La Industria, 1907), x.

The contract stipulated that the explorer was to map all of the waterways, and that his findings and photographs would belong to both the Casa Arana and the Peruvian government, including the publication rights. It also specified in great detail the geographic limits of the exploration. Interestingly, this included all of the area in which the Casa Arana was active by then (particularly the Igaraparaná and Caraparaná drainages), as well as adjacent ones that might be of future interest to Julio César.[273]

In 1904, Julio César reentered into partnership with Juan B. Vega (Arana, Vega y Cia), by then Colombia's viceconsul in Iquitos, to pursue further consolidation of his holdings in the Putumayo. If the Putumayo's Colombians thought that might improve their relations with Arana, they were mistaken. Indeed, one Colombian commentator opined

> [t]hat the conduct of the Colombian viceconsul in this epoch, Sr. Juan Vega, was most unpatriotic is undeniable, since he attended more to the interests of his partners, the Aranas, than to those of his country. It was he who astutely influenced the displacement of the Colombians from both rivers, the Caraparaná and Igaraparaná, through transactions that were nothing more than swindles. The Peruvian firm used every illicit means to appropriate the work of Colombians, but this is not just now rather it began several years ago, or since he [Vega] entered into business with them.[274]

There is apparently a public document in Iquitos dated April 8, 1904, regarding the founding of the Arana, Vega y Larraniaga company that states: "As for the Indians of the Putumayo they are obligated by the employees of the company to work by force."[275]

From a Colombian viewpoint, the Putumayo/Caquetá was essentially self-governed by vigilantes, since formal Colombian authority was absent. Until 1903, requests from the region's Colombians to Bogotá went unanswered. Rocha notes that there was considerable violence, including murder and arson, among the white settlers themselves.[276] In late 1903 and early 1904, there was a general uprising of the tribes of Aranalandia working with Colombians. The indigenes dug trenches and erected fortifications in the zone between Arana-dominated properties and those outside his control. A Colombian observer felt that this would have been

---

273. Ibid., xi–xiii.
274. Quoted in Pinell, *Excursión*, 218.
275. Quoted in Dominguez and Gómez, *La economía*, 183.
276. Rocha, *Memorándum*, 139.

impossible without a degree of collusion, indeed incitement, by the Casa Arana. It seems that at the same time the supply of ammunition for both pistols and carbines, heretofore provided by Arana, suddenly dried up. Whites living on the Igaraparaná were under particular assault and were faced with the choice of selling out or facing death at the hands of "savages." Once several had sold, José Gregorio Calderón bought out the interests on the Caraparaná of his two brothers and entered into a partnership with Julio César Arana. The bullets appeared once again "as if by magic."[277] However, the sword could cut both ways. It was Arana's practice to place his own people within the operations that he controlled—Calderón notes that Julio César's own brother-in-law, Néstor Zumaeta, was killed by Bora Indians at the Abisinia station.[278]

The Calderóns continued to petition the Minister of Foreign Affairs and the authorities in Mocoa for guidance, and it was then that (Basque-surnamed) Bernardino Ochoa showed up with some credentials and planted a Colombian flag on the Caraparaná. When he learned that the Peruvians had installed a garrison and customs post at the juncture of the Igaraparaná and the Putumayo rivers, Ochoa went there to assert a Colombian presence, but was rebuffed.[279] His stay in the region was brief, and he left behind as representative of Colombian hegemony (Basque surnamed) Bartolomé Guevara. However, that official quickly went over to the Peruvians by entering the employ of the Casa Arana. Then, in 1905, General Pablo J. Monroy and a few soldiers arrived to establish a Colombian customs house and military barrack on the Caraparaná, but they were not welcomed by the Calderón, Arana and Co. enterprise and departed quickly without leaving behind a trace.[280]

In 1905, Arana once again visited the Putumayo for a few days for the purpose of buying out some Colombian rubber operations.[281] A shiver was going through all of the Colombian *caucheros* on the Caquetá and Putumayo, since they had just learned that their own government was in the process of selling the region's rubber concession to the Colombian firm of Cano, Cuello, y Compañía. On January 27, it was announced that the Colombian government had ceded to Cano and Cuello no less than 30,000 square miles in the Caquetá.[282] While this initiative proved abor-

---

277. Pinell, *Excursión*, 217.
278. Calderón, "Importantes," 274.
279. Rocha, *Memorándum*, 139.
280. Calderón, "Importantes," 275–76.
281. Arana, *Las cuestiones*, 9.
282. Jorge Villegas and José Yunis, *Sucesos colombianos, 1900–1924* (Medellín: Universidad de Antioquia, 1976), 79.

tive, it certainly played into Arana's hands. Panicked Colombians were now willing to sell their operations to him for a fraction of their value.[283]

In short, between 1905 and 1908, Julio César acquired the properties of most of the remaining Colombians in the Putumayo,[284] including the critically situated La Chorrera station that would become the Casa Arana's field headquarters. The same Colombian observer noted that Benjamín Larrañaga was "an ignorant man" who was easily duped by Arana. Julio César exploited him in life and then made off with the major portion of the man's estate after his death.[285] Indeed, he exploited all of his Colombian employees and debtors, and even ordered the murders of holdouts José Francisco Gómez and Joaquín de Barros.[286] The commentator opined the Julio César's surname should have been Araña (Spanish for spider), given the complicated webs that he wove.[287]

José Gregorio Calderón remarked that his own partnership with Julio César, formed in the 1905 panic, proved disastrous. Their company became further indebted to the Casa Arana by buying supplies on credit at increasingly inflated interest rates. It also proved impossible to seek redress of disputes in Iquitos—given Arana's omnipresent prestige and political influence there. He summed up his departure from the Caraparaná as follows:

> Happily it was not too late for me to heed the warning from my heart that made me give up. I managed to get out by selling my shares to the other partners [Arana and Vega] in order to avoid total financial ruin, return to my family and recover the health and peace of mind that I had lost. And let it be said

283. Ibid., 216.
284. Arana, *Las cuestiones*, 14. Colombian writer Demetrio Salamanca T. believed that the takeover resulted from the shameful neglect by his country of its Amazon territory, the ingenuousness of the uneducated Colombian *caucheros* in their dealings with the wily Arana and the latter's willingness to employ violence against any holdouts. Salamanca T., *La Amazonia*, 91, 260, 556.
285. Larrañaga died suddenly in 1903, exhibiting all of the signs of arsenic poisoning. The following year, the *El Correo* newspaper in Cauca published an article saying that Larrañaga possessed "colossal wealth" when he died, including a paper factory in Manaos that employed four hundred workers, a fleet of river boats, and a bank account in the United States and England with half a million gold *pesos*—all of which seems preposterous (Dominguez and Gómez, *La economía*, 142).
286. Pinell, *Excursión*, 218. There follows a detailed account of how, with the cynical advice of Vega, Arana was able to turn two Colombian partners against one another and then acquire the property for a tenth of its real value from the victor (Ibid., 218–19). Virtually all of the Colombians were unlettered, and the shrewd Arana clearly played on their egos by insisting that their name precede his in their partnerships (Ibid., 220).
287. Ibid., 216.

that I did not sell land that was not mine nor Indians because I did not have them as slaves [being from] a Nation in which that traffic is immoral and forbidden. And if I was able to sell my improvements and active accounts to persons of Peruvian nationality, it was the same as selling them to Panamanians or Yankees under the legitimate [right] of universal commerce.[288]

Calderón received 12,000 pounds sterling (3,000 cash and three notes on London banks), or far less than the value of his interest. He had followed the advice of his brother-in-law, Mauricio Cuellar, a man "ignorant in such matters."[289] Then there was the case of Hipólito Pérez, the first operator on the Caraparaná to enter into partnership with Arana (Pérez & Arana), who, after ignoring the offer from the Colombian consul to assist him in finding a better buyer [no doubt Colombian], ended up selling his interest for 5,000 pounds sterling, or less than two years of the property's rubber proceeds:

> Not even the scene of hundreds of Indians wailing and trembling at the prospect of the Peruvian yoke, and on their knees before Señor Pérez pleading not to abandon them, was capable of moving him, notwithstanding that it is impossible to conceive of anything more terrorizing than the cries of these unfortunates.[290]

Then, too, there was the case of Justino Hernández, one of the Colombian *caucheros* associated within Calderón's orbit. According to the account of one former Colombian resident of Aranalandia, after the Calderóns sold out, and despite the condition that the Colombians abandon the area, Miguel Loayza[291] and Bartolomé Zumaeta exhorted Hernández to stay on until the new Peruvian supervisors were up to speed in terms of managing the local Indians. In May 1907, he decided to retire and went to El Encanto to collect the monies owed to him. There ensued an argument over the amount, and Hernández accused the Peruvians of trying to take advantage of him given that they had by then run off most of the Colombians. Zumaeta killed Hernández with his revolver, shattering the man's skull.[292]

Abel Calderón subsequently sent a report to Bogotá, dated September 30, 1909, in which he outlines the sad history of Colombians in Aranalandia. He claims to have established his own rubber enterprises in partnership

---

288. Calderón, "Importantes," 277.
289. Pinell, *Excursión*, 219.
290. Ibid., 220.
291. Rendered as Loiasa in some of the documents and publications.
292. Dominguez and Gómez, *La economía*, 187.

with three other Colombians, beginning in 1900. Their first three (named "Acacia," "Entre Rios," and "Atenas") were on rivers feeding the Caquetá. Taken together they employed 3,225 Indians. However, "by means of felony, violence and deceit," the Casa Arana relieved them of their properties, as well as all of their equipment and six thousand *arrobas* of rubber.[293] So, Abel and his partners decided to reestablish themselves several travel days distant from their original settlements, where, in the region of the Fahay River they established the stations known as "La Sabana," "Puerto Colombia," "Morelia," and "Puerto Tolima," all with agriculture and extensive rubber collection in the adjacent forests. They had three thousand Indian workers in all. But these were taken by force by Benjamín Larrañaga and his partner José María Moris Ramiréz. Abel's partners were taken to Iquitos and jailed. Abel himself was not, and he went to Bogotá to solicit the intervention of President Reyes. He was told by Reyes to remain absolutely silent about the matter and that he would resolve it by diplomatic means.

Meanwhile, the Colombian consul in Iquitos obtained the release of the prisoners, and they returned to the Putumayo/Caquetá to establish a new station. Soon thereafter, they, including Abel's brother-in-law Torres, were attacked and killed by Arana's men. Abel came to realize that his Bogotá mission had been pointless. The government knew what was going on. The newspapers, particularly Ecuadorian ones, were covering the events. He estimated that the number of Colombians killed now surpassed two hundred, the majority from Tolima. At least six hundred more Colombians had fled the area. However, there were many, mainly *Tolimenses*, prepared to resettle Aranalandia if only the government would expel the Peruvians and protect its citizens with a proper administration, three military garrisons, and a customs house, as well as missions and schools to educate the Indians.

Abel Calderón had been asked to become an amply rewarded instrument of the Peruvians as they established sovereignty over the region's immense resources. Unlike the traitorous *Pastuzo*, Benjamín Larrañaga, he had refused to sell his patriotic soul. Consequently, given that the Colombian government had failed to meet its responsibility to its own citizens, he was requesting indemnification of two million pesos.[294]

We might consider two Colombian overviews of the distinction between their treatment of the Indians and that of the Peruvians. José Gregorio Calderón, prime Colombian pioneer and player in the Putumayo, observed,

---

293. Cited in Gómez Lopez (ed.), *Putumayo: La vorágine*, 422.
294. Ibid., 423–27.

> The Indians ... esteem and obey the Colombians, because despite their ignorance they have the notion that the territory in which they live pertains to Colombia, and because the treatment that they are given by one and another [Colombian] is entirely distinct: our enterprise treats them gently whereas the colonizers or conquerors from Peru have made them feel and obey by means of cruel treatment, such as that at "La Chorrera," as recounted by the Indians themselves, for which reason they are constantly fleeing.[295]

Finally, Rafael Reyes, commenting retrospectively on the plight of the region's Indians in his *Memorias*, stated,

> These savage tribes tend to disappear, annihilated by epidemics, abused and sacrificed by those who practice the hunting down and trafficking in men, as in Africa, and for the rubber business. This trafficking or treatment of Indians was practiced in the time when, with my brothers Enrique and Nestor, we carried out the first explorations. The human traffickers penetrated the Putumayo and Caquetá Rivers in large canoes called *batelones* and stirred up the most powerful tribes to make war on the weakest in order to buy the prisoners, some of whom were kept to be sacrificed during their festivals, payment (for the remaining prisoners) being made in alcohol, tobacco, glass beads, mirrors and other trinkets. After the traffickers or buyers of Indians received their merchandise, they embarked in their *batelones*, men, women and children crammed together like human sardines, their hands and feet tied with cords, naked, devoured by mosquitoes and without protection from the sun's rays, where the temperature climbs to forty five degrees centigrade, and (without protection) against the copious rains. Food was scarce and in these conditions the human cargo took several weeks to arrive at the destined market in the margins of the Amazon. Many of these individuals died of hunger or because of the mistreatment. This barbarous commerce can be compared to that of the blacks in Africa. The human cargo, once it arrived in Amazonian waters, was sold in those remote farms and settlements for prices several times greater than their costs. Mothers were separated from children, husbands from wives who, destined to places so distant from one another, were never to see one another again and they were treated as slaves. Brazilian law did not condone this barbarous practice, but in those unpopulated areas official policy was easily mocked and this traffic in human

---

295. Calderón, "Importantes," 278.

flesh was at the pleasure of the majority of the marginal inhabitants along the great river who used the Indian-slaves to gather rubber and to fish and hunt. During the time that I was with my brothers in those explorations, we managed to completely destroy the trafficking in Indians in Colombian territory drained by the Putumayo and Caquetá Rivers and their tributaries, and in this endeavor the Brazilian government afforded efficacious support.[296]

Nevertheless, the depiction in the literature (particularly in Colombian sources) of Colombians in general as more humane than Peruvians in their treatment of the indigenes is somewhat shaky. There are "a great number or accounts and testimonies" preserved in the Colombian National Archive regarding maltreatment of Indians in Colombia before, during, and after the Putumayo scandal.[297] Others do, however, project an image of the relatively reasonable treatment of the indigenes compared with the barbarities of the Peruvians.

The Peruvians, particularly Arana and his defenders, disagreed with much of the foregoing. They contended that the Casa Arana extended credit to Colombians on request and serviced the Putumayo with the critical shipping without which the import of goods and export of rubber would have been intermittent and prohibitively expensive. Julio César maintained that he was a willing buyer of properties, at fair market value, from willing sellers. There is the succinct history of white settlement of the Putumayo in the document "Sketch of the Putumayo,"

> The Peruvian rubber dealers, since the year 1880, the period at which rubber was for the first time exploited in the west of Peru, have explored the river Putumayo on various occasions, the expeditions of Deofanto Reategui and Francisco Capa in the launch "Tahuayo," of Pablo Zumaeta in the "Galvez," and many others, having failed. The failure came from the limited resources with which they began that arduous enterprise, in which besides they had to subdue great numbers of cannibals, whilst in the neighborhood of Iquitos and in the rivers Purus, Javary, Jurua, Ucayali, etc. the rubber was found in abundance and without the drawbacks of the Putumayo. In the year 1896, Jose Maria Mori Ramirez, one of the clients of J. C. Arana, of Iquitos, formed a partnership with Benjamin Larraniaga, freighted a launch, loaded it with goods and started for the river Putumayo with sufficient men, and, following the affluent Igaraparana, they established the commercial house

---

296. Reyes, *Memorias*, 142–43.
297. Dominguez and Gómez, *La economía*, 193.

called "La Chorrera" at the foot of a beautiful cascade. The forests were immense, with abundant rubber, and everything promised the most successful working if sufficient civilized men could be brought to the place. But this was very expensive, and the house of Larraniaga, Ramirez & Co. had a very heavy outlay, and they became debtors to Mr. Arana for a very large amount. The difficulties increased with the obstacles that the Brazilian authorities put in the way of the free navigation of the Peruvian flag through the Brazilian Amazonas on the way to the Putumayo river. These circumstances discouraged Morey [sic], who transferred his share to Mr. Arana, and the house of Larraniaga, Arana & Co. began work. With the capital brought in by Mr. Arana, and with the credit which his name gave to the house, the working of the rubber trees began in a serious manner, and the Indians who were before so dangerous were used for the work. Strict orders were given to treat them well and to try and obtain their friendship, and this policy helped to obtain the submission of the Indians. They were made to understand that in exchange for the rubber they would get goods which would be useful to them, and advances were made to them in clothes, implements and so forth with the condition that they would pay in rubber or in work. In one of his voyages from Iquitos to Chorrera, freighted with merchandise and carrying on board a large number of men, the launch "Putumayo" was seized by the Brazilian authorities at the frontier and taken to Manaos, denying to the Peruvian flag the right of free transit by the Brazilian Amazonas on the way to the Putumayo. Then followed a long and costly litigation in which were invoked the existing treaties, celebrated in the year 1854 between Brazil and Peru with respect to the river Putumayo, and after many debates in Manaos and in Rio de Janeiro the Peruvian government obtained that Brazil would put into execution the treaties to which we have referred; since then the Peruvian flag encounters no obstacles in the Brazilian Amazonas on its way to the Putumayo. In 1903 Mr. Larraniaga died, and his heir was his son Rafael, who continued in the business, associating himself with Mr. Juan B. Vega, changing the name of the firm and commencing the present firm of Arana, Vega & Co. Since then Julio C. Arana & Hermanos have bought from Mr. Rafael Larraniaga his share. . . . The house of Arana has monopolized all the business of the region which we have already delineated; either for purchases made from other industrial concerns of their possessions or marked out according to the forest laws of Peru (Ley de Montanas) other points considered commercially strategic, and Mr. Arana is now obtaining a special concession by which the firm asks for a final title to certain lands and various privileges in the way of taxes, etc., in exchange for roads and other facilities that they would put at the disposition of the government . . .[298]

Then, too, we have the particularly fascinating account of Iquitos judge Rómulo Paredes regarding Putumayo history. Sent by the Peruvian government in 1911 to investigate the charges that Arana's firms had condoned and committed atrocities against the Indians of the Putumayo, Paredes would prove to be Julio César's most telling critic and nemesis. Nevertheless, the judge was a Peruvian nationalist to the core and assumed that the region belonged to Peru and that the Colombians were the usurpers. He was therefore told (and believed) an alternative historical narrative.

Regarding white settlement of the area, Paredes gives the founding date as 1895 when a group of Colombians descended from the Caquetá River and explored the Igaraparaná where they established the outpost of Último Retiro in its upper reaches. Their leader was Benjamín Larrañaga (there is no mention of the legendary Crisóstomo Hernández). Larrañaga then established a rubber gathering station at La Chorrera in partnership with Gregorio Calderón. The partners traveled together to Pará in Brazil with their first shipment where they dissipated all of its proceeds. The impoverished Larrañaga met in Pará the Peruvian José María Mori and the Jew Jacobo Barchillón, and convinced them to invest in his Putumayo operation. His new partners advanced the money for a cargo of trade goods and lease of a vessel that took Larrañaga to the Putumayo, whence he moved his merchandise by canoe up the Igaraparaná to La Chorrera. There, he found that several Colombians had recently arrived in the area and conquered Indians for rubber operations. Also, the Calderón brothers had taken some Colombian followers through the forest to the Caraparaná River, where they themselves engaged in conquering more tribes so that they could divide up the new work force.[299]

Then, in 1896, the vessel *Galvéz*, owned by the Peruvians Lizardo Arana, Pablo Zumaeta, and Abel Cárdenas (and commanded by Lizardo) went up the Caraparaná and purchased 6,000 kilos of rubber from the Colombians. The *Galvéz* was subsequently prohibited by Brazilian authorities from returning to the Putumayo. Larrañaga put together a shipment of rubber that he transported to Iquitos, paying customs duties on it

---

298. Memorandum No. 25. Sketch of the Putumayo, 1–2. Copy in the Bodleian Library S-22 MSS Brit. Emp. G338. [Memorandum 25]. This document is reproduced in its entirety as appendix 1 of the current text.
299. Chirif, "Los informes," 85–86.

to the Brazilians. He shopped it around, but finally sold the cargo to Julio César, receiving at the same time a considerable line of credit. Meanwhile, the Casa Arana negotiated free navigation of the Putumayo with Brazil. Nevertheless, the Colombians on both the Igaraparaná and Caraparaná (in particular, Rafael Tovar on the former and Gregorio Calderón on the latter) engaged in "crude warfare" against Larrañaga in an attempt to drive him out. They stole his rubber and his Indian workers. Calderón then organized a force of fifty "brigands" at El Encanto and sent them to attack La Chorrera. As Larrañaga prepared to defend himself, Arana's vessel, the *Putumayo*, arrived in the nick of time and its captain convinced the parties to negotiate. As a result, Tovar was to receive 50,000 soles and Calderón 14,000 in return for stations they had built on land claimed by Larrañaga, but with the promise to stay out of those areas in future. It was a promise that they subsequently ignored.[300]

Arana continued to deal with Larrañaga, but, as his debt with Julio César rose, Mori and Barchillón panicked and asked to liquidate the company. Larrañaga ended up as sole proprietor, but with an enormous debt to the Casa Arana. He therefore proposed that Julio César become his partner. Arana really didn't want to but had little choice. The firm was appraised, and Arana reduced its indebtedness in return for his share in it. It was then that the Casa Arana became active in introducing Peruvians into the area. It petitioned the government to establish a military garrison, a commissary and a customs house (at the confluence of the Putumayo and Cotuhé rivers). In 1900, the prefect of the Loreto sent an armed vessel to the Cotuhé to install the Peruvian presence there. The garrison and commissary were subsequently transferred to La Chorrera. Then, according to Paredes,

> It is said that Sr. Arana did even more: he extended navigation and commerce to the whole region of the Caraparaná; he developed a business of such magnitude that it increased tax proceeds [significantly], he protected the firms established there already, he nationalized [for Peru] an unknown territory through his efforts and money; he contracted the services of Sr. Robuchon to make a map of all of this region in order to complete the work of designating frontiers, a project frustrated when the engineer died during his risky undertaking. On the other hand, I was assured that the launches of Sr. Arana were always at the disposition of the Peruvian authorities, an attempt to serve the Nation and of guaranteeing its interests.[301]

---

300. Ibid., 87–88.
301. Ibid., 88–89.

After the death of Benjamín Larrañaga, Julio César and Rafael Larrañaga proposed to Juan Vega formation of a partnership. "Arana, Vega y Cia" was registered on May 15, 1904. It was then that the Peruvian garrison was transferred to La Chorrera. The following year, the Colombian general Pablo Monroy arrived in the area and established a forty-man garrison at San Gregorio [on the Putumayo upriver from Aranalandia]. He was accompanied by Rogelio Becerra, Colombia's newly named *intendente* in the Middle Putumayo/Caquetá. Soon thereafter, a Colombian general arrived with troops at Cotuhé, along with two Colombian customs officials who were supposed to exercise joint administration with the Peruvians in anticipation of the *modus vivendi* agreement that was under negotiation by the two countries [see below]. Paredes called this "an unexpected invasion" that "happily for us was of no effect."[302]

It was now that the Colombians on the Caraparaná, led by Gregorio Calderón, began harassing the Casa Arana, with the assistance of the Colombian garrison at San Gregorio. They encouraged the Indians around Último Retiro to rise up against the Casa Arana to rob it of its rubber and Indian gatherers. There were skirmishes in which three of Arana's men perished. Fortunately, the Peruvians were able to pacify the Indians, but at a cost of 70,000 soles. Furthermore, the theft of Indians and rubber continued until Julio César bought out Gregorio Calderón with the condition that he leave the area. Notwithstanding that the purchase price was considerable, after the sale Calderón remained on the Caraparaná. Arana had purchased the Indians' contracts, so he now sent an emissary to enforce them. Calderón hoodwinked Arana's representative into going to La Florida, where the Colombian general Durand was based. When they arrived, an armed force detained the Peruvians, took all that they had and then tortured them; one of the Peruvians died. The Casa Arana prepared to rescue its men by force, while an intense diplomatic effort convinced the Colombian *intendente* to free them. It was then that Julio César bought out the remaining Colombians in the area, at considerable cost, with the condition that they leave. Paredes concluded,

> So Arana, through licit means such as purchases, was finally able to rid this important zone of such bad people, and he had performed, in the view of many, a positive service to his country, nationalizing a territory that was almost lost through a foreign invasion.[303]

Paredes underscored in his report the international influences of the activities of the Casa Arana. First there was the network of "roads" [often

---

302. Ibid., 89–90.
303. Ibid., 91.

just rough trails] constructed by it throughout its domain that allowed the Peruvian military to deploy its forces effectively, often utilizing Indian bearers provided by the Company. Then, too, the Casa Arana encouraged local agricultural production by the Indians (over the opposition of its own section chiefs who believed that it detracted from rubber gathering). While giving the Putumayo a degree of autonomy from Iquitos, it also provisioned both the Indians and the Peruvian troops with abundant and healthy food. The Casa Arana's vessels and experienced crews provided Peru with a near invincible naval force in the Putumayo. What's more "nothing and no one in this vast zone moves without the all-powerful approval of said House." In Paredes's view,

> This power of the Casa Arana, undoubtedly created by the strength of its capital, with which it came to subject even those very Colombians who had established themselves in the Putumayo, has made it the owner of this great territory, which besides its natural wealth numbers thousands of Indians that here, where workers are so scarce, constitute incalculable wealth.[304]

Regarding the claims of Colombians and their country to the Putumayo, Paredes stated,

> The primitive dominators, removed from their possessions by the force of gold, have not been able to resign themselves [to their fate] after witnessing how the firm without indignation made great economic strides and developed its business with the same stations that they [the Colombians] had established and with the same tribes that they subjected; and after many of them left legally and willingly, they have not ceased to attempt to reconquer the area, at times presuming to attack the House in the same sold sections, at times . . . (illegible words in the original) complete Indian round-ups to carry them off to other places—principally to the left bank of the Caquetá—while at other times, out of their desperate spite, fomenting in their country ill-will towards Peru and achieving, finally, after their incessant campaigns, conversion into an international question what was simply a clash of private interests between Colombian adventurers, rubber men of the worst kind, and the Casa Arana, established in what is fully Peruvian territory, recognized by the enemies themselves when they exported their primary product through the customs house of Iquitos. Neither before nor after the House of Arana was established in the Putumayo did the

---

304. Ibid., 96.

Colombian State raise the question of its supposed domination in that territory. It has been certain ill-willed individuals or victims who thrown out of there in various ways went to their country to first glorify their conquests and later cry over their loss, finally stirring up national sentiment to such a degree that today there is talk of loss of a territory that [Colombia] never sincerely believed to be its own.[305]

Paredes noted that, while there had likely been some violence against Colombians, it had been perpetrated by lower officials of the Company. The most serious charge along these lines was an attack led by Bartolomé Zumaeta (Arana's brother-in-law) and Miguel Flores against La Reserva that allegedly resulted in twenty-five Colombian deaths. Paredes said that he was on his way back from the Putumayo when that accusation was first leveled by the Colombians during the negotiations between the two countries currently being held in Brazil, so he had been unable to investigate it. To be sure, none of the Colombians he interviewed in the Putumayo had even mentioned the incident, although he conceded that they regarded him as an enemy, rather than [an impartial] judge in whom to confide. Paredes did admit that there had likely been several abuses of Colombians by Peruvians, but labeled them as "of little importance" in and of themselves. They were perhaps understandable given the area's contested sovereignty. Unfortunately, the attacks gave rise to demands by the Colombian government for compensation of the victims—some justified, some not, while others were exaggerated.[306]

Paredes ended this section of his report by raising a question as his següe to what would become acerbic criticism of the Casa Arana and/or Peruvian Amazon Company:

> It is therefore of the greatest necessity and urgency to examine the Arana firm, with the goal that the Nation enjoy all of the benefits that it may bring to it, while diminishing as far as possible the causes of the hatred and resistance that its bad procedures have brought us, without it being determined whether this House is a strange organization, whether a saving grace or an evil that should disappear, because, on the one hand, it is a formidable, and even disinterested and patriotic, presence against the threat of any territorial usurpation, while on the other, it is death's harvester of lives, replete with national and international

---

305. Ibid., 96–97.
306. Ibid., 97–98.

enemies, abusive and cruel, as the sole cause of our international problems in these regions.[307]

If there was considerable disparity between the Colombian and Peruvian versions of white settlement of the Putumayo, we have yet another narrative that controverts most of the foregoing: the historical account of Aquileo Tobar. The *mestizo* son of a white father and an Indian mother from the Caraparaná, he was raised as a white. When interviewed as a middle-aged man in 1971 by a Colombian anthropologist, Horacio Calle Restrepo, he was working out of Puerto Leguízamo (on the Putumayo River) as a pilot of a tugboat. While raised a white, Tobar was fluent in one of the Huitoto languages and took great pride in his Indian heritage. He was born at El Encanto where his father was an Arana employee. The family moved to Iquitos when he was still small and it was there that he was schooled (he would eventually write two books). He also wrote an unpublished manuscript detailing his life and the oral history of Aranalandia that he learned from his ancestors. The following information is taken from that account[308] and regards the biographies of both Crisóstomo Hernández and Benjamín Larrañaga:

> During their time, it was the town of Florencia on the Caquetá that served as the main servicing center for the rubber districts of the Middle Putumayo/Caquetá. It received many migrants from Huila and Tolima intending to go downriver in search of the rubber trees that would allow them to become established as *caucheros*. The main rubber entrepreneur in the area was Don Francisco Gutiérrez and Crisóstomo Hernández was his employee. One night, in a *finca* located at some distance from Florencia, Hernández, and several of his *cauchero* friends were having a drunken party when one of them began harassing the mulatto. They began fighting, and before it was over, Crisóstomo had decapitated his adversary with a machete. Frightened and disgusted that he had killed his companion, Hernández seized a canoe and went several miles downstream to an outpost. A short time later someone from Florencia came to inform him that the authorities were preparing to arrest him, so the fugitive set out again overland through the forest, eventually reaching the shores of the Caguán River and then sailing in a makeshift raft downstream to its juncture with the Caquetá.

---

307. Ibid., 99.
308. Dominguez and Gómez, *La economía*, 201–26.

Weakened by hunger and close to death, he came across four Carijona Indians who fled from him into the forest to report to their *cacique* that they had seen their first black man. A search party was dispatched to bring him to be examined by the supreme *cacique* and his subordinate ones. They took pity upon Crisóstomo and nursed him back to health. He began learning their language and was eventually given a young girl as his wife. After spending some time there he became enamored with a different Indian maiden, but was told that she was off limits. However, they had an affair and Crisóstomo resolved to flee with her further downriver. The couple barely escaped their pursuers.

After seven days they saw a group of Indians on shore and, despite their fear of the unknown, he beached their raft and gestured that they should moor it. His wife spoke a few words of the Huitoto language and was able to assure them they had come in peace. The Indians were a fishing party of a major Huitoto settlement in the headwaters of the nearby Igaraparaná River. Their leader accepted the two newcomers and ordered his people not to harm Crisóstomo, pronouncing that even though his skin was black, Hernández was a white.

It came time to leave the Caquetá for the Igaraparaná; Crisóstomo and his consort walked through the forest accompanied by two hundred Indians and unsure of his fate. He noticed the abundant rubber trees along their way and that night could scarcely sleep, fearful that they were about to be killed. However, next day they were made welcome by the supreme *cacique*, Iferenanvique, who sent out word by *manguaré* summoning the region's headmen to a four-day festival. The fourth day, as was their custom, all of the *caciques* of the area (bearing gifts) assembled at the supreme leader's *maloca* to discuss the tribe's current affairs. Iferenanvique presented his two guests. Crisóstomo was asked to sit with them and observe the dancing that went on well into the night. About three in the morning, each embraced Hernández both as a sign of welcome and leave-taking.

In Florencia it was assumed that Hernández had been killed by hostile Indians. As the months and years went by, Gutiérrez and another rubber entrepreneur, Solís, gradually extended their trade and rubber gathering activities throughout the drainage of the Caquetá-Orteguasa and also established a presence in the headwaters of the Caguán.

Meanwhile, Hernández lived with the Huitotos and learned their language well. He was invited to their *cacique* councils and came to be highly regarded for his wisdom and knowledge of the outside world. He was given a second wife by the Huitotos and lived with both women in the house and garden plot provided by the tribe. He would have a child by each. He dedicated himself to hunting and planting crops. He was keenly aware of the abundant rubber in the area, but had no means of gathering or transporting it.

After four years, one day he told the Indians that he could not return to the world he had left, since he had nothing to offer it. Iferenanvique asked what he would like to take with him should he leave and Crisóstomo replied "rubber." The *cacique* asked him to specify how much and agreed to supply it within three months' time. Hernández promised that he would sell it to the whites and return with shotguns, machetes, fish hooks, knives, and clothing.

At the appointed time, the Huitotos delivered the rubber on the banks of the Caquetá and constructed three canoes to transport it under Crisóstomo's command. Each had five rowers, but they were inexperienced so upstream progress was slow. The naked Indians suffered and bled considerably from mosquito bites, but endured their "martyrdom." After several days, they spied two canoes descending the river loaded with trade goods. Don Francisco Gutiérrez was aboard one of them and Hernández greeted his startled former patron. Gutiérrez suggested that Crisóstomo come along to Florencia, but the mulatto was fearful that he might be arrested. So they all went ashore to weigh the Huitotos' rubber and transfer a considerable stock of trade goods to their canoes. Gutiérrez dressed all fifteen of the Indians (as protection against the insects) and what amazed them most were the shoes. The Huitotos remarked that whites wear "hats on their feet." The trade goods included metal tools, combs, needles, cloth, and six shotguns. There were also two arrobas of salt and two barrels of aguardiente. In parting, Gutiérrez gifted Hernández a musical watch.

On arriving back at his point of departure, Crisóstomo sent word to Iferenanvique to send bearers for the cargo and it was transported to his *maloca*. The *cacique* was intimidated and refused to open the packages, insisting that Crisóstomo do so. It was agreed that the distribution would transpire the next morning. That night, Iferenanvique assembled the *caciques* and told them, "this is the day in which we will receive things and articles that we have never had before, and if the white man who is with us continues to insist that we work rubber, well, we will do it, because the white man that we have here among us is true to his word."

Crisóstomo first opened a long and narrow package that contained the six shotguns and gave one to each of the *caciques*. They had never seen one before. He had the *caciques* give out the remaining trade goods to their people and everything disappeared quickly. It was then that Crisóstomo showed them how to aim and fire the shotguns, and they soon understood how to handle them. He amazed them by shooting a bird out of the sky.

Hernández was pleased to again have salt for his own food. As for the *aguardiente*, he waited until the next festival to bring it out. At first, the Indians found it distasteful and gagged on it. But that soon gave way to a major drinking bout that emptied both barrels in a few hours.

Of course, word spread quickly throughout the whole drainage of the Igaraparaná and other caciques came to Crisóstomo volunteering to collect rubber for sale to the whites. He agreed to receive some within four months time, after Iferenanvique's people had prepared their second shipment. Now that he had iron tools he was able to construct three large canoes for transporting the rubber, each of the vessels requiring eight rowers. They headed up river for about twenty days to a pre-arranged rendevouz with Gutiérrez. The Indians were clothed and it was out of season for the mosquitos, so the rowers did not suffer physical discomfort. The transaction with Gutiérrez was concluded, and Hernández asked him to recruit settlers to teach the large number of Indians in the vast territory how to collect rubber. Gutiérrez contracted ten such adventurers, including Gregorio Calderón, Jesús Antonio Calderón, Braulio Cuellar, Ildefonso González, David Serrano, and a Martínez and an Ordóñez. These would be Hernández's companions in the exploration and settlement of the Caraparaná River.

Crisóstomo traveled back downriver with a large cargo, and before transporting it inland he constructed a dwelling and warehouse at his launching point on the Caquetá. The site came to be called Puerto Yabuyanos. The demand for his goods among the Huitotos continued to increase, and more *caciques* volunteered to collect rubber. Gutiérrez now agreed to conduct their business at Hernández's port on the Caquetá. However, on his first journey downriver to it, he died of a tropical illness. His wife actually completed the mission and delivered a large shipment of goods to Hernández. Crisóstomo used some of the merchandise to entice tribes further down the Igaraparaná. He descended it accompanied by several Colombian intending settlers. Actually, they faced little opposition, indeed quite the opposite. News of the whites and their trade goods had spread throughout the entire Igaraparaná/Caraparaná, and their arrival was eagerly awaited. When they reached the site of the future El Encanto, Gregorio Calderón (who was an accountant by training) and three companions offered to settle there to administer the local tribe.

By the time that Crisóstomo returned to the Caquetá, there was an enormous shipment of rubber accumulated at his port. He resolved to take it himself to Florencia. When he arrived there, the affairs of his deceased patron were frozen, so the shipment was sold to Solís instead. The two men entered into an agreement to continue swapping rubber for trade goods, with Solís coming to Puerto Yubayanos regularly to do the exchange. By now, there were intending *caucheros* of modest means willing to settle in the future Aranalandia, as well as others who were exploring the Caquetá and its tributaries far below Puerto Yabuyanos.

Crisóstomo Hernández came to dominate the Caraparaná and part of the Putumayo. He always forbade his settlers to mistreat the Indians. He asked the *caciques* for a wife for each of them and the parents of the girls were given clothing in return. The women were proud to be married to a white man. Crisóstomo himself continued to live for years at Cabuyeno, or the domain of Iferenanvique, making trips to Puerto Yabuyanos with rubber and to retrieve merchandise. He had many settlers and employees dependent upon him, and consigned his affairs to lettered subordinates.

The other figure of note in this account is Benjamín Larrañaga. Larrañaga was residing in Florencia when he decided to explore the Caquetá for rubber, accompanied by six companions. They sought the patronage of Solís and were outfitted by his company. Rumor had it that the Middle Caquetá was populated by cannibalistic Indians that killed and ate whites, so many were afraid to travel there. However, Larranãga, along with his wife and young son, Rafael, set out with his companions in one large canoe. Four days below Port Yabuyanos, they set up camp and began exploring the nearby forest for rubber trees. They found many, but then stumbled across an Indian trail through the forest. That night Benjamín's wife was sure that they were doomed, but he dismissed the danger, observing that, "if it is our destiny to be devoured or killed by Indians so be it, but we have to resign ourselves to remain here because this place has much rubber." Over the next several days the Colombians began laying out *estradas* into the forest. No one appeared.

Then, one day, armed with shotguns, Benjamín and two companions followed the Indian trail and eventually came across three children at play. They were startled to see whites for the first time and ran into the forest. The Colombians continued on and arrived at a settlement of several huts that seemed abandoned. But then a man adorned with feathers, a *cacique*, came forward. The whites kept their weapons pointed at him, but prostrated themselves. The Indian then extended his arms in welcome and began talking in a happy tone of voice. Other Indians began to appear in their doorways. The *cacique* invited the visitors into his dwelling and gave them food and drink. Many curious Indians crowded inside to see them.

As it turned out, this tribe had already heard from the indigenes at El Encanto that the whites did not kill or steal women, and that they brought iron axes, sharp machetes and fine clothing. The three Colombians took off their outer clothing to give as gifts to the *cacique* and then returned to the river accompanied by two Indians sent along to assist them. The Indians spent the night in the whites' camp and were given shirts, pants, knives, fish hooks, and matches to take back to their *cacique*.

The whites continued working rubber, always going about two together as a precaution against being surpised by hostile Indians. They had also cleared some forest and planted corn that was already up when they were visited by the *cacique* and four men. The Indians were all adorned in feathers and body paint, and carried weapons. By means of gestures the *cacique* made it clear that he wanted to know how the whites made their balls of rubber. Benjamín took him through the process. The Indians spent the night and the next day Benjamín gave them pants, shirts, axes, and machetes. He also showed them a shotgun and demonstrated it by killing a bird. The Indians were amazed, and the *cacique* wanted one immediately. But Larrañaga made it clear with gestures that he would have to bring in some rubber first.

That night the *cacique* assembled several others at this *maloca*, including Nofigageré, leader of the cannibalistic Muinanes tribe. The *cacique* recounted how he had been received well by the whites and that he understood he was to bring rubber if he wanted a shotgun. He exhorted the others disposed to receive white goods to gather rubber as well. Nofigagaré wanted a shotgun and declared that no one who encountered a white in the forest should kill him. Should he do so, the *cacique* would personally kill the assassin.

Meanwhile, over the next four months Benjamín continued collecting rubber. He cleared more land and sent to Puerto Yubayanos for yucca, banana, and sugarcane seeds. Then, one day he observed a caravan of Indians led by Nofigageré coming out of the forest. They were bearing rubber that they had collected and processed without any prompting. It was then that Larrañaga decided that he could use the Indians as rubber gatherers.

The Indians spent the night, and the next day it was time to evaluate their rubber. Larrañaga and his companions were illiterate and innumerate, but they knew how to do accounting in their heads; the boy Rafael Larrañaga had three years of schooling and could write down the numbers he was given. The forty loads of rubber filled a large canoe. Two of the loads belonged to the *cacique* and were sufficient to get him his shotgun. Benjamín made it clear that he would sell the rubber and return with more goods, and Nofigageré promised to bring more rubber. He wanted ten more shotguns.

Larrañaga set out, and it took him forty-five days to reach Florencia. He sold the rubber to Solís, paying off the advance that had been given him and taking the remainder in trade goods. Solís had to order the shotguns, so Larrañaga awaited them for forty days in Florencia. He returned accompanied by six intending *caucheros*, some with spouses. So, now there were thirteen whites in all at Larrañaga's outpost. His crops were ready to harvest and he was well provisioned. He sent word for Nofagagiré to come for his trade goods.

It was then that Larrañaga proposed that they explore the Igaraparaná together. Six whites, sixteen Indian bearers, the *cacique*, and Benjamín set out through the forest. Along the way they distributed a few trade goods to other *caciques* and told them that they would be back within a month. Should the newly contacted *caciques* wish more goods, they should be prepared to hand over rubber in return. The expedition reached the rapids that would be the future site of La Chorrera. Its Indians were delighted that Larrañaga wanted to settle there and promised to introduce him to the fierce Bora tribes that inhabited the lower stretch of the Igaraparaná from the rapids to its confluence with the Putumayo. Larrañaga began the contruction work on a house and cleared land for some crops before heading back upriver (with Nofigagaré). Along the way he collected considerable rubber from the Indians for transport to Florencia. There, it took considerable time for Solís to assemble the cargo, given the shortage of goods due to Colombia's War of a Thousand Days. Indeed, Solís would soon cease descending the Caquetá on trade missions.

Larrañaga returned home and moved his family and a number of other whites to the La Chorrera area. According to Tobar, "Everything went along well, and the few whites that were there were like gods to the Indians." They all soon had two Indian wives because the women's parents pressed their daughters upon them. It was prestigious for an Indian family to have a daughter married to a white. And so passed the years of harmony between Indians and whites.

Thus, in Tobar's account of white settlement of the Igaraparaná/Caraparaná, there was no coercion or mistreatment of the Indians. The latter collected rubber of their own volition to sell to the whites in return for trade goods. Nor is there any indication that the whites took undue advantage in that exchange. The movements of Crisóstomo Hernández and Benjamín Larrañaga were highly circumscribed to the Caquetá River region downstream from Florencia—there was certainly no mention of long-ranging journeys to Pasto, let alone Manaos or Pará.

In sum, despite the differing versions of white settlement of Aranalandia, all concurred on one point—by about 1905–1906 the area was dominated by the Casa Arana. By then, José Pardo (1904–1908) served as president of Peru. So Julio César now had an established relationship with the man at the pinnacle of power. President Pardo was progressive and introduced major educational reforms that guaranteed primary schooling in any community with a minimum of two hundred inhabitants. Although a social reformer, President Pardo was also a firm Peruvian nationalist. Lagos believes that, without Pardo's steadfast policy regarding Peruvian sovereignty in the Putumayo, Arana would never have been able to acquire and consolidate his empire there.[309]

In the Putumayo the Casa Arana employed a cadre of hardened South Americans, some on the run from the law, to oversee its rubber collection stations. One, the Bolivian Armando Normand, had studied for a while in England and knew some English. In 1904, he was sent to Barbados to recruit workers for the Casa Arana along with Julio César's brother-in-law, Abel Alarco. They brought back thirty-four Barbadians with them. Between 1904 and 1906 nearly 200 Barbadians accepted employment in the Putumayo.[310]

Arana claimed ownership of nearly 6 million hectares between the Putumayo and Caquetá rivers (or a territory about twice the size of Belgium) and was easily in control of "the most extensive [private] empire ever known in Peru."[311] He was by then given to referring to the Middle Putumayo as simply "my river."[312] By 1906, Arana's Putumayo domain alone was yielding nearly 30 percent of the annual rubber exports from Iquitos.[313]

Lagos comments about those who tried to resist Julio César:

> The Colombians were captured on their plantations and those who were not assassinated on the spot, were transported on some boat by Arana to Iquitos, where they were placed in a cell of the local jail. In the Putumayo there was no Colombian authority to protect them. Desolate and imprisoned in Iquitos, and without anyone there willing to defend them, they were forced to sell their properties at whatever price [the Casa Arana] stipulated. Others, instead of enduring such a Calvary, agreed beforehand to sell out voluntarily.[314]

So, by 1907, Arana controlled almost everything on the Igaraparaná and about half of the rubber operations on the Caraparaná. The remaining three major Colombian enterprises there (El Dorado owned by Ildefonso González, David Serrano's La Reserva, and the La

---

309. Lagos, *Arana*, 85.
310. Stanfield, *Red Rubber*, 120.
311. Bákula Patiño, "Introducción," 85.
312. Stanfield, *Red Rubber*, 77.
313. Ibid., 105.
314. Lagos, *Arana*, 84.

Unión property of Antonio Ordóñez and Antonio Martínez) were interspersed among Casa Arana holdings (El Encanto, Argelia, and La Florida). The prescient commentator predicted that their days were numbered, and they would either have to sell out or be killed; it was "[t]he plan of the Peruvian firm not to leave a single Colombian in the Putumayo, and, to such effect, in all of the purchase contracts there was a clause whereby the seller promised to abandon the river and never again operate there."[315] The Casa Arana's purview was expanding to the Caquetá as well.

It seems that that same year, the Casa Arana apprehended and transported to Iquitos for trial several Colombians who were intending to establish operations in the territory of heretofore unsubdued Indian tribes.[316] According to one report, by 1907, the number of Colombians and their indigenes resettled in the vicinity of Puerto Pizarro on the Caquetá numbered about one thousand persons, and others were on the way.[317]

Peruvian machinations in the Upper Amazon did not go unchallenged. In 1887, Peru and Ecuador submitted their conflicting boundary claims in the Oriente to arbitration by the king of Spain, but little really happened. In 1903–1904, Ecuadorian and Peruvian forces clashed on the Napo River—twenty Ecuadorians died in one of the battles.[318] In 1904 (or 1905), there was a dispute with the Colombians over control of some Indians in the Middle Putumayo, and the Casa Arana sent out an expeditionary force of thirty-six persons—only six came back alive (one of Julio César's brothers-in-law died in the fighting).[319]

The conflict between Peru and Colombia was about to move center stage. After his visit to the Putumayo/Caquetá in 1902, Captain Espinar opined,

> The region of the Igara-Paraná River, tributary of the Putumayo River, is located in the area of undetermined frontiers with Colombia; but, paying attention to the tenor of the Royal Ruling of 1802, this zone belongs to Peru to which corresponds all the territory traversed by navigable rivers: nevertheless, the few Colombians with residence on these rivers believe [the territory] pertains to Colombia. It is high time that the respective governments try to define the nationality of this zone, given that within it there are to be found a considerabl num-

---

315. Pinell, *Excursión*, 220.
316. Ibid., 221.
317. Cited in Gómez López (ed.) *Putumayo*, 163.
318. Stanfield, *Red Rubber*, 99–100.
319. British Parliament, *Report*, 303.

ber of Peruvians and substantial capital in circulation pertaining to the citizens of both countries.[320]

Rafael Reyes, by then a war hero who had narrowly missed being a candidate for the Colombian presidency in 1900, became president of Colombia in 1904. He was declared the winner by the electoral court after a closely-contested election. Colombia's War of a Thousand Days had ended through sheer exhaustion, rather than sincere reconciliation between conservatives and liberals. The conservative Reyes believed that firm actions were necessary, and he instituted a series of autocratic measures. He suspended the fissiparous Congress and promulgated his own new National Assembly. It promptly abolished the vice presidency, extended the presidential term to ten years, and suppressed the Council of State.

From the beginning of his term, Reyes launched aggressive plans to affirm Colombian sovereignty in the Putumayo/Caquetá. He envisioned building a connecting road from the highlands to there and establishing river transport in his Oriente. He ordered that a new penal colony be established in the Middle Putumayo (thereby both sidelining some of his banished political opponents, while also flying the Colombian flag in the disputed territory). He created two new administrative divisions—one on the Caraparaná and the other on the Igaraparaná—the heart of Aranalandia. He dispatched his former opponent but now collaborator, liberal general (Basque-surnamed) Rafael Uribe Uribe, first to Ecuador to negotiate a commercial treaty and then to Rio de Janeiro to lobby for Brazilian recognition of Colombian claims in the Putumayo/Caquetá.[321]

However, much of this coordinated strategy foundered through lack of funding.[322] Rafael and other members of the Reyes family had a history

---

320. Espinar, "Viaje," 222.
321. While in Brazil he wrote his memoirs in which he opined that Colombia's three hundred thousand indigenes should be incorporated into the country as an even better source of labor for primary industries than foreign immigrants. Unlike his contemporaries, he argued that the Indians should not be forced into nucleated settlements. They could continue their hunting, fishing, and modest agricultural economy in the bush and sell their services to outsiders whenever they wished. Uribe did favor supporting missionary efforts that could entice Indians to settle voluntarily around the missions. They should all be evangelized in accord with Christ's mandate that all heathens be converted to Christianity. Peaceful assimilation of the indigenes into Colombian society, while respecting their cultures, would open up two-thirds of the national territory for settlement that was currently inaccessible due to the fear of the "barbarians." The Indians could also in time become the most efficacious defenders of Colombia's frontiers against foreign invaders and usurpers. Dominguez and Gómez, *Nación*, 39–40.
322. Stanfield, *Red Rubber*, 109–14.

of favoring concessions of Colombian resources to foreigners. There was the curious case, beginning in the year 1900, in which Florentino Calderón Reyes, Rafael's cousin and former partner in the Casa Elías Reyes y Hermanos firm, facilitated, for a payment of 25,000 pesos, the petition of a Colombian native of Popayán, Leopoldo Cajiao, for a thirty-year concession to exploit a vast territory (the future Aranalandia and then some) in the Putumayo/Caquetá region. The contract was to be transferred subsequently to Casa R. Samper, another Colombian-owned firm but based in Paris, and then to a French company, the real buyer. Florentino's brother, Carlos, was Colombia's minister of the treasury, and it was anticipated that he would be the conduit to President Sanclemente for ultimate presidential approval. Indeed, a presidential decree, facilitating the transaction and designed to permit leasing of all of Colombia's "vacant lands," was issued on February 9, 1900. After the declaration that the lands were deserted and unproductive for the nation, they were to be leased out by the state with the proceeds from an annual rent being used to civilize the indigenes, underwrite the defense of Colombian sovereignty, and offset some of the deficit that was being incurred by the government in the extant civil war.[323]

This was followed almost immediately, on February 19, by a petition from the major ministers in the government (Ministries of Treasury, Defense, Taxation, Foreign Affairs, and Public Education) to President Sanclemente requesting authorizing of the Cajiao concession. Its terms were quite reminiscent of the Peruvian legislation decades earlier designed to civilize and develop its Oriente. Cajiao was to establish steamer service on both the Caquetá and Putumayo rivers that would carry *gratis* government officials, missionaries, and mail. Colombian soldiers would be transported at half price. The concessionaire would fund any missions established in the territory by the ecclesiastical authorities. It would do all in its power to civilize the Indians, settling them along the riverbanks in permanent centers. It would build and maintain a road from Guineo to Pasto within fifteen years. Cajiao, or his designates, would receive a thirty-year concession during which in the first ten years it would pay an annual rent of 30,000 francs and then 50,000 francs yearly for the last twenty years. In return, Cajiao would be exempt fom import taxes on all materials used to develop and colonize his concession. He and all of his employees would be exempted from military service and personal taxes as long as they were resident in the territory. At the end of his lease, Cajiao would receive free title to all of the land he had cleared and any other improvements on it.

---

323. Dominguez and Gómez, *La economía*, 155–58.

He could also select from the remaining vacant lands additional hectares in the amount of twice what he had improved. The concession could be transferred to any private individual or company, but not to a foreign government. In the event that the new owners were foreign nationals, any disputes would be subject to Colombian laws and courts.

Before finalizing his application, Cajiao had second thoughts. He found it hard to believe that the Colombian Congress would approve such an arrangement—particularly the parts exempting his employees from military service. Florentino Calderón went so far as to offer Cajiao 35,000 francs and shares in the enterprise if he would proceed, but Cajiao refused. He was leery that the transfer of ownership to foreigners might become a scandal that would call into question his patriotism. While the petition called for respecting the rights of the existing white settlers in the territory (mainly Colombians, including investors in the established Compañía de Caquetá), Cajiao felt that it would be unpatriotic of him to introduce foreigners into their midst. He even went so far as to publish a rebuttal of the plan, while stating his reasons for not going forward with it. The Calderón brothers then tried to substitute for Cajiao an agent of the Casa R. Samper, but there was now too much public opposition and the project was dropped.[324]

Nevertheless, given the fact that the same rubber that was worth 1.25 pesos a pound in 1886 was now fetching 9.65 pesos per pound in 1900, the interest of foreigners in the Putumayo/Caquetá did not abate. In 1903, a French citizen bought a holding in the Caquetá from a Colombian and sold it the following year to the English firm Casa Chalmers, Guthrie y Compañía.[325] In 1905, W. G. Boshell, a representative of British interests in Peru, filed a complaint with the Bogotá government that his party that was intending to explore the Putumayo for rubber had been detained illegally in Florencia by the local authorities.[326]

By this time, the established Colombians in the Middle Putumayo/Caquetá were petitioning Bogotá with some regularity for formal infrastructure—including a road connection to the rest of the country, a local administration and security forces.[327] They underscored the peril of annexation posed by Peru and Brazil. Both were operating with relative impunity in the territory, which at times was depicted as god-forsaken but at others as an extraordinarily rich part of the Colombian nation.[328]

---

324. Ibid., 158–62.
325. Ibid., 163.
326. Ibid., 162.
327. Cited in Gómez López, *Putumayo*, 105–108.
328. One such document sent to Bogotá referred to the area as the richest part of the

There was discussion that the Colombians on the Caraparaná were organizing a commission to approach the United States with the proposal that they would sell their rubber exclusively to North Americans in return for a Yankee guarantee of their persons and interests.[329] There were also many rumors afloat among them of the penetration of foreign interests and even takeover. Some were downright bizarre, such as the testimony (1906) of several Colombians in Aranalandia to the effect that an American investor group, including President Theodore Roosevelt himself, was about to usurp their whole region.[330] There was also the fear that if Colombia prevailed over Peru regarding the sovereignty issue, it would, as with Panama, hand the Putumayo/Caquetá over to the North Americans.[331]

We have already noted that Julio César entered into partnership on two different occasions with the Colombian Juan B. Vega, the second time beginning in 1904, or the same year that Reyes became his country's president. By then, Vega was the Colombian consul in Iquitos and the relative of Enrique Cortés, a minister in the Reyes administration and confidante of the president.[332] It seems fair to say that Julio César now had a conduit not just to Lima but to Bogotá as well.

Then, in 1905, the Colombian government conceded for twenty-five years the area between the Caquetá and Putumayo below the Caguán River (precisely Aranalandia) to two concessonaires, a company formed by Cano, Cuello y Compañía and the enterprise of Pedro Antonio Pizarro. As we have seen, this was an area already inhabited by many Colombian *caucheros*, and it also contained the holding purchased by Chalmers, Guthrie y Compañía. The latter filed a lawsuit seeking at least indemnification from the Colombian government for its lost assets.[333] Shortly after receiving their concessions, both Cano, Cuellar y Compañía and Pizarro sold their interests to Arana, further grossly undermining Colombia's position in the region.[334] Some Colombians went so far as to accuse Reyes of treason.

---

national territory that is "capable of producing more [wealth] than all of the mines that Colombia has." Cited in Gómez López (ed.), *Putumayo: La vorágine*, 103. Or again the comment that if the government would protect and develop the region, while encouraging immigrant settlement there, it would be possible to create in the Putumayo/Caquetá two or three "Colombias." Ibid., 423.
329. Ibid., 32.
330. Ibid., 116.
331. Ibid., 118.
332. Dominguez and Gómez, *Nación*, 181.
333. Dominguez and Gómez, *La economía*, 163–64.
334. Lagos, *Arana*, 82–83.

As the tensions between Peru and Colombia heated up, in 1905 there were treaty negotiations in Lima that eventuated in a request by the two countries that Pope Pius X mediate their dispute. The tortuous nature of contending territorial claims in the Oriente was underscored by the fact that the border claims of Peru and Ecuador were still ostensibly being arbitrated by the Spanish monarch, and that outcome might influence the respective border claims in the Oriente. An interim agreement between them, dated in Bogotá on September 12, 1905, stated that both sides would provide the pope with its documentation and arguments, but that the pontiff's ruling would be final. Should the Holy Father decline the charge, the final arbiter would be the president of Argentina.[335] On July 6, 1906, there was another amendment, referred to as the *modus vivendi*, whereby both countries agreed to remove their authorities and military from the Putumayo, and both would enjoy equal free navigation on the river during the arbitration. Both would retain their existing administrations on the Napo and Yapurá [or Japurá] rivers, and remove them from everywhere else (including the Putumayo). In the interim, the customs house at Cotuhé was to be operated jointly by both countries, and its receipts divided evenly between them[336] (subsequently blatantly reneged by Peru). By the following year, the Colombian Congress had approved the amendment and the government had acted on it (removing Colombian officials from the Putumayo)—the Peruvian parliament had yet to consider the matter.

It would subsequently be argued that the *modus vivendi* had greatly favored the Peruvians, since they had a strong presence on the Napo and Yapurá that remained intact, whereas the Colombians were dominant on the Putumayo. In essence, the agreement was tantamount to a Colombian withdrawal from the latter. One Colombian official maintained that the Peruvians did not withdraw their military forces from the area at all; they just converted them into policemen (a *Fuerza Urbana*), while civilian administrators remained, but now in the employ of the company, and provided a Peruvian shadow administration. Both contingents were entirely at the disposition of the Casa Arana and there to protect its interests and claims.[337] So, temporarily at least, or until permanent resolution of the two countries' border dispute, technically the Middle Putumayo/Caquetá had become a no man's land—but one in which an ambitious Julio César Arana filled the void.[338] According to Stanfield,

---

335. Treaty of Arbitration for defining the frontiers between Colombia and Peru. Bogotá September 12, 1905. Bodleian Library MSS Brit. Emp. S-22 G332.
336. Agreement between Peru and Colombia. Lima, July 6, 1906. Bodleian Library MSS Brit. Emp. S-22 G332.
337. Reproduced in Gómez López, (ed.), *Putumayo: La Vorágine*, 267.
338. Hispano, *De París*, 231–35.

The agreement opened the Putumayo to both countries and gave each a sphere of influence—Colombia provisionally controlled the north bank of the river, Peru the south—and established a mixed customs house at Cotuhé near the Brazilian border. Nonetheless, conflict flared. Colombians ran into Peruvian resistance when they moved down river, especially near the Igaraparaná and Caraparaná. Moreover, Arana had plenty of reasons to rid himself of the mixed customs house: formerly, Arana had paid no export duties on rubber extracted from a contested frontier zone. The Cotuhé customs, ostensibly manned by both Colombians and Peruvians, rubbed Don Julio the wrong way.[339]

Peru and Colombia were both keenly motivated to establish their historical claims to the area. In 1905, Colombian writer Demetrio Salamanca T. published a book in which he reviewed the contending territorial claims of the two countries and rejected the Peruvian one to the Putumayo/Caquetá that was supposedly based upon Spanish colonial documentation. Salamanca argued that Colombia could rightfully claim everything north of the Lower Napo River. It, too, could play with the colonial sources to make its case that the Huallaga and Ucayali drainages were, in fact Colombian, but that would be as absurd as Peru's territorial grab in the Caquetá/ Putumayo.[340]

That same year, the head of the archives of the Peruvian Ministry of Foreign Affairs, (Basque-surnamed) Carlos Larrabure y Correa, began publication of a work with no fewer than eighteen volumes documenting his country's claims in the Amazon vis-à-vis all of its neighbors.[341] The Colombians responded in kind, if not quantity.[342]

The year 1907 would prove pivotal for the Casa Arana. In January, the Colombian consul in Iquitos, Germán Velez, made a trip to the Putumayo where he upbraided his fellow countrymen as traitors to their nation for selling their interests to Julio César. Velez was

> ... also said to have tried secretly to induce the Colombian overseers [employees of the Casa Arana] to join him in a raid upon the house at Encanto, bind (or kill, if necessary) the manager, also

---

339. Stanfield, *Red Rubber*, 112.
340. Demetrio Salamanca T., *Exposición sobre Fronteras Amazónicas de Colombia* (Bogotá: G. Forero Franco, 1905), 22–23.
341. Carlos Larrabure y Correa, *Colección de leyes, decretos, resoluciones, y otros documentos oficiales referentes al Departamento de Loreto* (Lima: Imprenta de "La Nación," 1905–1909). 18 volumes.
342. See (Basque-surnamed) Vicente Olarte Camacho, *Los convenios con el Perú* (Bogotá: Imprenta [*sic*] Eléctrica, 1911).

the captain of the launch *Cosmopolita*, which was in port loaded and about ready to return to Iquitos, take the launch with its rubber cargo and provisions up river to Colombian territory, loot her, and form an invading party of Indians and return and take possession. The Colombian overseers would not agree to such action, however, and afterwards told the manager of the plan. The consul is known to have talked a great deal with the Indian workers, and this fact is causing some uneasiness among the Peruvians, for the Colombians are recognized as having always treated the Indians with milder measures than the Peruvians, and as a consequence it is feared that the natives would be more inclined to aid the Colombians should any conflict arise between the two nations.[343]

On May 1, *The India Rubber World* published an article regarding formation of the Amazon Colombian Rubber and Trading Company, a North American firm launched in April of 1907 in New York and capitalized for $7.5 million. It had received a concession from the Colombian government to exploit rubber and timber on 47,000 square miles in the Putumayo/Caquetá, good until 1930. The enterprise would also be allowed to select 80,000 hectares of fee simple land anywhere within the concession. This triggered an exchange in the pages of the *New York Times* between Eduardo Higginson, Consul General of Peru in New York, and J. M. Pasos, Colombian chargé d'affairs in Washington. Higginson protested that Colombia had no right to approve such a concession in a disputed area that was undergoing international arbitration. Furthermore,

> That the Peruvian government has been and is in actual possession of said territory, maintaining customs houses and collecting taxes and that the territory is furthermore subject to a modus vivendi made some years ago between the government of Peru and Colombia, which provides that neither power should alter or disturb the status quo until the dispute between them as to the sovereignty of said region should be definitely settled.[344]

To which Pasos replied that the concession was actually made in January of 1905, or a year and a half before Peru signed the *modus vivendi*

---

343. In recognition of growing American interests in the Loreto, in 1906 the United States established a consulate in Iquitos headed by career diplomat Charles B. Eberhardt. Consul Eberhardt to the Secretary of State, Iquitos, November 30, 1907. United States Department of State, *Slavery in Peru: Message from the President of the United States, Transmitting Reports of the Secretary of State, with Accompanying Papers Concerning the Alleged Existence of Slavery in Peru* (Washington, DC: Government Printing Office, 1913), 116–17.
344. *New York Times*, September 6, 1907, published on September 18.

agreement in July, 1906. He further noted that he was informing his government of Higginson's contention that Peru still maintained customs houses and tax collectors in the area, given that the *modus vivendi* provided that both nations would withdraw all officials and military forces from it until the dispute had been resolved.[345]

At this same time, the Colombian minister for foreign affairs, Alfredo Vasquez Cobo, filed a complaint with the Peruvian government to the effect that Peruvians [Arana] were invading the area by taking advantage of the absence of Colombian authority there. The Colombian representative in Lima was ordered by Cobo to present a written complaint to the Peruvian minister for foreign affairs,

> You will point out to the Peruvian government that is impossible for our government to allow that our territory lying to the east of our Republic shall continue by virtue of the modus vivendi to be an asylum for brigands and converted into another Kingdom of Morocco where robberies and murders are committed with impunity and where the only law which obtains is that of the physically strongest.[346]

In the spring of 1907, the Colombian government recognized the need to strengthen its infrastructure in th Middle Putumayo/Caquetá. The governor of Nariño was ordered by Bogotá to create a police district whose jurisdiction reached as far downriver as the confluence of the Putumayo with the Campuya River, or where the upstream border of the *modus vivendi* began. Gabriel Martínez was appointed its first inspector of police and, in May, he was in Campuya. He returned to Mocoa where he was supposed to be given twenty men and rations for them. The latter were slow to materialize, so he bought victuals out of his own pocket. By August, he was back in Campuya with his troops, commanded by his *alférez*,[347] Sub-lieutenant Lisímaco Velasco. Martínez decided that it was a poor site for a fortification and so he moved fifteen of his men to El Remolino, a rubber operation on the Caraparaná (10 leagues outside the area of dispute as defined by the *modus vivendi*) owned by himself and Antonio Ordóñez, the remaining five were so ill that they returned to La Sofía for treatment. On October 11, 1907, Martínez began his official duties at El Remolino.

---

345. Ibid., September 18, 1907, published on September 21.
346. Despatch from the Colombian government to the Acting Minister for Colombia at Lima in reference to the murders that were being committed in the Putumayo. September 17, 1907. Bodleian Library MSS Brit. Emp. S-22 G332.
347. An expression for "aide-de-camp" with or without formal appointment.

On October 22, 1907, Colombian president, Rafael Reyes, repudiated the *modus vivendi*, seemingly making clear his intention to reassert Colombian hegemony in the Putumayo/Caquetá.

And thus there began a series of telegraphic exchanges between the Peruvian government in Lima and Julio César Arana in Manaos. On October 26, Lima informed him:

> Colombia rescinded *modus vivendi*. Necessary you should warn the Putumayo so as to avoid surprises. Tell us whether you consider possible repel any invasion by means of your own staff, and whether you fear that Colombia profiting by treaty with Brazil will introduce forces via Brazilian Putumayo. It is desirable government Peru be correctly informed advantages and dangers. Keep absolutely secret.[348]

Over the next several weeks there was a regular exchange of such telegrams between Lima and Manaos. To wit,

> October 28, 1907, Lima to Arana: Government repeated orders Prefect Loreto to act with you and take energetic measures defense territory.
>
> October 29, 1907, Arana to Lima: [Given] Ignorance proportion of aggression am unable to reply whether staff would repel; but in event of conflict government garrisoning south of Campuya, Florida, Último Retiro with outpost Delicias on Caqueta (see sketch) and our assistance will make large part of territory in dispute easily invulnerable.
>
> November 6, 1907, Lima to Arana: Government ordered dispatch forces places indicated by you. Necessary you order your staff support energetically troops any emergency.
>
> November 18, 1907: Arana to Lima: Manaos advise troops Colombia reached Putumayo.
>
> November 19, 1907, Arana to Lima: Ordered staff support action government repelling invasion with necessary rapidity and energy.
>
> November 20, 1907, Lima to Arana: Inform us places Putumayo occupied by troops Colombia. Government has given explicit orders. Has confidence resolute attitude yourselves. It has also ordered reinforce garrisons.

---

348. Copies of Cables passing between the Peruvian government and Mr. J. C. Arana (Thomson file-translations by Norman Thomson). October 26, 1907. Bodleian Library MSS Brit. Emp. S-22 G344a.

> November 22, 1907, Arana to Lima: Suggest government desirability gun boat be armed taking mouth Campuya, Caraparana to capture probable expedition troops Colombia proceeding upper Putumayo.
>
> November 26, 1907, Lima to Arana: Government given orders in accordance with your suggestions. Inform how many armed men available in Putumayo.
>
> November 27, 1907, Arana to Lima: We have at disposal five hundred men armed Winchester. Necessary government send Mannlicher rifles. Sufficient staff will be awaiting arrival troops to operate jointly in repelling invasion . . . leaving garrison in position ensuring rapid communications assisting garrisons whenever they are threatened. Inform as number of troops government ordered mobilise.
>
> November 29, 1907, Arana to Lima: We will advise invasion prepared long in advance. We think necessary to garrison weak points indicated and repel any attack, send a battalion with artillery thus carefully avoiding any contretemps. Am leaving direct Manaos.
>
> December 10, 1907, Lima to Arana: Government addressed to Prefect Loreto following cablegram: Send Putumayo forces of number necessary to ensure repulse Colombians and act in manner securing success in any event without exposing forces to unfavourable mishaps, which would have grave consequences.[349]

Immediately after establishing himself at at El Remolino, Martínez began receiving denunciations from several Colombians of the Casa Arana's outrages against them, including arson and murder. It was then that Inspector Martínez traveled with some men to El Encanto to investigate the killing of "Doctor" Justino Hernández and became convinced that it was carried out by Arana's agents. He demanded, and was given, five rifles that the former head of Colombia's forces had left there when he abandoned Aranalandia after the *modus vivendi* went into effect.

Upon arriving back in El Remolino, on November 15, Mártinez found most of his men ill with tropical diseases. The site was mosquito-ridden, so he moved them, accompanied by Velasco, to Serrano's La Reserva on the Caraparaná. It was there that the *alférez* learned that fifty-seven Huitotos (including women and children) from the Sebúa tribe were despairing due to their maltreatment by the Casa Arana and wanted to flee. Velasco encouraged them to do so by saying that he could protect them once they

---

349. Ibid.

were on undisputed Colombian territory. These Sebúas were now gathering rubber on the Buenavista station owned by Colombian Cornelio Sosa under the aegis of *Alférez* Velasco. Some of these Indians were gathering for other Colombians as well, including David Serrano. Martínez was appalled when he learned of the arrangement and warned his *alférez* that he was taking a great risk—in the Amazon you did not tamper with another man's indentured Indians.

Meanwhile, on November 28, 1907, Colombian consul general in Manaos, H. Jaramillo, informed both Rio and Bogotá that an Arana vessel loaded with merchandise, the *Rápida*, had departed Manaos for Nariño on the Colombian-Brazilian border. When asked to show documents approving the entry of the vessel and its cargo into Colombia, two Colombians on board, the brothers Julio and Cayetano Gómez, argued that they were citizens and had every right to enter their country and navigate its rivers as they pleased. According to Jaramillo, Cayetano had been accused of implication in the earlier plot to assassinate President Reyes and the two "traitorous" brothers had formed a partnership in Iquitos with Arana designed to exploit the Caquetá River. The Brazilian ship's captain, desirous of avoiding an international incident, decided to return to Manaos. Before leaving the area, the Gómez brothers recruited 30 men in Iça and Iquitos that they left in Teffé (where the Caquetá crosses into Brazil and is rechristened the Japurá River). In Manaos, the Casa Arana dispatched to Teffé regular supplies, 50 Winchester rifles, 30,000 shells, 36 shotguns, and two barrels of explosives.[350]

Velasco had ignored his commander's admonition, so Martínez went to see Miguel Loayza at El Encanto to try and work things out. Upon his return, at La Reserva, he ordered Velasco to release the Indians and accompany him to El Remolino. He argued that a Colombian force of fifteen ill troops was no match for the thirty or maybe up to a hundred veteran fighters that the Peruvians could put in the field. Not only was this ignored, Velasco told Martínez's troops that they should remain with him and they did. Back at El Remolino, the inspector wrote three letters to Velasco over the next eight days and received no reply. So, on November 29, he set out for La Reserva, but arrived there after a Peruvian force, under Bartolomé Zumaeta, had surpised some of the Indians, killing four of them and carrying off eight or ten others. The remainder had fled into the forest.

Martínez accompanied Zumaeta and his recaptured Indians to La Reserva and there found David Serrano and three other Colombians

---

350. Cited in Gómez López (ed.), *Putumayo: La vorágine*, 192.

bound up. The Buenavista house had been burned to the ground (out of spite once the Peruvians learned that Antonio Velasco, their real quarry, was no longer there). La Reserva had been sacked and Serrano had been forced to witness violations of his family that Martínez found too horrible to describe in his report. He then told Zumaeta that he would do all in his power to have the Sebúas returned to El Encanto, but that there was likelihood of a formal protest over the Peruvians behavior at La Reserva and a demand for retribution. Faced with the prospect of an "international incident," Zumaeta availed himself of Martínez's offer to be deposed regarding his side of the story. Meanwhile, on hearing the shooting, the *alférez* had fled with his men, proceeding on with two of them to Mocoa.[351] It was good for him that he had, since Zumaeta was under orders to find Velasco and bring him dead or alive to El Encanto.[352]

On December 12, Martínez descended to La Revuelta again with five men whom he still commanded and his secretary. He was intercepted by a force of thirty Peruvians who demanded that he return to La Unión to look for the scattered Indians in order to hand them over. On the way, Zumaeta informed him that he and the six persons with him were under arrest and they were being taken directly to El Encanto. When he arrived there, on December 22, the whole place was on a war footing. They stripped him of his documents, money, and letters of credit worth 10,000 pounds with which he had intended purchasing a steamship in Manaos for patrolling the Putumayo/Caquetá. He and his men were imprisoned.

During the last two weeks of December, Zumaeta dispatched men to attack a Colombian rubber station above La Florida on the Caraparaná, owned by Helidoro Moreno. In the fighting, Moreno's wife and a Peruvian were wounded. It all turned out badly for the Peruvians, commanded by Martinengui.[353] They were captured and sent to La Unión, where they were imprisoned by the recently arrived Colombian administrator of the Caraparaná, Jesús Orjuela. For his part, Zumaeta ascended the Putumayo to Campuya and attacked the Colombian N. Hernández and absconded with the man's wife. They also came across Bernardo Carvajal (a few leagues above Campuya and therefore in Colombian territory that was not part of the *modus vivendi* agreement) and grabbed his *peones* and paddlers, leaving the man stranded on a beach without anything.[354]

Meanwhile, by mid-October, Jesús Orjuela, the "honorary" (as in unsalaried) Colombian police inspector of the Caraparaná, had arrived on the

---

351. Ibid., 265–68.
352. Ibid., 275.
353. Ibid., 448.
354. Ibid., 276.

scene. He was a rubberman himself, intending to pursue his trade in the Putumayo. Now established on the Caquetá, on December 31, Orjuela received a letter from the Colombians at La Unión asking him to come immediately as they were about to be murdered or imprisoned by the Peruvians. He set out on a forced march and arrived at La Unión on January 2, 1908. By then, Zumaeta had attacked Serrano and carried off his belongings and wife, burned the nearby establishment of the Colombian Cornelio Josa, and murdered four Indians. Orjuela was told by one of the Colombians at La Unión that the Peruvians were holding Gabriel Martínez and his men as prisoners and that they would be back soon to transport all of the Colombians on the river downstream with "a chain around the neck." Rumors were rife that the Peruvians were about to attack.[355] Orjuela immediately declared there to be a state of war and ordered the Colombians in the surrounding areas to come to La Unión to constitute a defense force.[356]

At this time, the Peruvians captured by Moreno arrived at La Unión. Orjuela felt that, given his lack of military forces and the impossibility of receiving Colombian military reinforcements in timely fashion, he had no choice than to seek a peaceful solution. He sent two letters to El Encanto, one to Zumaeta asking that he desist in his attacks, and the other to Miguel Loayza requesting a meeting at El Dorado. The Colombians at La Unión were opposed and opined that Orjuela would be murdered or captured. In retrospect, he acknowledged that he had been incredulous, but he could not believe that the Peruvian government would back the excesses being committed by the Peruvians at El Encanto. He set out downriver for El Dorado and the anticipated meeting with Loayza. He was in the company of an American explorer, Walter Hardenburg, and a Colombian named Alfonso Sánchez (neither of whom was a party to the dispute).[357]

On January 3, 1908, the *Liberal* arrived from Iquitos with eighty-five soldiers headed by Captain Polack, as did the *Iquitos* five days later, under the command of Benito Lores, with twenty artillerymen and a cannon and machine gun on board five days later. On the 10th, the two vessels set out from El Encanto, their military contingent augmented by sixty Arana employees.

On January 12, 1908, the Casa Arana's steamship *Liberal* arrived at the rubber station La Unión on the Caraparaná River owned by the Colombians Antonio Ordóñez and Gabriel Martínez. It was accompanied by a Peruvian warship, the *Iquitos*, with eighty-five men from the military

---
355. Ibid., 297–98.
356. Ibid., 249.
357. Ibid., 300.

garrison in Iquitos, six cannons, and two machineguns. Between them the two vessels carried 140 men. La Unión's two owners were both absent. The assistant manager, Fabio Duarte, was nominally in charge of the twenty-odd[358] men present at La Unión.

The Casa Arana's representative, Benito Lores, and the *Liberal*'s captain, Carlos Zubiaur, presented a non-negotiable offer of 20,000 pounds for La Unión and insisted that all of the rubber on the premises be loaded aboard the Arana vessel immediately. When the Colombians unfurled their flag and warned the Peruvians not to land, the force on the *Liberal* opened fire. There was a fierce ninety-minute exchange. By then, the *Iquitos* had landed invaders downstream who were trying to encircle the Colombians. In any event, the defenders were out of ammunition and so they retreated into the forest, leaving behind five dead (including Duarte) and two wounded (they were dispatched with machetes). La Unión was then pillaged, its machinery and livestock seized, its buildings burned to the ground and its Indian women pursued into the forest, raped, and then boarded as "spoils of war."[359] Martínez subsequently valued the damages at a minimum of 300,000 dollars.[360]

We have the account of one Colombian eyewitness to the event:

> I am Manuel Cerquera, I am 38 years old, unmarried, a Catholic, a merchant and a resident of Puerto Córdoba. I arrived on the Putumayo River as an employee of Sr. Aurelio Gazca, in the year 1900 and lived in several places, especially that called "El Encanto," possession of the Commercial House Julio C. Arana y Hermanos, until the year 1908. I met Sr. Gustavo Prieto in the place known as La Unión, property of the Srs. Ordóñez and Martínez, Colombian businessmen established there. We were together in that place, La Unión, its owners Srs. Ordoñez and Martínez, Gustavo Prieto, Fabio Duarte, Artemio Muñoz, Miguel Antonio Acosta, I and some others whose names I do not recall at this moment, when we learned, from an Indian named Rosalía, who arrived fleeing from the Casa of "El Encanto," that the Peruvians had the intention of coming to attack us and burn down our dwellings. And that

---

358. There is disparity in subsequent testimonies regarding these statistics. Nevertheless, all agree that the Peruvians out-numbered and out-gunned the Colombians considerably. There is also disagreement over the Colombians' chain of command. By some accounts, former General Miguel Antonio Acosta was actually the commander at La Unión. Acosta later claimed that the armed Peruvians numbered 350 men and that he had 25, of whom 24 died. Ibid., 521.
359. Lagos, *Arana*, 86–87.
360. Ibid., 271.

in reality is what happened; at seven or eight o'clock one morning, more or less four days after the arrival of the Indian woman Rosalía, there turned up on the River Caraparaná, in front of the house La Unión, Julio C. Arana's launch "Liberal" and the warboat "Iquitos" of the Peruvian government. When we saw that they did not land we shouted to them to do so, the reply to which was to open fire against our people, some of whom were at the dock and others in the house. The shooting lasted for perhaps more than an hour. The "Liberal" fired its machine guns at the house and the launch "Iquitos" offloaded armed men a little below the house in order to encircle it from behind. As the number of attackers, more than three hundred, as I was told there, was very superior to the few men that we had, no more than fifteen persons, it was impossible for us to resist any longer and we had to flee into the bush leaving behind our wounded, among whom were Gustavo Prieto, Fabio Duarte, Hermógenes Correa, a boy Pedro (the domestic servant of Orjuela), and an Indian. The Peruvians then landed all of the forces that they had on both boats; they torched the houses after sacking them for all the merchandise and rubber that their owners had in them, along with some cows and poultry. They took several women, among them the Indian that I lived with and our two daughters. They finished off the wounded with a bullet to the head from their revolvers, while, after being soaked in petroleum, others were lit on fire. I saw the bodies, among them that of Sr. Gustavo Prieto, since, as noted, the fact that my woman and daughters had been taken prisoner caused me to return [in search of them] the next day to the scene of the events. For more than two weeks I was in the forest looking for my family, in the company of seven others who had survived the attack at La Unión. The circumstance that we lacked supplies entirely caused my companions to head for the house of David Serrano, more or less two hours distant from La Unión, in search of provisions. But shortly before arriving there they were surprised by a Peruvian patrol and taken to a place where they were murdered; among their number I remember the Srs. Melo Pulido, Juan Escobar, Fernando Quirajá, Félix Lemus, Rafael Cano and Francisco Ramírez. I stayed behind on a ranch in the bush because of an injured foot. But the same lack of supplies that caused my companions to go to the house of David Serrano, and the desire to find my family, forced me to give myself up voluntarily at the Casa "El Encanto," where, after managing to get them to turn over to me my woman and children, I remained until the time I moved here to Puerto Córdoba. The bodies of those killed at La Unión were left in the field where they were eaten by the dogs and other animals.[361]

When Loayza failed to either show or even acknowledge his letter, Orjuela decided to return to his men upriver. It was near Argelia that the travelers were ordered ashore by a Peruvian sentinel and Orjuela was arrested. Hardenburg and Sánchez were allowed to leave. Moored for the night, they were passed in the dark by the two vessels on their way to attack La Unión. After the assault, Orjuela was taken aboard at La Argelia by the triumphant Peruvians. He was beaten and thrown into the hold of the warship for transfer to El Encanto. Along the way, they came across Hardenburg and Sánchez. They were fired upon and then boarded brusquely.[362]

Orjuela was imprisoned with Martínez and his men in the same cell at El Encanto. He had a long discussion there with Commandant Polack in which the Peruvian asked him to recount frankly and in friendly fashion his understanding of what was going on. Polack was amazed. He could not believe that there were, in effect, no Colombian forces in the area. He had been told that Colombian General Monroy was on his way there with two thousand men. He had been told that he had been dispatched to the Putumayo after Colombia renounced the *modus vivendi* treaty because the Colombians were preparing to take the contested territory by force. That was Orejuela's first knowledge that Reyes had abolished the agreement, and he assured Polack that his mandate was to establish a small garrison, but under no circumstances below Campuya. Polack was clearly chagrined that he had been given false intelligence regarding an impending Colombian military invasion of the Putumayo/Caquetá. He asked if they could not resolve the matter by each returning to his original position? Orjuela demurred on the grounds that the Peruvian attack with regular military forces on La Unión constituted an invasion of Colombian territory. Resolution of such an international crisis was beyond their authority.

Polack informed Orjuela that the prefect of Loreto was aware of the detention of Martínez, and he felt that the best procedure at this point was to send all of his prisoners to Iquitos on the *Liberal*.[363] Orjuela and the Martínez contingent were placed together in a tiny and fetid cage below deck. Hardenburg and Sánchez had the run of the vessel.

When they arrived in Iquitos, the whole town was in an uproar, awash in false rumors such as one that the Colombians on the *Liberal* were pris-

---

361. Taken from a document in a Colombian archive quoted in Dominguez and Gómez, *La economía*, 187–89.
362. Gómez López (ed.), *Putumayo: La Vorágine* 301.
363. Ibid., 302–303.

oners of war and another that the Colombian warship *Cartagena* was currently in Manaos taking on supplies and munitions in preparation of an attack on the Peruvians in the Putumayo. According to Lagos, ". . . the incursion was coldly calculated by Julio César Arana and the government in Lima, dressed up as an heroic defense of Peruvian sovereignty."[364] Nevertheless, the Peruvian version of the confrontation at La Unión blamed the Colombians for initiating the violence prompting the Peruvians to fight back out of self defense. At the same time, an editorial by Rómulo Paredes in the Iquitos newspaper *El Oriente* (Arana-owned) called the attack at La Unión: "a patriotic act, energetic, manly and splendid."[365]

Enter the Peruvian Amazon Company (PAC)

Meanwhile, Julio César had embarked upon another strategy, one that would ultimately prove ill-fated. By the autumn of 1906, he was in London exploring the possibility of a forming a publicly traded company for his assets. He had dissolved his New York partnership, convinced that from Iquitos he could market his rubber directly and more economically in England via Booth Line ships (given the firm's exclusive right to circumvent Brazilian customs in Pará). Arana was also concerned by the rumor that an American, Percival Farquhar, was seeking a concession in the Putumayo/Caquetá from Colombia.[366] Farquhar had to be taken seriously, since he had received approval from the Brazilian government to construct the railroad from Madeira to Mamoré, and its construction was about to begin.

Arana's key business and personal connection in London was with José Francisco Medina, a director and shareholder in the Cortes Commercial and Banking Company, recently reconstituted as the Commercial Bank of Spanish America. For years, Medina had dealt in Arana's rubber consignments to England. Julio César now sought capital through a public offering with which to expand his rubber operations. This meant incorporation of an English company and finding an underwriter of its pubic offering. Medina and his co-directors were interested in helping the Casa Arana in order to retain their privileged situation within Julio César's expanding business.

---

364. Lagos, *Arana*, 86.
365. Ibid., 88. It should be noted that on January 21, 1907, Ordóñez and Martínez had signed a note for 5,500 pounds sterling with Arana, pledging La Unión and El Remolino properties as collateral. On June 16, 1910, the two properties were purchased by Arana (presumably the Peruvian Amazon Company) for 8,800 pounds (Ibid., 85).
366. Collier, *The River*, 60–61.

The first step was to retain London's Messrs. Deloitte & Co. accounting firm (paid by the Cortez Commercial Company) to send an auditor to South America to examine the books of the Casa Arana. Between early November of 1906 and early February of 1907, Mr. Leonard Wilkinson was in Manaos, and then Iquitos, but never in the Putumayo. By March, he had filed his favorable findings—"Report on accounts from the 1st December, 1900, to the 30th June, 1906."[367]

It was also time to cash in the Robuchon chip. The French explorer had spent extensive periods in the untamed parts of the Putumayo between 1904 and 1906. In a letter dated April 4, 1907, the Peruvian Ministry of Foreign Affairs had asked for the French explorer's maps and materials, as well as any land titles regarding the extent and configuration of the Casa Arana's holdings in the Putumayo/Caquetá. All of this documentation was to be used to further Peru's territorial claims in the ongoing negotiations with the Colombians. The requests were directed to Carlos Rey de Castro, Peru's consul general in Manaos.[368] On July 19, 1907, Rey de Castro forwarded the requested materials to Lima along with the news that Robuchon had disappeared several months earlier after sending out his Indian assistants for supplies. When they returned, their employer was simply gone. All that remained were his abandoned camp and some severely damaged and fragmentary research notes. It was likely that the explorer had been killed and eaten by cannibals. There would remain some skepticism about such speculation.[369]

By August 1907, Julio César was back in Manaos hard at work on the prospectus for the impending public offering. On August 23, 1907, the minister of foreign affairs asked Rey de Castro to prepare Robuchon's materials for publication.[370] Unsurprisingly, by this time Peru's Manaos consul general was receiving monthly payments from Arana.[371] The result was an extraordinarily rapid and self-serving exercise. Robuchon's "book" was in print by the end of that same year of 1907. The work was laced with photos of presumably well-fed and happy indigenes. It included portraits of Julio César, Lizardo Arana, Pablo Zu-

---

367. British Parliament, *Report*, 299–300.
368. Robuchon, *En el Putumayo*, xv–xvi.
369. Arana's future nemesis, Walter Hardenburg, later wrote regarding Robuchon's disappearance, "As he is known to have taken several photographs of the horrible crimes committed there [the Putumayo], it is thought by many that he was victimised by the employees of Arana." Hardenburg, *The Putumayo*, 219. The Colombian Cornelio Hispano (*De París*, 272–73) claims to have been shown in Iquitos three of Robuchon's damning photos after being sworn to secrecy as to the source by their cautious owner.
370. Robuchon, *En el Putumayo*, xviii.
371. British Parliament, *Report*, 471.

maeta, and Abel Alarco. There was even a full-page shot of the *Liberal*. The book appeared under the imprimatur of the Peruvian government and with a print run of an astounding twenty thousand copies. It seems likely that the costs were subsidized by the Casa Arana.[372] According to Stanfield, "Here then was the perfect vehicle to sell PAC stock, to quell disquieting rumors of abuse of Indian workers in the region, and to justify the conquest and Reduction of 'cannibalistic' natives."[373]

Meanwhile, beginning in late summer of 1907, an Iquitos newspaper owner and editor fired direct shots over Arana's bow. Benjamín Saldaña Rocca[374] was a former military officer with a record of distinguished service in the War of the Pacific. How he ended up in Iquitos is unclear. The man was committed to defending the interests of the *pueblo*, or "people," against those of the establishment. This would place him at odds with the Loretan elite, and particularly Julio César Arana. Before launching his two short-lived newspapers, Saldaña Rocca must have been in the city for a while, since he had already collected sworn statements from thirty-two of the Casa Arana's disaffected employees describing in gruesome detail the Putumayo atrocities. According to Hardenburg, Saldaña first worked in an actuary's office where he met the men.[375] Just why they confided in him is unclear. Nevertheless, it was in the summer of 1907, on August 9, that Saldaña Rocca filed criminal charges against eighteen current employees of the Casa Arana in Iquitos's criminal justice court and almost simultaneously began publishing a couple of four-page newspapers—*La Sanción* (August 22) and *La Felpa* (August 31).

Why he published two is also unclear, although they did differ somewhat in style and substance. The motto of the more serious weekly, *La Sanción*, was "To defend the interests of the people." The first page of its initial issue bore a poem (no doubt authored by its editor/publisher), entitled "Socialismo." Its last stanza states:

> Long live socialism. Its program is the precious motto of equality; [If] we pull up by the roots the bitter weed [of privilege] Society will surge upwards in all haste! The blood that is spilled for it will result in a virile and happy Peru. And finishing off so many vile Misters There will appear out of the people their Redeemer![376]

---

372. Ibid., 140.
373. Stanfield, *Red Rubber*, 55.
374. Rocca is rendered in some of the literature as Roca. I will employ the first iteration, since it was the form he used to sign his editorials.
375. W. E. Hardenburg, "Story of the Putumayo Atrocities," *The New Review* 1 (July 1913), 689.
376. *La Sanción*, August 22, 1907, 1.

La Felpa (The Reprimand), published initially as a bi-weekly, was far more acerbic. Its first number proclaimed the newspaper to be satirical and ironic. It bore a biting, center-page cartoon that lampooned Arana in almost every issue. It did this to the exclusion of vilifying (either by name or even class) the many other abusive rubber barons in Iquitos and the wider Upper Amazon. Saldaña Rocca's charges were picked up and echoed by Lima's *La Prensa* newspaper in its December 30, 1907 edition.[377]

Saldaña Rocca's attacks upon the Casa Arana in both of his newspapers could be characterized as unrelenting—even obsessive. *El Oriente*, owned by Iquitos judge Rómulo Paredes and subsidized by Julio César,[378] leapt to the defense of the employees of the Casa Arana. It also spearheaded a counterattack that impugned Saldaña Rocca's motives.[379]

Nevertheless, in the words of Goodman, Rocca,

> . . . was going to tell the truth about what was really going on in Iquitos behind the veneer of prosperity and progress. Iquitos was a dangerous place, especially for those on the wrong side, he told his readers, "Enough of mysteries and cover-ups," he declared; "enough of derision, abuse, prevarication, thefts, assassinations, embezzlements and fraud. In such a noble and worthwhile task nothing nor anyone can stop me. I will fight undaunted and without respite, scorning the switchblade of the assassin and the pistol of the thug. . . ." Over the coming months, Arana's Putumayo operations would continue to make *La Felpa*'s center spread; and sometimes so did Arana himself, always appearing perfectly turned out as a proper Edwardian gentleman, with a sharp, neatly trimmed beard and a smart, well-fitting suit adding to his distinction, but never as a man with clean hands. On more than one occasion Saldaña Rocca had him portrayed as a devil; on most occasions Arana and his men were surrounded by skulls.[380]

Printing his opinions prior to the subsequent formation of a *Pro-Indígena* movement in Peru (1909) and an Indian Protection Service in Brazil (1910), the audacious Iquitos journalist defended the character and dignity of the Indians, depicting them as the helpless victims of rapacious whites. The crusading editor also fueled the rumor in Iquitos that

---

377. Alberto Chirif, "Imaginario sobre el indígena en la época del caucho," in Alberto Chirif (ed.), *Cien años después del caucho: Cambios y permanencias en las relaciones con los pueblos indígenas* (Lima: Tarea Asociación Gráfica Educativa, 2009), 19.
378. Lagos, *Arana*, 85.
379. Goodman, *The Devil*, 163.
380. Ibid., 44–46.

Robuchon had been killed by Arana's men for taking incriminating photographs of abused Indians. *La Felpa* published cartoons on the topic: one showing the French explorer's death on Arana's conscience, and another depicting the flogging of an Indian that was reputedly based on a Robuchon image.[381]

By October 1907, it was evident to Saldaña Rocca that the criminal court would not act upon his charges. His editorial in the October 10 issue of *La Felpa* was headlined "Arana's Assassins" and stated:

> There continue to be repugnant murders of the House of Arana Brothers in their ominous and criminal activity; so far nothing has been achieved with the declarations that we have been making of the innumerable crimes that are carried out daily without justification and scandalizing the whole world. They keep killing, robbing, burning, etc., and our judicial authorities, in the more than a month and a half that we have been speaking of this, are doing almost nothing. The Señor Judge before whom there are the formal charges, doesn't bother to see that penal sanctions fall upon those who deserve them. Instead, with the scruples of . . . a little girl and inspired by God only knows what, he sends the matter upstairs to the Higher Court, so that it can resolve for him his *doubt* whether, given the present *modus vivendi* in effect between Peru and Colombia, that zone [the Putumayo] is neutral and whether or not he can exercise jurisdiction there. What a brave consultation![382]

Obviously, Rocca thought that the judge was prevaricating. So, under the rubric *Denuncias* (Denunciations), he began publishing in weekly installments excerpts from the texts of his thirty-two depositions. This evolved into a column headlined "The Wave of Blood." To avoid facilitating retribution against his informants, Saldaña Rocca noted, "I do not give the name of the signator in order not to orient the criminal firm."[383]

Saldaña Rocca's newspapers were little more than tabloids documenting inconvenient truths that were already pretty much common knowledge in Iquitos. However, in point of fact, he had become too much of a potential danger to be simply ignored. According to Collier, one man who distributed the newspapers in the streets was found dead in an alley with his eyes sewn shut and his ears filled with hardened beeswax.[384]

---

381. Ibid., 65.
382. *La Felpa*, Año 1, no. 1, October 10, 1907, 1.
383. Ibid., Año 1, no. 14, January 25, 1908, 4.
384. Collier, *The River*, 115–16.

Since 1903, Arana had maintained his primary office in Manaos, the hub of the Amazonian rubber industry. He had also cultivated the Brazilian city's newspaper editors. His influence over Rey de Castro, Peru's consul general there, gave him considerable access to Brazilian officials charged with formulating and implementing policy in the northwestern section of their country. In sum, regarding Brazilian public opinion and official policy, Julio César's eyes were not sewn shut nor were his ears plugged with wax.

The Casa Arana was pursuing a political strategy in Lima as well. Key to it was ensuring the election of at least one of the senator's from Loreto to the Peruvian parliament. In the 1907 election, Arana created the candidacy of his lawyer, (Basque-surnamed) Julio Egoaguirre. According to one subsequent critic, Judge Carlos Valcárcel, Egoaguirre was a "scrawny fifth-class attorney," but one with political connections.[385] These included (Basque-surnamed) Augusto B. Leguía, prime minister in the Pardo administration and future president of the country. The other senator from the Loreto elected that same year was Miguel A. Rojas, a business associate of Arana's partner Juan B. Vega.[386]

Then, too, there were Arana's ties from his Tarapoto/Yurimaguas days with the Morey's and Cecilio Hernández. All three had extensive political influence. In 1901, Luis Felipe Morey was elected to the Peruvian parliament as senator from the Loreto. He also served as house representative for the province of San Martín, as president of the Loreto's Junta Departament, and as president of the Iquitos Chamber of Commerce. Adolfo Morey had been sub-prefect of the province of Alto Amazonas in 1887. Also, he was elected alternate congressman from the Loreto for three successive terms (1901–1918). Cecilio Hernández was one of the original eighteen founders of the Iquitos Chamber of Commerce and in 1906 was elected to a term as the city's mayor.[387]

Meanwhile, in London, Medina introduced Arana to Henry Read, his personal friend and director and manager of the London Bank of Mexico and Peru, and endorsed Julio César's public offering. Arana was running low on cash, and Read arranged for a 60,000 pounds revolving line of credit in return for a 75,000-pound surcharge against future rubber consignments until the debt was satisfied. In September 1907, the

---

385. Judge Valcárcel would later (circa 1913) conclude, ". . . that the ex-President of the Republic of Peru, deceived by Don Julio Egoaguirre, one of the supporters of his policy, appointed authorities in the Loreto beholden to the Casa Arana, authorities who served to cover up the crimes in the Putumayo." Carlos A. Valcárcel, *El proceso del Putumayo y sus secretos inauditos* (Lima: Imprenta "Comercial" de Horacio La Rosa & Co., 1915), 308 [*El proceso*].
386. Ibid., 270–71.
387. Santos-Granero and Barclay, *Tamed Frontiers*, 59–63.

Peruvian Amazon Rubber Company (later to be shortened to Peruvian Amazon Company)[388] was registered in London, Medina serving as the signatory of the filing. On October 1, Arana y Hermanos sold its holdings (including the Putumayo) to the new Peruvian Amazon Company for 780,000 pounds.[389]

The firm of Bellamy and Isaac agreed to be the principal underwriter of the public offering. At the request of Bellamy, Medina served on the new company's Board of Directors as its first chairman. Initially, the two other directors were to be Julio César Arana and his brother-in-law Abel Alarco. Each Board member would receive an annual director's fee of 200 pounds. There would be four PAC managers, each with a 2,500-pound annual salary. Abel Alarco would run the London office, Pablo Zumaeta the Iquitos one, Lizardo Arana directed Manaos, and Julio César himself would move among all three.

On December 3, 1907, the Peruvian Amazon Company's Board of Directors was formally constituted.[390] A somewhat reluctant Henry Read had agreed to serve. Read possessed reasonable personal credentials in that he had been born in Peru, although he left there at age two, was raised and educated in England, and then returned to South America at twenty-one. He was then in business and banking for twenty years in both coastal Peru and Chile and was a fluent Spanish speaker. His shortcoming was a lack of any experience in the Amazon. Read purchased five hundred shares of PAC stock in the belief that a director should have a stake in the enterprise.[391]

In July of 1908, Medina became seriously ill and resigned from the Board of Directors. That same month, John Russell Gubbins accepted the invitation of Henry Read, "a very old and esteemed friend,"[392] to become an acting director in his place (Read wanted to leave on an extended vacation). Gubbins possessed very strong credentials, given his fluency in Spanish and residence of thirty-eight years in Peru—albeit

---

388. At times rendered henceforth in this text as PAC.
389. British Parliament, *Report*, 319.
390. The Colombian consul general in London filed a protest with the British authorities, noting that any such enterprise operating in the Putumayo needed Colombian governmental authorization and would have to conform to Colombian law. Gómez López, *Putumayo*, 151.
391. British Parliament, *Report*, 305. Read would later report that he agreed to serve only after considerable insistence by his friend, PAC Board Chairman Medina, who argued that the two of them represented the PAC's bank lenders and should be involved in order to protect their interests. Medina felt outvoted and ineffectual on a three-man Board that included two of the principals of the company.
392. Ibid., 229.

mainly in Lima and on the Pacific Coast (he had never been to Iquitos, let alone the Putumayo). Read could claim,

> I was in a good position there [Lima]; I was a member of the Chamber of Commerce; I belonged to the Benevolent Society; I was on a tariff mission for 16 years; I was a director of a large insurance company; and I came into contact with all classes of society from the President downwards, and I knew personally nearly every President in Peru.[393]

If Read and Gubbins were both conversant with Spanish and business, neither knew the first thing about rubber. Indeed, the only two members of the London Board who did were Julio César Arana and Abel Alarco. Then, in December of 1908, at the urging of Bellamy, Sir John Lister-Kaye, prestigious peer (and previous holder of a royal appointment) agreed to serve as a PAC director. While he later resisted the suggestion,[394] it seems clear that the underwriter wanted a preponderance of respectable English names on the Board as a part of the strategy to rescue the floundering stock offering.

Lister-Kaye, too, knew nothing about rubber, spoke no Spanish, and had never been to Latin America. He was vaguely aware of the scandal in the Congo but drew no conclusions from it. He did, however, possess business experience abroad (mainly in Canada). Before assenting to join the Board, Lister-Kaye ordered his solicitors to review the Peruvian Amazon Company's articles of incorporation and bylaws. He also checked out Arana's credit and reputation in London. He owned no shares of PAC stock, nor ever would.[395]

The public offering of Peruvian Amazon stock was launched on December 8, 1908. In all, there were to be one million shares (some preference and some ordinary) to be sold for one pound each. Initially, 850,000 shares were issued.[396] Julio César received 700,000 of the ordinary shares and 50,000 preference ones for £750,000, or the purchase price of Arana y Hermanos, and immediately pledged them to Read's bank.[397]

---

393. Ibid.
394. Ibid., 257.
395. Ibid., 252, 255.
396. There is some discrepancy in the various accounts of this. At times it is stated that the capitalization was for 1,000,000 pounds; at others the numbers seem to add up to 880,000 shares issued. It is not a critical point, since Arana's control of the PAC was paramount regardless of which is the correct figure.
397. Ibid., 319.

The prospectus was entirely in the hands of Arana and his friends.[398] There had been an earlier draft [see appendix 1] that was never shown to the public.[399] It had claimed that J. C. Arana and Hermanos had 1,500 civilized employees who maintained order and administered the Indians.[400] It further noted, "The occupation of the employees reduces itself to that of armed vigilantes, their general arms being Winchesters, which are being exchanged for modern automatic arms."[401] Seemingly, the words were meant to convey the notion that this was a "progressive" company. That statement was expunged from the final prospectus, which claimed that the main activity of the enterprise was its import and export businesses in Manaos and Iquitos. In effect,

> [t]he assets of the Company consist of the aforesaid important trading businesses at Iquitos and Manaos and in the upper tributaries of the Amazon, several freehold warehouses, the Iquitos tramways, the freehold estates of Pevas [Pebas] and Nanai, several valuable buildings in Iquitos, river craft for merchandise and passengers, stocks of goods, etc.[402]

The Peruvian Amazon Company's river fleet would include "fifteen steam vessels, eight iron lighters and other small craft used for conveying merchandise, collecting rubber and carrying passengers."[403]

Furthermore, the Company would have a half interest in eight other rubber holdings [Bolivia and Brazil][404] that had produced a total of

---

398. Ibid., 215.
399. Ibid., 321.
400. Memorandum 25, 1–2.
401. Ibid., 2.
402. Prospectus. The Peruvian Amazon Company Limited. December 8, 1908. Memorandum No. 7. Bodleian Library MSS Brit. Emp. S-22 G335, 2.
403. Ibid. There is also the preliminary document, "Sketch of the Putumayo" (see appendix 1), written in 1907, that differs on this point from the final prospectus. The earlier "Sketch" states,
> For the use of this business the firm has eight steamships, two of them being 120 tons and the other five [sic] smaller; they possess besides a great number of rowing vessels, and others that are towed by steamers. (Memorandum 25, 2)

There is the following parenthetical note at the beginning of Memorandum 25: (This document is translated from the Spanish. It is taken from a file of the papers belonging to the Company. It would appear to be a document used for the promotion of the Company. Some of its phrases can be traced to the Draft Prospectus of the Company.) It therefore seems likely that this unpublished document was generated by Julio César himself (and then translated into English—possibly by his secretary and brother-in-law Marcial Zumaeta) during his preparation of the prospectus for his public offering of Peruvian Amazon Rubber Company shares on the London stock exchange.
404. This included four estates (Puerto Carlos, Sindicato, San Francisco, and Colonia

190,000 pounds of rubber the previous year. It is unclear whether Arana had a half-interest share of each, or was holding back that percentage for his personal use.

According to another source,

> [t]he Manaos operation had many branches in Brazil, the principal ones being Puerto Carlos and Japury, on the Acre River; Sena Mandureira, on the Yaco River; Catahy, on the Purus River; Muru, on the Tarahuaca River; Jutahy at the mouth of the Jutahy River; Colonia Riojana on the Amazon, etc.[405]

The Putumayo was treated as a more speculative matter, but one with great potential. The intent was to send out "experts in agriculture and mining, to investigate the best way of developing the vast region of the Putumayo."[406] However, given the boundary dispute between Colombia and Peru—that is, the lack of clear land titles—the Putumayo was not being listed as a formal asset. Nevertheless, regardless of the outcome of that contestation "even if it should affect politically a portion of the Putumayo territory, [it] will not affect the legal rights of the settlers."[407] In other words, squatter's rights had value.

The prospectus listed "Arana's territory" in the Putumayo as 12,000 square miles inhabited by forty thousand Indians—implying that labor was abundant. More than 500,000 pounds had been expended by Arana there already, mainly from the region's cash flow.[408] Rubber production in the Putumayo had grown from 33,600 pounds in 1900 to more than one million in 1905. The combined 1906 and 1907 production was more than 2.7 million, and the first six months of 1908 had produced 883,012 pounds. In short, rubber production was up dramatically and would continue to increase.[409] Consequently, despite the recent drop in rubber prices,

---

Riojana) in Brazil with a total 850 square miles and six estates in Bolivia (Estela, Lucia, Belen 1, Belen 2, San Lorenzo, and Josefina) totaling 770 square miles. MEMORANDUM for the guidance of the Commission appointed by the Peruvian Amazaon Company, Limited, to inspect and report on their Business and Properties in Manaos, Iquitos, Putumayo, and the Amazon Valley generally. Memorandum 15. Bodleian Library MSS Bit. Empr. S-22 G332, 1.
405. Pinell, *Excursión*, 197.
406. British Parliament, *Report*, 619.
407. Prospectus, 3. In the event, wishful thinking.
408. British Parliament, *Report*, 123, 320. It had been expended in the main both for the acquisition of the Colombians' properties and the subjugation of Indians.
409. It should, of course, be noted that the lower numbers for earlier year's reflected the Colombian ownership of a considerable portion of Putumayo rubber production. By 1907, Arana had a virtual stranglehold on the area save for a few holdouts, and he was buying and transporting their rubber as well, given his monopoly on Putumayo shipping.

the Peruvian Amazon Company's income should continue to grow (from J. C. Arana and Hermanos' annual average of 50,000 pounds, excluding the Putumayo). In addition to rubber extraction, there were vast agricultural lands in the Putumayo, as well as deposits of auriferous coal and placer gold in all of its rivers.[410]

In the earlier draft or "Sketch" it was stated:

> In all the Amazon region there is not a better or healthier climate that that of the Alto-Putumayo, the best proof of which is the great number of Indians found living there. Whilst in the other rivers there are none, or if there are any, their numbers are but few. All the enterprises formed for the exploitation of rubber in the Amazonas have to face the grave difficulty of the scarcity of labour. In the Putumayo this problem does not exist, as the Indian is a splendid worker, sober, of very few needs, accustomed to the woods, and extremely cheap. Another difficult problem for those works is the food of the workers, food that almost always is of canned goods imported at some cost. This, however, does not apply in the Putumayo, where the Indian cultivates his vegetable garden. Suppose for a moment that there should be a great fall in the price of rubber, that the price should fall so low that it would be impossible to work it with any result. What would be the result? There would undoubtedly be a great crisis in the Amazonas which has the fictitious life given to it by rubber; but the region of the Putumayo, having navigable rivers which put it into communication with all parts, with a splendid climate, with thousands of square miles of land that Humboldt said would be the emporium of the world, with thousands of people to apply themselves to agriculture, woodcutting, etc., it would be, in all time, a stupendous store of wealth.[411]

Initially, Bellamy and Isaac sought to sell the remaining 130,000 preferential shares to the public, of which the underwriters would guaranteed placement of 100,000 in order to provide the company with a minimum of 100,000 pounds of working capital.[412] Arana was reimbursed 30,000 pounds for preparing the prospectus, payment of the registration fees for the offering and advertising (that likely included the costs of printing 20,000 copies of the Robuchon book).[413] However, the timing was atrocious. In October 1907, the United States experienced what is known as the Bankers' Panic or Knickerbocker Crisis, which in short order nearly

---

410. Ibid., 619.
411. Memorandum 25, 2.
412. British Parliament, *Report*, 216–17.
413. Ibid., 227, 237.

halved the values of the American stock market while rocking the global economy. Furthermore, it was now that a wave of Asian plantation-grown rubber hit the international market with a vengeance. The price of rubber plummeted.[414] Only ten thousand PAC shares were sold publicly,[415] leaving the underwriters with the remainder. Medina felt personally responsible and later stated, "it was in this way that I unfortunately became the owner of a large number of preference shares."[416]

Lister-Kaye subsequently testified that the three English directors (Gubbins, Read, and himself) met frequently to discuss PAC matters among themselves.[417] Since he was the only Board member ignorant of Spanish, and neither Alarco nor Arana knew English fluently, Board meetings were conducted in Spanish and translated for him by Read or Gubbins.[418] It was the English directors' intention to increase British control of the Peruvian Amazon Company over time.

Nevertheless, Arana and his circle held fully 83 percent of the outstanding shares and were therefore in a position to dictate policy.[419] It was quite evident that Arana and Alarco retained near total control over the flow of information to the Board. Indeed, the English directors were not even informed about Saldaña Rocca's formal legal complaint and revelatory newspaper articles denouncing the Peruvian Amazon Company's policies and behavior in the Putumayo.[420] Neither were they aware that their so-accused Putumayo managers had demanded an investigation of the charges against them, or that a concerned Peruvian government had dispatched Prefect Zapata to conduct an official "investigation" and prepare a report regarding the situation. To acknowledge such developments would have been to raise questions regarding the accuracy of the allegations. Then, too, most of the requests for funds from Zumaeta in the Iquitos office went unchallenged. In short, the Board of this particular rubber company was decidedly of the "rubber-stamp" variety.

In July 1909, while the intending whistle-blower, Walter Hardenburg, was en route to London, Medina's son joined the Board to represent his family's interests in the Peruvian Amazon Company. The senior Medina would subsequently cite that timing as demonstration that the directors knew nothing of the atrocities in the Putumayo, since no father would

---

414. Ibid., 321–22.
415. Ibid., 292.
416. Ibid., 619.
417. Ibid., 272, 275.
418. Ibid., 254. According to Lister-Kaye, Arana apparently understood considerable English but could not speak it (Ibid., 277).
419. Ibid., 255.
420. Ibid., 144–45.

encourage his own son to become a part of such an onerous scandal.[421]

The very same day that the Peruvian Amazon Company's Board of Directors had been constituted, or on December 3, 1907, the American consul in Iquitos, Charles B. Eberhardt, filed a report with the U.S. secretary of state. He had recently visited the Putumayo and was about to leave his Peruvian position. He observed,

> The [rubber] business is conducted from Iquitos, where a considerable office force is employed, though a resident manager is stationed at both posts, each of whom has complete charge of a given territory and to whom the foreman of certain specified tracts of territory, or sections as they are called, must report. These foremen, together with their assistants, all of whom are armed, number approximately 200, and they have control, by "rule of the rifle," over approximately 10,000 Indians—men, women, and children, principally of the Huitoto, Bora, Ocaino, and Andoque Tribes.[422]

The foremen all worked on commission and were capable of making fabulous sums. One man reported to him earning 80,000 soles (or 40,000 U.S. dollars) in a three-month period. He then told the consul that

> [h]is contract, however, is the most liberal of any of the employees, as he has charge of a district which has always been considered particularly dangerous, the Indians never having been completely cowed, always ready to revolt, and uprisings are to be expected at any time. Indeed, they have killed and eaten several white men during the last two years, and this foreman himself, after having been seized and bound to a tree, was only saved from a similar fate by the timely arrival of his armed assistants. Then, according to his own story, many of the Indians were killed in cold blood or tortured and put to death in ways exceeding for their sheer brutality the methods of the Indians themselves. When the Indians flee to the forests, expeditions headed by armed whites and made up of Indians of neighboring tribes towards whom the runaways have always been hostile go in pursuit, and so, hunted by the whites and surrounded on all sides by hostiles of their own race, they are eventually killed or brought back captives to work as slaves of the whites, though some of course do escape. The word "slavery" is used advisedly, for the condition of the Indians is in reality nothing else.[423]

---

421. Ibid., 619.
422. United States Department of State, *Slavery*, 112.
423. Ibid., 112–13.

Eberhardt commented that, beginning two years earlier, Barbadians were imported to work for the company. However, "... they soon sickened of the brutalities they were obliged to inflict upon the poor Indians and practically all have worked their way back to Iquitos or Barbados."[424] Then, too, many of the Indians had died of smallpox and venereal diseases. The American consul concluded,

> This decimation of the ranks of the Indians through the indifference of the whites toward the state of their health and by bullets from their rifles (to say nothing of those who successfully evade recapture) seems a short-sighted policy on the part of the management, inasmuch as they are entirely dependent upon the Indian for labor; but the exploiters do not care to look ahead, it seems, for, as one man expressed it to me, "My father left me nothing, and I do not care to look out for the future generation, either. I shall get all I can out of it for my own enjoyment and let those who come after look out for themselves."[425]

Finally, Eberhardt noted,

> The local papers, some of them said to be instigated by enemies of this company, and Colombians also, have recently had a great deal to say on the subject, publishing caricatures representing more or less accurately and truly many of the acts said to be perpetrated. However, I do not pose as a reformer in this matter, the instances I have cited above, both hearsay and what I have personally seen, being intended only for the department's information, and it is hoped that the dispatch, or, at least, the part pertaining to these cruelties, may be kept from the public. A great deal has been written in recent months about such acts in the Kongo rubber districts, and a similar state of things could, no doubt, be proven to exist here[426]

Enter Hardenburg

Unbeknownst to Arana and his minions, their fate was about to take a turn for the worse. For, beginning just before Christmas 1907, the Middle Putumayo would host a most improbable pair of visitors. Two young American adventurers, twenty-one-year-old Walter Hardenburg and his slightly older companion, Walter B. Perkins, had met while working for

---

424. Ibid., 114.
425. Ibid., 115.
426. Ibid., 114–15.

several months as surveyors on a railroad project in Colombia's Cauca Valley and were on their way to Manaos—where they hoped to find employment on Farquhar's Madeira-Mamoré railroad construction project. They had begun their journey on October 1 in Buenaventura (on the Pacific coast) and had crossed the Andes with considerable difficulty, given their large load of equipment and supplies. It was now early December as they departed Sibundoy for Mocoa at the headwaters of the Putumayo. Throughout their journey down the upper reaches of the river, the two Americans would be accorded warm hospitality (as well as critical logistical information and advice) by Colombian administrators and settlers.

Hardenburg kept a diary[427] that reflects his perceptiveness as amateur ethnographer[428] and biologist. It also demonstrates that he was interested in the area's economic potential:

> The region traversed by this magnificent river is one of the richest in the world. In the Andes and its upper course it flows through a rich mineral section. At its source, near Pasto, numerous gold mines are being discovered daily and are changing hands rapidly, and there are immense deposits of iron and coal.[429]

For Hardenburg, the Sibundoy Valley had "good, rich soil, quite suitable for agricultural purposes, and covered with a thick, short grass."[430] He underscored the hospitality that he received from the four Capuchin friars at Sibundoy, including the *Padre Prefecto* [Monclar] of all the Capuchin missions in the Putumayo and Caquetá. Father Estanislao de Los [sic- Las] Corts was particularly gracious.[431] At another point in the journey, they came across mineral deposits that appeared to be marble and took samples for later evaluation.[432] Upon arriving in Mocoa, Hardenburg observed:

> A good mule-road or highway connection with Pasto and La Sofía, the head of steam navigation on the Putumayo, would do much to awaken Mocoa from the torpor into which it is now plunged, for, in that way, this virgin region would have an outlet not only for the important forest products such as rubber, ivory, &c., but also

---

427. It would serve as the basis for articles published in the fall of 1909 in the London newspaper *Truth*, and then as part of a book that was published in 1912. Hardenburg and Enock, *The Putumayo*.
428. He published notes regarding the Indians and their customs. Ibid., 150–63.
429. Ibid., 54.
430. Ibid., 55. Hardenburg also reports meeting there "an old white hag who had been the compañera [mistress] of a certain ex-President in the eighties [Reyes]" (Ibid., 56).
431. Ibid., 61–62.
432. Ibid., 64.

for the valuable agricultural staples, as coffee, cotton, *yuca*, sugarcane, and the thousands of other products of the *tierra caliente*, which can be grown here. Besides, the opening of these means of communication would greatly facilitate immigration to this vast region, which is the most essential aid to its development.[433]

In Mocoa the two Americans called upon the *intendente* or mayor, (Basque-surnamed) General Urdaneta. It was then that they learned that Peru and Ecuador had submitted their dispute over sovereignty of the Putumayo to arbitration by the king of Spain. Once that was concluded, Colombia's claim to the area would have to be adjudicated. Until the month before, Mocoa had been an open-air prison for political dissenters, but Reyes had just pardoned all but nine of them. The latter were being escorted farther down the Putumayo and released so they could leave the country.[434]

Hardenburg and Perkins now obtained some Indian bearers who were willing to proceed part way downriver, but only to a certain point. Indeed, as they prepared for their journey, one young potential oarsman became alarmed over the rumors of dangerous cannibals in the Middle Putumayo and backed out. After buying a boat from "bilious-looking *aguardiente* merchant" (Basque-surnamed) Bernardo Ochoa,[435] as well as more supplies and goods to exchange for Indian artifacts, they set off. They were accompanied at first by Octavio Materón, a Colombian who that year, along with his partner González, had pioneered a new rubber station at La Sofía. On arrival there, they found that the former Reyes's outpost was abandoned and in ruins. The Colombians were now planting rubber trees to replace those that had been decimated by the earlier gatherers. Hardenburg noted,

> I was pleased to observe that strict morality was the rule, and that Gonzalez permitted no abuses against the aborigines either by taking away their women, by cheating them, or in any way at all. As to the *peons*, they seemed cheerful and contented.[436]

Since their original crew had reached the limit of a willingness to proceed, and the locals were about to engage in an important festival of several days duration, the impatient Americans set out alone from La

---

433. Ibid., 69.
434. Ibid., 67–68.
435. Ibid., 72–74. Possibly the same man mentioned as the Mocoa "official" in the Calderón account (see page 70–71).
436. Ibid., 94.

Sofía.[437] They paddled the craft themselves and endured a six-day delay after having been stranded aground during a precipitous drop in the water level one night while they were sleeping. Hardenburg had feared they might be stuck for several weeks or even months. They were therefore frustrated when a passing group of Indians in five canoes simply ignored their plight, but then a detachment of three Colombian police happened by and assisted them by dragging the stranded boat off the sand and into the main river channel.[438]

On December 23, or two weeks after leaving La Sofía, the Americans arrived at the Yaracaya rubber station. Its Colombian owner, Jesús López, welcomed them and then regaled his guests with seemingly exaggerated stories about Peruvian excesses. While the visitors were well aware of the border dispute from the Colombian viewpoint, they were skeptical that it was the whole story. Now, for the first time, they heard the name of the Peruvian Amazon Company. It was headquartered on the Caraparaná and Iguaraparaná tributaries of the Putumayo River. The two accessible (by steamboat) settlements there were El Encanto, on the former, under the surpervision of Miguel Loayza, and La Chorrera, on the latter, run by the "celebrated" (as in infamous) Victor Macedo. Both had several satellite stations in the interior (about forty in all) under the supervision of section chiefs who were a law unto themselves. Each chief had from five to eighty *racionales* under him whose job it was to keep his section's Indians in line. The latter were required to deliver a specified amount of rubber every ten days. If they failed to meet their quota they were flogged, or worse. The overseers were all paid on commission, rather than by fixed salary, an arrangement that led to considerable abuse of the work force.[439]

And then,

> López informed me that this company, planning to get possession of the rubber estates of the Colombians of the Caraparaná, had influenced the Peruvian officials at Iquitos, in open violation of the *modus vivendi*, to send troops up to help expel them, and that, moreover, these troops had just arrived.[440]

Quite taken aback, Hardenburg asked if there was some way that they could circumvent the trouble by trekking over to the Napo River. López suggested that they travel five days downstream to El Remolino, a rubber station owned by two Colombians—Ordóñez and Martínez. They

---

437. Ibid., 100.
438. Ibid., 11, 116–17, 125.
439. Ibid., 181–84.
440. Ibid., 132.

would surely provide the necessary Indian bearers and would likely purchase at a fair price anything that the Americans might choose to sell.[441]

It was then that several canoes arrived from downriver. It was the Colombian police force under the command of magistrate Don Gabriel Martínez (and co-owner of La Unión). He was no longer with his men. Rather, they had been confronted by a launch with a superior force of forty employees of the Peruvian Amazon Company who demanded that Martínez accompany them to El Encanto. When he refused, they brandished arms and threatened to take him by force. The magistrate then told his men to await his return for three days before heading upriver. He had not come back. The Americans decided to hasten to El Remolino in order to cross over to the Napo River, since "Above all, we did not wish to get mixed up in any backwoods frontier fighting."[442]

When they reached El Remolino they were informed by its single caretaker that it was just a shipping port for the real headquarters of the operation, the rubber gathering station called La Unión situated on the banks of the Caraparaná River about a three-hour hike away. They should find Ordóñez there. The travelers decided that Perkins would stay with their boat while Hardenburg went to La Unión. There, he was received by the assistant manager, Don Fabio Duarte. His employer was absent in the forest with some Indians but was expected back shortly. This company employed about two hundred Huitotos as rubber gatherers; however, the majority of the Indians in the area worked for the Peruvian Amazon Company, which "treated them very harshly, obliging them to work night and day without the slightest remuneration."[443] The Caraparaná area had been settled first by Colombians, but, after the strong-arm tactics of the Peruvian Amazon Company, only three remained. They had received frequent offers for their stations, but refused to sell. As a consequence,

> The autocratic Company commenced persecuting them in many ways, such as refusing to sell them supplies, buying their rubber only at a great discount, kidnapping their Indian employees, &c.[444]

Duarte believed that the arrival of the Peruvian force was prompted by the erroneous rumor that a large Colombian military contingent was proceeding down the Putumayo. He opined that, once they realized the truth of the matter, the Peruvian military would not countenance any attack upon the Colombian settlers—let alone participate

---

441. Ibid., 132.
442. Ibid., 133–34, 139.
443. Ibid., 144.
444. Ibid., 144–45.

in it. Martínez would be released. Nevertheless, "If, on the other hand, they (the Peruvians) did attempt any such iniquitous proceeding, . . . the Colombians would oppose them until the last extremity."[445]

Ordóñez sent word that he would be delayed for days in the forest, so Duarte suggested that Hardenburg proceed down the Caraparaná to the rubber station of another Colombian, David Serrano. He was assigned a Huitoto guide (presumably fluent in Spanish), and during the journey the American struck up a conversation with him:

> The guide seemed to be a fairly intelligent fellow, and gave me a quantity of information about the system of rubber-collection employed in this region. He also went on to inform me that the Peruvians treated his countrymen "very badly"; and when I asked him what he meant by this he gave me to understand that in case the Indians did not bring in a sufficient amount of rubber to satisfy the Peruvians they were flogged, shot, or mutilated at the will of the man in charge. When I asked if the Colombians also indulged in these practices he replied that they did not, for they always treated them well. It is unnecessary to state that I took all this information with a grain of salt, for it seemed to me very improbable.[446]

At Serrano's place, Hardenburg was introduced to two of the political exiles from Mocoa, General Miguel Antonio Acosta and Don Alfonso Sánchez. Their companions had gone on to El Encanto to catch a boat to Iquitos, but they had lingered behind to allow Sánchez to recuperate from a debilitating bout of malaria.

That evening, Hardenburg remarked that the Peruvians were probably not as bad as they were being depicted. An agitated Serrano then told his tale. He claimed that he owed a small amount of money to the El Encanto branch of the Peruvian Amazon Company. A month earlier, its manager, Loayza, had used this as an excuse to send a "commission" to "abuse and intimidate him so that he would abandon his estate." These employees of the "civilizing company," as they called it, chained him to a tree and then entered his house, seized his Indian wife, and raped her on the porch before his eyes. They then confiscated all of his rubber, worth about 10,000 soles, and embarked on their launch. They took his wife and son with them. He had not seen them since, but had heard that she was now the reluctant concubine, and his son the servant of the "repugnant monster" Loayza. Hardenburg wrote,

---

445. Ibid., 145.
446. Ibid., 145–46.

> This horrible story, in conjunction with the other accounts of the ferocity of these employees that I had been given and the treacherous kidnapping of the unfortunate Martínez, combined to make me think that we had stumbled upon a regular Devil's Paradise in this remote corner of the world.[447]

Serrano was awaiting "the arrival of Don Jesús Orjuela, the newly appointed Police Inspector and government agent from Bogotá," hoping that that official would do something to protect him. Next day, David Serrano took Walter on a tour of his property and then proposed selling a half interest in it to the startled American. After examining the books, Hardenburg realized that it was a very attractive offer. Serrano explained that he was buying insurance, since the Peruvian Amazon Company would have to think twice before harassing a foreign owner from such a powerful country.

Orjuela arrived, accompanied by a man named Sánchez. The Colombian administrator was certain that if he could but meet with Loayza they could work out an acceptable agreement to end the hostilities. Serrano remained skeptical and feared that the arrival of Peruvian troops meant that more trouble was coming. That afternoon, Perkins appeared and, with "several gasps of amazement" at the low asking price, agreed that they should consider purchasing an interest in Serrano's operation. Hardenburg then asked to accompany Orjuela to his intended meeting with Loayza in order to inform the manager of El Encanto that the Americans were contemplating entering into the rubber business. Serrano approved the plan.

The next morning, Hardenburg set out down the Caraparaná with the two Peruvians, their *peón* and three canoe paddlers furnished by Serrano. Their destination was El Dorado, another Colombian-owned rubber operation. They passed uninhabited Filadelfia, a former Colombian rubber property that was purchased, but then abandoned, by Arana. They also visited Argelia, one of the chief properties of the Peruvian Amazon Company, and had tea with its friendly caretaker, Don Ramiro de Osma y Pardo. After overnighting in their boat, they arrived at El Dorado and were received by Don Tobías Calderón. He had been left in charge by his boss, Señor González, who,

> ... tired to death of the continual raids, robberies, and other abuses of the Peruvians, had gone over on the right bank of the Putumayo to look for some other suitable place to establish himself, where he might be left in peace.[448]

---

447. Ibid., 148–49.
448. Ibid., 168.

Orjuela immediately dispatched a runner who would travel overland for several hours to El Encanto with the invitation to Loayza to meet. The following morning, several canoes arrived at El Dorado bearing several Colombian employees from El Encanto fleeing upriver out of fear. Sixty Peruvian soldiers had just arrived from Iquitos on the *Liberal*, and they believed that an attack on La Unión or La Reserva was in the works. Hardenburg wrote, "I began to wish that we had never set out on our trip down the Putumayo, if we were to be thus barbarously murdered by a band of half-breed bandits, as the employees of the 'civilizing company' now revealed themselves to be."[449]

After waiting for Loayza futilely for another day, they decided to return to La Reserva and started back up river. While camped on the shore that night, Hardenburg was awakened by Orjuela, who pointed out ascending lights on the river. Two boats passed in the dark, the *Liberal* and the *Iquitos*. Had the assault on the Colombians begun? Next afternoon, they arrived again at Argelia. De Osma and his *peón* were now under orders to arrest Orjuela and two of Serrano's Indian boatmen, but had no instructions regarding Hardenburg and Sánchez, so he allowed them to leave. They proceeded with the one remaining Indian and Orjuela's *peón* (who Hardenburg had passed off as his own man).

Having spent another night on the river, at about nine the next morning, while abreast of Filadelfia, they heard a firefight from the direction of La Reserva or La Unión that lasted for about an hour. They continued upriver and at about eight in the evening heard boats approaching. They hugged the shore to avoid being swamped by the wakes and their Huitoto fled into the forest, exclaiming that "Peruvians were very wicked."[450] Hardenburg wanted to run away as well, however Sánchez argued the Peruvians would not dare mess with an American. But then,

> ... turning a bend, two launches appeared, and as soon as we were perceived we hear a voice shout out: "Fire! Fire! Sink the canoe! Sink the canoe!" Before this order could be executed, however, the first vessel, the *Liberal* of the "civilising company," had passed us, but the second, the *Iquitos*, a sort of river gunboat, in the service of the Peruvian government, let fly at us, one of the bullets passing just between Sánchez and myself and splashing into the water a little beyond.[451]

---

449. Ibid., 169.
450. Ibid., 172.
451. Ibid.

And with the "most vile and obscene words," they were ordered to approach the launch. An argument broke out between two officers as to whether to simply execute them. So, as he stared down the barrels of 25 or 30 rifles, Hardenburg thought his number was truly up. Instead, they were jerked aboard the *Iquitos* and "kicked, beaten, insulted, and abused in a most cowardly manner by Captain Arce Benavides of the Peruvian Army, Benito Lores, commander of the *Iquitos*, and a gang of coffee-coloured soldiers, sailors, and employees of the 'civilizing company,' without being given a chance to speak a word."[452] The captives were placed under guard and were then told by Arce of his "victory" at La Unión. He claimed that the Colombians had opened fire on them, but they had killed them all—there being no prisoners or wounded. Hardenburg later learned that the Colombian owner, Ordóñez, had ordered the Peruvians to leave, and the unequal conflict began:

> There were less than twenty Colombians against about a hundred and forty Peruvians—employees of the criminal syndicate and soldiers and sailors with a machine gun. The Colombians resisted bravely for about half an hour, when, their ammunition giving out, they were compelled to take to the woods, leaving Duarte and two *peons* dead and Prieto and another *peon* severely wounded. The latter two were then dispatched most cruelly by some of the "civilising company's" missionaries. Then the thousand arrobas of rubber were carefully stowed away on the *Liberal*, the houses were sacked and burned, and several Colombian women, found hiding in the forest, were dragged aboard the two launches as legitimate prey for the "victors."[453]

That night Hardenburg witnessed the rape of one of the screaming women, who was in an advanced state of pregnancy, after she was assigned to a certain Captain _____.[454] At Argelia, Sánchez and Hardenburg were transferred to the *Liberal*, and he was reunited with Perkins. It seems that the raiders had stopped at La Reserva after destroying La Unión. Serrano and his men fled into the forest, but Perkins and one youth, Gabriel Valderrama, had remained behind. After loading all of La Reserva's rubber, the Peruvians "destroyed everything they could not steal."[455]

Then, according to Hardenburg, "The next morning, Monday, the 13th, I saw Loayza, a copper-complexioned, shifty-eyed half breed, who

---

452. Ibid., 173.
453. Ibid., 174–75.
454. His name was in Hardenburg's notes, but the publisher withheld it—probably out of fear of a libel suit.
455. Ibid., 175.

spoke a little pidgin-English, and [I] protested against our imprisonment and demanded to be instantly set at liberty."[456] Loayza replied that he was detaining them for their own good as the Colombians would certainly kill them if he left them behind—an observation that the American found to be preposterous. Then Orjuela was brought on board the *Liberal* under heavy guard and, after severe taunting, placed below in a cage. De Osma also appeared and shook the hands of Hardenburg and Sánchez. He had tried to intercede on their behalf with Loayza, but to no avail. De Osma was himself out of favor after having objected to the raid on the Colombians. The two "pirate vessels" then proceeded to El Encanto.

Hardenburg, Perkins, and Valderrama were taunted and then confined to a tiny hut (without furniture or food) where they passed a miserable night, fearing that they might be killed the next day. Hardenburg devised his plan—a bluff to be sure. He insisted upon meeting with Loayza and warned the man of the severe consequences should he and Perkins be harmed. He claimed they were the front men for a large American mercantile syndicate that planned to set up operations in the region.

The American government would become involved in punishing anyone responsible should the explorers disappear. He demanded they be released and allowed to return to Josa to retrieve their equipment. Loayza was clearly concerned. He would send for their belongings himself; meanwhile they were free to walk about El Encanto. It was then that they witnessed the plight of Martínez and his men, imprisoned in a tiny room and under guard. The prisoners were half-starved and abused daily. There was also the "pitiful sight of the poor Indians, practically naked, their bones almost protruding through their skins, and all branded with the infamous *marca de Arana* [scars from floggings], staggering up the steep hill, carrying upon their doubled backs enormous weights of merchandise for the consumption of their miserable oppressors...."

> ... But what was more pitiful was to see the sick and dying lie about the house and out in the adjacent woods, unable to move and without any one to aid them in their agony. These poor wretches, without remedies, without food, were exposed to the burning rays of the vertical sun and the cold rains and heavy dews of early mornings until death released them from their sufferings. Then their companions carried their cold corpses—many of them in an almost complete state of putrefaction—to the river, and the yellow, turbid waters of the Caraparaná closed silently over them. Another sad sight was the large number of involuntary concu-

---

456. Ibid., 176.

bines who pined—in melancholy musings over their lost liberty and their present sufferings—in the interior of the house. This band of unfortunates was composed of some thirteen young girls, who varied in age from nine to sixteen years, and these poor innocents—too young to be called women—were the helpless victims of Loayza and the other chief officials of the Peruvian Amazon Company's El Encanto branch, who violated these tender children without the slightest compunction, and when they tired of them either murdered them or flogged them and sent them back to their tribes.[457]

The Americans decided that Perkins should remain behind at El Encanto to await their equipment, while Hardenburg (with Sánchez) traveled to Iquitos on the *Liberal*. They boarded on January 15, and Hardenburg immediately had a run in with its captain, "the infamous" Carlos Zubiaur, brandishing a club. Although Loayza had promised them free passage, Zubiaur insisted upon being paid 17 pounds for each passenger. Once on board, the two travelers were denied supper and told there was no spare cabin, so they would have to string their hammocks on deck, which they did. Zubiaur reappeared and demanded the hammocks be moved elsewhere. At this point, Hardenburg lost his temper and threatened to kill his tormentor. After a pregnant moment, Zubiaur retreated; he left his two passengers alone for the rest of the trip.[458] Then Orjuela, Martínez, and their men were brought on board and put into a filthy cage that was so small the Colombians barely had room to sit down together.

Hardenburg became acquainted with the other passengers. (Basque-surnamed) Bartolomé Guevara ("to whom I took an instinctive dislike") was a former section chief in the El Encanto region. There was the Brazilian customs inspector who was a regular fixture to ensure that the *Liberal* did not discharge cargo illegally when passing through Brazilian waters. Another was a Peruvian lieutenant who was under arrest for refusing to take part in the raid at La Unión. He had resigned to return to Spain "to spend some of the money he had extracted from the tears, the bitter agony, the very life-blood of the unfortunate Indians under his control."[459] There was also the Peruvian commissary (policeman) of the Putumayo, "a miserable wretch who had a seven-year-old Huitoto girl with him for sale in Iquitos." According to Hardenburg,

---

457. Ibid., 180–81.
458. Ibid., 186–89.
459. Ibid., 190.

> His position was a sinecure, for, instead of stopping on the Putumayo, travelling about there and really making efforts to suppress crime by punishing criminals, he contented himself with visiting the region four or five times a year—always on the company's launches—stopping a week or so, collecting some children to sell, and then returning [to Iquitos] and making his "report."[460]

While most of the passengers ate miserable fare together, Zubiaur and the commissary dined privately (and well) in his cabin. The bug-infested trip to Iquitos took more than two weeks, and they arrived there on February 1, 1908. Hardenburg went immediately to inform

> ... the dentist Guy T. King, acting American Consul in this place, of the events already narrated to the reader, but this gentleman, considering solely and exclusively his own interests and forgetting the duties that his position as Consul incurred upon him, contented himself with congratulating me upon my narrow escape from death at the hands of the assassins of Arana and informing me that, owing to various circumstances, he could do absolutely nothing for us![461]

Dr. King's reticence may be appreciated if we consider what former prefect Hildebrando Fuentes had just stated in his recently published book (1908) on the Loreto:

> The notable patriot and rich merchant of Loreto, Don Julio C. Arana, who, for his personal qualities and civic virtues, is always referred to as the Abel of the Department, has been the civilizer of all the River Putumayo, and the one who with his talent and capital has made commerce flourish on its two most beautiful tributaries, which are the Cara Paraná and the Igara Paraná. He civilizes the Indians, he submits them to work, he creates among them needs, even if at the most elementary level of clothing, and he combats the ferocious instincts of this cannabilistic horde ... [462]

In light of the economic crisis that reduced trade between the United States and Peru, the American government had downgraded its diplomatic presence in Iquitos, and Dr. King had been appointed to replace Eberhardt. There was little doubt that King was apprised, at least in general terms, of the situation in the Putumayo. While there is no concrete

---

460. Ibid., 191.
461. Ibid., 195.
462. Fuentes, *Loreto* 2, 113.

evidence, it seems unlikely that King failed to receive a copy of Eberhardt's disturbing final report from either his predecessor or the U.S. State Department. Indeed, the abuses of the Indians, at least in general terms, were pretty much common knowledge in Iquitos. As early as 1907, some of the city's Catholic clergymen, Augustinian missionaries, had filed a report with Peru's minister of justice denouncing the *caucheros* of the Putumayo for serious abuses against the Indians, "whom they now ill-treat and murder for no reason, seizing their women and children."[463]

What was indeed curious was the feeble consular presence of both the United States and Great Britain in Iquitos. The city was at the vortex of the extreme tension (including the ruptured *modus vivendi*) between Colombia and Peru. The transition from Eberhardt to King coincided with the violence in the Putumayo that had resulted in several deaths. The Spanish monarch and the Vatican were supposedly trying to mediate the conflicting claims of sovereignty in the area. Brazil was intensely interested and would soon be hosting negotiations in Rio between the belligerents. Then, too, by December 1907 the U.S. State Department was in possession of a damning report on the Putumayo from a departing professional diplomat, yet chose to appoint a rank amateur as its new American consul in Iquitos.

For their part, the British had a resident English merchant, David Cazes, as their consul. This was an incredible posting for a city in which the English Booth Shipping Line exerted a virtual monopoly on direct import and export with Great Britain. The British influence[464] in Iquitos was so extensive that the pound was practically the local currency. A couple of hundred Commonwealth citizens (Barbadians) had worked in the Putumayo, and there was a smattering of European British nationals throughout the Loreto. Several of the rubber firms in the Oriente were British. Yet British diplomatic interests, too, were in the hands of another rank amateur—one who was more compromised than was the American consul by virtue of his extensive active commercial ties in the area—including with Arana's business entities.

It was then that Hardenburg received his object lesson in Iquitos political reality. When the young American announced his intention to confront Arana, Dr. King advised him heartily against such action. Iquitos was

---

463. British Parliament, *Report*, 461.
464. It might be noted, however, that this control was relative and varied over time and was declining during the first decade of the twentieth century—first in favor of France and then the United States. Indeed, some years between 1904 and 1914 French imports of Loretan rubber actually exceeded those of Great Britain. Santos-Granero and Barclay, *Tamed Frontiers*, 81–83.

simply Arana's town and the man was extremely dangerous if crossed.[465] It was doubtful that Julio César was even in Iquitos, since he was absent regularly for months at a time—whether in Manaos, Biarritz, or London. The consul did offer some assistance, however. The penniless Hardenburg could stay with him while he awaited transfer from the United States of 300 dollars that he had saved while working on the Colombian railroad. The American then quickly found employment teaching English in a private school for 10 pounds a month and another job designing a hospital building that paid 20 pounds.[466] He soon had fourteen private students in his English classes, including Julio Egoaguirre, who (unbeknownst to his teacher) was Julio César Arana's personal attorney.[467]

Hardenburg was now clearly in a position to remain in Iquitos, and it was then, in late February, that an event changed any thoughts that he might have had of leaving. According to Collier, one night Walter witnessed police raiding a nearby small printing shop while an angry mob milled about outside. The authorities led its battered proprietor to the dock and ordered him to board a vessel leaving for Yurimaguas. Dr. King informed Hardenburg that it was Benjamín Saldaña Rocca.[468]

Hardenburg later published his analysis of why Saldaña Rocca left Iquitos. Walter certainly did not mince words in his praise of the crusading editor:

> Imbued with a Socialist's hatred of oppression, urged on by an irresistible desire to serve humanity, the great soul of Benjamin Saldana Rocca rose in revolt. Single-handed, he pitted himself, a lone proletarian, against the Arana Company and its millions, together with the crooked and corrupt officials of Iquitos and the great business interests of the Peruvian Amazon which at once lined themselves up with the oppressors.[469]

Hardenburg then lists a series of graphic Putumayo accounts damning Arana published by Saldaña Rocca. Had he witnessed the mob violence and physical abuse reported by Collier that forced the newspaperman to depart Iquitos, Hardenburg would have certainly said so in his first-person account. What he does tell us is that, prior to founding his newspapers, Saldaña Rocca was employed in Iquitos's actuary's office where, in the course of his duties, he met many of Arana's disgruntled

---

465. Collier, *The River*, 112.
466. Ibid., 114.
467. Ibid., 135.
468. Ibid., 115.
469. E. A. Hardenburg, "Story of the Putumayo Atrocities 2," 689.

former employees, men who, "had been persecuted and tortured by the Company's officials because they would not murder, flog and mutilate Indians for the benefit of the Arana gang."⁴⁷⁰ And so it began. However, Saldaña Rocca's departure from Iquitos was considerably more prosaic than Collier would have it. According to Hardenburg,

> As to the Arana gang, they were but little disturbed by Saldana's exposures. By means of their subsidized press, they denounced Saldana as an agitator and accused him of having tried to blackmail them. This and a transparent denial of the charges on the part of some of the chiefs of sections was the only reply they made. They were not worried. Why should they be? They knew they were safe. They had nothing to fear, for did they not hold in the hollow of their hands the puny officials who conducted the affairs of government in their interest? Not daring, however, to take libel proceedings against Saldana, they nevertheless soon secured his downfall. All they did was to pass the word to their apologists and to the other interests to refuse to advertise in or subscribe for his papers. This was sufficient to bring about the desired result, for while many workers bought the papers eagerly, the lack of advertising matter and the diminished circulation resulted in each issue being published at a loss. As Saldana's capital was limited, he soon had to yield, and the papers were both suspended in February of 1908.⁴⁷¹

And so Saldaña Rocca left Iquitos, presumably of his own free will.

On February 1, 1908, *La Felpa* had published an article claiming that Bartolomé Zumaeta, under orders from Loayza, took more than a dozen Peruvian soldiers to La Unión where they tricked Martínez and seven persons into surrendering.⁴⁷² The following week, or precisely a month to the day of the raid on La Unión, *La Felpa* published a cartoon on the violation of Peru (depicted as a comely blindfolded woman). In the background, the *Liberal* is anchored off of La Unión, which is going up in flames. The caption reads: "The Aranas and *El Oriente* pressure the Pe-

---

470. Ibid.
471. Ibid., 693. Nevertheless, both newspapers had considerable advertising from Iquitos restaurants, stores, and professionals. Clearly, not everyone in the town was afraid of Arana.
472. As we have seen, Martínez was actually already in Arana's custody, and his partner Ordóñez fled successfully into the forest. A Colombian source later claimed that *La Felpa* accused the Peruvians of firing some bullets into the *Liberal*'s hull as "evidence" that they had been attacked by the Colombians. Pinell, *Excursión*, 223.

ruvian Republic without pity in [plain sight] of our government."[473] The February 22 issue of *La Felpa* carried a cartoon labeled "In search of the criminals" with the caption "When a society gilds over its blemishes and is anxious over pursuing the criminal, and kneels before the murderer who has a fistful of gold, the judge kisses that hand." It was signed by (Basque-surnamed) Uribe. The image is of a blinkered official in coat and tie, holding a lantern out of which gold coins fall into a sack. He is oblivious to the skulls and rifle under his feet as he probes the distant darkness with his beam. In the background a *peón* holds out the head of an Indian and more gold coins fall from its bloody severed neck into sacks that sprout wings and fly away.[474] This was likely the last issue of the newspaper.

Hardenburg resolved to pursue his own investigation of the Putumayo abuses. Collier depicts the young American as a stubborn crusader in his own right by virtue of his "Currier and Ives print come to life" upbringing, his childhood reading habits—books like *Two Years before the Mast* and *Uncle Tom's Cabin*—and his Methodism.[475]

There was now a breakthrough. A young man, Miguel Gálvez, Saldaña Rocca's natural son, came to see Hardenburg. He had in his possession all of the original accounts of Putumayo abuses that had informed his father's articles in the broadsheets, as well as paternal permission to pass them on to anyone who might continue the campaign. Miguel claimed to have overheard the American when he had asked a bartender if he knew of the publications. Hardenburg was somewhat suspicious that it might all be some kind of Casa Arana trap, but he decided to take the risk. Soon he was in possession of extensive incriminating testimony that documented the abuses on the Putumayo as being far more extensive than he had imagined. Hardenburg now believed all of the allegations that he had heard from the Colombians.[476]

Months after Hardenburg reached Iquitos, Perkins finally managed to exit the Putumayo and the two Americans were reunited. To wit,

> Towards the end of April Perkins arrived without our baggage, for the miserable murderers of El Encanto, their cupidity aroused by the idea of getting something for nothing, had stolen it while Perkins was held prisoner at that place. Thus they became aware of the deception I had practiced upon them in regard to the Ameri-

---

473. *La Felpa,* Año 1, no. 16, February 8, 1908, 2–3.
474. Ibid., Año 1, no. 18, February 22, 1908, 2–3.
475. Collier, *The River,* 124–25. Writing more than half a century after the Putumayo affair, Collier dedicated his book: "To the memory of WALTER ERNEST HARDENBURG, Son of Liberty, 1886–1942."
476. Ibid., 130–35.

can syndicate, and so great was their anger that they were upon the point of murdering Perkins, but their fears getting the better of them, they contented themselves with keeping him a close prisoner and abusing him, as is their custom.[477]

Perkins wished to leave for the United States as soon as possible, and the determined Hardenburg advanced him passage money, declaring that it was his own intention to remain. Now that the Americans' equipment was definitely gone Dr. King agreed to initiate the process whereby the U.S. State Department might petition the Peruvian government for compensation of them for their loss.[478]

Hardenburg asked Perkins to send him an account of what he had learned and experienced at El Encanto. He later received that undated missive sent on the letterhead of the Lowndes County Well Drilling Company of Crawford, Mississippi. It stated that there were about forty Colombians at Serrano's place when the assault on it began. The Peruvians outnumbered the Colombians about ten to one and struck from both the river and the land. Four of the defenders were killed outright and the remainder fled into the jungle, leaving behind several wounded companions who were dispatched on the spot by the attackers.[479] Perkins considered himself to be a part owner of the Serrano property and believed that his American passport would protect him. He approached the Peruvians with his terrified bodyguard, and both men were immediately taken prisoner. Serrano's rubber (worth about 2,500 dollars) was then expropriated.

Perkins was taken to El Encanto, where he met up with Hardenburg. He decided to remain behind while Hardenburg traveled to Iquitos. Perkins recounted,

> I stayed to recover our baggage and to suffer the role of the helpless prisoner for 72 days. I was promised freedom and permission to look for our baggage, but was held strictly a prisoner the whole time. During my stay there I suffered a good deal in various ways, sickness without medicine or even drinking water, daily threats of death, etc., no clothes and all that, but I fared better than any other prisoner in that I was the only one not murdered after I came there. While there I saw many things which filled me with horror and disgust and gave me an insight into the ways

---

477. Hardenburg and Enock, *The Putumayo*, 195–96.
478. British Parliament, *Report*, 503.
479. Appendix 2 is a translation of one of the many accounts in Colombian archives of Peruvian abuses in Aranalandia. Its author provides an eyewitness account of the assault on La Unión.

of the Peruvian Army and the Peruvian Amazon Co. After I had been there about three weeks, Colombians were captured by a large force of Peruvians after being tricked into laying down their arms. These prisoners, unarmed and tied, were then butchered with pistols and machetes. The bodies were mutilated and some decapitated . . .[480] Indians were frequently tied to stakes driven in the ground and whipped on their backs until the flesh was a quivering jelly like mess and then turned loose. In a few days if the Indian was not already dead he was generally shot, because the maggots in his back, together with the stench and the improbability of his further usefulness, made him undesirable. Several cases of wholesale murder of Indians occurred in or near Encanto while I was there—the hut surrounded, torch applied and the poor devils shot as they ran out. Castration of the men was another form of amusement that generally resulted fatally and putting Indians in stocks for several weeks at a time until their legs were useless or they had died of starvation also helped to pass the time. The tribes were searched thro' and all the young girls of pleasing appearance were brought up to the house to serve as wash women, servants and prostitutes for the officials. Also every boat that went to Iquitos carried off several of these girls, to say nothing of small boys that were sold as slaves or sent as presents to friends. To see the mothers of these children in their sorrow over the loss was enough to stir the heart of anyone except those cowardly murderers. Why are these cruelties committed? The Peruvians say they punish the Indians because they fail to bring in the required amount of rubber. That is true in some cases, but at other times there was not even this excuse. I tried to get at the cause and found none other than that a coward when he has one at his mercy turns into a savage beast. These Indians are great fighters among themselves but they are afraid of the white man and the Peruvians were quick to see this. By beating and killing a few every now and then they have managed to keep the Indians in a state of terror and this terror, visible to the Peruvians, has invited more atrocities, the Peruvians themselves being as cowardly as the Indians, but better organized. In closing I will state that in Iquitos, Peru, I met a few Peruvians I believe were gentlemen and I do not care to make sweeping assertions against the race as a whole, but I do say that the crew of mongrels the Peruvian Amazon Co. has in charge of their business and

---

480. This may be reference to the fate of Erazo and his recruits.

that part of the Peruvian Army they have leased to protect them and help them to steal are a set of cowards, murderers and thieves without an exception. When they find an exception to this rule they discharge him as being useless.[481]

A key issue remained: the extent of Julio César Arana's personal involvement in (or even knowledge of) the situation in the Putumayo. In April, Arana was back in Iquitos, having just visited the Putumayo. In early May, Hardenburg met Julio César in the Arana residence and there ensued an exchange of frigid courtesies.[482] Arana claimed to have been ordered by the Peruvian government to accompany a two-hundred-man force to the Putumayo that it was sending there to secure the border against Colombian incursions. Julio César tried to feel out the American regarding what he had witnessed on the river, but a cautious Hardenburg gave vague replies. He then asked Arana to do something about retrieving the two Americans' missing equipment and scientific instruments (or reimbursing them for their loss). Julio César promised to look into the matter and the two men parted.[483]

In the event, Arana had misrepresented to Walter his recent Putumayo mission. Saldaña Rocca's campaign had stirred the Peruvian pot sufficiently to prompt adverse articles in certain Lima newspapers.[484] Then, too, Colombia denounced the events at La Unión and La Reserva resulting in the deaths of several of its citizens, and was demanding satisfaction. So, the Peruvian government had ordered an investigation of the charges by Loreto's prefect, Carlos Zapata. Zapata was also an Iquitos merchant who owed a considerable sum of money to J. C. Arana y Hermanos (a debt by now transferred to the books of the Peruvian Amazon Company).[485] Julio César had returned to Iquitos from London to accompany the prefect to the Putumayo, along with Peru's consul general (and Arana employee) in Manaos, Carlos Rey de Castro. According to subsequent analysis by Iquitos judge Carlos Valcárcel, Arana's initial fear was that the prefect would reveal the truth about the atrocities committed

---

481. In Bodleian Library MSS Brit. Emp. S-22 G338. Goodman claims that Serrano was killed during the attack as well. Goodman, *The Devil*, 35.
482. Goodman reports that Perkins was present at the meeting (*The Devil*, 35). However, in their later accounts neither Hardenburg nor Arana confirmed (or denied) this.
483. Collier, *The River*, 128–29.
484. Both the American Consul King and the British Consul Cazes stated their belief that the Saldaña Rocca broadsheets were simply ignored by the public. However, Cazes would subsequently testify that he banned them from his home because his wife found them to be disturbing (British Parliament, *Report*, 159, 162).
485. Ibid., 333.

by his overseers (practices that Arana believed were essential for the success of the operation).[486]

On their arrival in the Putumayo, the party found that Peruvian military commandant Polack had detained the alleged overseers of a subsequent attack on La Reserva a few days after the raid at La Unión. Bartolomé Zumaeta, Arana's brother-in-law, was accused of having been a leader of the force. Bartolomé was also said to have ordered the execution of a Peruvian sergeant, Ricardo Caceres, for having refused to attack the Colombians.[487]

The prisoners were released, and there would later be an unexplained entry in the books of the Peruvian Amazon Company for payment of 7,000 pounds to Prefect Zapata.[488] His 9,000-pound debt to the Company also disappeared from the ledgers.[489] Unsurprisingly, the Zapata report would exonerate the Peruvian Amazon Company of abusing either Colombians or Indians. It had responded patriotically to Peru's appeal for assistance in the defense of its borders, on the one hand, and its treatment of the Indians was entirely humane, on the other.

Judge Valcárcel believed that the initial concern in Lima had been genuine, but that Arana was essentially able to deceive the Peruvian government through a combination of his control of the Iquitos judiciary and administration, his influence in the Congress (especially through Senator Egoaguirre), and his ability to garner media sympathy as a civilizer of cannibals and defender of Peruvian sovereignty in the Loreto (orchestrated by Manaos Consul-General Rey de Castro). Egoaguirre's (and, consequently, the Loreto's) power and influence were certainly waxing in the Peruvian capital.

However, should anyone doubt (and Judge Valcárcel surely did not)[490] that the Zapata investigation was orchestrated by Arana, we have the latter's testimony to that effect. There is his letter (dated July 5, 1908, and signed in Manaos), sent to (Basque-surnamed) Andrés A. Aramburú[491] in gratitude for his defense of the Casa Arana against its accusers. The letter stated,

---

486. Valcárcel, *El proceso*, 262–63.
487. British Parliament, *Report*, 141.
488. Ibid., 218.
489. Ibid., 333.
490. Valcárcel, *El proceso*, 237–38. Valcárcel called the trip and its report "an infernal confabulation" by Zapata, Arana, and Rey de Carlos—while accusing the consul general of being the key offender (Ibid., 263.).
491. Aramburú, owner of Lima's most prestigious newspaper, was descended from an illustrious Peruvian publishing family and was the grandson of a Lima mayor.

My Dear and very appreciated Sir—I have just read in the April 9, 1908, edition of *La Opinión Nacional*, the profound defense of the house of Julio C. Arana and Brothers against the imputations of *"La Sanción"* and *"La Felpa,"* two adventitious and ephemeral handbills in Iquitos. Your gallant pen, exalted by your generous just spirit, has annihilated our firm's detractors. . . . During my lengthy business career, it has been my sole concern to acquire a patrimony that would allow me to collaborate modestly in the project of making our country great and happy, through honorable work; and the observance of its laws and institutions. . . . Due to the heavy-handed interventions of *"La Sanción"* and *"La Felpa"* I have undertaken a journey in the Putumayo and its tributaries, accompanied by the gentlemen Carlos Zapata, Prefect of the Department of Loreto and Carlos Rey de Castro, Consul General of Peru in Manaos, and these distinguished functionaries of our country will say that there are no indications of violence and cruelties in the rubber collection establishments of the region. Indians, who until five or six years ago lived in the furthest reaches of the jungles, fleeing from the white man or trying to eat him, today come willingly and satisfied to the work centers, remain there for as long as is necessary and engage in the lives of its civilized beings without hatred or protests. Those unfortunates, who were ignorant of the most rudimentary notions of their responsibilities and rights, are now beginning to have an idea of life's worth; and of the meaning of country. And this dawning of culture will continue to grow, undoubtedly, in relation to which our sovereignty in these regions is financed and the degree to which civil authority is allowed to be felt.[492]

The language of this particular letter is vintage Carlos Rey de Castro. Indeed, Rómulo Paredes, publisher of the newspaper *El Oriente*, claimed that Rey de Castro regularly wrote and submitted articles to him and other newspapers in Iquitos, urging that they be published under the editor's name. In one, the phantasmagoric claim was made that Arana had converted eleven thousand Indians into enthusiastic Peruvian patriots who paraded before Prefect Zapata to show their resolve to take up arms against the Colombians in defense of a Peruvian Putumayo. Arana then used his influence to have these favorable (to the Peruvian Amazon Company) Iquitos reports reproduced in sympathetic Lima newspapers.[493]

---

492. Valcárcel, *El proceso*, 368–69.
493. Ibid., 263–64. This particular one was reproduced in Aramburú's *La Opinion Nacional* on September 14, 1908 (Ibid., 266).

Nor was his manipulation of the media limited to Peru. After the Colombian version of the events at La Unión were published in the Manaos newspaper *Journal do Comercio*, Rey de Castro gave its reporter an interview laced with disinformation. He stated that various Peruvian rubbermen had been active for nearly two decades in a Putumayo that was clearly Peruvian territory. The Casa Arana had consolidated this reality by purchasing many properties of willing Colombian sellers. The recent conflict was due to the fact that certain Colombian *caucheros* had convinced President Reyes that they were capable of retaking the Putumayo if given the chance. So Colombia had torn up the *modus vivendi* agreement. The Colombians, commanded by General Acosta,[494] started the conflict at La Unión by opening fire on the Peruvians. General Acosta and his men had already captured many Peruvians and murdered several of the bound prisoners. Rey de Castro insinuated that events at La Revuelta were confusing. It was likely a dispute between individuals over their private affairs, although he didn't rule out the possibility that General Acosta had a nefarious role in the matter. Loreto's prefect was taking measures to ensure that the peaceful Peruvians in the Putumayo could pursue their enterpises without interference. Rey de Castro underscored the good relations that existed between Peru and Brazil, and the former's commitment to improve them. The reporter concluded his article with a paean to the Peruvian consul's powers of persuasion:

> Such was the information that the distinguished Dr. Carlos Rey de Castro deemed to provide us, to whom we once again thank for his characteristic amiability. Publishing this [article] we consider the matter closed.[495]

In Valcárcel's view, therefore, two of its key officials (Egoaguirre and Rey de Castro) criminally deceived the government in Lima—while using their own credibility to undermine that of Saldaña Rocca. Valcárcel

---

494. Recall from the Hardenburg account that he met General Acosta at the Serrano property. Acosta was a convicted "felon" who had been released recently from his political exile in the Putumayo and allowed to leave the country. He was en route to Iquitos. It is certainly unlikely that Bogotá placed him in command of anything at La Unión. It seems that he went to La Unión when he learned that it had become the rallying point for Colombian resistance to an order from Miguel Loaysa that they should all abandon Aranalandia within five days or suffer the military consequences. Acosta claimed that he was appointed commander by the "government of President Reyes" (cited in Gómez López [ed.], *Putumayo*, 521–22) when, in fact, he was asked by Orjuela (an honorary Colombian official to be sure) to take charge of the defense while he went to El Encanto to try and reason with Loayza—a logical move given Acosta's military experience. Cited in Ibid., 195.

495. Gómez López, (ed.), *Putumayo: La Vorágine*, 346–48.

claimed that the only reason he did not later indict Carlos Rey de Castro for his complicity and iniquity was the immunity afforded the man by his status as a Peruvian consul general.[496] In the aftermath of the Zapata "investigation," Rey de Castro provided the eyewash and Arana the influence to ensure that the Lima press published articles favorable to the Casa Arana.

At the end of May, Perkins departed Iquitos for the United States via Brazil and happened to coincide on the same vessel with Arana. During their conversations about the rubber industry, Julio César complained about the inefficiency of his Peruvian overseers in the Putumayo. Once in Manaos, Arana presented the startled Perkins with a written offer to employ him and Hardenburg. The American did not regard it as an attempt to head off their request for compensation for their lost possessions,[497] but it could certainly have been designed to coopt them. If so, Arana was clearly aware by then of Hardenburg's investigations.

As he pursued his sleuthing in Iquitos, Hardenburg still did not have concrete evidence of Julio César's involvement in the Putumayo/Caquetá atrocities. None of the testimonials mentioned him directly. It was then that Walter learned of a promising lead in Manaos and went there. After considerable difficulty, he located Aurelio Blanco, a man who had been contracted to do carpentry in the Putumayo. However, at one point Blanco had been ordered to join a patrol that was setting out in pursuit of fugitive Indians. When he refused, he was ostracized and even fired upon. He then managed to escape downriver, but along the way lost his tools when his canoe was swamped. In Iquitos, Blanco confronted Arana in person and demanded his back pay and retribution for his losses. Julio César promised to look into the matter. When nothing happened within the agreed time frame, Blanco took his contract to a notary public to initiate formal proceedings against the Casa Arana. Instead, the notary sold the document to the firm (for a third of its claim).

An irate Blanco again confronted Julio César and informed him of what was going on in the Putumayo. Arana remained impassive and unsurprised by the "revelation." He told the carpenter that his contract with the company had never been notarized and was therefore invalid. He then handed Blanco 15 pounds sterling to go away. Blanco counted out the bills and then threw them at Arana's feet with the admonition that he add them to the pile of his millions—ill-gotten as it was through the murder of helpless Indians. Being no fool, Blanco fled Iquitos immediately for Manaos and once there went underground. While the carpenter refused

---

496. Ibid., 267.
497. Goodman, *The Devil*, 43–44.

to have a statement drawn up and notarized (he feared it, too, might be sold to Arana), he agreed to put his story into a letter that the American could use in any fashion that he wished.[498]

Once back in Iquitos, for the next eight months Hardenburg tried to gather additional evidence from former Arana employees. Miguel Gálvez put Hardenburg in touch with Julio Muriedas, a Spaniard who was an Iquitos accountant and the owner of a small (financially challenged) rubber station. At one point, Muriedas had been in Arana's employ and spent six months in the Putumayo—three each at El Encanto and La Chorrera. According to the witness,

> At both depots . . . it was common to torture Indian children to force them to reveal where their fathers were hiding in the forest. One of the worst offenders, he swore, was the agent Bartolomé Zumaeta, Arana's brother-in-law, a hard-drinking syphilitic who the Indians had sworn to kill."[499]

As it turned out, Muriedas was narrating a death foretold. It regarded the vengeance of a rebel Bora Indian *cacique*, Katenere, who roamed the forest within the Matanzas section headed by Armando Normand. When Katenere decided to stop collecting rubber and fled with his followers, he was captured and put into a cage. His wife was then raped before him by one of the high officials [possibly Bartolomé Zumaeta]. A young Indian girl helped Katenere to escape and secure some Winchester rifles. He and his followers spent the next two years on the run, periodically attacking both whites and Indians who continued to collect rubber. In May 1908, Bartolomé Zumaeta was overseeing the washing of rubber at an outpost in Bora territory when Katenere and his men fell upon them. The *cacique* shot and killed Zumaeta. This triggered an intense manhunt that failed to capture the wily Indian. However, Katenere made the fatal mistake of launching an attack against the Abisinia station and was shot dead during the altercation.[500]

Little by little others came forward to Hardenburg with tales that were almost too grim to believe. Arana's overseers come across as psychopathic killers. One informant, (Basque surnamed) Aurelio Erazo, reported that Indian children were used for target practice—the forehead counting as the bull's-eye. According to him, overseer Rafael Calderón, "a Colombian mule-driver-turned-bandit who had fled his country to duck a murder rap," commented how it was interesting to see the wounded children

---

498. Collier, *The River*, 136–47.
499. Ibid., 148.
500. Lagos, *Arana*, 280–81.

jump about when the shot failed to kill them outright. As for the women, Calderón was reported to have said, "I had as many as I could handle. Our motto was, kill the fathers first, enjoy the virgins afterward."[501] Hardenburg was able to collect eighteen such accounts by late spring of 1909. In every case he insisted that the witness appear before Frederico M. Pizarro, an Iquitos notary public (and a man who was disaffected with Arana) and swear to the truth of his statement. They were informed that they could actually face legal liability in the event that they were simply defaming Arana.

Initially, Hardenburg could not understand why the overseers were so brutal with regard to the seemingly scarce, and therefore valuable, Indian labor force. But he then learned the key. Most of the overseers received a commission on the rubber extracted in lieu of a salary. Under such a system, they could aspire to amassing considerable wealth quickly. It was in their interest to maximize gains over the short term—intimidation thereby becoming part of the business strategy.[502]

Hardenburg's every move in Iquitos was now being monitored. Julio Egoaguirre even asked the American outright if he was lingering in the city for the purpose of researching a book. His nationality was possibly the only thing that had saved Walter's life up to this point. If Hardenburg now had his smoking guns of sorts, it was equally obvious that it would be impossible to pursue justice in Arana's Iquitos. For openers, Germán Alarco, Julio César's brother-in-law and brother of Abel (manager of the Peruvian Amazon Company's London office), was the mayor.[503]

Nevertheless, as agreed, Blanco and Perkins had sent their letters. Hardenburg now felt that, along with them, his own recollections, Saldaña Rocca's materials and the eighteen[504] sworn testimonials he had collected, he now had more than enough to challenge Arana openly. On May 21, 1909, Walter asked Dr. King to forward his evidence to the American embassy in Lima, but the consul refused to do so. It was then that Hardenburg resolved to take the matter to London, since, after all, it now housed the headquarters of Arana's Peruvian Amazon Company. He would leave Iquitos in early June. Meanwhile, on June 6, 1909, the Iquitos

---

501. Collier, *The River*, 149.
502. Ibid., 150–51.
503. The Casa Arana's stranglehold on this office speaks volumes regarding Julio César's influence upon Iquitos and Loretan politics. He was mayor in 1902; Germán Alarco was elected in 1909, 1911, and 1912; Pablo Zumaeta assumed the post in 1912, 1914, 1922, and 1923; and Arana's close business associate, Victor I. Israel was mayor in 1917, 1918, and 1930. Bákula Patiño, "Introducción," 86.
504. Collier, *The River*, 151. Goodman places their number at twenty. Goodman, *The Devil*, 47.

firm of Arana, Alarco and Co. (another of Julio César's businesses) sent a letter to the Peruvian Amazon Company stating:

> We have knowledge that a person called Hardenburg is traveling towards England, who has been preparing a business of blackmail against you, as it is said. We mention this to you for your guidance."[505]

Hardenburg experienced a last, seemingly trivial, adventure on his way out of South America—one that would come back to haunt him. As he prepared to leave for Manaos and then Pará, the American was informed by Muriedas that he had just sold his rubber station and was departing at the same time for Spain. However, there was a difficulty. He had in his possession a bill of exchange (rather than money) from the buyer. The instrument could be cashed at the Bank of Brazil in Manaos. Meanwhile, an illiquid Muriedas could not afford his passage. Hardenburg considered the Spaniard a friend and was grateful to him for the testimonial he had supplied regarding the Casa Arana's Putumayo operations. So a financially strapped Walter bought Muriedas's ticket and leant him 20 pounds, which was half of the American's travel stash. In early June they left together for Manaos.

There they shared a room and Muriedas went on a binge. The last day before they were to continue on to Pará, the Spaniard was desperately ill and hung over, but refused all medical attention. It looked to Hardenburg like he would be proceeding alone at dawn the next morning, and so he requested repayment of his loan. Muriedas, seemingly incapable of getting out of his bed, signed over the bill of exchange. A harried Hardenburg, after enlisting the help of the Colombian consul in Manaos who had befriended him, rushed to the bank and was able to complete the transaction a few minutes before it closed. Back at the hotel, Muriedas repaid the loan out of his proceeds. He was suddenly chipper and pronounced that he might be up to making the voyage after all.

The next morning they boarded the steamship to Pará, where each was to go his separate way—Hardenburg to Liverpool and Muriedas to Lisbon.[506] By now, the two men had become estranged. When they arrived in Pará, the Spaniard disappeared for a period and then returned to announce he had accepted a position as an accountant in the Mato Grosso. He would not be going on to Europe.[507]

In late June, Hardenburg boarded his ship to England, determined

---

505. British Parliament, *Report*, 504.
506. Collier, *The River*, 151–60.
507. Ibid., 162.

to expose the atrocities in the Putumayo. He arrived there in mid-July. Curiously, on July 3, 1910, the American minister in Lima wrote to Hardenburg,

> I enclose a copy of the reply of the Peruvian government to my representations on behalf of Mr. Perkins and yourself. You will receive from the firm of Arana, Alarco and Co., for yourself and your friend, through the next Budget, £500 in full quittance of any and all claims. Under the circumstances of the case, I think you should be very thankful for this settlement.[508]

It seems that, once back in the United States, Perkins had pestered the State Department with a demand for $20,000 in compensation for their lost possessions, but to little avail. The U.S. officials found that to be exorbitant and reduced the request to the Peruvian government to 5,000 dollars. It, in turn, agreed to 2,500 dollars (500 pounds) in retribution (or an eighth of the amount that the two Americans had originally demanded).[509]

Was this coincidental timing or was the offer part of a preemptive strategy designed to mollify Hardenburg? Whatever the case, the letter was sent to the American's Iquitos address and therefore had to be forwarded to him in England. It did not arrive until late September, or after the appearance of the first newspaper article in *Truth* that triggered what eventually became an international scandal.

Modus Vivendi Revisited

After off-loading their booty from the La Unión raid, the Peruvian vessels proceeded up the Caraparaná, where they established garrisons at La Florida and Argelia, as well as fortifying those at the mouths of the Caraparaná and Iguaraparaná, all to prevent any Colombians from returning to the area. Along the way, they destroyed the El Dorado settlement, the only Colombian one remaining in the Caraparaná drainage. The Peruvians also established a fortification just below Campuya.[510]

In mid-January of 1908, Loayza dispatched Zumaeta to extirpate the remaining Colombians from Aranalandia. Two Colombians (who were passing through El Encanto to collect on outstanding debts before abandoning the area definitively) informed Loayza that Zumaeta had captured the Colombians Primitivo M. Pulido and La Unión employee Carlos Mejía, along with two of their men, a few days after the assault

---

508. Ibid., 504.
509. Goodman, *The Devil*, 43.
510. Gómez López, (ed.), *Putumayo: La Vorágine*, 279.

on La Unión in a nearby Indian settlement. As it happened, there was also a group of twenty-two recruits from Tolima at Serrano's La Reserva who had been contracted to work for the Casa Arana. Loayza had sent the (Basque-surnamed) Colombian Manuel Erazo in search of them, obviously some weeks before the events at La Unión. Erauso had left his cohort at La Reserva behind and traveled to El Encanto to secure permission to bring his recruits there as originally agreed. But there was now an order from Company headquarters in Iquitos that no new Colombians should be permitted into Aranalandia. Erazo returned to his men to devise a plan.

Loayza then dispatched his aide, Miguel Flórez, with the order to kill all the Colombians, beginning with Pulido. When Zumaeta resisted initially, Flórez threatened to kill the five men in custody himself. After a night of much drinking and discussion with Flórez, Bartolomé agreed to the assassinations. By one account, Pulido and his companions were killed and their bodies thrown into the river.[511] By another, they were taken into the forest, tied to trees and then shot.

Next, Flórez and Zumaeta proceeded to attack La Reserva, killing Serrano and his people—and, after torturing them, dispatching Erazo along with all twenty-two of his Colombian recruits. Zumaeta, Flórez, and their men lingered for a few days at La Reserva, slaughtering its livestock for nightly banquets and drinking sessions with Serrano's liquor, while combing through the property for valuables that they then divided among themselves as booty.

It was then that Commander Polack learned of the killings and dispatched a force to detain the perpetrators. Loayza forewarned Zumaeta and Flórez, and they fled into remote parts of Aranalandia. The others were imprisoned at El Encanto awaiting the "investigation" some months later of Prefect Zapata, accompanied by Julio César Arana.[512]

The anti-Colombian campaign in Peru continued unabated. In February 1908, the Colombian Ildefonso González, ensconced at El Dorado, decided to abandon that area and resettle near Campuya. Loayza sent his employees to take over El Dorado and ordered Mariano Olañete to murder González. This was done and Olañete was given the El Dorado station as his reward.[513]

---

511. Ibid., 486–87. Again, there is some confusion in the sources. By another account, there were five victims in all, killed by "Paulo Zumaheta," and this did not happen until early February. Ibid., 505.
512. Arana paid Zapata 8,000 pounds, and the men were ordered released—to resume their posts within the PAC. Ibid., 506.
513. Ibid., 507.

Meanwhile, in Iquitos, Martínez and Orjuela were left to their own devices, basically abandoned in the streets and without resources—fearful of mob violence and even that they might be assassinated as the Colombian eyewitnesses in what would likely become an international investigation. The newspapers, with one exception, were inciting the public. *La Felpa* was denouncing, in strong terms, the actions of the Casa Arana and the Peruvian military.[514]

Fortunately, Martínez found a Colombian friend in the city, the *pastuzo* Emilio Cabrera, who loaned him enough money to secure health care for himself and his companions and purchase for everyone tickets to Barbados. Out of funds there, he left his men and proceeded to Bogotá, where he filed his report (April 23, 1908) and a request for governmental assistance for the men stranded in Barbados.[515]

Orjuela sought a meeting with the prefect. The Peruvian official admitted that there had been excesses, and he was planning to send an investigative commission to the Putumayo to determine the facts and punish the guilty. Meanwhile, he suggested that he could arrange to send the Peruvians back to the area to resume their posts. They might be indemnified for all their suffering. Orjuela responded that it was not a matter of indemnifying individuals; Colombia's sovereignty had been violated and its national honor impugned. He had no interest in returning to the Putumayo, fearing that he might be murdered once there like many of his compatriots.[516]

Meanwhile, the Colombians in and around Aranalandia were quick to denounce the attack on La Unión to their own government. On February 12, 1908, twenty-eight of them signed a letter in Güepi to the governor in Pasto accusing the Peruvians of regularly violating the *modus vivendi* line (they were obviously unaware that Reyes had abrogated that agreement several months earlier) and of "abusing the Colombians, killing them and sacking their properties leaving them in misery, just as they have done last January 12."[517]

On April 4, José Ignacio Neira, Colombian consul in Iquitos, reported to Bogotá that the town was awash in rumors and hysteria. Residents were clamoring for Lima to send more gunboats to the Putumayo/Caquetá and were volunteering to enlist in a military force. The press was stirring up much public resentment and "insulting"

---

514. Orjuela characterized *La Felpa* as a "newspaper that even though it says falsehoods approximates the truth [more than the other Iquitos newspapers]." Cited in Ibid., 306.
515. Ibid., 271–72.
516. Ibid., 306.
517. Ibid., 215.

Colombia with regularity. In sum, ". . . in this part of Peru they are at war with Colombia, which, so far, remains ignorant of what has happened."[518]

Orjuela eventually made his way to Bogotá where, in early May, he filed his report on the events with the Colombian government.

On May 19, 1908, (Basque-surnamed) Colombian Minister of Foreign Affairs Francisco José Urrutia wrote to his representative in Rio de Janeiro[519] with details regarding the Peruvians' capture of police chief Gabriel Martínez (and his men) on December 14, 1907, the fatal attack upon David Serrano at La Unión on January 12, 1908, and the detention about the same time of Jesús Orjuela, Colombian police inspector in the Putumayo. The Colombian officials were, "marched through the streets [of Iquitos] in a famishing and nearly naked condition, after being brutally taken there." They were released only through the intervention of some influential Colombians in the town [untrue]. It was also believed that the Peruvians were now invading the Caquetá. Urrutia added,

> Apart from the deeds which I have briefly sketched, this government is aware of others, equally cruel, which have been perpetrated on Colombian citizens and their property, sometimes by the civil and military authorities of Peru, and at other times by the employees of the Arana Brothers, who are under the authority and the absolute and unconditional protection of the Peruvian government. We should also bear in mind the persecution, or rather the extermination which is being carried on against native Colombian tribes; outrages which even surpass those of a similar nature that were committed in ages gone by and that are still a disgrace in the history of mankind.[520]

Urrutia left it up to his subordinate to frame the protest to be sent immediately to the Peruvian government. He noted that his representative in Lima had been preparing one in February when he died suddenly. Urrutia was then told by Peruvian officials that their representative in Argentina was embarking from there and would soon be in Bogotá to negotiate a new *modus vivendi*. That was two months earlier, and he had failed to show. Since then, there had been total silence from the Peruvi-

---

518. Ibid., 244.
519. It would seem that both the pope and the Argentine president had declined to mediate the dispute, and it was now under informal Brazilian arbitration pending formal approval by the Peruvian and Colombian congresses.
520. Despatch from the Colombian government to the Colombian Minister at Rio de Janeiro in reference to the extermination of native tribes in the Putumayo. Bodleian Library, MSS Brit. Emp. S-22 G332.

ans. Urrutia wanted the protest to be framed as a blatant violation of the world's Hague Convention of July 29, 1899, governing the rules of war. Both Colombia and Peru were signatories to that treaty.

On August 6, 1908, Urbano Gutiérrez wrote to President Reyes in the name of his fellow Colombians, stating the the crimes against Colombians in the Putumayo/Caquetá were continuing Some of his employees had been killed horribly. When Justino Hernández was dispatched with eleven shots, the comment by his murderers while he was dying that there was "one fewer dogs." Reyes sent a copy of the letter to his minister of foreign affairs with the charge that that he enlist the Brazilians and then, through the Colombian ambassador in Lima, coordinate action by all three countries against the "bandits."[521]

Nevertheless, by April 1909, tensions between Peru and Colombia were abating. Perhaps both feared that they could easily spin out of either's control. They signed an agreement whose preamble stated, "The governments of Peru and Colombia respectively desiring to put an end in a friendly manner to the differences which have arisen between them, and to abolish in future all possibilities of conflicts on the frontier, as well as to place their amicable relations on a thoroughly satisfactory footing, have resolved to draw up an agreement which shall faithfully interpret their desires." It was agreed to set up an International Commission to investigate the claims of both countries and issue a report of its findings. If either party disagreed with them, the disputes would be submitted to binding arbitration. They were still awaiting the decision of the king of Spain regarding the boundary dispute between Peru and Ecuador.

In any event, the final agreement between Peru and Colombia would call for punishment of anyone found guilty of crimes in the Putumayo, as well as compensation for their victims. Finally, there was a desire to conclude a Treaty of Commerce and Navigation that would facilitate trade relations between the two countries.[522] Shortly thereafter, this April 21 agreement was amended and amplified to create the International Commission that would meet regularly in Rio de Janeiro and consist of a designated representative of both countries, and the Brazilian minister of foreign affairs. In the event that said minister declined, the third member would be the "British Minister at Rio Janeiro," followed by the "Minister for the German Empire at Rio Janeiro." In the event that all of the aforementioned were unavailable, the Peruvian and Colombian delegates

---

521. Cited in Gómez López (ed.), *Putumayo, La Vorágine,* 357–58.
522. Agreement between Peru and Colombia (April 21) 1909. Bodleian Library MSS Brit. Emp. S-22 G332.

would appoint a third mediator at their first meeting in Brazil. In any event, the third member would be the president of the joint commission and cast the deciding vote in the event of disagreement between the two parties to the disputes. Article 11 states, "The decision of the Joint International Commission shall be final. There shall be no appeal therefrom, and execution of the same shall take effect from the day on which the said decision is made."[523] It was later decided that the International Commission would determine what compensation one country would pay to the other (or its citizens) for any damages caused by its policies or actions up until the time of the agreement.

This did not mean, of course, that there was no longer mutual suspicion between Colombia and Peru. In October 1909 (or after Reyes left office), Leopoldo Triana, resident in Pará and Colombia's consul there, wrote Bogotá about the absolute necessity to clarify through executive order and an act of Congress Colombia's sovereignty in the Putumayo/Caquetá. He opined that talk of missions and schools should be relegated until a future date after Colombia had established a real infrastructure in the region and extirpated all Peruvian officials from the contested [*modus vivendi*] territory. He pointed out that the honorary consul in the Caquetá, Pedro Antonio Pizarro, was effective and should be retained. Nevertheless, he needed real resources to establish and staff a customs house. Colombia should situate "confidential consuls" [spies] in both the Putumayo/Caquetá and Iquitos to be well informed of Peruvian machinations in the region. Then, too, he said that there should be an immediate propaganda campaign in Europe, the United States, and Brazil exposing the atrocities of the Casa Arana and denouncing the public offering of the Peruvian Amazon Company (informing potential shareholders that the Colombian government would refuse the PAC's territorial claims). He requested funds to launch a publicity campaign in Brazil in which there would be at least one article documenting Casa Arana excesses in the Putumayo/Caquetá (such a campaign would cost money as the newspapers would expect compensation, but he was optimistic of access since the Brazilians were very ambivalent regarding Peru in general). There should be a paid Colombian minister assigned to Rio (and others in Manaos and Belem as well).[524] The Manaos one, Triana maintained, should keep

---

523. Agreement concluded between Colombia and Peru (April 13, 1910) to reform that which was concluded on April 21, 1909. Bodleian Library, MSS Brit. Emp. S-22 G332.
524. There is evidence that the Colombian consul in Iquitos had been both unpaid and ineffective as well. The Colombians in that town and the Putumayo/Caquetá periodically petitioned Bogotá for a better qualified and motivated paid replacement. Gómez López, (ed.), *Putumayo: La Vorágine,* 139–40. Triana would later inform Bogotá that they

close track of the large numbers of Colombians who had entered Brazil to work on the Madeira-Mamoré railway. They would be a willing patriotic military force if one needed to be dispatched quickly to the Putumayo/Caquetá to protect Colombian national interests there.[525]

The foregoing had transpired during and against the backdrop of the Reyes presidency. It should be noted that he had domestic political problems throughout his term. While a conservative, he held out an olive branch to the liberals in the interest of national conciliation. He was also a reformer at heart, sponsoring legislation to protect Colombia's minorities, including its Indians. He thereby managed to alienate hardliners within the conservatives' ranks and, on February 10, 1906, he and his family were the targets of a failed assassination attempt. Then, in 1909, he angered many Peruvians with an initiative to sign a treaty of reconciliation with the United States and Panama. Ironically, it had been Rafael Reyes who commanded the outgunned Colombian forces posing feeble and futile resistance to the powerful U.S. military supporting Panamanian secession.

The treaty was signed by Colombia's secretary of state on January 9, 1909. In reality, both Colombia and Peru were prime beneficiaries of the canal. It provided the Pacific coasts of both nations with far shortened sea connections with parts of their own national territories, not to mention better access to the Atlantic seaboards of both the United States and Europe.

However, the treaty struck a sore nerve in Colombia, triggering violent street demonstrations by Reyes's opponents. It was rejected by the National Assembly and, on June 20, 1909, the president reconvened the Congress that he himself had suspended in 1905 and submitted to it his resignation to avert a possible civil war.[526]

In Iquitos, the summer of 1911 was crucial, particularly regarding Peru's ongoing conflict with Colombia. In January of that year, no doubt emboldened by the budding Putumayo international scandal, Bogotá had sent more than a hundred troops to La Pedrera on the Caquetá with orders to retake it, which they did. Their mission was also to determine the possibility of wresting the entire Putumayo from Peru and the Peru-

---

should be wary of any formal communications with Rio since such correspondence was obviously being intercepted by the Peruvians in some fashion.
525. Ibid., 461–66.
526. Reyes was himself an admirer of American ingenuity and idealism, although critical of its materialism. He predicted that the former would prevail over the latter. General Rafael Reyes, *The Two Americas* (New York: Frederick A. Stokes Company, 1914), 22–77. He would spend part of 1913 in the United States involved in organizational meetings for a Pan-American Congress and Union to be held the following year

vian Amazon Company.⁵²⁷ During May and June, the Peruvian gunboat *América* patrolled the Putumayo River, and then, on June 28, a Peruvian force in four vessels, commanded by Lt. Col. Oscar R. Benavides, left Iquitos to attack the Colombians at La Pedrera. Initially, the entrenched defenders, under the command of (Basque-surnamed) General Isaac Gamboa, resisted the assault. But, on July 12, 2011, Benavides was victorious (giving him a hero's national fame that would enable him to become the Peruvian president in 1914). Colombia suffered 150 fatalities in the fighting.⁵²⁸ In Bogotá there was patriotic outrage, fueled in part by a pamphlet written and published by Dr. Eduardo Arias questioning if this would be another Panama.⁵²⁹

Ironically, while both countries were still unaware of the definitive battle's outcome, Peru and Colombia had reached an agreement, on July 19, whereby Peru would have to return to Colombia any "war" booty, and particularly La Pedrera itself. On August 3, 1911, there was a huge public demonstration in Iquitos against the Lima government's concessions to Colombia. The outraged citizenry denounced the "national disgrace" of Lima's treachery against its own patriots and armed forces.⁵³⁰

Enter Casement

Roger Casement, born September 1, 1864, was the son of Captain Roger Casement, a British soldier and Protestant Ulsterman, and a Catholic mother. She had him baptized secretly as an infant and then died in childbirth when Roger was nine. Her children were raised as Protestants and in relative comfort in Ulster by their uncle, John Casement. Four years later their father died, so Roger was fully orphaned by the age of thirteen.

John Casement was a director of the Elder Dempster Shipping Company of Liverpool. Roger's aunt on his mother's side was married to Edward Bannister, who was in charge of Elder Dempsey's West African affairs. By 1880, Roger Casement was living in the Bannister household, and shortly thereafter he went to work as a clerk in the shipping firm. He was bored by the desk job and dreamed of African adventures. In 1883, at age nineteen, he went to West Africa as a purser on one of the Elder Dempsey ships and fell in love with the continent. The following year,

---

527. Fernando Romero, *Iquitos y la fuerza naval de la Amazonía (1830–1933)* (Lima: Dirección General de Intereses Marítimos, Ministerio de la Marina, 1983), 110.
528. Villegas y Yunis, *Sucesos*, 155.
529. Las guerras con el Perú, Credencial historia no. 191 (November 2005), http://www.banrepcultural.org/blaavirtual/revistas/credencial/noviembre2005/guerras_peru.htm.
530. Stanfield, *Red Rubber*, 159.

Roger secured employment with King Leopold's African International Association.[531] By this time, the notorious Henry Stanley ("Dr. Livingstone, I presume?") was in Leopold's employ as well, all part of the king's attempt to garner international respectability,[532] and Casement became reasonably well acquainted with the famed explorer. Meanwhile, he witnessed many human abuses in the Congo and heard harrowing tales of others.[533] Casement also shared a room for ten days with Joseph Conrad and provided the future famed novelist with much information about the Congo—some of which likely informed *Heart of Darkness*. But, in 1890, or after six years in various capacities with King Leopold, Casement returned to England. The Congo was simply becoming too much of a "Belgian enterprise" for his taste.[534]

Nevertheless, Africa was now in his blood, and Roger returned to it (this time Nigeria), in 1892, for a three-year stint with the Oil Rivers Protectorate. It was administered through the British Foreign Office and its consuls and vice consuls. At age twenty-eight, Casement was now the assistant director-general of customs at Old Calabar. It was all an excellent apprenticeship for his entry into the consular service. In 1892, Casement's uncle, Edward Bannister, was Great Britain's assistant consul in Loanda (with consular jurisdiction over all of Leopold's Congo). In that capacity, he reported that British subjects were being employed in the Congo and were sometimes flogged—and was told by London to stick to commercial affairs. Then, in 1895, there was a further diplomatic incident when an Englishman who was engaged in illicit trade in the Eastern Congo was captured by Leopold's *Force Publique* and summarily hanged on the spot.[535] At the same time, the complaints about conditions in the Congo were growing louder.

---

531. Founded in Brussels in 1876, the African International Association was to have national chapters in the several countries engaged in the imperial scramble for Africa. Ostensibly, its national representatives were to meet periodically to develop a humanitarian agenda for Central Africa. In fact, it proved to be extremely ineffectual and ultimately an obfuscation for King Leopold's real agenda—control and exploitation of the Congo. Casement became an employee of the AIA at the time that its failure was a part of the reason for the 1884–1885 Congress of Berlin in which Leopold was essentially granted a personal fealty in the Congo.
532. In 1898, Stanley would write in the Introduction to his friend Guy Burrow's book *The Land of the Pygmies*, "what great gratitude the world owes to King Leopold—for his matchless sacrifices on behalf of the inhabitants of the region ... (W)ho can doubt that God chose the King for his instrument to redeem this vast slave park?" Quoted in William H. Bryant, *Roger Casement: A Biography* (Lincoln, NE: iUniverse, 2007), 39.
533. Hochschild, *King Leopold's Ghost*, 196.
534. Bryant, *Roger Casement*, 6.
535. Ibid., 37–38.

Meanwhile, King Leopold's costly personal empire and lifestyle were straining his finances. In 1890, he negotiated a 30 million franc loan from the Belgian government and, in 1895, he agreed that Belgium should annex the Congo Free State as one of the conditions of granting his request for another 7 million franc loan. Furthermore, annexation might serve to shield Leopold's personal fief in Africa from too much international scrutiny, although in practice he intended that little would change on the ground. The Belgian parliament refused to approve annexation, although it did go along with the loan.

Then, in 1896, E. V. Sjöblom, a Swedish Baptist missionary assigned to the Congo, denounced the "rubber terror" there in articles in his home country that were soon translated and published in other European newspapers. King Leopold's administrators counterattacked in both the British and Belgian press. In London, the Anti-Slavery League and the Aborigine's Protection Society were demanding an investigation. If Leopold and his apologists assailed their critics in the Belgian and British press, the monarch tried to appease them as well. In 1896, he founded the Commission for the Protection of the Natives, comprised of six Congo missionaries (half of whom were Catholic and the other half Protestant). It all proved to be eyewash, since there was no budget for them to even meet; nor were they to serve in any other capacity than an advisory one.[536]

The most serious challenge to Leopold came out of the Elder Dempster Shipping Company itself in the person of its young (mid-twenties) employee, Edmund Dene Morel. The company held a monopoly on the sea traffic between Europe and the Congo. Morel spent considerable time in Antwerp supervising the arrivals and departures of Elder Dempster vessels and was therefore privy to the records regarding what was being imported and exported. He was struck by the quantities of arms and ammunition heading to Africa (for what purpose?). It was also plain to him that only a small portion of the rubber and ivory coming out of the Congo was being reported in the European press (was someone pocketing a fortune?). Finally, there was a huge imbalance in the flow of goods, suggesting that the wealth of the Congo was not being obtained by means of fair and legitimate trade (then how?). He was able to collect evidence from various contacts and whistleblowers within Leopold's organization from which he began to piece together the full story of the incredible exploitation and atrocities.

Morel took his findings and suspicions to his employer, Sir Alfred Jones, head of Elder Dempster—not to mention president of the Liv-

---

536. Hochschild, *King Leopold's Ghost*, 173–74.

erpool Chamber of Commerce and honorary consul of the Congo in the city. Rather than receiving satisfaction, let alone sympathy, Morel was immediately proscribed within the company. If he went public, Elder Dempster would most certainly lose its most lucrative client. There were blatant attempts to purchase his silence—but all to no avail. At age twenty-eight, Morel left his employer, filled with "determination to do my best to expose and destroy what I then knew to be a legalized infamy ... accompanied by unimaginable barbarities and responsible for a vast destruction of human life."[537]

Over the next several years, Morel launched a torrent of books and articles. He founded the *West African Mail* periodical in which he regularly published the latest revelations provided by disenchanted Congo missionaries, civilians, and even anonymous Leopold staffers. Despite King Leopold's extensive efforts to convey the impression of his benign civilizing mission through various displays at world fairs, botanical gardens, and museums,[538] a very different image of the Congo was emerging. Linked with those of British humanitarian groups, Morel's campaign began to gain traction in the European press—including even the Belgian media.

In July 1898, Roger Casement was named consul of Loanda, a posting in the very consulate in which his uncle had preceded him. But then, or but a few months later, and given that a Boer War seemed imminent, he was reassigned to Southern Africa as consul in Lourenço Marquez. While there, he would volunteer for a British military operation in South Africa and was awarded the Queen's South African Medal for valor.

On August 20, 1900, Casement was appointed British consul in Kinshasa. King Leopold invited Roger to stop in Brussels during his journey back to England on home leave. It was there that Casement attended a private lunch and meeting with the monarch himself, during which Leopold tried to explain away the negative rumors and reports coming out of the Congo—to a very unimpressed British consul.[539] From Kinshasa, during the next year, Casement conducted some informal investigations into the workings of Leopold's empire. By then, his purview as consul had been broadened to the adjacent French Congo, and he was appalled by similar stories of atrocities being committed against the natives there. By this time, he was regularly ill from various tropical ailments and required periods of recuperation in Europe.

---

537. Ibid., 186.
538. All three institutions were prime means of informing the citizenries of imperialist European nations about the world's exotica and the Occident's civilizing mission within it.
539. Bryant, *Roger Casement*, 41.

While the British government remained circumspect in its condemnation of Leopold, Morel and the Aborigines Protection Society were turning up the heat. In May, the Society sponsored a massive rally in London, demanding that the signatory nations to the Berlin accord redeem their pledges to look after the welfare of the Africans.[540] Casement went back and forth between Europe and the Congo over the next several months, and, in February 1903, he left England for what would be his last trip to Africa. Then, in May 1903, hearings on the Congo were held in the House of Commons that resulted in the unanimous resolution that Congo natives "should be governed with humanity."[541] There was also the issue of the treatment of British subjects, and particularly Zanzibarians, who had been contracted to work for Leopold.

It was already a delicate moment in the diplomatic relations between Belgium and Great Britain. Belgians had naturally sided with the defeated Boers, and they harbored considerable anti-British sentiment in the aftermath of that war. Despite his kinship with King Edward VII, Leopold was wary of British intentions. He despised Morel, and believed that he was being paid by Liverpool merchants to undermine Belgian hegemony in the Congo.[542]

The British government was compelled by its own public's opinion to do something, so, the day after Casement arrived back in Kinshasa, he received an order from the Foreign Office to venture into the interior on a fact-finding mission in order to prepare a report. For most of the summer, he traveled the country taking testimony from missionaries and natives alike regarding indescribable barbarity. He then peppered both the Foreign Office and Leopold's administration with dispatches detailing the abuses. Casement sent a critical letter to the Congo's governor general denouncing his administration, since: "Instead of lifting up the native populations submitted to and suffering from it, it can, if persisted in, lead only to their final extinction and the universal condemnation of mankind."[543]

Clearly, Roger Casement was posing a big problem for the Belgian monarch. He was a twenty-year veteran of Africa, including serious time in the Congo. He had worked for both King Leopold and the Elder Dempster Shipping Company. In short, he was an observer with great credentials and credibility. In late 1903, he was recalled to London, where he began work on his final report. It was then that he encountered considerable frustration with his own Foreign Office. It wanted him to tread lightly. Con-

---

540. Ibid., 41–43.
541. Hochschild, *King Leopold's Ghost*, 194.
542. Bryant, *Roger Casement*, 91.
543. Quoted in Hochschild, *King Leopold's Ghost*, 201.

sequently, Casement began leaking some of his information to the press, a move that forced the Foreign Office to agree to publication of his report.[544]

It was then that Casement and Morel actually met. From Africa, the British consul had followed Morel's publishing avidly, and Roger now shared parts of his own (as yet unpublished and still evolving) work with him. In the latter's words,

> I was mostly a silent listener, clutching hard upon the arms of my chair. As the monologue of horror proceeded . . . I verily believe I *saw* those hunted women clutching their children and flying panic stricken to the bush: the blood flowing from those quivering black bodies as the hippopotamus hide whip struck and struck again; the savage soldiery rushing hither and thither amid burning villages; the ghastly tally of severed hands. . . . Casement read me passages from his report, which he was then writing, whose purport was almost identical with oft-repeated sentences of my own. He told me that he had been amazed to find that I, five thousand miles away, had come to conclusions identical with his in every respect . . . An immense weight passed from me.[545]

The two men had their differences. Casement was becoming increasingly disillusioned with the British imperialism that he served. He was gradually developing an interest in his Irish roots, and increasingly regarded the plight of his native land as one more example of British colonialism. Morel was not against colonialism per se, only against the abusive variety perpetrated in the Congo. He believed that Great Britain, in particular, was on a civilizing mission, and regarded its treatment of indigenous peoples as qualitatively superior to that of King Leopold. Both were moralists who, like Hardenburg, were profoundly influenced by their Christian upbringing. All were certainly determined and uncompromising crusaders on behalf of the oppressed—damn the costs.

Their main playing field, of course, was Great Britain. There, in Morel's opinion, British foreign secretary Edward Grey would act only "when kicked, and if the process of kicking stopped, he will do nothing."[546] In February, 1904, Grey ordered publication of Casement's report. While it censored (to Casement's disgust) the identities of some of the culpable, it had devastating effect. It estimated that no fewer than three million Congolese had died for rubber.[547]

---

544. Ibid., 203–204.
545. Quoted in Ibid., 205–206.
546. Quoted in Ibid., 209
547. Goodman, *The Devil*, 9.

Despite their strong and stubborn personalities, as well as their differences, Casement and Morel were now fast friends and firm allies. Rather than continuing to work through the Aborigenes Protection Society, with the danger of having their campaign subordinated to its wider agenda and fund-raising goals, Casement urged Morel to start his own Congo defense organization. As a British consul, he could not participate officially; however, he gave the impecunious Morel a 100 pounds sterling donation (the Congo Reform Association's first) out of his own limited funds. He also promised to advise Morel on how best to press the cause within British officialdom. Morel founded the CRA, and it attracted a crowd of more than a thousand persons to its first meeting on March 23, 1904—held in Liverpool's Philharmonic Hall.[548]

And so it began. In the words of Hochschild,

> The crusade of E. D. Morel now orchestrated through the Congo Reform Association exerted a relentless, growing pressure on the Belgian, British, and American governments. Almost never has one man, possessed of no wealth, title, or official post, caused so much trouble for the governments of several major countries.[549]

It all became a magnificent test of wills. Morel's publications and innumerable public appearances throughout the British Isles kept the pressure on. Nevertheless, Leopold was after all a king with vast connections and resources. The details of the contest are beyond the scope of the present book;[550] however, there are the highlights. There was Morel's successful campaign across the Atlantic—where he even managed to recruit to the cause such formidable allies as Mark Twain and Booker T. Washington. The former lobbied Washington, DC, on three occasions, including President Roosevelt himself during a lunch. Regarding Twain, Booker T. Washington observed, "I think I have never known him to be so stirred up on any one question as he was on that of the cruel treatment of the natives of the Congo Free State."[551] Meanwhile, in Great Britain many powerful industrial magnates and personalities joined the cause. Chief among them was Sir Arthur Conan Doyle.[552] He barnstormed personally around Great Britain with Morel—voicing his opposition to King Leopold. He wrote an introduction to Morel's book, *The Crime of the Congo*, in which

---

548. Hochschild, *King Leopold's Ghost*, 206–207.
549. Ibid., 209.
550. I would refer the interested reader to Ibid., 235–74.
551. Quoted in Ibid., 241
552. Sir Arthur Conan Doyle maintained a friendship with both Morel and Casement. For instance, one evening the three men attended a London performance together of a new play based on the Sherlock Holmes work *The Speckled Band* (Ibid., 270–71).

he even exceeded its author by pronouncing the situation in the Congo to be "the greatest crime which has ever been committed in the history of the world."[553]

For his part, Leopold bribed key newspapermen and politicians in both Europe and America to at least obfuscate and blunt the practical consequences of the crusaders' revelations and criticisms. He hired his own publicists and published and distributed his version of the Congo's history and current state of affairs. They configured the narrative in terms of a sinister plot contrived by vindictive Protestant missionaries against a Catholic monarch, on the one hand, and the power play of British political and commercial interests to control the Congo, on the other. Leopold ultimately hired Henry I. Kowalsky, an influential and flamboyant American attorney/lobbyist who also gained access to the president, ". . . to whom he gave a photograph of Leopold in a silver frame, an album of photos of the Congo, and a memorandum asking him not to be deceived by jealous missionaries and Liverpool merchants."[554]

It was to no avail. In Brussels, on three occasions, the British and American ambassadors met with the minister of foreign affairs to urge Belgian annexation of the Congo. In London, Morel had begun to advocate that a British warship be sent to the Congo to abolish its system of enforced labor. Foreign Minister Edward Grey declined, but then pressured Belgium by refusing to recognize the new Belgian Congo. Morel organized a mass demonstration at the Royal Albert Hall that was endorsed by 20 bishops and 140 members of parliament.[555]

Finally, resigned to the loss of his African fiefdom, King Leopold resolved to sell it to Belgium. There ensued lengthy negotiations, begun in 1906 and not concluded until early 1908, which at times turned testy. The March 1908 agreement called for Belgium to assume the Congo's 110 million francs debt, forgive Leopold's outstanding 32 million francs loans, consign 45.5 million francs to certain pet projects that Leopold had underway and to pay him ("as a mark of gratitude for his great sacrifices made for the Congo") 50 million francs. It really mattered little to the seventy-three-year-old Leopold, since fewer than two years later, or on December 17, 1909, he was dead—probably of cancer.

So, now Belgium was in possession of the Congo. Morel, stripped of his foil, continued to monitor the situation with considerable skepticism. However, as early as autumn 1909, the Belgian government had

---

553. Quoted in Ibid., 271.
554. Ibid., 247.
555. Ibid., 272.

implemented many reforms in the Congo, and trustworthy observers were reporting "immense improvement."[556] Morel continued his vigilance over the next few years, but then it became time to declare victory and wind down the Congo Reform Association. On June 16, 1913, the CRA held the last meeting of what had been the first major human rights movement of the twentieth century. Many dignitaries spoke and others sent congratulatory telegrams and letters that were read to the assemblage. Morel was praised as the man of the hour. When, at last, it was his turn to speak, he deflected the greatest credit to someone else,

> While I was listening to all that was being said, I had a vision. The vision of a small steamer ploughing its way up the Congo just ten years ago this month, and on its decks a man that some of you may know; a man of great heart . . . Roger Casement.[557]

In the aftermath of the publication of his Congo Report in early 1904, Casement had an international reputation as a humanitarian. During most of 1904 and 1905, he lived in Ireland and England while on sick leave from the consular service. He actively recruited for Morel's Congo Reform Association, and, on April 24, 1904, wrote a letter to Alice Strepford Green soliciting her support. She was an influential London socialite; many celebrities and intellectuals (e.g., Florence Nightingale and Winston Churchill) frequented her salon. One of her major interests was the Irish question; she was an advocate of home rule. Casement's growing reservations about British imperialism predisposed him for much of the discussion and debate in the Green residence. Roger became Alice's close friend, and she would eventually mentor him in Irish history and current events. He even undertook (unsuccessfully) the study of Gaelic. He would write to her that he had discovered his Irish identity while comparing the plight of the Congolese and Irishmen during his investigation of King Leopold's abuses in Africa:

> I realized then that I was looking at this tragedy with the eyes of another race—of a people once hunted themselves, whose hearts were based on affection as the root principle of contact with their fellow men and whose estimate of life was not of something eternally to be appraised at its market price. . . . And I said to myself then, far up the Lulanga river, that I would do my part as an Irishman, wherever it might lead me personally."[558]

---

556. Ibid., 273.
557. Quoted in Ibid., 274.
558. Quoted in Goodman, *The Devil*, 91.

In 1905, Casement was given the prestigious Companion of Saint Michael and Saint George award reserved for those who distinguished themselves in Great Britain's foreign service. In light of his budding commitment to Irish nationalism, he considered refusing the distinction until convinced by his friends that it would be the death knell of his career.[559] Roger Casement simply had to earn a living. He was never good at handling money, and both his brother and sister asked him periodically for financial support. He considered leaving the consular service, but, after so many years of experience, it was really his only true profession. Nevertheless, by this time he was totally disdainful of the Foreign Office—and Grey in particular. He was equally critical of Theodore Roosevelt. He denounced both British imperialism and American materialism, and admitted that in his country's service, "I was well on the high road to being a regular Imperialist jingo."[560]

His celebrity gained him a prestigious posting in Lisbon, but he resigned after one month due to illness and returned to Great Britain. Once recovered, he was in the running for a position in Haiti, but ended up instead as the (reluctant) British consul in Santos (southern Brazil). By sheer coincidence his fellow passenger during his journey on the SS *Clement* to assume his post in Brazil was none other than Julio César Arana—the two men exchanged pleasantries at the captain's table. After an unsatisfactory stint in Santos, Casement was transferred first to Pará and then, in 1909, he became British consul general in Rio de Janeiro.[561] Fate was about to seek him out once again.

---

559. Ibid., 86.
560. Ibid., 89–91.
561. Ibid., 88–89.

Part 2
The Scandal

Blowing the Whistle

Walter Hardenburg arrived in London in mid-July of 1909, determined to confront the British directors of the Peruvian Amazon Company with his evidence. He soon thought better of it, as they might well try to suppress his facts. He then made the rounds of the London press, but met with indifference (particularly given his insignificance), incredulity, and the fear of libel suits. After a few weeks of frustration, relieved only by a romance with Mary Feeney, a young Irishwoman (and his future spouse) he met at his lodgings, Hardenburg decided to take his manuscript to the newly merged Anti-Slavery and Aborigines Protection Society. He left his own written summary with its director, John Harris, a former missionary in the Congo—and one of Casement's key collaborators when Roger was researching his Congo report. Walter also flashed the dossier containing his supporting evidence (pronouncing that he would not let it out of his possession) and promised to return in two days' time. Harris read the Hardenburg précis, and, at their next meeting, urged him to take it to *Truth*, a newspaper that specialized in investigative reporting. He also thought that the publisher Fisher Unwin might be interested in doing a book. At no time did Harris suggest that the young American approach the British Foreign Office, concluding from his Congo experience (like Morel) that Sir Edward Grey was only likely to act once there was public pressure.[1]

Thanks to Harris's introduction, Hardenburg gained immediate access to *Truth*'s assistant editor and chief investigative reporter—G. Sidney Paternoster, the man who would author the first *Truth* article (September

---

1. British Parliament, *Report*, 65.

22, 1909) under the provocative headline "The Devil's Paradise: A British-Owned Congo." There would be others in the coming weeks, all based on the materials provided by Hardenburg (his written account of the trip down the Putumayo, translations of the Saldaña Rocca articles and depositions and copies of his own testimonials by former Arana employees). Hardenburg subsequently published his own book, in which he stated:

> In making these exposures I have obeyed only the dictates of my conscience and my own sense of outraged justice, and now that I have made them and the civilized world is aware of what occurs in the vast and tragic *selvas* of the River Putumayo, I feel that, as an honest man, I have done my duty before God and before society and trust that others, who are in a position to do so, will take up the defence of these unfortunates and the prosecution and punishment of the human hyenas responsible for these crimes.[2]

London's more respected newspapers initially ignored *Truth*'s exposé. Nevertheless, Horace Thorogood, reporter from *The Morning Leader*, was sent by his editor to the office of the Peruvian Amazon Company to interview the directors regarding the allegations. He met with three individuals. Two were introduced as Board members, while the third was Germán Alarco, former mayor of Iquitos. All were foreigners, so the other two were likely Abel Alarco and Arana himself. A black-bearded man (probably Julio César), who was "fairly" fluent in English, did most of the talking. Thorogood was told that the charges were baseless. Hardenburg had a bad reputation in the Amazon and was a blackmailer and forger to boot.[3]

The reporter was asked to return the next day, and he would be given a detailed statement refuting the charges. When he did so, he was informed that the Board was too busy to meet with him and to come again the following day—which he did. Told then that the manager was absent, and that he should return later, again the persistent journalist complied. Late that afternoon, the company's secretary, a Mr. Smith, took him into a private room and announced that the directors were not interested in any more articles. He handed Thorogood an envelope. The journalist noticed that it contained money and handed it back without removing the contents. He informed Smith that it was highly improper to offer money to a reporter and left. Shortly thereafter, Smith turned up at the office of *The Morning Leader* to assume personal responsibility for the envelope. He claimed that he had misunderstood his instruction, given as it was in poor English by a native Spanish speaker. In that gentleman's country, Smith ex-

---

2. Hardenburg and Enock, *The Devil's Paradise*, 196.
3. British Parliament, *Report*, 49–50, 54.

plained, it was customary to provide "tips" to journalists for their trouble.[4]

Thorogood's article ran on the front page of the newspaper and could not have been worse from the Peruvian Amazon Company's viewpoint. Not only did its headline echo "Our Congo," it disclosed the attempted bribe.[5] Oops!

As summarized subsequently (1912) in his own book, Hardenburg was alleging that the conditions in Aranalandia were comparable to those obtaining in the Belgian Congo during its darkest days. To wit,

> 1. The pacific Indians of the Putumayo are forced to work day and night at the extraction of rubber, without the slightest remuneration except the food necessary to keep them alive.
>
> 2. They are kept in the most complete nakedness, many of them not even possessing the biblical fig leaf.
>
> 3. They are robbed of their crops, their women, and their children to satisfy the voracity, lasciviousness, and the avarices of this company and its employees, who live on their food and violate their women.
>
> 4. They are sold wholesale and retail in Iquitos, at prices that range from £20 to £40 each.
>
> 5. They are flogged inhumanly until their bones are laid bare, and great raw sores cover them.
>
> 6. They are given no medical treatment, but are left to die, eaten by maggots, when they serve as food for the chiefs' dogs.
>
> 7. They are castrated and mutilated, and their ears, fingers, arms, and legs are cut off.
>
> 8. They are tortured by means of fire and water, and by tying them up, crucified and head down.
>
> 9. Their houses and crops are burned wantonly and for amusement.
>
> 10. They are cut to pieces and dismembered with knives, axes, and *machetes*.
>
> 11. Their children are grasped by the feet and their heads are dashed against trees and walls until their brains fly out.
>
> 12. Their old folk are killed when they are no longer able to work for the company.

---

4. Ibid., 54.
5. Collier, *The River*, 168–72.

13. Men, women, and children are shot to provide amusement for the employees or to celebrate the *sábado de gloria*, or, in preference to this, they are burned with kerosene so that the employees may enjoy their desperate agony.[6]

The American's *cri de coeur* shouted,

This state of affairs is intolerable. The region monopolized by this company is a living hell—a place where unbridled cruelty and its twin-brother, lust, run riot, with consequences too horrible to put down in writing. It is a blot on civilization; and the reek of its abominations mounts to heaven in fumes of shame. Why is it not stopped?[7] And all this, let us remember, is done by a gang of human beasts, who, consulting exclusively their own evil interests, have had the audacity to form themselves into an English company and put themselves and their gruesome "possessions" under the protection of the English flag, in order to carry out more conveniently their sanguinary labours in the Putumayo and to inspire confidence here. People of England! Just and generous people, always the advanced sentinels of Chrisitianity and civilisation! Consider these horrors! Put yourselves in the place of the victims, and free these few remaining Indians from their cruel bondage and punish the authors of the crimes![8]

Not surprisingly, from the outset there was pushback by Arana in the form of an *ad hominem* attack on the accusers—Saldaña Rocca and Hardenburg (and soon thereafter Captain Whitten). Two days before the first article in *Truth*, in a letter from the Manaos office dated September 20, 1909 (Arana was still there at the time), it was claimed that a certain Englishman in the city named Garnier[9] had accused Hardenburg of attempted blackmail and forgery. Julio César certainly had no way of knowing the imminent timing of the *Truth* disclosures, but was preemptively posturing himself in anticipation of Hardenburg's potential impact. Then, three days after publication of the first London newspaper article, on September 25, Señor Eduardo Lembcke, chargé d'affaires of the Peruvian legation in London, sent a response to it. He denied the allegations categorically and denounced Hardenburg as a

---

6. Hardenburg and Enock, *The Putumayo*, 184–85.
7. Ibid., 186.
8. Ibid., 214.
9. British Parliament, *Report*, 212–13. Garnier is an ambiguous figure who is also sometimes referred to as a newspaper editor and at others as an Englishman and/or Manaos businessman. His name is rendered in some texts as "Guarnier."

notorious blackmailer. That same day, another letter was written in Iquitos and sent to *El Comercio* newspaper in Lima by Loreto's sub-prefect Juan Tizón accusing Hardenburg of asking for 7,000 pounds to suppress his book project.[10] Then, on October 4, *Truth* published another letter from Lembcke stating:

> My government had been for some time aware of the object that the prosecutors of the present defamatory campaign have in view. They also know the worthlessness of the declarations obtained [Hardenburg's informants], with the same object, in Iquitos, from individuals entirely undeserving of confidence on account of their antecedents; many of them have undergone terms of imprisonment in Peru and the neighbouring countries, where also several of them still remain completing their sentences. The Peruvian courts of justice have dismissed as unfounded most of these declarations, and you give publicity to some of these which have no legal value. The said papers [*La Felpa* and *La Sanción*] published fantastic crimes alleged to have been committed by the employees of the firm of J. C. Arana and H'mos, and when the employees of this old and well-known house commenced proceedings for libel against said editor, he disappeared, in order to evade the grave responsibility he had incurred"[11]

One might suspect a certain amount of coordination of these initiatives by Arana and Zumaeta.

The PAC directors' first reaction was to ignore Hardenburg (he remained in London and could have been invited to meet with the Board). On September 27, they sent their own categorical denial to *Truth*, and then included a copy in their letter of October 6 to the Anti-Slavery and Aborigines Protection Society, while refusing the latter's request for a meeting to discuss the allegations. They underscored that the abuses described in *Truth* antedated formation of the Peruvian Amazon Company, and therefore they were not responsible in any way for them.[12] In point of fact, Pasternoster would later testify that he purposely avoided blaming the directors because he could not imagine Englishmen being complicit in such atrocities.[13]

---

10. Ibid., 281.
11. Ibid. In point of fact, neither Saldaña Rocca's nor Hardenburg's depositions had ever been admitted by any Peruvian court. Nor did the Putumayo's managers file a libel suit—although there was speculation that they might do so; rather they requested an official investigation of the charges against them.
12. Ibid., 55.
13. Ibid., 98.

While initially the British directors did not question Arana's word that the charges were baseless, they had some evident concern. Director Gubbins called on David Cazes and was assured the allegations in *Truth* were untrue and defamatory. Arana was a good and highly respected man.[14] On October 11, 1909, Director H. M. Read asked a famed American author,[15] Charles Reginald Enock, to meet with the Board to discuss his terms for conducting an investigation in Peru of the charges. Read had clearly read Enock's book (which contended that enslavement of forest Indians was common in many parts of the country).[16]

Chairman Gubbins was an old Peru hand, and he had a personal relationship with President Augusto B. Leguía. On October 8, he wrote a letter to the Peruvian president, enclosing the *Truth* articles, while contending that their allegations against the Peruvian Amazon Company were groundless. Nevertheless, Gubbins questioned the integrity of some of the local governmental officials in the Loreto. Presumably without the knowledge of Alarco, he sent the letter to his brother in Lima with orders that it be hand delivered to Leguía.[17] Unsurprisingly, when he learned of the exercise a few days later, Julio César was not amused. He particularly felt that the aspersions cast on Loreto officials were impudent and imprudent.[18]

Hardenburg was not the only potential whistleblower. Of related concern for Julio César was British Captain Thomas Whiffen, another would-be amateur anthropologist (shades of Robuchon). In May 1908, the wounded Boer War veteran, remittance man, and adventurer had gone to the Putumayo to live with the Bora and Resigero Indians. While in Arana's Putumayo territory, Whiffen had witnessed some atrocities and received an earful of others as recounted by his Barbadian guide. Whiffen testified that he had arranged for his own travel to the Putumayo and adjacent areas, journeying through Arana's holdings both at the beginning and end of his essentially year-long exploration. During his first visit, the Peruvian Amazon Company assigned Normand as Whiffen's handler (the captain spoke no Spanish and Normand was fluent in English). At

---

14. Ibid., 189, 224.
15. Charles Reginald Enock, *The Andes and the Amazon: Life and Travel in Peru* (New York: C. Scribner, 1907).
16. British Parliament, *Report*, 445. Enock was not contacted again until the following May, when he was asked to head up a PAC investigative Commission that would leave soon for the Putumayo. As it turned out, he requested a considerable fee and was passed over.
17. Ibid., 231.
18. Ibid., 190.

La Chorrera he hired a young "Barbadian,"[19] John Brown, to accompany him throughout as his guide. The captain believed that he himself was likely suspected at the time of being a spy sent to the Putumayo by the shareholders of the PAC.[20]

Whiffen later concluded that evidence of the atrocities was being hidden from him systematically (the stocks, whipping posts, and scarred indigenes). However, Brown began to recount his own knowledge of the abuses and, on his unannounced return visit back through the Putumayo, Whiffen saw much of the evidence—although he only witnessed one actual whipping and demanded that the young woman be cut down and released (which was done).

Brown decided to leave the Putumayo with Whiffen. He was owed 30 pounds in wages, and was given a chit and told to present it for payment at the company's headquarters in Iquitos. There, Pablo Zumaeta told Brown that he was not authorized to honor the instrument. Whiffen himself met with Pablo over the matter and was informed that he would have to collect the money in Manaos from Julio César. It was then that Whiffen recounted what he knew about the atrocities in the Putumayo. Zumaeta seemed surprised, but noted that he had heard similar reports before.[21]

It was now July 1909, and Whiffen was in Manaos for two days. He met with Arana and Rey de Castro. Whiffen recalled,

> The subject discussed was the atrocities which had taken place. Mr. Arana seemed horrified at the idea that any atrocities had taken place. However, he appeared to accept my word for it; he promised immediately to clear away Macedo, the manager at headquarters at Chorrera, and to make a clean sweep of the personnel.[22]

But, according to Julio César, that reform would require a certain amount of time.[23]

After his arrival in England the captain underwent medical treatment for a tropical fever and then reposed for a couple of months. In late September or early October, he received a letter from Julio César sent from Paris. Since Whiffen planned to visit a friend there anyway, he arranged to meet Arana in the French capital. The first couple of Putumayo articles in *Truth* were out by then, and Julio César was clearly concerned.

---

19. He was actually from Montserrat.
20. British Parliament, *Report*, 526.
21. Ibid., 316–17, 526–27.
22. Ibid., 317.
23. Ibid., 517.

He asked Whiffen if he had contacted the newspaper or intended to? Whiffen was not interested in such notoriety. Arana then asked about his report to Cazes, and Whiffen replied that it was practically written and that he planned to submit it soon. Arana asked him if he knew Hardenburg? The British captain had met the American briefly at a social event in Iquitos and was told by his acquaintances that Hardenburg was a Colombian spy. They had never corresponded before or since.[24] Arana and Whiffen agreed to meet next in London within a couple of weeks.

After he returned to England, Whiffen received a formal request from the British Foreign Office for a report on the Putumayo. Since he was finalizing the text of his Cazes's report, he resolved to send along to the British Foreign Office a modified version. Meanwhile, he'd still had no satisfaction regarding the 30 pounds owed him by the Peruvian Amazon Company (he had advanced the wages to Brown). Whiffen now arranged to host Arana for dinner on October 12, 1909. After the Peruvian's hospitality in both Manaos and Paris, the captain felt that it was his turn to pay. While they dined, the captain informed Julio César about the request from the Foreign Office and how he planned to handle it. Arana asked "Are you bound to do it?" and was told by Whiffen that he was. They then went out on the town together and ended up about midnight at the captain's private club—the Motor Club. Whiffen, in particular, had been drinking quite a lot of champagne.

The subject of possible payment for his materials came up, and Whiffen stated that his expenses in Peru were about 1,400 pounds, but that he would consider 1,000 pounds pounds for them. Arana wanted him to put that request in writing—in Spanish. Whiffen knew practically none, but was fluent in French (having spent a year studying in Switzerland). He and Arana spoke to one another mainly in that language. Sometimes Arana would slip in the odd Spanish word that Whiffen could understand from the context. He noted, "Arana speaks French in some sort of way." He also felt that Arana understood, but did not speak, English.[25]

Arana insisted that the letter must be written in Spanish, since first it would be sent to Abel Alarco (presumably a non-English speaker),

---

24. Ibid., 519.
25. Regarding Julio César's language skills Collier remarked on an exchange between Hardenburg and Arana,

"English, no" he [Arana] declared, holding up the fingers of his right hand. "One, two, three, four, five—that is my English." As Hardenburg, apologizing, switched back to Spanish, he could not know that Arana's grasp of English was prodigious. And though he had an equal command of Portuguese and French, he would rarely admit to any language but his own. Convinced of his ignorance, people lowered their guards, often letting fall facts he could use against them. (Collier, *The River*, 126–27)

before being forwarded with the Company's recommendation to Lima. Whiffen began writing phonetically as Arana dictated. However, when he had finished and tried reading his text, he became suspicious that Julio César was preparing some sort of "trap." So he tore the paper into pieces and later could not remember what he did with them.[26] Whiffen claimed that he was not drunk, but subsequently, in his formal testimony before the British Parliamentary Select Committee, conceded that he was quite hungover the next day and had trouble recalling what had happened at the club.[27]

The other subject of conversation that evening was the 30-pound debt. Arana had told him to come by the office the next day for his check. Whiffen did so, and Julio César then said the money would have to be forwarded to him by mail. When it was not forthcoming, he met one last time over the matter at Arana's London hotel. He was again assured that the check would be mailed shortly, and the two men parted on a somewhat "frigid" note. Whiffen never did receive payment.[28] Shortly thereafter, Arana provided his directors with the reconstructed draft of Whiffen's letter, having retrieved the scraps from the wastebasket at the Motor Club. It appeared to be a demand for 1,000 pounds in return for the captain's silence.[29]

Meanwhile, *Truth* had published additional articles in late September and into October. They detailed too many atrocities to simply be ignored—especially for a country coming off press coverage of the Congo scandals. Furthermore, they were being picked up by the international press in Germany, Belgium, and the United States.[30] Many prominent persons sided with Reverend Harris in demanding that the Foreign Office investigate.

Once back in London, in mid-October, Arana assured his Board of Directors that there was absolutely no "truth" to the charges in *Truth*, leveled, as they had been, by a forger and intending blackmailer. All of Hardenburg's informants were criminals of the lowest type themselves and possessed no credibility.[31] Arana would soon provide his Board written evidence of the several attempts to blackmail him. Julio César also sent a

---

26. British Parliament, *Report*, 520.
27. Ibid., 519.
28. Ibid., 521.
29. Ibid., 482. At a later point, we will consider both men's positions regarding what transpired between them that evening. Whiffen's subsequent testimony before the Select Committee introduced the possibility that the letter was not only reconstituted but also doctored.
30. Ibid., 191.
31. Ibid., 314.

letter to David Cazes at his personal residence asking to meet with him. For the Foreign Office this suggested too close of a relationship between the two men, and it denied Cazes's request to proceed with the meeting.[32]

While Arana retained the full confidence of his directors, clearly they had growing reason for concern. At their insistence, on November 3, 1909, Alarco sent a letter in the name of the Board to the section chiefs in the Putumayo ordering that the Indians be treated humanely:

> We give you these instructions not because we think the publications in "Truth" have the slightest vestige of truth, since we know with what particular object Hardenburg came to Europe, but with the idea of insisting that in future any new employees who may be engaged may follow the same line of conduct as that followed by former ones in regard to the proper treatment which it has always been customary to give to the natives.[33]

Given that London had virtually no impact over activity on the ground in the Putumayo, as well as the likelihood that Alarco knew very well that the admonition to mete out "treatment which it has been customary to give to the natives" was a coded way of stating that business should be conducted as usual, this was a cynical exercise on his part. Arana and Alarco were also working closely with the Peruvian legation and chargé d'affairs in London on damage control.[34]

Then, too, the directors wanted changes in management and unsurprisingly, on November 4, Juan Tizón, a former sub-prefect in the Loreto, went to work for the Peruvian Amazon Company in Iquitos. Arana was opposed to granting Tizón too much authority. Julio César was an admirer of Macedo and was concerned that an empowered Tizón could undermine both him and Pablo Zumaeta.[35] The directors also demanded that any Europeans employed by the company in the Putumayo would have their contracts annulled. They should be offered the option of returning home or renegotiating the terms of their employment.[36]

In early November, twenty-eight-year-old Deloitte accountant, Henry Lex Gielgud, returned from South America after several months there examining and reorganizing the books of the new company. He had spent time in both the Manaos and Iquitos offices, as well as two months in the Putumayo. While on his way to the latter, he learned that

---

32. Goodman, *The Devil*, 70.
33. British Parliament, *Report*, 204.
34. Ibid., 394.
35. Ibid., 225.
36. Ibid., 191.

Bartolomé Zumaeta and two other company employees had just been killed by the Indians.[37] Gielgud had seen the *Truth* articles while passing through Manaos on his homeward journey, and was appalled by them.

He first met with Julio César, and then the entire Board of Directors, stating his positive opinions of the Peruvian Amazon Company's activity in the Putumayo. Abel Alarco requested that he write a letter to the Board to that effect, which he did (dated November 24, 1909). It praised the two main station chiefs—Victor Macedo and Miguel Loayza.[38] On December 22, the Manaos office dispatched to London a packet of testimonials and newspaper articles (six from *La Felpa* and *La Sanción*) denouncing Hardenburg (clearly Rey de Castro was in the loop because he was said to have opposed forwarding the documentation, given that some of it might actually help Hardenburg's case).[39] It included a letter from Egoaguirre that claimed he was the one approached by Hardenburg with the demand for 7,000 pounds to drop his book project.[40] Gielgud believed him because, "Dr. Egoaguirre is a man of good standing in Peru. He is a Senator."[41]

On November 3, 1909, Peruvian senator Ward demanded an investigation of the charges in *Truth*. On November 22, the Senate passed a motion to constitute an investigative commission to be sent to the Putumayo. Senator Egoaguirre was able to stall the matter by invoking the recent Zapata one that had exonerated the Peruvian Amazon Company entirely. By now, Egoaguirre was certainly in favor in Lima, since he was named minister of development by President Leguía on December 17, 1909 (a post that he would hold until August 31, 1911).[42]

Then, the Peruvian Amazon Company received a letter from the British Foreign Office, dated November 24, 1909, noting that the practice of providing commissions to the chiefs of sections and sub-managers was conducive to abuse of the Indians.[43] It stated that the Foreign Office now

---

37. Ibid., 378. It happened in May of 1909 (Ibid., 189).
38. Ibid., 102, 126, 388, 390.
39. Ibid., 31, 426.
40. Ibid., 327.
41. Ibid., 395.
42. Within days after his appointment, or on December 31, 1909, the Peruvian Congress passed *La Ley General de la Montaña*, a bill introduced by himself and Rojas in 1907 right after they entered parliament. It was designed to stimulate the economy of the Loreto by facilitating the sale and concession of tracts of public land (Indian territories) in the forests. Unsurprisingly, while designed in part to foment agriculture, it favored "especially the rubber industry." Under its terms, initially, rubber interests could acquire title of up to 30,000 hectares, a limit that could be lifted on a case by case basis by the Congress (García Jordán, *Cruz y arado*, 187–88).
43. British Parliament, *Report*, 361.

believed in the accuracy of the allegations in *Truth* regarding atrocities committed against the Indians and the murders of Serrano and other Colombians, and that Julio César Arana was fully knowledgeable and therefore complicit. On December 11, Grey also ordered his Washington ambassador, James Bryce, to apprise the Americans of the situation and solicit their cooperation in demanding that the Peruvian government intervene.[44] It included excerpts of Whiffen's report, copies of the *Truth* articles and the letter that had been sent to the PAC's Board of Directors.[45] Arana then produced for his Board the reconstructed blackmail letter drafted by Whitten.

The PAC annual shareholders meeting was scheduled for soon thereafter, or on December 31, 1909. At it, Julio César wanted to denounce Hardenburg and Whiffen as blackmailers, but his directors refused to endorse that strategy. While they believed his allegations, they also felt that they were predicated upon Arana's personal experience (and evidence) rather than their own. They likely were acting on the sage advice of their solicitor—concerned as he probably was over the possibility of a libel suit. Arana was free to make his own personal allegations and did so in a private letter sent by him to shareholders on the eve of their annual meeting.[46]

On December 30, the Peruvian Amazon Company's Board of Directors finally responded to the Foreign Office. In a carefully crafted letter clearly written with legal advice, Gubbins underscored that crimes in the Putumayo, if any, were under active investigation by the Peruvian government and antedated the existence of the Peruvian Amazon Company.[47] Consequently, its directors had no culpability. Arana's letter to the shareholders was included as demonstration that Hardenburg and Whitten were both blackmailers without any credibility whatsoever.[48] Neither man was ever able to secure a copy of Arana's accusatory circular—although Whiffen subsequently attempted to do so.[49] Furthermore, the contention that compensation of managers in the Putumayo involved a commission system was erroneous. Some few of them were owners of a tiny piece of the company and were therefore entitled to dividends.[50]

---

44. Goodman, *The Devil*, 72–73.
45. British Parliament, *Report*, 281.
46. Ibid., 243.
47. Ibid., 356.
48. Goodman, *The Devil*, 76.
49. British Parliament, *Report*, 522.
50. This proved to be inaccurate in that on June 17, 1909, Pablo Zumaeta had written to Arana (a letter never shared with the other directors) requesting funds to buy out the existing contracts of the section chiefs. Some received 7 percent commissions on the rubber produced and others 20. Then there was the egregious case of the two Rodríguez brothers, each of

At the shareholders' meeting, a brazen owner of 100 shares, Mr. Morgan Williams, demanded to know if the allegations in *Truth* were accurate. He was alone in his audacity, as his fellow shareholders preferred to accept, uncritically, the representation that the charges were baseless. Gubbins informed the gathering that six of the company's employees had been killed by cannibals in the past year, but that, "We have not chastised the Indians beyond what was proper." He also noted that some employees had been used as soldiers in the ongoing frontier dispute with Colombia. Williams found these to be strange activities for a rubber company and demanded a proper inquiry.

Chairman Gubbins further noted that the matter of any past criminal activity was properly under investigation by the Peruvian authorities. Meanwhile, the Peruvian Amazon Company planned to form its own commission that would look into the present (not the past) with an eye to improving conditions in the Putumayo—if warranted. Gubbins promised that the report of the commission would, of course, be shared with the shareholders. He agreed to meet privately with Williams and did so on two occasions over the next few months. According to Williams, those proved to be courteous exercises in stonewalling.[51]

Then, too, there were letters from Englishmen—notably Mr. Philips (sometimes spelled Phillips), an Australian who worked on a Peruvian Amazon Company steamship in the Putumayo, and an American, Herbert Spencer Dickey (sometimes rendered as Dickie), a physician who had worked during the latter part of 1908 and early 1909 at El Encanto. Neither one had witnessed either atrocities or signs of them. Dickey stated,

> I am ready to certify upon oath if necessary that, during the ten months' permanency in the Encanto, I have never seen an ill or debilitated Indian compelled to work, nor have I seen a single Indian bearing the marks of scars or floggings on his back. On the contrary, I was struck by the extremely gentle methods used in the management of the Indians.[52]

---

whom received a 50 percent commission. Since their sections were among the Company's most productive, it was a business (rather than humanitarian) decision. In the event, it took 9,000 pounds to buy out the Rodríguez contracts (Ibid., 312). There is no record in the London PAC files of a response; however, the contracts of many section heads were bought out at considerable expense. Zumaeta's letter would prove a double-edged sword. On the one hand, it showed that the commissions' system had been abolished, while, on the other, it was proof that it had existed until quite recently.

51. Ibid., 292–94.
52. Ibid., 416.

Philips not only saw nothing unusual in the Putumayo, in Iquitos he talked with Hardenburg,

> Mr. Hardenburg at that time told me he was writing a book upon the Putumayo, and that, unless the Company paid him well, he would publish it and make them suffer in England. He also said that he had paid various ex-employees of the Company to give affidavits, and that he thought he had a case strong enough to make the Company pay a few thousands rather than stand for publication.[53]

The Philips and Dickey letters were dated identically (January 8, 1910), which raises the question of their likely orchestration by someone.[54]

On January 26, 1910, Gielgud was hired as the Company's secretary. Then, in February, the Board resolved to terminate Abel Alarco's services. While it had contributed to his problems, Alarco was not dismissed solely for the attempted bribe of Thorogood. Director Read noted, "He was a difficult man, a man of rather violent temper; he quarrelled with people."[55] According to Collier,

> For months the British directors' relations with Alarco had been at the snapping point; his vile temper had provoked countless stormy scenes. Then, too, he neglected the business, making off for weeks at a time to his rented house in Geneva, to strut the boulevards in Inverness cape and deerstalker with his two prize German wolfhounds.[56]

Arana concurred in the firing of his brother-in-law, and Gielgud was made both secretary and manager of the Peruvian Amazon Company with a salary of 1,000 pounds a year—or seven times his former compensation (150 pounds) at Deloitte.[57] The young man had proposed the amount and refused to budge when Arana and Medina tried to get him to accept 800 pounds instead. He believed that his knowledge of Spanish (albeit limited), his training at Deloitte, and his first-hand experience in South America made him worth it. Besides, he was assuming Alarco's

---

53. Ibid., 244.
54. Ibid., 243. To wit, on May 16, 1938, in a sworn deposition, Dickey noted, "I myself was offered £500 if I would pen one letter simply stating that I as a company doctor had seen no mistreatment of the Indians." Angus Mitchell (Introduction, Commentary and Footnotes), *Sir Roger Casement's Heart of Darkness: The 1911 Documents* (Dublin: Irish Manuscripts Commission, 2003), 735 [*Sir Roger*].
55. British Parliament, *Report*, 350.
56. Collier, *The River*, 190.
57. British Parliament, *Report*, 149.

responsibilities at considerably less salary than the former general manager. Gielgud also refused their proposal that he move to the Putumayo to serve for three years as the Peruvian Amazon Company's on-site manager.[58] Gubbins was receiving 100 pounds annually for chairing the Board, in addition to his 200-pound director's fee. Beginning January 26, 1910, he would be given a 600-pound yearly salary in recognition that he would be working more intensely with the inexperienced general manager.[59]

Meanwhile, for his part, Grey decided to follow the strategy of his Latin American affairs manager, Louis Mallet. In a letter dated February 8, 1910, the Foreign Office insisted the Peruvian Amazon Company constitute an impartial investigative committee that would visit the Putumayo and prepare a report on the allegations, as well as recommendations for redressing any abuses. There seemed to be growing frustration on the British government's part, since, should the Board remain "obdurate," the Foreign Office intended to summon Sir Lister Kaye for an explanation.[60]

Nevertheless, the Board remained solidly behind Arana. According to Director Read, when it came to Hardenburg versus Arana, "We had to choose between the two." The directors "knew Arana; we had been with him some time; we were very much impressed with his quiet manner, but when we asked questions he appeared to speak with the greatest frankness, and Alarco as well."[61]

That same month of February, Hardenburg left London for Canada with his new bride. *Truth* had been providing him with some living expenses—maybe about 20 pounds in all. However, that was only because the American had exhausted his personal funds, and the newspaper wanted him to remain in England to testify in its defense should the Peruvian Amazon Company (or anyone affiliated with it) choose to file a libel action.[62] Since none was forthcoming, Hardenburg's continued presence in London was no longer required. He moved to Toronto, where he would live for a year, surviving mainly on the proceeds of his wife's sewing. They then resettled in Red Deer in western Canada, where Walter entered the employ of the Canadian Pacific Railroad. He built his wooden house there with his own hands.[63]

---

58. Ibid., 391, 393.
59. Ibid., 192. Gubbins was not paid from March 1911, until liquidation of the Company the following September (Ibid., 195).
60. Goodman, *The Devil*, 78.
61. British Parliament, *Report*, 315.
62. Ibid., 96–97.
63. Lagos, *Arana*, 337.

Then, on March 8, 1910, Chairman Gubbins wrote the Foreign Office, "My directors must once more emphatically repeat that they uphold most absolutely the written word of their co-director, Senor Arana, that the alleged atrocities related in 'Truth' are entirely unfounded."[64] They cited Gielgud's earlier eyewitness report that conditions were fine in the Putumayo.

As the public pressure continued to mount, British parliamentarians demanded to know what was being done. On March 14, the Foreign Office informed the House of Commons that Peru had launched an official investigation.[65] The Foreign Office's short response (March 31) to the Peruvian Amazon Company's most recent letter was to simply reiterate the insistence upon an impartial investigation.[66]

Meanwhile, the Peruvian Amazon Company's directors were interviewing candidates for its investigative commission. Whether enthusiastically or not, Arana and Alarco had been the ones to propose forming it. Clearly, the Foreign Office was under considerable pressure and might very well mount its own inquiry—better that they attempt to spike that particular cannon by constituting a PAC team. Quite probably the two Peruvians thought that they could control the selection process and/or the Commission's agenda and experience once it was in the Putumayo.

Arana continued to assure the Board, and Chairman Gubbins in particular, that there was simply nothing to be found. After all, given the value of each Indian to the Company, wasn't it simply absurd to believe that they were being killed deliberately by their employer?[67] In the interim, his agents in the Putumayo would have ample time to hide any incriminating evidence—the stocks, whipping posts, human cages. Any badly scarred Indian would be removed from sight.

In the event, the PAC Board had become so thorough in its vetting and, then rejection, of flawed candidates, that the months had dragged on (the PAC would later be accused of strategic foot-dragging in the hope that the scandal would simply blow over). Director Lister-Kaye eventually proposed that his friend, Colonel Bertie, be appointed head of the Commission—arguing that the man was eminently qualified, thorough and thoroughly incorruptible.[68] Bertie was hired at a salary of 2,500 pounds, and it was decided that Gielgud should accompany the group. While abroad, his salary would be doubled to 2,000 pounds. In Gielgud's ab-

---

64. British Parliament, *Report*, 326.
65. Ibid., 371.
66. Goodman, *The Devil*, 79.
67. British Parliament, *Report*, 213.
68. British Parliament, *Report*, 253–54.

sence, Gubbins would serve as PAC general manager in London, and his salary was doubled accordingly to 1,200 pounds.

Julio César wrote to British consul Cazes to inform him that the Commission would soon be on its way and that its scope was to enquire into the present situation in the Putumayo and not the past. Investigation of former crimes (if any) were the purview of the Peruvian government. Interestingly, he added,

> There have been difficulties in constituting this Commission given that the repeated insinuations of the English minister and of the Anti-Slavery Society, motivated by the constant publications against the Company, has pressured [us] to send out the best personnel that it has been possible to assemble.[69]

Then, on April 11, 1910, Gielgud wrote a letter to the shareholders of the Peruvian Amazon Company that stated,

> I am to inform you that the Board absolutely decline to attach any credence to the allegations that have appeared in "Truth." The evidence on which those allegations are based is, in their opinion, of such a character that no reliance whatever can be placed on it either as to its impartiality or its general trustworthiness. On the other hand, Mr. J. C. Arana, founder of the business in the Putumayo, a man in whose integrity the Board have the utmost confidence, asserts most positively that the allegations in "Truth" are gross misrepresentations. I may add that my own experience during a visit to the Putumayo last year certainly does not bear out the horrible stories told in "Truth." While I was in that region I visited various centres of rubber collection and saw many hundreds of Indians. These had not the cowed and down-trodden appearance one might have expected to find in the victims of brutal savagery. . . .[70]

Gielgud would regret those words soon enough.[71]

Meanwhile, that same day Ambassador Bryce informed London that the Americans were dragging their feet. Peru and Ecuador were practi-

---

69. J. C. Arana to David Cazes. July 25,1910. Bodleian Library MSS Brit. Emp. S-22 G338.
70. British Parliament, *Report*, 143.
71. In his diary Casement quotes the Barbadian Frederick Bishop stating that in the case of the stays of both Captain Whiffen and Henry Gielgud the place was "cleared up" prior to their arrival. Prisoners were marched into the forest by the *muchachos de confianza* and kept out of sight until the visitors left. Angus Mitchell (ed.), *The Amazon Journal of Roger Casement* (London: Anaconda Publications, 1997), 98–99 [*The Amazon Journal*].

cally on a war footing after the recent collapse of the attempt by the king of Spain to arbitrate their border dispute. Both countries had amassed thousands of troops on their common border, and the American government was trying to involve Argentina and Brazil in the attempt to get the belligerents to the negotiating table. In short, it was an extremely delicate moment and the Americans were unlikely to embarrass Peru at this juncture.[72]

About this same time, the Peruvian Amazon Company's directors received a letter (written April 23) from Mr. Meech, their former bookkeeper in Manaos. He had visited the Putumayo and reported, "My impressions of this region are generally very good, and I anticipate a large and prosperous future for it under good administration." He had read the *Truth* articles and dismissed them as "merely the result of unsuccessful attempts to blackmail the firm of Arana."[73]

Then, too, Tizón had just been dispatched to the Putumayo to observe conditions there and report back to the Board. His initial take could scarcely have been rosier. He toured much of the region and saw only contented, well-fed Indians working hard to supply the Company with rubber in exchange for goods. His only concern was the frequency of illnesses. He noted that Indians who had to carry rubber from the interior to shipping points were sometimes weakened to the degree of becoming susceptible to diseases. In sum, far from alarming, Tizón's first report (in June of 1910) would be most reassuring news to the Board.[74]

Like Gielgud, Tizón too would soon come to regret his words. Meanwhile, in June the Company paid its fourth straight semi-annual dividend on preference shares (throughout its history the Peruvian Amazon Company never paid a dividend on the ordinary ones).

Nevertheless, the recent past could be compromising, since evidence was produced to show that in that same month of June, 1910, a group of fifty-seven Indians had absconded to the Colombians, and a PAC expeditionary force was dispatched to bring them back.[75] There were other disturbing entries in the Company's books, like the 11,420 pounds expense the year before (signed by Macedo) for *gastos de conquistación de Indios* (charges for the conquest of Indians), for "104 Indians; 55 males and 49 females, ransomed from the power of various persons on various rivers" or for "40 Indians carried to the River Caquetá by Colombians implicated in disturbance at the beginning of the year, and those Indians being ran-

---

72. Goodman, *The Devil*, 81.
73. British Parliament, *Report*, 246.
74. Ibid., 237–38.
75. Ibid., 306.

somed by pacific arrangements."⁷⁶ Miguel Loayza had written a letter on December 22, 1909, stating that the trouble at La Reserva and La Unión had stemmed from a quarrel over runaway Indians.⁷⁷

A couple of months earlier, the Colonial Office had forwarded to the Foreign Office a deposition by John Brown detailing the ill-treatment of its West Indian employees by the Casa Arana, and claiming that those who remained there were virtual slaves of the Peruvian Amazon Company, while noting that British consul Cazes was in cahoots with the Peruvians.⁷⁸ Grey now had his excuse for direct intervention—his obligation to investigate the welfare of British citizens. Meanwhile, in London, the Anti-Slavery and Aborigines Protection Society pressed for inclusion in the investigation of a British consul. On May 11, Secretary Travers Buxton wrote to Sir Edward Grey:

> It has been stated that the condition of affairs in the Putumayo valley is identical with those prevailing at the worst period of the Congo State administration. The Society has for several years given much attention to that question, and therefore, feels competent to compare the two questions. It has come to the conclusion that in many of its features the system of enforced rubber collection closely resembles those of the Congo State; moreover, my Committee does not hestitate to say that nothing reported from the Congo has equaled in horror some of the acts alleged in detail against this rubber Syndicate. The nature of the evidence is indeed too revolting to permit of full publicity, but the documents are at the disposal of His Majesty's government, should they desire to examine them. In view of the fact that this Syndicate is not only a British Company, but that several of its Directors and principal shareholders are British subjects, this Society would urge His Majesty's government to request the Peruvian government to permit the presense [*sic*] of the British Consul during the proposed enquiry in order that a full report may be made to the British government.⁷⁹

---

76. Ibid., 309.
77. Ibid., 426.
78. Goodman, *The Devil*, 79–80. According to Goodman, "Yet prejudices at the Foreign Office almost dismissed Brown's testimony when Sir Charles Hardinge, the permanent undersecretary, cast doubt on the reliability of the evidence, arguing that the writer would have needed a diary of events to recall dates and names so precisely, the implication being that black people didn't keep such things" (Ibid., 80).
79. Travers Buxton to Sir Edward Grey. May 11, 1910. Bodleian Library MSS Brit. Emp. S-22 G322 (file 2 of 2).

The Peruvian Amazon Company agreed that a British consul could accompany its commission—no doubt hoping that it would be David Cazes. However, he quickly disqualified himself on the grounds of his association with Arana. The Company's one condition was that the Peruvian government agree and that approval was quickly forthcoming.[80]

The Foreign Office received a steady stream of correspondence urging it to intervene—including a petition from a group of concerned citizens in Dublin. From Canada, Walter Hardenburg continued his input into the process. On June 28, 1910, he sent his advice to the secretary of the Anti-Slavery Society:

> 1. Watch them at every turn, so that they cannot bribe the Peruvian witnesses nor appeal to a false sense of patriotism.
>
> 2. The best witnesses are the Indians themselves, but great care must be taken in selecting interpreters. I recommend Colombians on general principle, but they too may be bribed. Hundreds of Indians—unless they have all been despatched—will be seen covered with enormous scars.[81]

He also opined that, since several months had passed, the Peruvian Amazon Company has "probably had time to clean things up a little." He urged that Saldaña Rocca and Iquitos physician (Basque-surnamed) Dr. Urdaneta be interviewed. Finally, he provided contact information for two Colombians in Pasto who might give evidence and noted, "I have always refrained from attacking Peru itself as much as possible for Peruvians are extremely patriotic, and if it were represented that the good name of Peru were being attacked, some witnesses might not care to give evidence."[82]

The PAC commission had been named finally on June 10, 1910. Reverend Harris of the Anti-Slavery and Aborigines Protection Society met with Grey on July 1, 1910, demanding an investigation and pressuring the British government to send its own investigative team to the Putumayo. How could a commission appointed and paid by the Peruvian Amazon Company be expected to come up with an impartial report? The Foreign Office should constitute its own team or at least send someone credible to accompany the Peruvian Amazon Company one. Roger Casement's

---

80. Ibid., 84–85.
81. Walter Hardenburg to Travers Buxton, June 28, 1910. Bodleian Library MSS Brit. Emp. S22 G322 (file 2 of 2).
82. Ibid. Hardenburg also notes, "I have repeatedly written to Iquitos, but have rarely received answers. I expect my mail was probably tampered with."

name came up for the first time (Rio's British consul had arrived back in England in May on leave).[83]

On July 13, Peruvian approval of inclusion of a British agent in the Putumayo investigation was received in London. So, Grey once again gave an order to Casement. He was to depart immediately for Iquitos and the Putumayo with the PAC commission (it would leave England on July 23), albeit not as one of its members. Officially, Casement's brief was to investigate the circumstances of Barbadians (British subjects) in the employ of the (British) Peruvian Amazon Company. However, in private conversation, they agreed that he would probe the full range of the allegations against Arana. Casement was to maintain his independence from the PAC commission to the degree that such was practicable.

Coincidentally, Casement and the Commission traveled on the same vessels from Southampton to Iquitos with Lizardo Arana.[84] Between Manaos and Iquitos, their ship's captain, Buston, noted that the Amazonian indigenes were all enslaved: "The Indians and the rubber trees are equally reckoned as personal property of the estate owner." Then, too, according to Casement,

> He says that Indian children are stolen or bought constantly up the rivers and brought to Iquitos for sale or gifts. That he has often carried them on his ship with Peruvian families that as long as things go well with the family they are treated all right—but if they lose money these slave children are at once sacrificed. In no case do they ever, according to him, get back to their native homes—but if the family cannot maintain itself they would be cast adrift or sold possibly.[85]

Casement also met Lizardo Arana again in Iquitos, where, according to Bryant,

> Lizardo, representing Julio César Arana, his brother, showed a good-humored willingness to help them in their work. The Peruvian Amazon Company would offer a company launch, the *Liberal*, to take them to the various stations in the area. Expenses would be borne by the company. Casement received a visit from the late Acting French Consular Agent, Vatan, who told him that "the condition of things on the Putumayo had been disgraceful—that the existing method was slavery pure and simple—but that it

---

83. Edward Grey to John Harris. July 13, 1910. Bodleian Library MSS Brit. Emp. S-22 G322 (file 2 of 2).
84. Goodman, *The Devil*, 106.
85. Mitchell, *The Amazon Journal*, 86–87.

was the 'only way' in Peru as she exists." It was necessary to civilize the Indians, he added.[86]

Iquitos British consul Cazes had generally kept the Foreign Office in the dark regarding the Putumayo and probably hoped that there would be no official British investigation at all, given his belief that there were no longer Barbadians working there. But now that Casement, Grey's appointee, was in Iquitos, Cazes took pains to cooperate. He invited Casement to be his houseguest, and then served as interpreter at a meeting with Prefect Francisco Alayza y Paz Soldán. Roger was favorably impressed with the man's sincerity, but noted that,

> The local prefect, Dr. Paz Soldan, knows nothing about the Putumayo. He believes simply the stories told him by the interested people, and he has been in Iquitos, coming from Lima, only a year, and not on the Putumayo at all.[87]

As for his own position, Casement observed, "I am viewed with grave suspicion already, I think, but as I have got the Commission with me we are all right."[88]

Cazes then introduced Casement to Carlton Morris, an influential Barbadian living in Iquitos who was anxious to meet up with Roger. Morris proved to be the key to securing several frank interviews with former employees of the Casa Arana.[89] There was Frederick Bishop, who told Casement that, "we shall have great difficulty in his seeing active wrongdoing, such as flogging or anything of that kind, because they will be far too careful, but that the Indians will tell, if a good interpreter is found, and in any case we shall see their backs and buttocks cut and scarred."[90] There was also the problem of the statements of the Barbadians themselves,

> Some of these men, I believe, have committed grave crimes, much worse than Bishop owns up to of flogging, and there may be dif-

---

86. Bryant, *Roger Casement*, 123. Pearson, after visiting Manaos in 1910, notes,
    A young Englishman whom I met had spent some months up in the Putumayo district and brought down with him a nine year-old boy as body servant who was a veritable little savage. Friendly and smiling he was when all went right, a murderous little tiger if things went wrong. He would accept reproof from his master but from no one else. One day a man servant struck him and his master returned two hours later to find the boy sitting in the courtyard, a loaded Winchester across his knees, and all the servants hidden in a hastily barricaded room from which they dared not emerge. Had the offender shown himself the boy would certainly have shot him. (Pearson, *Rubber Country*, 102; 104)
87. Mitchell, *The Amazon Journal*, 108.
88. Ibid.
89. Goodman, *The Devil*, 103–105.
90. Mitchell, *The Amazon Journal*, 107.

ficulty in getting them to speak. A man came in today, Adolphus Gibbs, and told me his story. He was very frightened, and refused twice to come, but probably thinks it safer to keep in with me than with the company.[91]

Casement was disappointed to learn that John Brown (who had been both Whiffen's guide and the man who had denounced the Casa Arana's treatment of West Indian employees) was ill and unavailable to act as his translator. The Montserratian, married to a Huitoto woman and fluent in that language, would have been ideal. However, through the intervention of the former French consul in Iquitos, Casement was able to retain a former Arana employee, a Peruvian named Viacara—a man fluent in both the Huitoto and Bora languages. Viacara had left the Putumayo after refusing the order to shoot two Indian chiefs.[92]

In the event, the difficult logistics in accessing the Putumayo meant that Casement traveled there aboard the *Liberal* in late August with the members of the Company's Commission. While his investigation would be independent of theirs, and he was often separated from them while in the field, they shared and discussed their findings. Casement held all of the commissioners in high regard (excepting Gielgud). Indeed, with the exception of Gielgud, those appointed were relative PAC outsiders and possessed impressive credentials.

Colonel Reginald Bertie was to head the Commission. Along the way, however, he had become critically ill. Under doctor's orders, he remained behind when the group departed Manaos for Iquitos. Then there was Walter Fox, a tropical botanist and rubber specialist who had been recommended by the director of London's Kew Botanic Gardens. Seymour Bell was a merchant with considerable experience in Latin America and knowledge of Spanish. Finally, there was Louis Barnes, a tropical agriculture specialist with many years of practical experience in East Africa and Brazil. After Bertie's resignation in Manaos, Barnes became the Commission's leader.

At La Chorrera, once a Barbadian, Joshua Dyall, admitted to killing five Indians, including two by smashing their testicles under orders from Armando Normand, Casement decided to let the entire commission, Juan Tizón, and a nervous Victor Macedo sit in on the interrogations. By and large, most of the Barbadians cooperated with the process, although Roger would suspect throughout his investigations that there were attempts by section chiefs to bribe, intimidate, and otherwise suborn the witnesses.

---

91. Ibid.
92. Ibid., 102, 107.

Casement summed up the situation as follows:

> One moves here in an atmosphere of crime, suspicion, lying and mistrust in the open, and in the background these revolting and dastardly murders of the helpless Indians. If ever there was a helpless people on the face of this earth it is these naked forest savages, mere grown up children. Their very arms [slightly built] show the bloodlessness of their timid minds and gentle characters.[93]

Their testimony was astounding—it described regular resort to inhumane practices to enforce slavery. Initially, Tizón was incredulous, but that quickly changed. Sobered, he promised to institute reforms immediately and resolved to cooperate with the investigation in every way. Both he and Casement agreed that, for the Indians' sake, it was better to have a reformed Peruvian Amazon Company in place, rather than a scramble within a void of aspiring new rubber men.[94] Tizón confided that there was a piece of him that wanted to take the *Liberal* back to Iquitos and simply resign his post, but he had resolved to stay on and effect reforms to save Peru's national honor.[95] He was effusively grateful to Casement and the Foreign Office for bringing the atrocities to light; he also believed that nothing would be served by confronting the worst of the blackguards. Indeed, if they felt cornered, anything was possible, including violence against Tizón and/or the foreigners. Tizón believed,

> If the Company can be kept going and he can have his hands strengthened, then he is prepared to make a sweep of all the criminals "prudently" but as quickly as possible. They would be called down to La Chorrera, one by one, and then got rid of—paid off and sent away, and not allowed to return. Better men would be obtained and put in their place.[96]

Casement understood the logic, but was frustrated by his inability to confront the criminal overseers with their crimes,

> The whole air of the place is atrocious. The one preoccupation now of even ourselves—this Commission and myself——is to see that nothing of the truth shall leak out. We, in search of the truth, proceed as if we were the liars and wrongdoers, and hold our meetings in secret, and constantly pledge ourselves to be "prudent" and "silent," and to show no one of our local hosts what we think of them

---

93. Ibid., 124.
94. Ibid., 125–27.
95. Ibid., 130–31.
96. Ibid., 127.

or of the things we are hourly looking at. We play a part the whole day, and when investigating (as far as we can) a most appalling crime like that told this morning, pretend to be butterfly catching. And all this caution now is enjoined on us for the sake of . . . the Indians![97]

Then, too,

> We cannot escape from our surroundings, and I said that it was all very well for Tizon to say I was his guest, or the Company's guest, I was really the wretched Indians' guest. They paid for it all. The food we eat, and the wine we drank, the houses we dwelt in, and the launch that conveys us up river—all came from their emaciated and half-starved, and well-flagellated bodies. There is no getting away from it, we are simply the guests of a pirate stronghold, where Winchesters and stocks and whipping thongs, to say nothing of the appalling crimes in the background, take the place of trade goods, and a slavery without limit the place of commercial dealings.[98]

Some non-Barbadian PAC employees approached Casement as well, wanting to testify. However, he urged them to travel to Iquitos to inform the authorities there instead. It was beyond his purview to work openly with non-British citizens.[99]

Casement and the commissioners eventually traveled out to several sub-sections and continued their interviews. He became increasingly concerned for his own safety and that of the Barbadians who had openly cooperated. He began keeping a loaded firearm in his room at night and scheduled his and their moves to avoid providing opportunities for ambushes on the trail.[100]

One can legitimately question whether Casement entered the Putumayo with an open mind; if so, it certainly closed during his stay there. His diary is filled with running observations about his companions and interlocutor—all of whose standing rose or fell according to the degree to which they shared his belief in the atrocities, and his crusader zeal in redressing them. Tizón was the only Peruvian that he admired and trusted. As for the other "whites" in the Putumayo, Casement noted that, in their view,

> The Indians who actually prefer their forest freedom to the whip, the *cepo* [stocks], the bullet and the raping of their children are

---

97. Ibid., 162.
98. Ibid., 161–62.
99. Ibid., 168, 173.
100. Ibid., 170, 375.

spoken of in terms of reprobation as lazy, idle and worthless—and this by men who never leave their hammocks all day, and whose only "work" is to work crime. They have not cultivated a square yard of ground or done one useful thing with their hands since they came here. Their only use—their sole purpose—is to terrorise and rob. And this is the function of the paid employees; the higher staff of a great English Company! Mr. Arana has planted a strange rubber tree on English soil![101]

Casement was given to loathsome descriptions of the "criminals." Upon his arrival at Matanzas, its infamous section chief, Armando Normand, was away on a *correría*. He was shown Normand's sitting room, where there were many photographs on display of "brutal faced South American people." And then, "one in particular I should imagine to be Normand himself 'when a boy'—it looks like a low typed East End Jew, with fat greasy lips and circular eyes."[102] When Casement later met Normand in person, he found Armando to possess, "a face truly the most repulsive I have ever seen, I think. It was perfectly devilish in its cruelty and evil. I felt as if I were being introduced to a serpent."[103] Or again, while contemplating the fate of the tortured Indian, Casement observed in his diary,

> His manhood lashed and branded out of him. I look at the big, soft-eyed faces, averted and downcast and I wonder where that Heavenly Power can be that for so long allowed these beautiful images of Himself to be thus defaced and shamed. One looks then at the oppressors—vile cut-throat faces; grim, cruel lips and sensual mouths, bulging eyes and lustful—men incapable of good, more useless than the sloth for all the work they do—and it is this handful of murderers who, in the name of civilisation and of a great association of English gentlemen, are the possessors of so much gentler and better flesh and blood.[104]

---

101. Ibid., 250.
102. Ibid., 255. All parties to this story, whether Anglo-American or South American, held in varying degrees their era's negative stereotypes of Jews that would culminate but a few decades later in history's most infamous holocaust. Casement also wrote in his diary, Every man flies to get rubber—by hook or by crook; and all who can try to get the labour of someone else. The petty trader is the next step in the commercial ladder. He gets the rubber from the vicarious collectors by comprehensive swindling it would be hard to match anywhere else in the world. Iquitos, which has no church, has a large colony of Jews. It also has a big colony of Chinese, half-castes—that is to say a Chinese cross with the Cholo Indians, and quite a good physical type it is. The Jews are the predominant business factor in Iquitos. . . . (Ibid., 441)
103. Ibid., 256.
104. Ibid., 335.

As for the commissioners, they all lacked in varying degrees backbone and a proper humanitarian vision, concerned as they were primarily with the Peruvian Amazon Company's financial challenge. However, Casement won them all over, excepting Gielgud. He reserved choice criticisms for that individual:

> I had a long talk with Gielgud, just before lunch. His mental attitude (or moral?) is a strange one. Frankly at heart I think he is as bad as any of these Peruvian scoundrels. He is a cold-blooded, selfish guzzler, who thinks of himself first and always.[105]

Casement feared that men like Macedo and Normand might try to prevent the Barbadians from leaving. Alternatively, they could be counting on transportation of the Barbadians back to Iquitos, where they might be bribed and/or intimidated into repudiating their testimony. After all, most had implicated themselves in crimes and could be imprisoned readily for them. It might become a choice between freedom and repudiation.[106]

In the event, however, Tizón agreed to let the Barbadians leave the Putumayo with Casement on condition that their debts on the Company's books were satisfied. Indeed, he and Macedo authorized a discount of 25 percent of the indebtedness in order to facilitate matters (although Casement felt that it was to blunt the full impact of the accusation that both the indentured *peones* and Indians were grossly overcharged in the Company's stores).[107] Tizón confided to Casement that he himself was now *persona non grata* within the Peruvian Amazon Company for having joined "the enemy" and was uncertain of his future.[108] Casement reported:

> Tizon tells me privately that he has written to the Editor or proprietor of the Lima *El Comercio*, who is his cousin, saying the government must take action. He is also writing to the Prefect at Iquitos, begging him to come round here as soon as he can. He tells me the Prefect has written him too—and he begs me to go and see him and speak frankly to him. This I have said I shall do privately. He says it will be a friendly act—an act of goodwill and kindness. Our conversation was long and very friendly, as indeed my conversations with Tizon always are. He has repeated all his former assurances, with emphasis—that the gang at Iquitos must be broken up and Arana eclipsed, and the London Board take complete control. He says the criminals here will fly to Brazil as

---

105. Ibid., 399.
106. Ibid., 298–301, 385, 415.
107. Ibid., 351–54, 358–59.
108. Ibid., 305.

soon as the Judge appears—and that that will be the best settlement. There they will drift to the Purus and other rivers to repeat their crimes![109]

Casement arranged for fourteen men, four women, and four children to be off-loaded from the *Liberal* in Brazilian waters. He then left word that any passing English vessel willing to transport them to Barbados would be compensated for the passages by the British vice-consul in Manaos.[110] Casement was also to be accompanied to Iquitos (and eventually Europe) by an orphaned Indian lad, Omarino. The plan was to display him in a campaign to seek funding from wealthy English sympathizers to finance reform in the Putumayo. Arédomi, a young Andoke Indian, also pleaded successfully to come along.[111] Roger decided to drop them off in Barbados for a little schooling and acclimatization in western ways before bringing them to London.[112]

Meanwhile, Casement was more than a little concerned about his own situation,

> Then there will remain my few days in Iquitos, when I must be hourly on my guard. I shall not feel safe or happy until I see Tabatinga and the Brazilian flag waving over its mulatto-troop of soldiers. Brazil and freedom are synonymous up here in this beknighted region.[113]

Indeed, his friend the former French consul Marius Vatan opined that it was only because Casement had come in an official capacity that he was allowed out alive: "Your death would have been put down to Indians—I know what I am talking about."[114]

Back in Iquitos, Casement was again the house guest of Cazes. There, his suspicions increased that the city's British consul had deluded his own government to protect his close association with Arana:

> Long talk with Cazes about Putumayo—his mental attitude is not a desirable one. Every time he opens his mouth on the subject he shows how very much more he knew about it all than he admitted to the FO when asked for information. He was cheek by jowl with Arana in London at that time. He knew heaps and heaps of things, and yet in his letter to FO he pretended he knew practi-

---

109. Ibid., 391, 408.
110. Ibid., 326–28.
111. Ibid., 130–31.
112. Ibid., 340–42.
113. Ibid., 394.
114. Ibid., 472.

cally nothing—even that Huitoto slaves were brought and sold here in Iquitos. Why, I already know of several.[115]

Casement also wrote, "Every fresh conversation I have with the man [Cazes] convinces me that to be a trader in Iquitos is a very perilous occupation for one's integrity or sense of right and wrong."[116] There was also the astounding fact that Zubiaur was now commanding Cazes's launch the *Beatriz*, after having been dismissed as captain of the *Liberal*, presumably for horning in on Macedo's and Pablo Zumaeta's personal profit from inflating the costs of the goods shipped from Iquitos to the Putumayo.[117]

As for another key player in Iquitos, "The Judge Paredes, editor and leader writer of El Oriente, is another of the scoundrels here. This paper is the organ of the Arana gang."[118]

As promised to Tizón, Casement did visit the prefect with Cazes along to serve as translator. Accordingly,

> He begged me again and again that there should be no publicity, that I would not write a report for publication, as that would be a "crushing weight" on the "guiltless" shoulders of Peru, that the Company for its "criminal negligence" deserved punishment, that the Indians should be protected in future and all possibility of recurrence of these things removed.[119]

The prefect even asked Casement if he could suppress some of the most damning testimony of the Barbadians in his report to the Foreign Office. No, Casement replied, but suggested that he might write two reports, one for internal use at the Foreign Office and the other more general in which he would avoid criticism of Peru. The prefect seemed relieved by that suggestion. Casement also believed that Cazes editorialized at times, while ostensibly simply translating.[120]

The missions of Casement and the PAC commission were not the only concerns in Iquitos. Pressure was coming from other quarters as well. Lima's *El Comercio* newspaper published a letter sent from Barcelona, Spain, by Señor Enrique Deschamps stating that there was investigation in progress in London into the "various horrible crimes" committed in the Putumayo. This caused "a great sensation" over the matter in the Peruvian

---

115. Ibid., 454.
116. Ibid., 457.
117. Ibid., 421, 453.
118. Ibid., 482.
119. Ibid., 458–59.
120. Ibid., 459–60.

capital.[121] Then, too, as early as 1909, the city's concerned university students and others had constituted the Asociación Pro-Indígena, committed to publicizing Peru's mistreatment of its Indians. Its main focus was upon the exploitation of Indians as miners in the Peruvian highlands with a system of indentured labor that was practically tantamount to enslavement. Joaquín Capelo,[122] co-founder and president of the Pro-Indígena and senator in the Peruvian parliament, gave an address (in October of 1911) to Lima's College of Engineers, in which he stated:

> The fact is that after the Republic's ninety years of existence the Indian of Peru has returned to being the domestic sheep whose wool, meat and blood shelter, enrich and feed the avarices of those who hold power over this race, and [who are without] conscience and with a spirit lacking the ideals of justice and freedom. Today, in the twentieth century, the Indian lives just as the independence campaign found him in 1821, living as a pariah deprived of all human rights, even that of his own person that anyone was free to enslave or imprison for indebtedness, and deprive of whatever he might possess, as if the Peruvian constitution had made of the Indian an exception [to the law]. Those who believe that the simple fact of their having been born into a distinguished social status, or of having achieved this condition by some other means, creates in their person a privileged right, an efficacious title, that confers upon them supremacy and command over other citizens less favored by destiny. These are those who believe that the fact that they find themselves in a political situation that makes it possible for them to abuse their strength and political power constitutes

---

121. Mitchell, *Sir Roger*, 270.
122. The British chargé d'affairs in Lima, Lucien Jerome, wrote to Casement on June 9, 1911:
> *Entre nous* the Sociedad Pro-Indígena is a farce. It is more or less a political society whose real object is to *embête le gouvernement!* (which, seeing, as how you and I are both paddies, sh'd appeal to us!). Capelo is an old rascal and has a bad reputation as one of the worst slave-drivers in the country. (Mitchell, *Sir Roger*, 463.)

This judgment would prove to be both harsh and inaccurate. In point of fact, Capelo was a sociologist and civil engineer, serving as well as a professor of sociology at the University of San Marcos. He had extensive experience in the Loreto, having overseen construction of the highway from the highlands to the Pichis River, thereby providing the Loreto with its first land link to the rest of Peru. In 1900, he served briefly as the Loreto's interim prefect, based in Iquitos. Capelo was the first minister of Development and Public Works when that ministry was created, in 1896, under President Pierola. He also occupied that post under President Oscar Benavides in 1914. Capelo was a proponent of economic development of the Oriente, its transportation infrastructure, and its military defense. He had considerable personal experience in the Oriente and extensive knowledge of the social and political climate in the Loreto, as well as the plight of its indigenes.

the right to do so; those who, in the final analysis, in whom the ability to impede that the free exercise by others of their rights, empowers them to do so and constitute themselves into the arbitrators and lords over the destiny of peoples and of each citizen. All of these men are backward beings in term of human evolution, they are in error and proceed badly. These men are not only against the public good, that they might not even perceive, but are also against anything contrary to their personal venture. They have strong-willed commitment to this egotistical goal that they pursue and believe can be realized at the expense of others.[123]

The purposes of the Pro-Indígena were to

1. Attract well-intentioned persons to the cause of defending the vulnerable Indians.

2. Foment laws protecting Indian rights, while founding a newspaper[124] to raise Indian self-consciousness and a staff to confront administrative and juridical abuses of Indian rights.

3. Educate the Indians to read, write and count so they could become aware of, and defend, their constitutional rights. They needed to learn Spanish to be effective. This was relatively simple to accomplish if only the national law requiring establishment of a school in every community with a minimum of 200 residents (systematically ignored in the case of the indigenes) was enforced. [The idea was to work outside the framework of the religious missions—viewed by then as collaborators in the exploitation of the Indians].[125]

The Pro-Indígena was keenly aware of the scandal in the Putumayo and, from the outset, entered into regular correspondence with London's Anti-Slavery Society. Lima's newspaper *La Prensa* editorialized against the abuse of the Indians "everywhere" in Peru, noting that at the time of the Independence (from Spain) there had been 2 million indigenes in the country and now there were half that many. It labeled the trend "permanent suicide that Peru is putting into practice through the slow extinction of its Indians" and the fact that "even the sacred undertaking of this new Society Pro-Indígena is a motive for disdain or anger."[126] A year later, *El*

---

123. *La Prensa*, October 14, 1911.
124. It would be called *El Deber Pro-Indígena* and the first issue appeared in October of 1912.
125. Ibid.
126. *La Prensa*, November 18, 1911. About this same time, or in the year 1910, Brazil charged national hero Colonel Cândido Rondon, with creating an Indian Protection Ser-

Comercio, reporting on the forced labor imposed on five hundred Indians of the Shanuzi drainage, referred to them as having been "Putumayoed" (*putumayizados*), with the added comment "the Putumayo exists here as well."[127]

In mid-summer 1910, Lima had dispatched a new judge to the Loreto, Carlos A. Valcárcel. In November, Judge Valcárcel was ordered by the Peruvian Supreme Court to conduct his own investigation in the Putumayo—although it was stalled when the acting prefect (an Arana man) failed to authorize the funding for him to do so.[128] The judge initiated proceedings against the local *vocales* (overseers of Loreto's judicial process) for dereliction of duty. He was also astounded to learn that, four years earlier, Saldaña Rocca had filed charges against Pablo Zumaeta over the alleged atrocities in the Putumayo and, not only had they never been acted upon, the very accusatory documents were actually in Zumaeta's possession.[129] Valcárcel now issued arrest warrants for several persons, including Zumaeta. Despite this, Casement had a low opinion of Valcárcel. On December 3, just before leaving Iquitos himself, Roger wrote in his diary, "Valcárcel is going down to Manaos on a launch, good riddance! Altho' I liked the rascal's bright face and Indian skin and splendid teeth."[130]

On September 22, 1910, Zumaeta prepared a demand in the amount of 888,934 pounds on behalf of the Peruvian Amazon Company, requesting compensation from Colombia for its actions in the Putumayo/Caquetá since 1901. On October 4, 1910, or while the Company's Commission and Casement were still in South America, Julio Egoaguirre filed the claim on the Peruvian Amazon Company's behalf in Lima with representatives of the Colombian government. The amount was more than the valuation of the Peruvian Amazon Company's total assets. When his directors found that to be absurd, Arana responded that it was only the

---

vice. Working from "attraction posts" in the heart of the jungle stocked with ornaments, it quickly overcame the suspicions of a dozen tribes. Rondon's order to his men was "Die if you must, but never kill" https://en.wikipedia.org/wiki/Fundaçao_Nacional_do_Índio. The phrase was adopted by John Hemming as the title of volume three of his encyclopedic work on Brazilian Indians. He begins the book with a comprehensive treatment of Rondon and the creation of the Indian Protection Service. John Hemming, *Die if You Must: Brazilian Indians in the Twentieth Century* (Oxford: Pan Macmillan, 2004), 1–23.
127. *El Comercio*, November 26, 1912.
128. Válcarcel would later lament the fact that, had the Supreme Court issued such an order when it should have, in 1907–1908, as many as ten thousand lives of indigenes might have been saved. Valcárcel, *El proceso*, 15.
129. Ibid., 16–17.
130. Bryant, *Roger Casement*, 135.

opening gambit in what would likely become a prolonged negotiation.[131]

Throughout the autumn of 1910, Loreto's prefect, Francisco Alayza y Paz Soldán, sent a series of damning reports to Lima.[132] He was opposed to a proposal of creating a new province in the drainage of the Yavarí River. It would simply create an expensive and redundant new bureaucracy without solving any problems. There was also a major population drain of *peones* attracted by other areas, some in neighboring countries, since the rubber resources of the Loreto had been overexploited and were in severe decline. Then, too, there was the common practice of enslaving Indians and selling them to foreigners. The prefect estimated the population loss to be forty to fifty thousand persons in a region already plagued with an acute labor shortage. *La Prensa* reported shortly thereafter:

> It happens that 3 or 4 thousand men, Peruvians by birth, have gone to the mouth of the Purús to establish a town that they call Nuevo Loreto, in Brazilian territory, because in their own country they find nothing but ill-treatment and oppression.[133]

It would seem likely that acculturated Indians skilled in rubber collection were in the ranks of the migrants. Then, too, we have the provocative account of British traveler W. O. Simon, who visited the Madre de Dios river drainage about this time and there was the guest of a *cauchero* named Torres,

> The bulk of Torres's workmen, and those who gave their name to the station ["Puerto Huitoto"] were Huitoto cannibals, whom he had purchased wild and cheap, on the Putumayo river, a thousand and more miles away. The Huitotos had been with Torres four years when I had arrived, and in that time had been taught to work satisfactorily. He had endeavored to make them subsititute alcohol drinking for their former man-eating orgies.[134]

The prefect's own attempts at fomenting immigration and the settlement of new colonists in the Loreto had failed for several reasons. The cost of living in Iquitos was about twice that in Lima. The town lacked an

---

131. British Parliament, *Report*, 359, 427.
132. Cited in Luis Alayza y Paz Soldán, *Mi país: Algo de la Amazonía Peruana* (Lima: Librería e Imprenta Gil, S. A., 1960), and elsewhere [*Mi país*]. Luis was brother of Loreto's prefect, Francisco, and published many of his sibling's comuniqués to the Peruvian government in his book.
133. *La Prensa*, December 3, 1912.
134. W. O. Simon, "Frontier Life in South America," *The Wide World Magazine* (April 1913), 92.

adequate school and had no hospital, despite the constant influx of persons with tropical diseases seeking medical assistance. Rubber gathering dominated the economy.[135] It created a floating population little given to the sedentary settlement required by other forms of agriculture. Communications were abysmal—there was no railroad, no proper roads, nor adequate telegraphic service.[136]

Finally, there was a security issue, given the inadequacy of the Peruvian military force in the region. The prefect complained that the number of troops sent by the central government was insufficient to meet the external threats, and the training of most of the men was inadequate. He warned that Iquitos itself might be in danger. He was greatly concerned over the incursions of Ecuador in the Upper Napo River and Colombia in the Upper Putumayo. Nor was Brazil to be trusted. He urged establishment of a serious military garrison and customs house in Leticia to countermand the influence of the latter. Furthermore, given all of the foregoing, Loretans had a very jaundiced view of the country's authorities and believed that the only viable measures were local initiatives.[137]

That same December, the Peruvian Amazon Company held its annual shareholders' meeting in London. Investor Morgan Williams again demanded to know what the Peruvian Amazon Company was doing about *Truth*'s allegations. He even made a motion to dissolve the Company (it failed to receive a second). He then asked what the Board of Directors intended to do about the scandal, and was told that an investigative commission had been dispatched and, of course,

---

135. Pineda Camacho, *Holocausto*, 184–85. The rubber industry was in the midst of its own profound crisis, particularly given the budding international competition from Southeast Asian plantations. According to Fuentes, even during his time in office (1904–1906), there were signs that the resource was disappearing through abusive harvesting practices. The reserves had been exhausted in most of the regions within easy reach of Iquitos, and the *caucheros* were forced to go ever farther into the remote forests. Nor were Peruvians sensitive to the need to both conserve wild rubber trees and establish plantations. He knew of only two of the latter (Fuentes, *Loreto* 1, 220–21). He called for a reform in the land laws that would allow *caucheros* to gain title to their holding. Without such a guarantee, the temptation was to maximize the short-term returns in order to liberate oneself of the indebtedness from one's *enganche* contract. This meant felling the trees for a one-time bonanza or bleeding them excessively (i.e., without proper resting for recovery) (Ibid., 223–24). Then, too, there was the problem with the Indians. By obliging them to meet quotas, but without supervising them directly, the forest Indians also felled trees. However, in their case they delighted in exterminating the resource, as they viewed it as the reason behind their enslavement by whites and believed that, once the rubber was gone, the whites would disappear as well (Fuentes, *Loreto* 2, 118).
136. Alayza y Paz Soldán, *Mi país*, 124–28.
137. Ibid., 128–29.

its report would be shared with the shareholders. Questioned by Williams whether the Peruvian Amazon Company actually had freehold land titles in the Putumayo, a flummoxed Chairman Gubbins had to concede that, in such places, there simply were none—only squatter's rights. He also promised greater transparency in the future. Williams then made a motion to remove Arana from the Board, one that he agreed to withdraw only after it was demonstrated to be at cross purposes with a technicality in the Company's by-laws.[138]

In January of 1911, Valcárcel was forced to leave Iquitos for several months of treatment in New York for beriberi. While Senator Julio Egoaguirre had managed to delay constitution of the Peruvian government's investigative team, he was ultimately unsuccessful in heading it off altogether. Nevertheless, it seems that his tactics provided Zumaeta with the opportunity to rush to the Putumayo in advance of the investigators to make certain that the most egregious suspects could make good their escape.[139] Many did so by fleeing to Brazil, some accompanied by Indian slaves to be sold there and to prevent them from testifying about the destruction of their settlements.[140] Rómulo Paredes, an acting Iquitos judge and editor of the *El Oriente* newspaper as well, was appointed to head up Peru's own commission that would conduct an on-site investigation in the Putumayo. Given that Arana had once owned *El Oriente*, one can only assume that he preferred that Paredes, rather than Valcárcel, head the investigation (if there must be one). In the event, he was mistaken.

By now, Casement was secretly back in London ahead of the Peruvian Amazon Company's own commission, and hard at work on a report that substantiated Hardenburg's charges and even elaborated upon them. Meanwhile, in the Putumayo, Macedo resigned his post and fled. Tizón replaced him at El Encanto and informed London, "As you are already aware, many changes in the chiefs of section have taken place."[141] This was not without cost to the Company, however, since seven employees now on the lam had drawn down a total of somewhere between 10,000 and 20,000 pounds owing to them.

On January 8, 1911, Casement wrote to Louis Mallet, Edward Grey's private secretary at the Foreign Office:

---

138. British Parliament, *Report*, 294.
139. Valcárcel, *El proceso*, 19, 272.
140. Mr. Jerome to Sir Edward Grey. Lima. April 27, 1911. Miscellaneous, No. 8. *Correspondence Respecting the Treatment of British Colonial Subjects and Native Indians Employed in the Collection of Rubber in the Putumayo District* (London: Her Majesty's Stationery Office, 1912), 143 [Miscellaneous].
141. British Parliament, *Report*, 246.

The Prefect is honest but weak. All round him at Iquitos are rogues and villains. The truth about the abominable crimes on the Putumayo have been known to scores and scores of people in Iquitos, including the actual Prefect's predecessor Sr. Zapata, who was unquestionably bribed and in the secret pay of Arana. The present Prefect, Sr. Alayza Paz Soldan, promised me to clean out the whole pig stye. He is terribly afraid of my report being published—our chief weapon for good is that I have got the evidence of so many men—given under such circumstances too—and that the Peruvian govt. know this. I may add that Sr. Pablo Zumaeta, the director of the company in Iquitos (J.C. Arana's brother-in-law) called on me in Iquitos on my return with all sorts of heartfelt regrets and promises. He said that he had learned from the Prefect of the deplorable state of things I had found to exist and wished to assure me that he and J.C. Arana were entirely innocent and that if I would give him a list of those agents of the Company on the Putumayo whom I believed to have committed crimes he would have everyone instantly dismissed and handed over to justice. This, I told him, I could not do—but I said that the Company's representative in Putumayo, Sr. Tizon, and the members of the Coy's commission were all cognizant of the facts within my knowledge and the names and crimes of the agents implicated and that he could get any information from that quarter. I said that the information in my possession was the property of H.M. Govt. who would doubtless convey it to the proper quarter. Zumaeta, I believe, left Iquitos the day of my departure, 7 Dec., for the Putumayo. I wrote fully to Mr. Barnes, the leading member of the Commission, giving him all the facts I thought might be of further service to him and the Commission in view of the approaching visit of the judge [Paredes]—supposing the judge wants to be convinced. Of this latter point I heard conflicting views. Mr. Cazes our Consul (who was very helpful to me throughout) said he heard this judge well spoken of. The Prefect himself told me the judge was a new comer and had not been long enough at Iquitos to get corrupted or locally tainted—on the other hand I heard privately from a Peruvian naval officer (in strict confidence) that the Judge was open to bribes and Cazes himself had very little confidence as to the genuine result of his journey. He thought that, for form's sake, and to keep us quiet, they would arrest and imprison (for short periods) some of the worst men—such as Montt, Aguero, Normand and Fonseca to name a few. I told the Prefect that every one of them should be hanged—and I begged him to go himself to the Putumayo. Tizon (who was sub-prefect formerly) after he saw that he could not hide things and that I was bent on getting to the bottom and would find out, threw up the sponge and completely joined us—and when I left Chorrera on 16 November he begged me to see the Prefect, to tell him everything and to do all I could to smash the Zumaeta gang in Iquitos. Zumaeta is a rascal, but not a stupid one. They have a far worse blackguard in their pay there, named J.C. Dublé—a clever and dangerous rogue. I fear he accompanied Zumaeta to Putumayo on 7 Dec. The Commission know his character and it is obvious (to my mind) that the visit of these two men to the Putumayo a week or more before the judge started was most improper and should not have been allowed. While pretending that they were going to assist justice the real object of their journey was to see how much they could plan with Macedo (& Loayza probably) to defeat the end of justice, and to forestall the investigation of the examining judge. Tizon is a weak man too, but he became completely convinced, and, as I told you, promised me most emphatically and again and again that he should do his duty—since his duty (as he would interpret it) is to save the "honour" of Peru to prevent any public exposure—and he hoped, so he told me, that Aguero, Montt and Co would bolt down the Japurá to Brazil before the judge arrived. He told me that would be the 'best solution'—the company would be rid of these scoundrels & 'scandal' would be avoided.[142]

Casement noted that just the day before he had received a letter from Arana, posted erroneously to his cousin's address in Ireland and then forwarded to him in London (Julio César was obviously being kept well informed of Casement's movements). It was in Spanish and Casement provided Mallet with the following translation:

> I've heard of your return and should very much like you to tell me when you will be in London in order that I may visit you and exchange views on the reforms to be effected in the Putumayo, and if possible to have an idea of the impressions you have derived from your recent visit to that region, and also to obtain any suggestions you may have to offer upon the better development of the Company's affairs.[143]

Casement commented, "It is rather cheeky of this gentleman to write to me. I will not acknowledge his letter until I have seen you or heard

---

142. Mitchell, *Sir Roger*, 20–21.
143. Ibid., 23.

from you."¹⁴⁴ Mallet was against the meeting, but advised Roger that, if he proceeded with it he must take someone along, in case the Foreign Office at some future date needed a witness to what had transpired. Casement then received a second letter from Arana requesting the meeting and informing the consul that the Peruvian Amazon Company's commission was still in the field, and that his Iquitos general manager [Zumaeta] had left for the Putumayo to effect reforms. On January 10, Casement replied to Arana that he was in receipt of the two letters, but that "As my recent journey was undertaken solely by discretion of His Majesty's government I regret that I am not at liberty to comply with your request or discuss with you any matter connected with it."¹⁴⁵

Casement was also in continual contact with Bertie, the original chairman of the PAC investigative commission who left the mission for health reasons, but who obviously remained in the mix. On January 17, Bertie wrote Roger a letter disclosing that he and Chairman Barnes had met with PAC Board member Lister-Kaye and briefed him on the forthcoming report that was in substantial agreement with Casement's assessment. According to Bertie,

> I pressed the imperative necessity of the British members of the Board seeing the matter through, coûte que coûte, & insisting, hereafter, on ousting the common enemy & cleansing the Augean stables by a clean sweep. . . . Mr. Read alone shd. be taken into confidence—as being, in my judgment, the brain and motive power of the Board—but the utmost secrecy must be maintained 'till all is ripe for action. At the opportune moment, the bomb shd. be thrown which ought to blow the enemy out of his seat & both countries. Sir John pledged himself to do his utmost to secure the adhesion of his colleague to the proposed plan of campaign, & to sit tight himself.¹⁴⁶

On January 31, Casement informed Grey that the Barbadians had been recruited in 1904 and 1905 by Abel Abarco and signed to two-year contracts. Some were forced to take part in the atrocities, despite their reluctance. While there had been thirty thousand Huitotos at one time, there were no longer more than ten thousand left. Many had died from smallpox and other illnesses, but the majority of deaths were due to

---

144. Ibid.
145. Ibid., 28–29.
146. Bertie to Casement, January 17, 1911. Bodleian Library MSS Brit. Emp. S-22 G344a.

maltreatment.¹⁴⁷ Roger named the worst offenders among Arana's overseers—including (Basque-surnamed) Elías Martinengui.

On March 17, Casement sent his completed report to Sir Edward Grey. He invoked two prime examples of abusers:

> Of Martinengui, the worst things were alleged to me by those who had served under him. During his term of service at Atenas he had wasted that region, and so oppressed Indians that they were reduced to a condition of wholesale starvation. . . ; owing to the strain put upon them by Martinengui, the Atenas Indians had been unable to cultivate their own clearings, women as well as men being compelled to work rubber.¹⁴⁸

Then there was Aurelio Rodríguez at La Sabana and Santa Catalina, where,

> Wholesale murder and torture endured . . . and the wonder is that any Indians were left in the district at all to continue the tale of rubber working on to 1910. This aspect of such criminality is pointed to by those who . . . assert that no man will deliberately kill the goose that lays the golden eggs. Their argument would have force if applied to a settled country or an estate it was designed to profitably develop. None of the freebooters of the Putumayo had any such limitations. . . , or cares for the hereafter to restrain him. His first object was to get rubber, and the Indians would always last his time. He hunted, killed, and tortured today in order to terrify fresh victims for to-morrow. Just as the appetite comes in eating, so each crime led on to fresh crimes, and many of the worst men of the Putumayo fell to comparing their battles and boasting of the numbers they had killed. Everyone of these criminals kept a large staff of unfortunate Indian women for immoral purposes—termed by a euphemism their "wives". Even "peons" had sometimes more than one Indian wife. The gratification of this appetite to excess went hand in hand with the murderous instinct which led these men to torture and kill the very parents and kinsmen of those they cohabited with.¹⁴⁹

Casement observed that young Indian *muchachos*, or trusted boys, were recruited, trained, and armed as enforcers and then assigned to plac-

---

147. Consul-General Casement to Sir Edward Grey. London. January 31, 1911. Miscellaneous, 4–12.
148. Consul-General Casement to Sir Edward Grey. London, March 17, 1911. Miscellaneous, 43.
149. Ibid., 44.

es where they had no relatives or fellow tribesmen. He then asked the rhetorical question of how such a small band of oppressors could intimidate and control a large number of indigenous men who were not individual cowards. His answer was rather complex. He noted that the whites exploited existing divisions and enmities among the tribes. They splintered members of the same tribe into separate settlements. The whites also possessed superior weapons and then systematically stripped Indians of theirs. He further commented,

> Perhaps a greater defense than their spears and blow-pipes even had been more ruthlessly destroyed. Their old people, both women and men, respected for character and ability to wisely advise, had been marked from the first as dangerous, and in the early stages of the occupation were done to death. Their crime had been the giving of "bad advice." To warn the more credulous or less experienced against the white enslaver and to exhort the Indians to flee or to resist rather than consent to work rubber for the newcomers had brought about their doom. I met no old Indian man or woman, and few had got beyond middle age. The Barbados men assured me that when they first came to the region in the beginning of 1905 old people were still to be found, vigorous and highly respected, but these had all disappeared, so far as I could gather, before my coming.[150]

Paredes arrived at La Chorrera on March 27, 1911, to begin a more than three-month investigation:

> Paredes carried confidential but somewhat ambiguous instructions from the prefect. He was to follow the letter of the law and fulfill his duties as a judge; to observe the behavior of the Indians and their exploiters and to report secretly the same to the government and to punish the guilty and prosecute the murderers with the utmost tenacity. However, he was "not to damage the Arana company or interfere with our garrisons which are performing a genuinely patriotic duty in defending those faraway frontiers of our territory." Accompanied by a scribe, a medical doctor, two interpreters, and a small guard, Paredes set out to gather hard criminal evidence.[151]

Arana clearly had some notion of the disastrous content of Casement's report, since, on March 22, 1911, he urged his Board to write to

---

150. Ibid., 45.
151. Stanfield, *Red Rubber*, 152.

the Foreign Office to enter into discussions of the consul's findings.[152] The Peruvian Amazon Company's commission was still writing its own report. While there was no outright collusion with Casement, the commissioners had worked closely with him in the Putumayo and therefore could have given Arana a preview of what was likely to come in both reports. Then, on March 30, 1911, Grey sent Casement's document to the Peruvian government, as well as to his counterpart in the United States. As yet, it would remain unpublished in order to give the behind-the-scene pressure on Lima a chance to work.

Casement remained skeptical of Paredes. On April 27, he wrote about his concerns to Grey. There was the possible problem posed by Aurelio Lopez, a clerk of the Iquitos judicial court, who was accompanying Paredes. The implication was that he might influence adversely the investigation. Casement included an *El Oriente* editorial from November 15, 1910, that seemed to have been written by Paredes himself. It was a "disreputable" attack on the colonizing scheme of a Franco-Dutch company (to be serviced by steamship from Iquitos) on the Pastanzas River claimed by Ecuador. It was obviously opposed by certain local vested interests, and the local port captain was said to have declared that any Iquitos business assisting the foreigners would be burned down. The *El Oriente* editorial itself suggested that colluders in Iquitos would likely be lynched. The initiative quickly stalled. Then there was the false accusation, in another editorial in the newspaper, to the effect that Booth Company ships were deliberately creating wakes in the river to capsize local boats for malicious amusement.

It was also to be remembered that *El Oriente* was a strong defender of Arana in its treatment of the 1907–1908 violence between Peruvians and Colombians in the Putumayo. Indeed, Paredes was said to have close ties to many of the Arana supervisors now under investigation. Casement was unsure whether *El Oriente* was owned by Paredes at the time of the *La Felpa* and *La Sanción* campaigns against Arana, but the newspaper was clearly Saldaña Rocca's most outspoken foe. In short, Casement believed that Paredes was a strong anti-foreigner Peruvian nationalist, and any investigation by him of Arana would be suspect.[153] Acting British Consul Cazes shared Casement's skepticism of Paredes.[154]

Meanwhile, Peru's requests to Brazil for extradition of the accused had been denied for lack of such treaty arrangements between the two

---

152. British Parliament, *Report*, 248.
153. Mitchell, *Sir Roger*, 270–76.
154. Ibid., 292.

countries.¹⁵⁵ Lucien Jerome reported that he had been assured by the president himself that Augusto Leguía would take a personal interest in resolution of the scandal. The British chargé d'affairs had requested that Peru pass additional legislation banning slavery, and Grey forwarded that suggestion to the Peruvian government. He also recommended that Lima should institute and fund Christian missionaries in the Putumayo as an act of good faith.¹⁵⁶

On April 27, 1911, Valcárcel returned to Iquitos. When a new judge arrived on May 17 to replace acting Judge Paredes on the Iquitos bench, the absent Paredes was relieved of his assignment in the Putumayo as well by Peru's Supreme Court (a removal likely orchestrated by Senator Egoaguirre). According to Valcárcel, even while in the Peruvian parliament, Egoaguirre continued to work as Julio César's personal attorney and remained on the Company's payroll.¹⁵⁷ One can only assume that word of Paredes's in-depth probe had reached Iquitos, causing consternation in the Arana circle. Paredes later noted that if the arrival of Casement had frightened many of the culpable managers of the Peruvian Amazon Company in the Putumayo, just the rumor of his coming provoked veritable panic that caused many of the criminals to flee. There was the case of Abelardo Agüero and Augusto Jiménez who were escaping on the *Liberal* when a vessel approached and someone said that the judge was on board it (he wasn't). The two fugitives insisted that the crew hide them below deck.¹⁵⁸

In any event, the ostensible retraction of the judge's credentials posed the concern for Valcárcel that, upon his return to Iquitos, Paredes's findings would be ignored. So Valcárcel again asked the prefect for funding to undertake his own investigation in the Putumayo. When it was not forthcoming, the judge offered to pay most of his own expenses if only he could be accorded transportation on a government vessel. The prefect guaranteed him space on the next boat leaving Iquitos. Two would depart without the judge ever being informed.¹⁵⁹

On May 13, the British Foreign Office sent a copy of Casement's devastating 135-page report to each director of the Peruvian Amazon

---

155. Mr. Jerome to Sir Edward Grey. Lima. May 4, 1911. Miscellaneous, 145. It seems that a year later Brazil promised to dispatch Colonel Rondon to Northwest Amazonia to search for the wanted Putumayo criminals, but nothing ever came of it. Mitchell, *The Amazon Journal*, 71.
156. Mr. Jerome to Sir Edward Grey. Lima. April 27, 1911. Miscellaneous, 143.
157. Valcárcel, *El proceso*, 274.
158. Chirif, "Los informes," 135.
159. Valcárcel, *El proceso*, 18.

Company. The Commission had had little difficulty in determining that the forests of the Putumayo were a "battlefield of bones," and that, within the previous five years, the indigenous population had declined precipitously.[160] May 25, the Peruvian Amazon Company Board of Directors wrote the Foreign Office requesting a meeting with Casement to discuss his findings and possible reforms.[161] On May 31, the undersecretary of state for foreign affairs told the House of Commons that Consul-General Casement had confirmed fully the accuracy of the reports of Putumayo atrocities.

Meanwhile, on May 16, 1911, the Peruvian Amazon Company's own Commission of Enquiry submitted its report to the chairman and directors. Regarding the Indians it opined:

> In the Commissioners' opinion the natives are far from being the blood-thirsty and treacherous savages they are often said to be. Cannibals they undoubtedly were, and in outlying parts still may be—that, however, appeared from enquiries in various parts to be more a religious or magical rite, the eating of a slain enemy, than the expression of a gross ferocity—and they would certainly have been less than human if they had not occasionally broken out in resentment of the treatment they received. To the Commissioners they appeared to be a simple people of naturally friendly disposition, whose confidence and affection it would not be difficult to gain. Much bad work remains to be undone, and it will take time to eradicate the memories of past ill-treatment and the distrust engendered by it, but it will be time well spent.[162]

Concerning administration of the enterprise:

> Misunderstanding and ill-feeling between the local managers and the Iquitos office appear to have been unceasing, and the managers both in Putumayo and at Iquitos were at best criminally ignorant of the state of things obtaining in the huge territory that was subject to their administration. There can be no doubt that the great distance and lack of regular communication between Iquitos and the Putumayo rendered it extremely difficult for the General Manager [Pablo Zumaeta] to keep in touch with his subordinates in an effective manner; still, the Commissioners are of opinion that it could have been done, and still more strongly that it ought to have been done, especially by personal visits to the Pu-

---

160. Collier, *The River*, 214.
161. British Parliament, *Report*, 247.
162. Ibid., 610.

tumayo, not merely to Chorrera and Encanto, but to the outlying sections.[163]

Zumaeta had only actually visited the Putumayo once since formation of the Company in 1907 and the visit of the Peruvian Amazon Company's commission in mid-1910. Furthermore, the section chiefs were not only "practically free from all effective control"; they were totally ignorant of rubber and, "with a few exceptions, unfitted for the work they should have done, the large majority being men of practically no education and absolutely no character." The only demand placed upon them by headquarters was to produce as much rubber as possible, "and the methods of collection were left to themselves." Consequently,

> The Indians came to be regarded by the Chief as rubber-producers existing solely for his benefit; disinclination to work rubber became a heinous offence and was not infrequently treated accordingly. The Chief's object was only too often simply to get the greatest quantity of rubber in the shortest possible time. The ultimate welfare of the Indians did not interest him, and such important matters as the planting of "chácaras" [sic—*chacras* or the indigene's food gardens] and conservation of rubber trees were allowed to go by the board. The Putumayo early acquired an unenviable reputation as an asylum for undesirables, a class from which, unfortunately, too many of the Company's employees were recruited. Labour supervision as carried out by such men under the descriptions described above and in the absence of any governmental control rapidly became indistinguishable from slave-driving. . . . Unfortunately, it has frequently been the custom to chastise the Indians with whips, sometimes unmercifully, when they did not deliver what the Chief of Section considered a sufficient quantity of rubber. This has been the cause of many Indians running away rather than face the ordeal. When this happens, an armed force is sent out to hunt them down, and they are more or less severely punished when caught.[164]

Additionally, men, women, and children were forced to transport "excessive" loads of rubber, the diet in general was inadequate and made the Indians more susceptible to disease and medical care was practically non-existent. The Indians were considered to possess none of the ordinary rights of humanity, particularly the women, who were simply assigned to employees of the Company. In sum, "The general conditions that have ob-

---

163. Ibid., 608.
164. Ibid., 609.

tained in the Company's territory can only be characterized as disgraceful." As to "the allegations against the Company's agents published in "Truth" and other papers, the Commission are [sic] convinced that they are substantially correct."[165] Indeed, the commissioners learned of hitherto unreported atrocities and underscored that the evident marks they observed on the "backs of men, women, and children" could only have been made with a hot iron. Even if from time to time orders were issued from Iquitos to treat the Indians differently, there was no follow up. As to culpability,

> The Managers, whether in the Putumayo or at Iquitos, cannot shift from themselves the responsibility for this state of things; if they were unaware of what was going on, their ignorance admits of no excuse and can be attributed only to the fact that no proper visits of inspection were made; if they were acquainted with the facts, no condemnation is too strong for them.[166]

Such systematic abuse was only possible in the absence of government control in the Putumayo. While there were two Peruvian officials present there during the Commission's visit, they simply were not doing their job. On a more positive note, the Company had placed the whole Putumayo under the administration of a sympathetic and humanitarian general manager at La Chorrera [Juan Tizón], and he now was fully empowered to clean up the mess.

The commissioners underscored mismanagement of the Company's funds. The system of paying commissions to section chiefs led to "resultant evils." Not counting commissions, salaries alone totaled more than 10,000 pounds per annum—including the Iquitos manager's pay of 2,500 pounds. The broad authority accorded the general manager in Iquitos was excessive—particularly as regards to the payment of large sums of money. They recommended that there should be a general review.[167]

The report concluded with the following:

> Mr. Rey de Castro is, the Commissioners understand, the Peruvian Consul-General at Manaos. The Commissioners have been unable to learn that Mr. Rey de Castro has been of service to the Company. The Company on the other hand appear [sic] to be acting as his Bankers [sic], inasmuch as they [sic] are paying his house, and other expenses, and receiving in return small repayments; at June 30th, 1910, Mr. Rey de Castro stood in the books

---

165. Ibid., 610.
166. Ibid.
167. Ibid., 611.

as a debtor for about £4,600. The Company does not appear to derive any benefit from this account, which is outside of legitimate trading business, and is running the risk of a bad debt.[168]

On May 31, the Peruvian Amazon Company Board forwarded its Commission's report to the British Foreign Office, along with a statement that the worst offenders in the Putumayo had been removed. They requested that Casement attend their next meeting to discuss further reforms with them.[169]

The Peruvian Amazon Company's directors were soon to be accorded another shock. In early May, Zumaeta, acting on behalf of his sister and with her power of attorney, had filed a lien against the Company's assets in Iquitos. It seems that Eleonora had been accorded ownership valued at 40,000 pounds (and bearing a guaranteed 6 percent annual return)[170] in Arana y Hermanos when it was incorporated in 1903—an obligation that was transferred to the Peruvian Amazon Company upon its incorporation. Her collateral included the Peruvian Amazon Company's extensive assets and real estate holdings in Iquitos, as well as the Company's two main launches—the *Callao* and the *Liberal*.[171] Given the compound interest upon her investment, she was now owed 60,000 pounds. While Julio César disclaimed any knowledge of Zumaeta's actions and contended that Pablo was acting independently on his sister's behalf, the effect was to ensure that the Peruvian Amazon Company was now defunct—unable to either raise fresh capital or to meet its obligations out of existing funds. Indeed, a subsequent dispatch from Zumaeta to Arana tipped their hand. In it, Zumaeta approved Commissioner Fox's suggestion that the Peruvian Amazon Company might plant two million rubber trees in the Putumayo, since that could serve as the basis for founding a new company once the old one was totally back under their control.[172]

Casement was far from convinced that the Foreign Office would act decisively, whether on its own or in concert with the Americans. On June 6, he wrote to one of his humanitarian benefactors, William Cadbury, expressing his disappointment with the United States. Grey had tried to enlist the U.S. Department of State earlier "but they were no good then"

---

168. Ibid., 611.
169. Ibid., 248.
170. Mitchell, *Sir Roger*, 476–77. It seems that this represented half of the slightly less than 80,000 pounds' worth of Julio César's holdings in the business in 1903. He left the balance in it as working capital, much of it in the name of his three partners—Lizardo Arana, Pablo Zumaeta, and Abel Alarco.
171. Ibid., 293.
172. Ibid., 455.

and were unlikely to be so in future"especially when it is a question of 'injuns.'"¹⁷³ Nor did he think Peru would heed criticism from a country in which Negroes were burned alive and lynched as a matter of course. Casement further opined:

> I think the Monroe Doctrine is at the root of these horrors on the Amazon—it excludes Europe (the mother of Western civilization with 500,000,000 white people against the U.S.A. with less than 80,000,000 whites) from her proper correcting and educating place in the whole of S. America. . . . Northern Europe—especially Teutonic Europe—might redress the wrong—even today—but all effective European influence is shut out by the Monroe Doctrine—or its modern interpretation. That interpretation in U.S.A. statesmanship is that some day all these misgoverned Iberian "republics" must come to be incorporated. I have talked it over with U.S.A. diplomats and ambassadors & I find the underlying belief always the same—viz. that Uncle Sam is to inherit the New World. I don't agree—and I don't think that belief or its fruition good for mankind. Europe is a bigger thing than Uncle Sam—& must remain so.¹⁷⁴

Casement went on to note that South America suffered from the barbarism imposed by centuries of Spanish and Portuguese rule:

> It is true the native life of S. America has been destroyed—practically. There are no longer any Indians to speak of on the Amazon—and the region where I have been, the Putumayo, is the last stronghold (or was) of the primitive native race. Inside of twenty years of the vegetable filibusters rule of Iberian civilization that large reservoir of human life has been almost drained dry. There were at least 50,000 Indians in J.C. Arana's pirated territory 15 years ago, and now there are certainly not 20,000—and so with other neighbouring districts. All those Iberian republics of northern S. America are not only pirate states—but they are slave states. Slavery is rampant—and with the recent extension of the rubber trade has taken on a large extension in the Upper Amazon valley. The whole history of slavery needs, as you say to be rewritten. It is more deeply written in vice than we conceived. It can evade the law—and flourish in defiance of law. Legalized slavery is far better than the lawless slavery of Peru, Colombia, Bolivia, etc. etc. and doubtless of Mexico and Central America. What law established,

---

173. Ibid., 360–61.
174. Ibid., 361.

law can assail, law can modify. But this lawless, ruthless enslavement of millions of our fellow men, founded on a lie & on denial of its own existence is harder to uproot. You can refuse to receive a slaver at court—but who rejects a minister of Peru or Mexico etc.?[175]

In a June 9 letter to Casement, Lucien Jerome (in Lima) wrote:

As for the Peruvian government, you probably will have read between the lines of my despatches and seen that I have never had much confidence in the sincerity of the officials. I have to pretend to have, but I can tell you I have none. The Aranas have bought up the government is only too plain. They have been stuffing me with the Franco-Dutch Company being a military expedition of the Ecuadoreans and with yarns as to possible war with Colombia. I have to make believe that I take it all seriously. The President of this Republic [Leguía] is one of the worst Presidents South America has ever had and I have reason to believe that he is himself interested in the Arana undertaking.[176]

On June 17, Casement wrote to Gerald Sydney Spicer, his superior and confidante at the Foreign Office:

The crux of the situation is 1st finance and 2nd the Reduction of Arana and his pals out there—the Zumaetas to impotence. Here your official help may always be needed with the Peruvian Govt.—for if Arana resists reform or cannot be reduced, then I think he should be indicted. I can give you, if that should ever be needed, quite enough prima facie evidence to show his knowledge of what took place which would justify your suggesting that he might be put on trial. Let us hope it won't come to that and that the [PAC] Board will be able to terrorize him into complete acquiescence, or better still impotence.[177]

Obviously, Casement harbored deep reservations that the "system"—defined broadly as the British, American, and Peruvian governments—was capable of rectifying, or even disposed to rectify, the situation in the Putumayo. Nor did he doubt the guilt of the Aranas and Zumaetas, let alone that of their henchmen.

It was now that Consul Casement began communicating with Chairman Gubbins in private and was invited to attend future sessions of the

---

175. Ibid., 362.
176. Ibid., 463–64.
177. Ibid., 384.

Board of the Peruvian Amazon Company. After obtaining the necessary approval of his superiors, Roger appeared at three of their meetings:

> The intention was that the directors should devise reforms to end this intolerable state of things in Putumayo and set up a decent commercial administration; but I very soon realized by my attendance at these meetings that the directors were much more preoccupied with their financial position. Their view was that unless the company could be saved—it was threatened with dissolution through lack of funds—there was no possibility of the company conducting any reforms; so that the situation really resolved itself during these meetings into discussing means of saving the company, keeping it in effective existence, and making it a really decent and honourable British company that should have effective control on the spot.[178]

Casement brought a stenographer with him to the first meeting, as previously advised by Mallet. In the event, the British consul was underwhelmed by the directors, who were "all at sea" and without any concrete plans for reform. Casement concluded, "They are all in the hands of Julio C. Arana and his informal gang of robber cousins and brothers-in-law out there."[179] Arana was back in England, but did not attend this first meeting. Casement read the Board a letter that he had just received from Arana in which he claimed to be astonished that "there existed a state of affairs which one could hardly expect of mankind." He thanked Casement for all he was doing to protect the natives. Arana claimed to have taken the necessary reform measures and excused himself for not having exerted better oversight, attributing the failure to the Company's financial difficulties that had precluded him from making more "visits of inspection" to the Putumayo.[180]

Casement also faced a new personal dilemma. King George V had just ascended the English throne after his father's death, and the coronation's honors' list included a knighthood for Casement in recognition of his great humanitarian work. As with his earlier honor, Casement was ambivalent about the award and considered refusing it (in particular out of concern for what his fellow Irish nationalists might think). In a letter to Gerald Spicer, he stated, "Thank you for your congratulations and good wishes, altho' if you ever attempt to 'Sir Roger' me again I'll enter into an alliance with the Aranas and Pablo Zumaeta to cut you off someday in

---

178. British Parliament, *Report*, 17.
179. Goodman, *The Devil*, 148.
180. Ibid., 148–49.

the woods of St. James' Park, and convert you into a rubber worker to our joint profit."[181]

Arana was absent from the second Board meeting as well, given that the other members believed his presence might be offensive to Casement. This prompted the British consul to insist on Julio César's presence at future ones—after all he *was* the key component within the Peruvian Amazon Company. Earlier on, Julio César had offered to forego most of his director's salary for the duration of the crisis. Casement urged that Zumaeta be replaced as manager of the Iquitos office and that Seymour Bell, former member of the Company's investigative commission, be sent to take his place. Bell was willing to go. Since there was no money to pay him, the possibility of each director contributing 200 pounds was discussed.[182]

Casement believed that the former abuses would resume if the Company reverted to Peruvian control. He therefore undertook a private initiative to try and find new British capital willing to invest in the Peruvian Amazon Company, but it had debts in the amount of 247,000 pounds. Casement tried to raise funds through his connections with wealthy philanthropists like chocolate-magnate William Cadbury,[183] Charles Booth of the Booth Steamship Company and Andrew Carnegie. However, Asian plantation rubber had clearly eclipsed the collection of the wild South American product in the world market. The prospects of the latter seemed dismal.[184] According to Casement, in July of 1911,[185]

> As the company, through their chairman, Mr. Gubbins assured me that they could not get the money necessary to ensure their own future, and as I had quite failed to get any money for them through private efforts I had made, entirely on my own responsibility, it seemed to me useless to continue discussing academical reforms which these gentlemen were not in a position to undertake, and I therefore, wrote, with the sanctions of the Secretary of State, to

---

181. Quoted in Ibid., 149–50.
182. British Parliament, *Report*, 221–22.
183. In a July 11, 1911, letter to Cadbury, Casement noted that Arana had written him offering to sell his 608,493 ordinary shares in the PAC and his 50,000 Preference ones. Sir Roger believed that only a Carnegie could resolve the situation through such a purchase. Lacking this,

> The fear is that Arana and Co. will sail for the Putumayo and re-establish their pirate camps in the forest and complete the destruction of the Indians within the next six years. I know the Iquitos head of the firm—Pablo Zumaeta—contemplates this and is even suggesting it to Julio Arana here in London. How to save the remnant of the Indian only providence can show. Mitchell, *Sir Roger*, 461.

184. Goodman, *The Devil*, 152–53.
185. British Parliament, *Report*, 17, 33, 108–109, 114, 121.

the company, discontinuing my attendance at those meetings, and at the same time I wrote to the chairman, Mr. Gubbins, urging upon him in his own interest, the absolute importance I attached to Mr. Bell being sent out without delay.[186]

The pressure was building. A July 1, 1911, article appeared in *The Nation* with the title "The Curse of Rubber." It concluded that a tenth of Britain's income came from foreign investments, a significant portion of which was in tropical agriculture. It speculated that the Peruvian Amazon Company's main mistake was to have been exposed (there were likely many similar extant enterprises functioning within the empire). The newspaper asked, "Who knows, and who cares? Certainly, finance cares little." The real problem was,

> We sometimes think of slavery as a thing of the past. We pride ourselves on our country's emancipation a century ago, or we think the atrocious system died on the plains of Gettysburg. It is not true. The problem of slavery is still before us. Of all the great problems in the world, there is none more urgent. Speaking last Saturday at a Welsh chapel in London, Mr Lloyd George said that "if the Christian Church were destroyed, the country would be turned into a burnt up wilderness." Well, we have seen a heathen land converted into what has rightly been called a Devil's Paradise under a nominally Christian government, Christian directors and Christian agents. And, we ask, what feeling but execration have those tormented Indians for the name of Christianity, or with what thought but terror does the idea of the white man's civilization inspire them in their anguish?[187]

Then, in mid-July, Casement received information that the British chargé d'affairs in Lima believed that the Vatican was prepared to increase Catholic missionary work in the Putumayo—but there remained the critical question regarding its expenses. The Catholic Church was strapped, as was the Peruvian government. There was the suggestion that the Foreign Office should lean on Peru to come up with the financing as clear evidence of its commitment to reform.[188] Any mission would have to be Catholic, as the Peruvian constitution ruled out Protestant ones. No matter, since,

> The real point is that any mission is better than none—that all Missionaries are better than J. C. Arana and Pablo Zumaeta—and

---

186. Ibid., 109.
187. Mitchell, *Sir Roger*, 435.
188. Ibid., 486–87.

that no London Company and only Pablo Zumaeta and such birds of prey in the Putumayo (which will be the case in future I fear) the presence of any good man of any denomination or Church in Christendom is likely to be the surest check on criminality we can introduce."[189]

At this point, the U.S. government informed the British that it had instructed its ambassador in Lima to side with Jerome in insisting that the Peruvian government become pro-active in reforming the abuses.[190]

Roger attended his last Peruvian Amazon Company Board meeting on July 17, and two days later the Company filed for bankruptcy. The Board designated Julio César Arana as its liquidator, and Zumaeta remained at his post.[191] Casement agreed to continue his search for private funds to bail out the Peruvian Amazon Company, but he made it clear to Gubbins and the Board that it was his private initiative and not that of the Foreign Office. He shared with the Board the concerns of his potential backers. They were disturbed by the Peruvian Amazon Company's lack of clear title to the Putumayo lands and would likely insist that the Peruvian government issue one. They also believed that the Company's creditors should be prepared to take a haircut of some magnitude (to be determined) on their outstanding debt. It was only then that Sir Roger learned that the Company owed debts in South America amounting in the hundreds of thousands of pounds. Finally, he could not assure the directors that any restructuring of the Company would necessarily head off publication of his reports. That decision was yet to be made by the Foreign Office and was outside of his control.[192]

On July 19, Casement wrote to Spicer reporting on his meetings with the Peruvian Amazon Company Board. Arana was duly subdued and seemed disposed to replace his brother Lizardo and his Iquitos manager Pablo Zumaeta ("quietly").[193] He had even offered to finance, out of his own pocket, the trip of Horace Bell to assume oversight of PAC affairs in Iquitos. It seemed to Sir Roger that Arana should be kept in the PAC mix as it confronted its financial difficulties, since,

> The fate of the Company is the prime consideration. If the company goes to pot, Arana takes the Putumayo—that is absolutely certain. Mrs. Arana, under the mortgage she holds will seize all

---

189. Ibid., 521.
190. Mr. Bryce to Sir Edward Grey. Washington, D.C., July 21, 1911. Miscellaneous, 147.
191. Ibid., 121.
192. Mitchell, *Sir Roger*, 476–82.
193. British Parliament, *Report*, 221–22.

outside the Putumayo—steamers, train at Iquitos, offices, etc. etc. While J. C. Arana as the original squatter will certainly walk off with the Putumayo and no one can possibly contest his claim. He could then work the Indians and rubber out in the shortest time. If the company is saved it will be reconstructed with fresh capital and fresh directors—men who can be trusted.[194]

Then, on July 24, the Parliamentary Committee of the Anti-Slavery and Aborigines Protection Society passed a formal resolution calling upon the British government to initiate an enquiry into the operations of the Peruvian Amazon Company to determine its responsibility for the Putumayo atrocities documented in Casement's reports. It resolved to continue to raise public awareness regarding the Putumayo in both Great Britain and the United States. It encouraged joint action by the British and American governments in the matter and called upon Argentina, Brazil, and Colombia to apprehend and extradite back to Peru the Peruvian Amazon Company's wanted fugitives. The Society also went on record favoring creation of non-denominational missions in the Putumayo, particularly medical ones.[195]

Meanwhile, in the Putumayo, the Paredes mission was clearly unimpeded by the Peruvian Amazon Company and ranged throughout the region between the Igaraparaná and Caraparaná rivers—or the very heart of Aranalandia. It also enjoyed the full and sincere cooperation of the Peruvian Amazon Company's section chief at La Chorrera—Juan Tizón. On July 15, Paredes arrived back in Iquitos from the Putumayo, having confirmed in his capacity as investigator both the atrocities and the fact that most of Arana's overseers had fled. Whatever the status of his credentials at that point, he was preparing an incredibly detailed report.[196] He had interacted with members of seven Indian tribes, and among the Huitotos alone observed the horrible scars (*marca de Arana*) on the bodies of about three thousand indigenes. Tizón seemed committed to genuine reform, but (despite the contention in the PAC commission's report to the contrary) had questionable authority to implement it.[197] Nevertheless, Paredes was of the opinion that, while still falling short of the Peruvian government's "noble aspirations," the situation regarding the treatment of the Indians in the Putumayo had improved:

---

194. Mitchell, *Sir Roger*, 494.
195. Bodleian Library MSS Brit. Emp. S-22 G323 file.
196. Both before and after his stay in Iquitos Paredes was a well-published novelist and playwright. He was adept at both writing and description—the latter sometimes lurid. Chirif, "Los informes," 77–78.
197. British Parliament, *Report*, xvii.

The true hecatombs, the horrible killings of the Indians happened until 1906, a period in which they flared up in a frightful manner. Since 1907 they have attenuated somewhat, although there are still murders and flagellations, the general level of criminality diminishing up until the time of my arrival in the Putumayo, the 26th of March of 1911, a date in which the crimes against the savages were rare and isolated.[198]

In short, no doubt to Casement's amazement and satisfaction, Paredes proved to be anything but an establishment dupe or puppet. Nevertheless, while clearly an acerbic critic of the Peruvian Amazon Company, Paredes was also a confirmed Peruvian nationalist. As editor of *El Oriente* newspaper, he had applauded the Peruvian victory at La Pedrera and even praised the role of the Casa Arana in the campaign. That history notwithstanding, the other newspapers (controlled by Julio César),[199] and particularly *El Heraldo* (now owned by Zumaeta), skewered Paredes for being unpatriotic. There was evident nervous anticipation in the Arana camp of what his Putumayo report would contain and contend.

Paredes responded in the July 21, 1911, issue of *El Oriente*:

> This deals with a delicate and grave question. It deals with hecatombs that have gained world attention. They aren't legends; they are tangible facts, they are real facts. The crime is not a single one, but a multiple one. . . . To apologize for the criminals, forgetting the prestige and renown of the country, is to wrap oneself in the same crime. . . . Justice has no country . . . for what is crime here, is a crime in England and in all parts of the world."[200]

After previewing the Paredes report, on August 4, 1911, Judge Valcárcel issued 237 arrest warrants[201]—including ones for Pablo Zumaeta, Juan B. Vega, Martín Arana,[202] Victor Macedo, and (Basque-surnamed) Luis Alcorta. Then, on August 5, Valcárcel met with the prefect to com-

---

198. Chirif, "Los informes," 137.
199. Apparently Zumaeta, while in "hiding" after Judge Valcárcel issued a warrant for his arrest, purchased the *El Heraldo* newspaper to oppose that of Paredes. The situation was exacerbated by the fact that Paredes headed the Iquitos branch of a faction within the governing party that was in active opposition to President Leguía's administration. (Department of State, *Slavery in Peru*, 135).
200. Cited in Stanfield, *Red Rubber*, 154–55.
201. Valcárcel, *El proceso*, 342–46.
202. He was accused of ordering thirty lashes for a five- or six-year-old Indian boy (Valcárcel, *El proceso*, 60–61). According to Lagos, Martín was Julio César's illegitimate half brother. He worked at Entre Rios where he was paid eight pounds sterling monthly to do household chores and prepare cocktails for visiting guests (Lagos, *Arana*, 281).

plain that Zumaeta had been under indictment for a long time yet remained free. It was common knowledge that he was running Casa Arana affairs from his residence. When Valcárcel pressed the point, he was informed that the order had been to arrest Zumaeta in the streets—but didn't specify doing so in his house![203] As Paredes declared in his report, for the majority of people in the Loreto, "a crime committed in the Putumayo region is not the same as a crime committed at Iquitos."[204]

Paredes's candor, and the subsequent Valcárcel indictments, had nullified any possibility that Sir Roger's private search for fresh capital to save the Peruvian Amazon Company might bear fruit. On August 3, 1911, Sir Roger was informed by Morel that Andrew Carnegie had turned down the request to invest in the Peruvian Amazon Company, convinced that, as presently constituted, it was a lost cause and should be allowed to fail.[205]

Ironically, one of the fatal blows to the Company's finances was the flight of section heads accused of atrocities. They were owed considerable funds and had to be paid off. This was only possible by exhausting the London accounts. By September, or the month the Peruvian Amazon Company entered into formal liquidation, some salaries had not been paid since March, and the bank balance stood at 2 or 3 pounds.[206] Director Read's bank held 300,000 shares of PAC stock and was still owed about 20,000 pounds on its line of credit.[207] The Board had resolved to send Abel Alarco (who continued to receive the annual 2,500 pounds on his five-year contract) to collect some money owed to the Peruvian Amazon Company on Brazil's Purús River (in the event, Alarco never went).[208] Arana was to leave for Manaos to collect other debts that were outstanding there.

---

203. Valcárcel, *El proceso*, 23. Later, when none of the accused had been detained by mid-1912, the judge noted,
> In effect, Pablo Zumaeta was not captured . . . because it did not please the Iquitos authorities to do so, and he is still not in custody because the Tribunal of this city has declared that a *gentleman* like Zumaeta cannot be in jail; Victor Macedo is not in prison because the secretary of the ex-President Leguía intervened on his behalf; Luis Alcorta has not been captured because the authorities of Loreto do not wish it either; neither have Singer King, Gregorio Oliveros and Belisarios Suárez been detained because they had fled the Putumayo before Dr. Paredes arrived at that river; and Martín Arana, who is still in the Putumayo, is not in prison because that would not please the authorities of the Loreto. (Ibid., 314)

204. Department of State, *Slavery in Peru*, 161.
205. Mitchell, *Sir Roger*, 538.
206. British Parliament, *Report*, 248–49.
207. Ibid., 335.
208. Ibid., 206.

Meanwhile, on August 1, *The Daily News* published interviews with Arédomi and Omarino. The twenty-year-old Arédomi, said "They say that our country was very happy before the white man came for rubber. Now it is very unhappy." He went on:

> Now we are slaves. We are sent far, far into the forest to get rubber, and if we do not get it, or if we do not get it quickly enough, we are shot. That is how I lost my eldest brother. He was called Morujigi, and he was shot by a Viracocha [white man] called Montt. Our father was flogged to death by a Peruvian called Sarote. . . . They took me to help manage the other Indians. So I became a *muchacho* (Spanish for "boy") and lived in a house with other young men who served the Viracocha. We had to beat the Indians who did not collect enough rubber. We had rifles. Some of us were sent into the forest with the Indians who were getting rubber, and we had to whip them to make them go quickly enough when they were bringing back the rubber. I was a bad *muchacho* because I would not beat the Indian. They flogged me when I refused to ill-treat a kinsman.[209]

The twelve-year-old Omarino said:

> My mother was beheaded by a muchacho. I saw it done. The Indian was ordered to cut off her head by a man called Fonseca. She had not collected rubber. There were possibly a hundred people in my clan, and we all lived together in a great house. There are now only five left. All the rest have been killed. The Chief of my clan, Gwamay, was shot. . . . One old woman was very brave, she told us not to collect rubber for the Viracocha. Therefore they cut her head off. . . . Before the Peruvians came we were never hungry, because we had time to grow food and to hunt. The men we work for promise to give us food, but often they do not. So some people have got very thin and they have fallen down under the weight of the rubber they had to carry through the forest, and have died.[210]

It was in early August that Casement convinced Gerald Spicer that he should be permitted to return to Iquitos, unannounced, to observe first-hand how the Peruvian government was implementing its investi-

---

209. Mitchell, *Sir Roger*, 534–35.
210. Ibid., 535. Michell notes that the reporter knew no Spanish and that he, Casement, conducted the interviews and thereby became "the self-appointed mouthpiece for the Amazindian people." Ibid., 534.

gation and reform in the Putumayo. Casement's former colleague in the Congo, George Babington Michell, had been appointed to the newly created Loreto British consular post.

Roger left England in mid-August, accompanied by Arédomi and Omarino. He had changed his mind regarding their future, and now felt they would be better off in Peru rather than as celebrities (and oddities) in Europe.[211] Their first stop was in Barbados, and Casement found O'Donnell there in a rather abject condition. While the former section chief of Entre Rios was still owed a considerable sum of money by Arana, he held out little hope of collecting it. Casement wrote Spicer that the man was "the least criminal of the chief agents," but nevertheless culpable. He wanted the Peruvian government to know of O'Donnell's presence in Barbados, should it wish to seek his extradition.[212]

On September 4, the three travelers boarded a ship for Pará. It was then that Casement met the American medical doctor, Henry Spencer Dickey, who was returning to South America after recuperating from yellow fever. He had worked at a mine in Colombia for seven years after his year of service in the Putumayo. He was then employed by a rubber firm in Brazil on the Javarí River before taking ill and was now returning to his post. While on the Javarí, Dickey had learned that José Fonseca and Alfredo Montt, both "wanted" in Peru, were working at a nearby rubber station.

Casement and Dickey traveled together from Barbados to Pará, and then up the Amazon and Maranhão. Dickey would later write of his eccentric companion. To the horror of the captain on the vessel between Pará and Manaos, Casement carried a shillelagh and dressed in Irish wool, but then insisted on going barefoot because of the heat. When they later reached the closest town to the whereabouts of Fonseca and Montt, Casement asked the local official to arrest them and turn them over to Peruvian authorities. In the event, the man did nothing. Dickey believed that he had been bribed, and the two fugitives warned by Peru's Manaos consul Carlos Rey de Castro.[213]

There was no reception committee for Sir Roger in Iquitos. By this time, rumors and fears of his as yet unpublished reports from his first investigation in the Putumayo were rampant in the city and region. Fur-

---

211. Curiously, a century later, a Huitoto woman from La Chorrera, Fany Kuiru, would learn about the two boys and, believing that they disappeared in Europe, launched a search to determine the ultimate fates of Omarino and "Ricudo" [sic], http://fucaicolombia.org/blog/colombian-woman-hopes-to-solve-century-old-mystery.
212. Goodman, *The Devil*, 159–60.
213. Ibid., 161–63.

thermore, the price of Brazilian rubber had plunged from its speculative peak of 15 milreis in April of 1910 to but 6 milreis at present. The Amazon economy was now in full-blown economic crisis. Casement was perceived to be the enemy and a likely English spy to boot. Mitchell finds this assessment to be reasonable and adds that part of Sir Roger's brief likely included reporting upon the penetration of the Amazon by German influence and interests.[214]

Meanwhile, Judge Paredes met in Lima with a concerned President Leguía and then, on September 30, 1911, produced the 1,242-page report that he sent to Loreto's prefect.[215] In attributing the abuses to the commissions' system, whereby section chiefs received no salary but participated in the profits, Paredes claimed to have examined a contract between J. C. Arana & Co. and Arístides Rodríguez (brother of Aurelio) conceding him 50 percent of the profits on the collected rubber. Paredes contended,

> Once at the head of the sections, these soulless creatures gave free vent to their evil instincts. They were genuine dictators, without morals and without God. They enacted terrible laws and created shameful institutions, such as that of the "trusted boys." They legislated in regard to the gathering of rubber, imposing on the Indian labor which exceeded his strength in order to obtain the greatest possible yield of rubber in the shortest possible time. They considered the Indians as chattels and disposed of their lives by a simple imperative mandate which was irrevocable. They respected neither women, the aged, nor the children, and all, without exception, were subject to the audacious rules laid down for the work. Being interested solely in the profits, they did not concern themselves with the wages, which, if paid at all, were so miserable and ridiculous that they caused a clamor, hence it was that hunger claimed more victims among the unfortunate Indians than diseases themselves or the whip and the plummet of their taskmasters, who, being absorbed in getting profits for the business never took the trouble to learn whether those whom they were exploiting ate or not. With a can of sardines, another of salmon, a pasteboard strap, a cap or Cushma, they demanded comparatively impossible things. The Indians who earned a shotgun labored for

---

214. Mitchell, *Sir Roger*, 567.
215. It was also sent to Lima and later forwarded by the minister of foreign affairs to the American ambassador to Peru. An abbreviated version of the Paredes report was later published in the U.S. Department of State's *Slavery in Peru*, and is the source for what follows.

many years, unless they were so fortunate as to belong to the invidious group of "trusted boys."[216] According to the strange idea of these chiefs, the Indians had no right to live unless they worked for them, and this demand went to the inconceivable extreme of prohibiting them to cultivate farms, for the time which they spent in agriculture was lost to the business. For this reason there were chiefs who desolated cultivated fields and burned houses in order that the Indians might not settle in particular spots. . . . Being genuine autocrats, they pronounced the sentence of death with the greatest coolness, and once an order had been given it had to be executed. Morbid, degenerate criminals, of a sensual nature, they lived surrounded by women, selected Indian girls, most of whom were under age, from whom they exacted fidelity and of whom they were very jealous, going so far as to kill them if they ever caught them smiling at an employee. There was one chief who had 20 concubines, and so great was their power that they went so far as to kill Indian chieftains in order to take coveted women away from them. As an example, of this type may be cited one Armando Normand, who assassinated four wives out of jealousy, first torturing them in them [*sic*] most fearful and dastardly manner.[217]

Another problem was the *modus vivendi* agreement that created a power vacuum in the Putumayo. With no real authority on the ground, the overseers could perpetrate atrocities with seeming impunity.[218] Then, too,

> Everyone [in Iquitos] made efforts to keep the chiefs untouchable, as if their disappearance meant the disappearance of profits. They were considered indispensable and irreplaceable, for they held the key (and we know what it was) to the flourishing state of the business, and if they had been restrained in crime they might have ruined the enterprise; at least this must be assumed if we think of the unanimous concealment of their crimes, which were never stopped or punished, and if we remember this secret approval on the part of all who were determined to conceal, defend, and deny.[219]

In sum, Paredes not only confirmed Casement's accusations, he embellished upon them with additional evidence and testimony. His view-

---

216. Rómulo Paredes, "Confidential Report." U.S. Department of State, *Slavery in Peru*, 145. The "trusted boys" were hated more than the white oppressors and were targeted by the Boras (the least subjected tribe) for killing. Ibid., 153.
217. Ibid., 145–46.
218. Ibid., 162–63.
219. Ibid., 163.

point regarding the Indians, and the way they had been depicted, was telling:

> I think differently than Engineer Robuchón. In the latter's report we discover a marked tendency to represent the Indian as a detestable, bad, treacherous, morally monstrous, and dangerous creature, and, finally, as a fearful cannibal. According to these fantastical paragraphs regarding the customs of the Indians, regarding their strange manner of being and their macaberesque mode of living, it would seem that no one, unless he were a daredevil, would venture to come in contact with them, for, as they are painted, it is impossible to establish relations of labor with this kind of people, who live in human orgies and assassinate for the sole pleasure of eating their fellow men. These pictures of horror, while they give an idea of the wrought up imagination of the person who conceived them, deserve rather to figure in a blood-curdling novel, but by no means in a serious dissertation by a man of science, unless he pretended to have different object, the scope of which we do not pretend to know, or unless Mr. Robuchón, without well knowing the element which he so gloomily represents, was carried away by exaggerated reports of the interested parties [Casa Arana], determined that the Indian should be considered as a depraved creature, dangerous and incapable of being subdued, in order to extenuate the crimes which they have committed against him. There is no other explanation possible.[220]

Paredes provided new insight as well into a previously unmentioned abuse when condemning Arana's business practices:

> The evil started with the manager's office at Iquitos. I am assured that a special business was carried on there with the goods sent to the managers' offices at El Canto [sic] and La Chorrera. It is

---

220. U.S. Department of State, *Slavery in Peru*, 150. It might be noted that Paredes was anticipating by several decades the late twentieth-century debate within anthropology regarding cannibalism. In 1979, William Arens published his book entitled *The Man-eating Myth: Anthropology and Anthropophagy* (Oxford and New York: Oxford University Press, 1979). In it he underscored Paredes's point that accusing a people of cannibalism dehumanizes it as part of the justification for the usurpation of its territory and destruction of its way of life. Arens acknowledged that, in rare cases, there may been ritual cannibalism related to funerary practices and warfare against avowed enemies, but went so far as to question whether cannibalism as a food source ever existed anywhere at any time. Arens's work has been criticized by other anthropologists. In particular, see Lawrence R. Goldman (ed.), *The Anthropology of Cannibalism* (Westport, CT: Bergin and Garvey, 1999).

a public and notorious fact at this commercial town that every time a steamer left for the Putumayo certain employees of the Arana firm ransacked the commercial establishments of the various ports in search of cheap merchandise, even though it was bad, the only thing required being that it should be cheap, for it was for the Indians, and therefore its bad condition made no difference, the result being that a veritable lot of refuse canned goods, cloth and groceries, almost useless and unserviceable, were sent to the region. The majority of the employees there and many merchants have informed me of this fact, and it may be proved from the correspondence of the firm, . . . So far did the abuses go in this regard that the Indians often refused to accept the goods, and the employees themselves would not accept them, preferring to do without them and suffer privations rather than submit to having heavy items charged against their accounts, to the risk of their health and their life. The cloth would unravel on being touched, the canned goods were rotten, and the rest of the goods detestable. But this was doing business, and nothing else made a difference.[221]

Finally, the judge made some interesting observations about the behavior of insecure civilized men once placed in the wilderness:

This lack of guarantee creates a peculiar psychological condition in the minds of those who experience it, especially in the midst of the forest. They are always seeing dangers. They feel brave. Living out of necessity or habit with their weapon in their arm, even when they sleep, they become wayward, authoritative, autocratic, domineering. They know that they can not ask for help of anyone in the moment of a struggle in which their life is to be lost, and they think that the only salvation is in their weapon. They think of nothing but self-defense. These ideas of death, constantly striking their imagination, make them timid and cowardly and they are capable of any act, however reprehensible; and as they consider the Indian an inferior being they assassinate him without the slightest scruple. They think that the Indian is a product of the forest. They are creatures who live in constant alarm and continuous alertness. They are unable to think of peaceful composure, for they consider themselves to be in the midst of war. The constant struggle with nature, seeking food and wealth with the machete and bullet, imparts to them a certain ferocity, and they think solely of the fact that they live surrounded by vipers, tigers, and cannibals.

---

221. Ibid., 154–55.

Like children who read the Arabian Nights, they have nightmares of death, treason, and blood. This is a phenomenon which I have often observed. The solitude of the forest produces this disease, a mixture of morbid valor and timidity which perturbs the imagination and corrupts every human sentiment. In the Putumayo region this phenomenon was developed enormously, perhaps owing to the nature of the men in charge and to the character of the forest itself, which is so remote from the world, so dense, dangerous, and gloomy. The reputation given to the Indian by considering him a cannibal (an absolutely false accusation) caused these so-called chiefs to suffer still more acutely this disease of the mountains [forests] and to implant more firmly in their minds the idea, stoutly adhered to and constantly put into practice, that the only way to live there was to enforce respect by inspiring terror.[222]

Jerome informed London that Judge Valcárcel was now "on vacation," having left Peru for a four-month absence in Europe. His substitute in Iquitos was much more compliant and sympathetic to the interests of the Peruvian Amazon Company. Zumaeta had been detained (not true), but was then allowed to escape from the Iquitos jail.[223] What was true was that Zumaeta was openly thumbing his nose at the "system." Nevertheless, in a November 24 missive to Grey, Casement reported that special treatment of Zumaeta by the judicial and civil authorities, and particularly the annulment of the warrant against him, ". . . is regarded by many people in Iquitos as a scandal."[224] In short, while the Casa Arana was predominant in local affairs, it did not enjoy absolute control over them.

When Valcárcel was in Paris, he learned that the local court in Iquitos had fired him and he was now under indictment for abandoning his post, so he was being treated as a common criminal on the run. He would also subsequently be accused of having trafficked in illegal favors from the bench.[225] Clearly, the Judge was being sent the strongest of warnings not to return to the Loreto. Once back in Peru, he went straight to Lima to straighten out the matter. Since he had requested and received permission to take the leave of absence, the Peruvian Supreme Court quickly dismissed the spurious charges.

On October 16, 1911, or about three months after Valcárcel had issued his arrest warrants for several of the officials of the Peruvian Amazon Company, Casement arrived in Iquitos. He reported a couple

---

222. Paredes, "Confidential," 158.
223. Mr. Jerome to Sir Edward Grey. Lima. August 28, 1911. Miscellaneous, 149.
224. Mitchell, *Sir Roger*, 640.
225. Valcárcel, *El proceso*, 317.

of months later that most of the indicted overseers were still at large and that Zumaeta was hiding (in plain sight).[226] Meanwhile, British first minister in Lima, Charles Des Graz, informed Grey that he had met with President Leguía and minister of foreign affairs, Don Germán Leguía y Martínez (the president's first cousin). The two Peruvian leaders brought up the recent Colombian incursion into the Putumayo/Caquetá and promised to pressure Brazil further regarding the extradition of the Putumayo fugitives there.[227] The many indictments and several detentions (of small fry) showed that the Peruvian government was acting in good faith—despite considerable local opposition.[228]

While Valcárcel was away defending himself, a former high official of the Casa Arana in the Putumayo, (Basque-surnamed) Isaac Ezcurra had filed a declaration with the Iquitos court seconding the Paredes report and claiming to have witnessed many atrocities. He had resigned his post with the Peruvian Amazon Company when he became convinced that its administrators in Iquitos were fully aware of the abuses in the Putumayo and condoned them. Escurra added an accusation against two section chiefs, Alfredo Montt and Andrés O'Donnell. They had ordered that any Indian women found to have venereal disease (imported into the region by the *racionales* themselves) should be killed. Escurra testified that he had read a written order to that effect.[229] He further complained that when he returned to Iquitos with an Indian girl who had willingly agreed to enter his employ as a household servant, during the final accounting of the wages owed to him, he was charged 50 gold pounds for her.[230]

Escurra was frustrated in all of his attempts to formally denounce certain Putumayo section heads—particularly Victor Macedo. Macedo warned him against going public with his information. Then, in Iquitos, Dr. Cavero, head of the superior court, also told him to keep quiet, while adding that Arana was due in town shortly and would discuss the matter with him. Ezcurra next went to the prefect, who twice made excuses for not receiving him. In the event, it was Dr. Cavero who filed the brief that led to annulment of the warrant against Zumaeta. Shortly thereafter, Cavero became Peru's prime minister during the Leguía administration. According to Casement,

---

226. Sir Edward Grey to Mr. Des Graz. London. November 29, 1911. Miscellaneous, 149.
227. Mr. Des Graz to Sir Edward Grey. Iquitos. November 17, 1911. Miscellaneous, 151.
228. Mr. Des Graz to Sir Edward Grey. Iquitos. December 3, 1911. Miscellaneous, 150.
229. Valcárcel, *El proceso*, 69.
230. Ibid., 241.

> Sr. Escurra has frankly no belief in any official, high or low, of his country. He declares that they are dishonest and corrupt to a degree and that for the Putumayo it would be far better in the hands of the Colombians, its rightful owners, who treat the Indians more justly than the Peruvians do; and that the best course is to make public everything connected with the scandalous crimes his country has not only tolerated but for far too long has taken an active part in abetting.[231]

Escurra contended that Abelardo Aguero, section chief of Abisinia station, fled the Putumayo accompanied by "others of the gang of ruffians who had so long terrorized the district."

Aguero carried away with him fifty Indian families for sale in Brazil. He was offered 2,000 pounds by a firm in Manaos that wanted to place them as gatherers on the Acre. However, Aguero declined and took them up the Purús River instead where he was now overseeing operations.[232] Fonseca and Montt had also taken Indians to Brazil and were now gathering rubber openly on the Javarí River. Indeed, Cazes (probably informed by Captain Zubiaur) estimated that the Putumayo absconders had carried away at least five hundred Indians.[233]

As for Macedo, according to Casement's sources, he left the Putumayo in possession of a large payment from the Peruvian Amazon Company, including a 2,000-pound bonus, before Paredes's arrival. He went to Lima, where he believed himself to be safe. Paredes received two telegrams, one from the prefect of Callao and the other from Lima's mayor, urging him to annul the arrest warrant against Macedo. While some said he had been detained and was on his way to Iquitos to face trial, Casement was told that the man was in Barbados, entirely free, and awaiting the outcome of events in Iquitos (particularly the fate of the Zumaeta warrant).[234]

On November 27, Casement wrote to Grey with news that he had learned that the Barbadian Armando King, wanted by the authorities for alleged crimes in the Putumayo, was actually in Iquitos throughout the ostensible police search for him. He was sheltered by Veliz de Villa, former officer in the Peruvian Army, and was now employed by Villa on his rubber estate on the Tapiche River (tributary of the Ucayali) along with another Barbadian who was also wanted by the Iquitos police. This was but another example of the ineptness, or connivance, of the Peruvian

---

231. Mitchell, *Sir Roger*, 679.
232. Ibid., 645.
233. Ibid., 646, 652.
234. Ibid., 642, 680, 685.

authorities. Nor did Casement believe that Tizón would last long with the Peruvian Amazon Company, now that he was unwilling to further Arana's and Zumaeta's agenda. It was also unseemly that Consul Cazes had just leased the *Beatriz* to the prefect for the purpose of transporting supplies and troops to the Putumayo, yet it would be returning to Iquitos with Arana's rubber. This Peruvian force would be sheltered and fed by the Company. Meanwhile, the civil authorities in the Putumayo were its employees as well. The whole thing was incestuous beyond belief. The welfare of the Indians was of little concern.[235] This was simply continuation of an ongoing collaboration,

> The arrangement no doubt worked to the satisfaction of both parties. Messrs. Arana and Co. (it is really absurd to use the name of the ineffectual London Company at any stage of the business) obtained the military help and prestige of Peru in attacking, murdering and pillaging the Colombian settlers on the Caraparaná in securing their rubber and enslaving fresh tribes of Indians, while the Peruvian government through this "patriotic action" of the Aranas extended the frontiers of the "national territory" and became possessed of regions to which it had not any moral or lawful claim.[236]

As for other Putumayo criminals, even after the warrants for their arrest had been issued, several continued to work out of La Chorrera. Then, when the *Beatriz* arrived there in October, or when any other detachment of troops from El Encanto showed up, they took flight into the forest until the coast was again clear.[237]

Paredes warned Casement not to go to the Putumayo (as did Escurra), as his life might well be endangered there.[238] Then, on November 30, 1911, Paredes gave Casement a copy of his own confidential report that he had submitted to the prefect two months earlier. It would inform the consul's own reporting to the Foreign Office.[239] Casement and Paredes were now mutual admirers and intimate friends. Indeed, Roger proposed Rómulo for membership in the Anti-Slavery Society, and, by June of the following year, the Peruvian was soliciting London for copies of the Congo's reform legislation and any advice that the Society might wish to proffer regarding possible changes to Peruvian law in order to protect the country's indigenes.[240] Furthermore, Paredes was now a strong supporter

---

235. Ibid., 653–55.
236. Ibid., 654.
237. Ibid., 655.
238. Ibid., 658–59.
239. Goodman, *The Devil*, 166.
240. Rómulo Paredes to President of the Anti-Slavery Society, June 3, 1912. Bodleian

of the Pro-Indígena Society in Lima, as Peru's best hope for reform of its Indian policy.[241]

Casement was stunned by the Paredes report that contended at least one thousand Indians had been murdered at Normand's headquarters alone. The judge believed that the heaps of bones lying about during his visit had been buried in anticipation of Casement's arrival. Then, too,

> ... the outraging of children, of even very small children, was frequently practiced by these men and that these innocent victims of this atrocious lust were killed or died from the effects of outrages committed upon them.[242] Then, on December 4, 1911, Paredes published an editorial in *El Oriente* stating that, "while we are not descended from this land" he could no longer remain silent in the face of the propaganda of another newspaper (likely Zumaeta's *El Heraldo*) on behalf of the "conquistadors and heroes of the Putumayo." He noted, For no one is it a myth that the author of those present publications and of the active propagandas ... is our celebrated Consul in Manaos, Don Carlos Rey de Castro, the evil man who has just officially handed over [to the Colombians] La Pedrera and Córdoba, and who is the very one that for many years has been hoodwinking our government, causing through his reproachable conduct immense damages to the Country ... [243]

The *ad hominem* attack on Rey de Castro concluded with the comment that certain judges (Paredes himself) went on foot into the bush while in the Putumayo in search of truth, rather than hanging out in La Chorrera or drinking champagne on board the *Liberal*. The battle was clearly joined.

On December 5, 1911, Zumaeta sent a belated letter to the Peruvian Amazon Company's London directors that made an improbable claim that "almost all the heads of sections have been accused of acts of cruelty which have remained a secret both to our Manager at La Chorrera and ourselves."[244] A letter written by Pablo four days later informed the Board that one S. R. Zumaeta [Pablo's brother? Son? Cousin? Nephew?] had been dispatched to the Putumayo "with the

---

Library, MSS Brit. Emp. S-22. G322 (file 2 of 2).
241. He later forwarded to the Anti-Slavery Society in London a number of Peruvian newspaper clippings—including several documenting the activities of the Pro-Indígena Society and its president, Joaquín Capelo. The clippings are in the Bodleian Library MSS Brit. Emp. S22 G334 file.
242. Mitchell, *Sir Roger*, 660.
243. *El Oriente*, December 4, 1911.
244. British Parliament, *Report*, 237.

main object of removing the section employees who have been for years serving the Company."²⁴⁵

Casement left Iquitos, bound for Europe, on December 7, 1911. While en route, he forwarded to Grey several letters. In a particularly detailed missive, dated December 11, Sir Roger told his superior,

> I am now convinced that no serious steps are in contemplation in Iquitos and that none will be taken there unless the Lima government itself strongly takes the matter in hand. The Superior court of Iquitos is on the side of the assassin, and the prefect, while personally honest, has by his futile attitude towards the criminals shown a lack of foresight and interest in his duty quite contemptible. . . . The accused now turns into the accuser and Dr. Paredes is assailed openly in the columns of *El Heraldo* the daily paper that Zumaeta has bought as the personal property of his firm, while the prefect and government of Peru are also referred to in terms of growing contempt. His Majesty's government is charged with improper action and the Secretary of State with having 'issued secret instructions to the Consul' (myself) to withdraw from the jurisdiction of Peru the "sole criminals" in the Putumayo region, the Barbados negroes, who alone had committed reprehensible acts, while the gang of murderers headed by Señor Zumaeta are referred to as "heroes and patriots who only advance the interests of their country." At the Municipal Election, Zumaeta, but just absolved by his judicial friends, was put up as a candidate [for mayor] and although he and his partisans were beaten at the polls, the elections were annulled through the influence of Dr. Borda, a member of the Court. A "Social Club" has been formed in Iquitos, the first and only of its kind, and Zumaeta has been elected President. The steamer *Liberal* is announced to return to the Putumayo at once for more rubber and it is widely stated that the riches of that region are by no means exhausted. The Judge, Dr. Valcárcel, who ordered this man's arrest, on evidence collected by the investigating Judge Paredes, is dismissed from his office (on a pretext), and Zumaeta taking the place of the Public Prosecutor brings further criminal charges against him. Dr. Paredes has more than once assured me that did he not own a paper with a circulation outside Iquitos, in which he is able to defend himself he would be in jail and this for having told the truth to his government.²⁴⁶

---

245. Ibid.
246. Mitchell, *Sir Roger*, 674–75.

Paredes characterized the ostensible investigation in Iquitos of the alleged atrocities in the Putumayo to be a "farce." He also noted to Casement "I ordered the arrests of more than 200 murderers, and I have made 200 enemies any one of which is capable of shooting me when he gets a chance." The ten men [all small fry] currently in jail in Iquitos would be released after things had blown over.[247] Then, too, Casement opined,

> The Prefect and his local authority may be dismissed as of little or no account in the present fight. He equally is assisted in the *El Heraldo* and when he sought to replace the criminal comisario, Amadeo Burga [Zumaeta's brother-in-law] with a new official (directly chosen by Dr. Paredes I learn and who impressed me very favourably) the 's steamer the *Liberal* leaves Iquitos without the new official, accelerating her departure in order that he should be left behind. As a consequence he had to walk some hundreds of miles thro' the forest to reach his post. Meanwhile the former comisario Burga, having at length (after five months delay) been ordered to arrest as many of the remaining agents of the Company (a list of whom to the number of 18 Dr. Paredes handed to me), warned these men of his intention, so that when he came to La Chorrera with troops they had all had time to get away. . . . The old Juez de Paz, Señor Coloma is now replaced (altho' still of course one of the Company's agents on the Putumayo) and the announcement was made in Tuesday's press of Iquitos that the new Juez de Paz for the Putumayo is Sr. Manuel Torrico. This man is another agent of the Company—and Dr. Paredes denounces this appointment as a scandal and a clear proof that the local judicial body intend to perpetuate so far as they can, the old state of things on the Putumayo. When I was on that river last year Sr. Torrico was a sub-agent of the Company, a subordinate at Occidente to the atrocious criminal Fidel Velarde (one of the first to 'escape'). He has since been promoted to be a chief of section and in this capacity he will now be the new and sole representative of Peruvian justice on the Putumayo. In the *El Heraldo* of the 5th instant it is triumphantly announced that "the infamous process against the servants of the firm of J.C. Arana instituted by the weak government of Señor Leguia (the President of Peru) at the instigation of a foreign government shows signs of dying "locally" and Dr. Paredes and his investigations are referred to in terms of contempt and ridicule.[248]

---

247. Ibid., 675.
248. Ibid., 677–78.

In short,

> Of the 237 warrants issued by Dr. Paredes only ten appear to have been executed, although a great number of the men denounced were at the time within the jurisdiction of Peru and easily accessible to the police. In many cases the criminals charged remained on for several months in the employ of the Company and no effort was made to execute the judge's warrant until the date of my return to Iquitos, when a pretended attempt was made to apprehend some twenty individuals of La Chorrera District, who were all reported as having escaped a few hours before the arrival of the soldiers. I have proof that in several cases these men did not even run away, but arranged to return to their posts directly the magistrate and soldiers quitted the district, and further proof that in these cases, the magistrate gave warning of his coming.[249]

Finally, Sir Roger's overall assessment of Paredes was that,

> Whatever his former opinions may have been Dr. Paredes since he went to the Putumayo has been an earnest friend of right. His report is an able and complete exposure of the atrocious state of things he found there. He hides nothing and rarely seeks to minimize in the supposed interest of Peru, the crimes of his countrymen. . . . Dr. Paredes' feeling for the Indians is warm and sincere. In his report he does ample justice to them, and his projected scheme of reform for the Putumayo would leave little to be desired. I hope at an early date to send you, very confidentially a précis of Dr. Paredes' report which will permit you to judge of how faithfully he did his duty on the Putumayo and how fully he deserves our support at Lima.[250]

Meanwhile, Paredes wrote of his frustration in carrying out his new assignment to return to the Putumayo. He blamed the influence of Egoaguirre for his problems. However, he added, "Happily, we have a new government, and it seems that this Senator has lost all his influence; so much so that some days ago I received a fresh order from the Minister for Foreign Affairs to go to that region as soon as possible." He predicted that it would take him until the following summer to complete his new investigation and report, adding, "It has not been my fault, but that of the influences of the House of Arana with the government of Sr. Leguia."[251]

---

249. Ibid., 685.
250. Ibid., 675–76.
251. Translation of Paredes letter of December 10, 1912. Bodleian Library MSS Brit. Emp. M-22 G322.

When Casement reached Manaos, Julio César, recently arrived from Europe after being appointed liquidator of the Peruvian Amazon Company by its Board of Directors, tried to meet with him. However, Sir Roger avoided Arana before continuing his journey.[252] From Manaos, Casement wrote to Spicer that Macedo and Normand were recently seen walking the city's streets together and were likely on their way back to Peru to bring criminal libel charges against Paredes before being reinstated as section chiefs in the Putumayo where they would earn "50% of the gross output of Indian corpses." He promised to send copies of Zumaeta's "charming" newspaper, "with quite a Leopold touch about them."[253] He told Spicer that he left Iquitos when he did so as not to fuel the new party line there that Peru's government was being pressured unduly to act by London. Rumors were again making the rounds that Hardenburg, Whiffen and Casement himself had tried to blackmail Arana in return for their silence. He had a sense of *déjà vu* in Manaos, since Arana was now accusing his recently fired Swedish accountant there of stealing cash and records that he turned over to the Colombian consul ("The old story—exactly the same old yarn he trumped up against Hardenburg & Julio Muriedas").

Meanwhile, letters and telegrams of support of the Peruvian Amazon Company's accused, demanding that the charges against them be dropped, were pouring into Iquitos—sent by influential people in Lima. These included two deputies to Congress: Hildebrando Fuentes and Ingoyen Canseco.[254]

Sir Roger noted that the Colombians had just published a book on the Putumayo scandal that praised the British Foreign Office for its determination, "I am very sorry for the Colombians—they have been treated infamously in this whole business & from all that I can learn on every side (including from Paredes himself) they are far better men with the Indians than the Peruvians." Casement concluded, "We should back up Paredes all we are worth & inform those wretched creatures in Lima that unless we get satisfactory proof of reforms on Putumayo on his lines we publish everything."[255]

Casement was convinced that the success of any attempt to pressure Peru into genuine reform in the Putumayo would require American participation. Therefore, without informing the British Foreign Office, on his own initiative he detoured to Washington, D.C. There, British ambassador, James Bryce, arranged for him to meet with officials of the U.S.

---

252. Goodman, *The Devil*, 168.
253. Mitchell, *Sir Roger*, 680.
254. Ibid., 681.
255. Ibid., 681–82.

Department of State and with President Taft. Bryce wrote about the visit:

> He [Casement] states that officials committed with judicial investigation, though in possession of conclusive and sufficient evidence against well-known rubber gatherers, have been forced to drop action by corrupt local influence, and entire case threatens to terminate to great discredit of the good name of Peru, with perfunctory punishment of a few underlings detained at Iquitos, while those responsible for iniquitous system are seen daily on streets and remain unpunished.[256]

Casement clearly impressed the American government, and it acted quickly on his suggestion that it open a consulate in Iquitos. Secretary of State Philander Knox ordered the American ambassador in Lima, Henry Clay Howard, to communicate his government's concerns over the situation in the Putumayo to the Peruvians. Ambassador Howard was reticent to place undue pressure upon the Peruvian government, particularly since he viewed his primary mission to be the advancement of American business interests in Peru.[257] In his view, the Peruvian government was trying to do its best in the face of difficult physical and political obstacles (unlike Casement, who believed that the Peruvian government itself was a big part of the problem). According to Howard (in a letter to the U. S. Secretary of State dated February 19, 1912), during a meeting that he had with the two Leguías concerning the Putumayo,

> The President especially alluded to the great difficulty, in that remote section, in finding men of sufficient or any legal education who had also sufficient character to withstand the temptations offered by the large corporations in the interior, who seek to get such petty officials upon their own pay rolls, under one device or another, with a view of absolutely dominating their official acts. He instanced a case that had come to his knowledge of an official who received something like £30 per month in salary from the government, and in less than two years had returned with a fortune of £20,000, when it was known that said sum could not have been legitimately made by him.[258]

Meanwhile, despite their debacle at La Pedrera, the Colombians would continue to press for an advantage in the Putumayo/Caquetá. By

---

256. Mr. Bryce to Sir Edward Grey. Washington, D.C. January 17, 1912. Miscellaneous, 151.
257. Goodman, *The Devil*, 174–75.
258. Mr. Howard to the Secretary of State. Lima. February 19, 1912. U.S. Department of State, *Slavery in Peru*, 134.

January 1912, the Peruvians had five gunboats and a thousand troops stationed on the Putumayo River and five hundred men at Puerto Pizarro on the Caquetá. Nevertheless, Bogotá sent orders to its administrators in the region that they should introduce spies (disguised as aspiring *caucheros* or rubber gatherers) into the disputed territory to ascertain Peruvian military weaknesses, while at the same time seeking to foment rebellion among the Indians. Over the next several months both initiatives would fail.[259]

In February 1912, now back in London, Casement informed Grey that the proceedings in Iquitos were a total sham. Valcárcel's replacement had refused to try the accused over a technicality in Peruvian law. He ruled that until all of the more than two hundred indicted individuals had been detained, none was required to stand trial (the indictments against Zumaeta and Arana would be dropped entirely a few weeks later on the grounds of insufficient evidence). Magistrate Amadeo Burga had been sent by the authorities to the Putumayo to serve warrants but always managed to arrive just a few days after the accused had escaped. Regarding the abusive collection of rubber on the Putumayo, it was business as usual. A nascent program to plant rubber trees around the stations had been abandoned, and the Indians had been ordered back into the forests to gather latex.

Casement reported the rumor that Arana was telling associates in Peru "that as soon as what he called 'the fuss' was over, the natives would be set to work again."[260] Furthermore, the critical parts of the Paredes report were being kept under wraps in Lima more than four months after its submission, clear evidence that the Peruvian government was engaged in a cover up.

Casement forwarded the text of his extensive report to Grey on February 5, 1912, accompanied by a lengthy cover letter in which he stated:

> It was abundantly clear that the company, or those who locally controlled the Putumayo in its name, having recovered from the shock of exposure and fear that followed the visit of the commissioners and myself in 1910, had determined to retain forcible exploitation of the Indians as their right by conquest and their surest means of speedy gain. . . . The fate of the Indian supporter of this fabric of civilised society is of no account. The short-sighted policy which ends in working him to death, and denuding whole regions of their entire population, is only what has been settled custom and practice of well nigh 400 years of Iberian occupation of that

---

259. Stanfield, *Red Rubber*, 161.
260. Quoted in Goodman, *The Devil*, 175.

part of the world. It was not ever a fact, and is not now a fact, that the presence of the Peruvian or Amazonian Indian is incompatible with the existence or civilisation of the white man. It was not ever a war of plough against tomahawk, of colonist and cultivator against barbarism and warrior hunter. On the contrary, the Peruvian Indian is a being of extreme docility of mind, gentleness of temper, and strength of body, a hardy and excellent worker, needing only to be dealt with justly and fairly to prove the most valuable asset the country possesses. Instead of this he has been from the first enslaved, bent by extortion and varying methods of forced labour to toil, not for his own advantage or the advancement of his country, but for the sole gain and personal profit of individuals who have ever placed their own desires above the common welfare. . . . To these remote people civilisation has come, not in the guise of settled occupation by men of European descent, accompanied by executive control to assert the supremacy of law, but by individuals in search of Indian labour—a thing to be mercilessly used, and driven to the most profitable tasks—rubber getting—by terror and oppression. That the Indian has disappeared and is disappearing rapidly under this process is nothing to these individuals. Enough Indians may remain to constitute, in the end, the nucleus of what is euphemistically termed a civilised centre. . . . In this instance the force of circumstance has brought to light what was being done under British auspices—that is to say, through an enterprise with head-quarters in London and employing both British capital and British labour—to ravage and depopulate the wilderness. The fact that this British company should possibly cease to direct the original families of Peruvian origin who first brought their forest wares (50,000 slaves) to the English market will not, I apprehend, materially affect the situation on the Putumayo. The Arana Syndicate still termed itself the Peruvian Amazon Company (Limited) up to the date of my leaving Iquitos on the 7th December last. The whole of the rubber output of the region, it should be borne in mind, is placed upon the English market, and is conveyed from Iquitos in British bottoms. Some few of the employés in its service are, or were when I left the Amazon, still British subjects, and the commercial future of the Putumayo (if any commercial future be possible to a region so wasted and mishandled), must largely depend on the amount of foreign, chiefly British, support those exploiting the remnant of the Indians maybe [sic] able to secure.[261]

---

261. Mitchell, *Sir Roger*, 725–26.

Grey had had it. He informed Washington through Ambassador Bryce of his belief that Peru would do nothing unless the Casement report was published.[262] On February 26, 1912, Grey ordered Great Britain's Iquitos consul, George B. Michell, to conduct an on-site investigation of the situation in the Putumayo.[263] By April 6, 1912, and particularly because of the growing concern over the Putumayo scandal, Washington reopened its consulate in Iquitos and appointed Stuart J. Fuller to the post. He was ordered to accompany Michell on the British consul's upcoming investigative trip.

While awaiting Washington's response, on March 5, a member of parliament demanded that the Foreign Office publish the Casement report. And then, on March 25, 1912, London's *Morning Post* newspaper published an interview with Seymour Bell of the PAC investigative commission. Bell was perplexed that Casement's report remained unpublished, and he underscored that his team was in substantial agreement with it. Three days later, there was good news from Washington. Despite Ambassador Howard's opposition to publication, the State Department had decided to defer to Grey's judgment in the matter.[264]

Howard now urged that the U.S. government should be patient in order to give the Peruvians a chance to place their own house in order. Surely, Peru had good intentions. Hadn't it just announced its plan to initiate and fund missionary activity in the Putumayo?[265] The delays with the Peruvian government's own investigation might in large part be due to the difficult logistics that impeded speedy communications between Lima and Iquitos. Then, too, Bryce underscored "the almost insurmountable obstacles that confront the Central Administration in the peculiar character of local conditions at Iquitos." (The reference, of course, was to Arana's dominance of the Loreto's political and juridical institutions.)

On April 22, 1912, the Peruvians responded to the growing international pressure by constituting a new investigative commission on the Putumayo (as well as relations with Indians in other regions of the country) that would make recommendations for policy changes by the end of the year. Its members included (Basque-surnamed) Dr. Javier Prado y Ugarteche (the former president of Leguía's cabinet and ex-minister of foreign

---

262. Sir Edward Grey to Mr. Bryce. London. February 23, 1912. Miscellaneous, 160.
263. George B. Michell, *Report by His Majesty's Consul at Iquitos on His Tour in the Putumayo District* (London: His Majesty's Stationery Office, 1913), 2 [*Report*].
264. Goodman, *The Devil*, 180–81.
265. Mr. Bryce to Sir Thomas Grey. Washington, D.C. March 28, 1912. Miscellaneous, 161.

affairs who had negotiated border matters with both Brazil and Ecuador) and none other than Julio Egoaguirre. Casement was appalled by that appointment (given the senator's known ties to Arana). In short, foxes charged with guarding the chicken coop. The process was further impeded when, on May 4, 1912, Brazil ruled that there would be no extraditions to Peru of Putumayo fugitives (again citing the lack of an extradition treaty between the two countries).[266] While technically true, one suspects that Brazil had little stomach for shedding further light on Arana's activities and those of its own rubber barons.

On May 31, 1912, Casement sent his translation of the Paredes Report that he had made before leaving Iquitos six months earlier. He had underscored all of the discrepancies (and omissions) between it and the version forwarded by the Lima government to Great Britain and the United States. He used the term "suppression" in referring to the exlusion of sentences, whole paragraphs, and all of chapter three.[267]

Despite Howard's lobbying to the contrary, the United States opted to follow Great Britain's lead.[268] Sir Edward Grey argued (somewhat cynically) that publication of the Casement report would assist the Peruvian commission in its work.[269] In the event, the U.S. government concurred with publication of Casement's findings, but insisted that all of its correspondence with London regarding the matter be appended to it (i.e., the Americans' cautions and reservations).[270]

Stuart Fuller sent his first report on May 31, 1912, in which he noted that he and Michell would likely travel to the Putumayo together in July or August. He underscored that they wanted to avoid going on a Peruvian Amazon Company vessel, if possible.[271] Then, on July 1, he sent a more detailed report. By then, Fuller was convinced that real reform would be impossible without the support of the Loreto's inhabitants, and that seemed unlikely:

> In the first place, those in control of the Putumayo concession are among the wealthiest and most influential men in this part

---

266. Document in the Bodleian Library MSS Brit. Emp. S-22 G321.
267. Mitchell, *Sir Roger*, 744.
268. Mr. Huntington Wilson [Acting U.S. Secretary of State] to Sir Edward Grey. Washington, D.C. Miscellaneous, 162. Sir Edward Grey to Mr. Mitchell Innes. London. July 27, 1912. Miscellaneous, 164.
269. Mr. Mitchell Innes to Sir Edward Grey. Washington, D.C. May 25, 1912. Miscellaneous, 163.
270. Ibid., 164.
271. Consul Fuller to the Secretary of State. Iquitos. May 31, 1912. U.S. Department of State, *Slavery*, 12–13.

of Peru, and in fact in the whole country. Their influence in Lima is great, and locally they could bring pressure to bear on many people who might otherwise strongly support a movement to protect the Indians and improve their conditions. An indication of the state of local public opinion in regard to these men is to be found in the Iquitos attitude toward Pablo Zumaeta, the moving spirit in the Peruvian Amazon Co., who is still under indictment and for whose arrest a warrant was at one time issued. He is in enjoyment of most of the local honors, vice alcalde (vice mayor of the municipality), vice president and acting head of the chamber of commerce (an influential organization), president of the benevolent society, etc., to all of which he was elected subsequent to his exposure. He is well respected in the town and stands high, the charges under the shadow of which he rests being entirely disregarded. In the second place, for a full comprehension of the existing situation it is necessary to examine into the general labor situation throughout this part of Peru. An important factor in this phase of the situation is found in the ancient, deep-rooted, and almost universal attitude of the Peruvians, who, while they may not approve of cruel and inhuman treatment, generally regard the Indians as placed here by Providence for the use and benefit of the white man and as having no rights that the white man need respect.[272]

Fuller elaborated by describing the debt peonage system that trapped illiterate and innumerate Indians in eternal servitude tantamount to slavery. Indeed, their contracts (and therefore their persons) were bought and sold among the white employers. Yet, they were in scarce supply and absolutely critical to the Loreto's existing economy. Were said labor practices to be abolished, the region's entire economic system, particularly the structuring of credit and debt, would simply collapse. Then, too, there was the parallel system of the domestic slavery of children and young women prevalent in the majority of Iquitos households. The wards were generally not maltreated, but were bought and sold. That traffic was so blatant that

> The crews in launches operating in this river all expect to make something by trading in girls and children. The practice has repeatedly been complained of by the clergy, but without result."[273]

The prefect of Loreto, Señor Francisco Alayza y Paz Soldán, had tried to see that Judge Valcárcel's arrest indictments were enforced. He was

---

272. Ibid., 14.
273. Ibid., 16.

apparently under a certain amount of pressure from Lima to do so. He was believed to favor proper punishment of those guilty of crimes on the Putumayo. He wrote a letter in which he summed up the situation there as follows:

> In the Putumayo there exists no real settlement whatsoever. Its principal centers, "La Chorrera" and "El Encanto" are the exclusive property of the old commercial house Arana y Hermanos, today the firm called the "Peruvian Amazon Company." The dwellings, warehouses, industry, commerce, vessels, in a word, all that civilization has taken to these immense regions belong in an absolute manner to said Company. It operates with 400 employees and servants between whites and civilized indigenes, 20,000 semi-savage indigenes and 5,000 that remain to be conquered, counting men women and children, and constituted by the Huitotos, Boras, Andoques, Ocaínas and other tribes, who were cannibals, and if they do not practice cannibalism today it is because of the constant precautions of the Company.[274]

In the prefect's view, neither the vastness of the area nor its rubber production justified ever converting it into a formal district under its own sub-prefect. Its only link to the outside was the steamship from Iquitos that paid a visit about once every six weeks. The Putumayo had a commissary that reported directly to him. The commissary had at his disposal between sixty and eighty Peruvian soldiers, the number varying according to the circumstances of the moment. This military presence was justified by the importance of the region's strategic position with respect to the Colombian and Brazilian borders.[275]

Prefect Alayza y Paz Soldán had just departed Iquitos and planned to be away for three months; it was common speculation in the city that he would never return. The acting prefect was Señor Estanslao Castañeda (appointed through Egoaguirre's influence), and the acting sub-prefect was an employee of the Peruvian Amazon Company.[276] There was a single justice of the peace for the vast Putumayo territory, who was also Arana's employee.

Fuller further commented:

> Pablo Zumaeta, in whose case nothing further has been done, continues to stand high in the esteem of the local public. He took

---

274. Alayza y Paz Soldán, *Mi país*, 110.
275. Ibid., 110–11.
276. Consul Fuller to the Secretary of State. Iquitos. July 1, 1912. U.S. Department of State, *Slavery*, 16–17.

a prominent part in the official ceremonies of July 12 connected with the mass celebrated for those who died in the Battle of the Caqueta last year in the troubles with Colombia. He and his friends blackballed, at the Iquitos Club, the judge, Dr. Valcárcel, who had issued the warrant against Zumaeta, and that in a club where a majority is necessary to shut out a proposed member. Zumaeta may not be guilty of all that is laid at his door; he may have been accused and the warrant issued against him unjustly, but if as innocent as he claims to be, it is strange that he does not go into court and vindicate himself once for all.[277]

Nor was Arana without his Peruvian detractors. On July 6, 1912 a Peruvian physician, Dr. Olivares, now resident in Paris, wrote to Dora Mayer of Callao, a longstanding German resident of Peru and an activist in the protection of Indian rights. She was an outspoken critic who published frequent articles in the Peruvian press. Olivares stated:

> I address myself to the Apostle of the helpless native race, to the talented and large hearted writer, with the conviction that you will take up the cause of the poor Indians of the Putumayo and that you will know how to rouse public opinion from the tribune of the press, which will form an overwhelming current forcing the public authorities to put an immediate end to so much wrong and to free more than 5000 Peruvian Indians, our fellow countrymen, otherwise condemned to disappear, victims of the rapacity and boundless greed of a handful of scoundrels.[278]

He included in the envelope press clippings demonstrating that the Putumayo scandal was all over the French and English press. He had served as a medical doctor for three years in the Loreto and seven months in the Putumayo and witnessed much himself. He stated that, in the Putumayo,

> A not insignificant portion of that native race is groaning in the horrors of slavery, victims of the insatiable rapacity of a band of pirates, the Peruvian firm of Arana, Zumaeta and Company, whose Agents on the River Putumayo have achieved by their abominable crimes a detestable reputation in the civilized world, arousing a general sentiment of indignation and protest, and thus ultimately affecting the prestige of Peru as a cultured and civilized country for tolerating and leaving such atrocities unpunished.[279]

---

277. Ibid.
278. English translation of Salvador Olivares to Dora Mayer, July 6, 1912. Bodleian Library MSS Brit. Emp. S-22 G335.
279. Ibid.

Nor was the issue of Indian enslavement in Peru limited to the Putumayo. In his 1913 article, W. O. Simon stated:

> The aboriginal forest savages are bought and sold freely. When I was in the Madre de Dios the market price of a man was sixty pounds, a woman forty pounds, and a child ten pounds; although, for good men used to collecting caucho (low-grade wild rubber), much higher prices prevailed. One landowner I know had just bought twenty families, say, eighty persons in all—for five thousand pounds. I myself was offered a hundred people by their master, who was retiring from business. He appeared quite surpised when I told him that Englishmen did not deal in human flesh.[280]

Olivares praised Valcárcel, Paredes, and Casement, while describing Normand as,

> The most accomplished type of bandit that can be found; he was degenerate and realized the type of born criminal of Lombroso, for he committed unheard of cruelties and murders for sport, from sheer Sadism, surpassing Nero and Caligula; he caused hecatombs, almost depopulating the section of which he was chief; nevertheless in the eyes of the Company (who could not ignore such atrocities) Normand was a model, and ideal employé, the object of all kinds of attention, for he made his section produce much rubber.[281]

On July 8, 1912, Lima's *La Prensa* newspaper published an article announcing the death "in a cruel state of abandonment" of Benjamín Saldaña Rocca. In sum,

> His efforts were completely wasted because, in those quarters where he did not find the gold of the Firm of Arana in his path sealing the lips of the government officials, the circumstance that the frontier controversy with Colombia as to the ownership and possession of the Putumayo had at that time been rekindled, forced honourable men to observe a patriotic silence. And the humble defender of Right and Humanity had to fly from the Loreto and died miserably in a wretched hole in order that he should not in his turn be the victim of the criminal exploiters of the river.[282]

---

280. Simon, "Frontier Life," 87.
281. Ibid. Olivares signed his letter by stating "As anonymity is repugnant to me" and thereby gave Mayer permission to quote from it.
282. Quoted in Goodman, *The Devil*, 200.

The article mentioned the death of Robuchon, whose photographs documenting the, "macabre orgies of the servants of the Casa Arana still pass from hand to hand."[283]

Meanwhile, Eduardo Lembecke, Peruvian chargé d'affairs in London, peppered the British Foreign Office with claims that Peru had instituted most of the necessary reforms, and was dispatching an investigative commission to the Putumayo, so publication of the Casement report was unnecessary. However, it was quite clear that any new report (even allowing that the investigation was thorough and genuine) would not be available for many months. And it was at this point that Grey informed the U.S. government of a shipment of 75 tons of rubber that had just arrived in Iquitos from the Putumayo, one of the largest consignments in years. Furthermore, "the amount exported from January 1st up to the end of April of this year equaled, three-quarters of the total output for 1911, figures which can only have been rendered possible by a continuance of the old system of enforced labour."[284] Grey was therefore authorizing expeditious publication of the Casement investigations (which came out on July 13, 1912). It contained both of Casement's reports and a list naming the Putumayo's worst offenders.

The so-called "Blue Book" (dubbed thusly because of its deep blue cover) caused an immediate furor. Canon Hensley Henson sermonized at Westminster Abbey, denouncing the British Board of the Peruvian Amazon Company,[285] and the Dean of Hereford gave a speech in which he stated:

> It was found there were men connected with the Peruvian Rubber Company [sic], men bearing titles and honourable positions in our country, who actually had shares in this company, whose abominable and horrible treatment of the natives was quite as bad as anything which had been done in the Congo.[286]

On July 15, *The Times* of London published a scathing article under the headline "The Putumayo Atrocities: A South American Congo—Sir Roger Casement's Report Published." It recounted floggings, torture, and murder. Furthermore, the readership was being treated gingerly, since,

> Far more terrible examples of cruelty are quoted on the direct evidence of witnesses of the crimes. We cannot reproduce them. It

---

283. Ibid.
284. Sir Edward Grey to Mr. Mitchell Innes. London. July 27, 1912. Miscellaneous, 164.
285. Collier, *The River*, 220.
286. British Parliament, *Report*, 255. Lister-Kaye had the PAC's attorneys send the canon and the dean letters demanding a retraction (or face possible legal action). The dean complied; the canon did not (Ibid., 255, 270–71).

must suffice to say that all of the Indians of such a district, without discrimination of age or sex, became the helpless victims of the lust, cupidity, and savage cruelty of its conquerors.[287]

The article further stated that many of the worst offenders were at large. There was also the anomaly that Peru excused its lack of policing the Putumayo by underscoring its isolation while at the same time claiming sovereignty over it. In the following weeks, similar articles appeared throughout much of Great Britain's popular press (e.g., *Truth*, *The Saturday Review*, *The Spectator*, *The Economist*, *The Times* and *The Nation*). Both the *New York Sun* and *The New York Times* followed suit.[288]

It was also at this point that prominent Britons like Sir Arthur Conan Doyle, Lord Rothschild, and the Duke of Norfolk set up a Putumayo Mission Fund (in the belief that Christian missionaries would make a difference), and Pope Pius X exhorted all Latin American leaders to protect their Indians.[289]

There was an immediate reaction by the Peruvians. In a July 16, 1912, article, Lima's *El Comercio* editorialized that the crimes reported in the Blue Book should be investigated, while noting that the British, too, had their historical failings in their dealings with native peoples. The Putumayo was exceedingly remote, and therefore proper administration of it was extremely difficult. The newspaper praised the efforts of a leading Peruvian PAC administrator (it is unclear whether reference was to Arana or Zumaeta—probably the former), "a man of renowned moral culture and humanitarian sentiments," for his attempts to reverse the excesses of the section chiefs on the ground. Nevertheless,

> to explain the evils does not mean to accept them, and our government should take measures to see that legal sanctions for crimes reach even the far off regions of the Putumayo, since only in this fashion will it really become possible to eliminate them.[290]

Nevertheless, *El Liberal* weighed in with an article under the title of "Barbarians of Civilization," by Spanish national V. Romero Fernández, the medical doctor on the Paredes investigative mission in the Putumayo (and at that time living in Barcelona). He declared that the Arana forces used violence whenever necessary to displace the Colombians and de-

---

287. Quoted in Goodman, *The Devil*, 185.
288. Ibid., 187.
289. Appeals for the initiative appeared widely throughout the English-speaking world. For example, the *Singapore Free Press and Mercantile Advertiser* published an appeal in its August 20, 1912 (page 6) issue.
290. *El Comercio*, July 16, 1912.

scribed the cold-blooded dispatching of one of them with a machete.[291] In a successive interview, he noted that he was asked by the prefect of the Loreto to accompany Paredes, since he was the only foreign doctor in Iquitos at the time (the implication being that the Peruvian ones would be compromised when it came to criticizing Arana). Romero Fernández had agreed on humanitarian grounds. He had met Roger Casement in Peru and was now prepared to confirm without any reservation everything that had appeared in the London press based on Sir Roger's reports. Indeed, the reality in the Putumayo was even worse. The commissioners were told by credible witnesses that, at La Chorrera, the whites had corralled three hundred Boras and a few Huitotos into a structure that they then burned to the ground. For amusement, they shot any Indians that tried to escape. The investigators had visited the site and it was a veritable boneyard— many of the bones exhibiting bullet holes.[292]

On July 22, 1912, Lima's *El Comercio* published an interview with Abel Alarco in which he claimed that Rafael Reyes had approached him and Julio César in London with a proposition—probably a guarantee to respect the Peruvian Amazon Company's position in the Putumayo in return for their support of Colombia's sovereignty claim there. It was rejected and it was then that Reyes recruited Hardenburg and Perkins with an offer of 4,000 pounds each to make their journey down the Putumayo and then proceed on to London to denounce the Peruvians. Hardenburg later proposed to Egoaguirre that he would write a report favorable to Peru, and the Peruvian Amazon Company, in return for 8,000 pounds. Hardenburg went to Great Britain to launch his anti-PAC campaign, beginning with his article in the *Trust* [sic] newspaper. In short, the American was a blackmailer in the service of the highest bidder (Reyes).

Alarco concluded on a brash note. Invoking the history of atrocities committed by both Great Britain and the United States while expanding their respective imperial domains, he observed that, "One does not conquer with caresses." Indeed,

> When I went to the Amazon basin a little more that twenty years ago, no one could penetrate those unknown and mysterious jungles without grave personal risk: the savage saw in the explorer his natural enemy; and he killed him. This lasted for a long while. The explorer, who was after rubber, lowered his own psychology to that of the forest dwellers; and saw in him his natural enemy. And he killed him: the one who (killed) through the natural instinct to

---

291. *El Liberal*, July 21, 1912.
292. Ibid., August 9, 1912.

defend the land where he was born, while refusing to submit this defense to the courts; the other, driven by the higher civilizing instinct, went to those lands to extract a product that in Liverpool was esteemed like gold. The latter had right and justice on his side: he was a Peruvian and these regions were Peruvian as well. The extracted rubber paid export taxes in the Peruvian customs houses, contributed to the public wealth. The civilized man should have let the savage kill him; that anonymous and unproductive adversary, living beyond the law. What would the Englishman have done faced with the brutal and aggressive forest dweller. Before I became a partner in the Casa Arana and engaged in formation of the Peruvian Amazon Company, I was a boss of rubber gatherers and, with my own eyes, and almost without being able to avoid it, preoccupied as I was with defending my own life, I have seen innumerable of my companions fall one after the other, at times in open combat with the savage tribes and at others in those shrewd ambushes that only the forest dweller knows how to prepare for his enemy . . .[293]

In short, it was egregious for the international do-gooders to proclaim that "beneath the rays of the same sun that illumines civilized Europe," there is a country (Peru) in South America where the inhabitants kill one another for pleasure, innate cruelty, and with impunity—given their impotent government. Nevertheless, Alarco asked (rhetorically) if the average Englishman sipping his cup of tea and riding down the road in his car appreciated the great daring and sacrifice of the Peruvians who obtained the critical rubber for its tires. He also complained that, "The issuance of arrest warrants against innocent, brave men was bleeding the Putumayo dry, as these heroes of commerce are forced to disappear into the 'depths of the forest' to escape rough justice."[294]

While outwardly the Peruvian government appeared to be promulgating genuine investigation of the alleged Putumayo abuses, there is the interesting communication dated August 16, 1912, sent by Vatican apostolic delegate in Lima, Angelo Scarpadini, to Cardinal Merry del Val at the Vatican Secretariat of State:

> According to what the president [of Peru] assured me, the Casa Arana is not in bankruptcy, unlike the English Corporation that

---

293. Ibid., July 22, 1912.
294. Ibid. The following month, when Casement forwarded the article to Grey, he expressed his skepticism that it had been an interview, noting instead, "it is clear, both from the language employed and the signature appended (*Máximo Pesar*) that it is the work of some Peruvian journalist [or Rey de Castro?]."

had assumed the firm; and he added that all of the publications issued in recent times regarding the horrors in the Putumayo, were above all a stock market ploy, that is to say, they were designed to scare Peruvian capitalists, and in particular Arana, into selling their shares at a bad price, and, in this fashion, the English Corporation would not only resolve its bankruptcy, it would become the proprietor of that vast holding in the Putumayo for the purpose of rubber gathering, a goal that England had not realized because Peru has taken great care in the matter.[295]

On August 1, 1912, the U.S. House of Representatives passed a resolution directing its Secretary of State to provide it with all documentation in his possession regarding the alleged existence of slavery in Peru ("if not incompatible with the public interest"),

> ... and especially all information tending to show the truth or falsity of the following statement made in an editorial in the London Times of July fifteenth, nineteen hundred and twelve: "The bluebook shows that in an immense territory which Peru professes to govern the worst evils of the plantation slavery which our forefathers labored to suppress are at this moment equaled or surpassed. They are so horrible that they might seem incredible were their existence supported by less trustworthy evidence"[296] A few days later, Consul Fuller sent the State Department a dispatch from Iquitos stating,
>
> A short time ago I was privately informed by a Peruvian military officer, who held a command in the Putumayo region up to May, 1912, when he was transferred to Iquitos, that reprehensible practices were still being carried out there when he left. He also said that he did not believe that the local civil authorities were trying to do anything at all in the matter further than to throw dust in the eyes of the central government in Lima.[297]

---

295. Quoted in Pilar García Jordán, "La misión del Putumayo (1912–1921). Religión, política, y diplomacia ante la explotación indígena," in Pilar García Jordán, Miguel Izard, and Javier Laviña (eds.), *Memòria, creación e historia: Luchar contra el olvido/Memòria, creació I història: Lluitar contra l'oblit* (Barcelona: Universitat de Barcelona, 1994), 257.
296. Wm H. Taft, "Letter of Transmittal." Washington, D.C. February 4, 1913. U.S. Department of State, *Slavery*, 3.
297. Stuart J. Fuller to the Secretary of State, "Labor Conditions in the Putumayo River." Iquitos. August 5, 1912. U.S. Department of State, *Slavery*, 42. There was also the likelihood that the military were in cahoots with Arana. It was said that military officers also enriched themselves in the Putumayo by facilitating Arana's agenda, particularly regarding commerce and contraband—becoming in effect *"cauchero* soldiers." Pineda Camacho,

He showed the American consul a copy of a report he had filed with the prefect (after his military commanders told him it was none of his affair), denouncing the treatment of the Indians and advocating strengthening civil administration of the Putumayo, as well as introduction of missionaries there. Prefect Alayza y Paz Soldán turned it over to the military commander while "intimating that it would be better to place in command an officer who would attend to his military duties and leave civil matters alone."[298] Fuller further advised Washington that he had received a very equivocal answer from Loreto's prefect to his written request for a report on what the authorities were doing regarding the Putumayo allegations. The American consul concluded:

> In fine, the letter is really an admission that the responsible government of the Department of Loreto can not point out anything that has been done to better conditions, and a warning that any attempt to get at the facts on the ground will not meet with their assistance or cooperation, as well as an effort to hold off any inquiry until the matter blows over, or until they have at least had time to start something in the Putumayo. It is a good example of the shifting equivocation that meets one throughout in the attitude of the local authorities.[299]

Rómulo Paredes happened to be in Washington, D.C., and he joined forces with the minister of the Peruvian legation, Federico Alfonso Prezet, in denouncing the House resolution as precipitous while urging patience. They both believed that there had been many recent reforms, and Paredes was on his way home to join the Peruvian government's latest investigative team on its pending trip to the Putumayo.[300]

The British and American Iquitos consuls were poised to leave on their fact-finding trip and had retained John Brown, a British subject fluent in Huitoto, as their translator (he had formerly been Whiffen's translator as well). The problem was that Arana controlled completely the physical access to the Putumayo. Their efforts to gain independent trans-

---

Holocausto, 179.
298. Ibid., 43.
299. Stuart J. Fuller to the Secretary of State, "Labor Conditions in the Putumayo Region." Iquitos. August 6, 1912, U.S. Department of State, *Slavery*, 44–45.
300. The Peruvian Minister to the Secretary of State. Washington, D.C. August 7, 1912. U.S. Department of State, *Slavery*, 178–81 and elsewhere. Nevertheless, the following month Paredes wrote to the secretary of the Anti-Slavery Society, "I learn that in Lima the representatives for Loreto . . . and above all Dr. Ego Aguirre, who is Arana's lawyer, are opposing my return to the Putumayo." Rómulo Paredes to Travers Buxton, September 21, 1912. Bodleian Library MSS Brit. Emp. S-22 G322 (file 1 of 2).

portation there were unsuccessful, and the two consuls were reduced to setting out from Iquitos on the *Liberal*.

They embarked on August 7, 1912, and then, four days into the journey, at a place called La Colonia Riojana (named after Arana's birthplace and located where the Putumayo entered Brazil) the *Liberal* rendezvoused with another vessel. It was there that the consuls were joined by Julio César Arana and his young (early twenties) brother-in-law Marcial Zumaeta.[301] The Arana party included Carlos Rey de Castro, who had in tow a Peruvian agronomist, (Basque-surnamManuel Reátegui,[302] and a Brazilian photographer (Silvino Santos).[303] The Peruvian consul from Manaos produced a letter from Acting Prefect Alayza y Paz Soldán authorizing him as Peru's representative. Casement had been openly concerned that consuls Michell and Fuller would be duped by Arana and Rey de Castro during their impending journey to the Putumayo.[304]

The Colombian consul in Manaos shared this concern. On September 2, he wrote to Michell and Fuller in Iquitos, saying that he had just been authorized by Bogotá to accompany the consuls if they so desired. But he realized it was too late and warned them against the danger of being hoodwinked by Arana and Zumaeta, the criminals themselves. He protested that Colombia should have been represented in the mission, since it was to incontrovertible Colombian territory. He noted that the previous March, at just one Arana station in the Putumayo, 122 skulls of Indian men, women, and children were counted. The trafficking in Indian slaves continued, and they were being sold in Iquitos "like donkeys." Eight indigene boys and four girls had just been boarded on a steamer at Cotuhé, dispatched for sale by the commander of the Peruvian military detachment there. He noted that, in 1909, or after the tragic events of

---

301. Marcial had resided in London for several years and was reasonably fluent in English. He assisted Arana with translations in the capacity of personal secretary.

302. Likely shortened from Larreátegui. This man may very well be the one described in Turvasi:

> This Manuel Reategni [sic?], an old man living in Iquitos, was rich and a great scourge of the Indians. Three times he succeeded in avoiding a trial by means of money. The fourth time he tried in vain because he found an incorruptible judge [Valcárcel?]. He had a stone house, which served as a harem, with fifty women or more living under strict rule and kept for lust and for market. He had a large cage on his grounds in which he kept a tiger [jaguar]. A houseboy who did not come immediately when called was thrown into the cage before his father's eyes. The father was threatened with a gun when he fell on his knees to beg for mercy. Turvasi, *Giovanni Gennochi*, 76.

303. Consul Fuller to the Secretary of State. Iquitos. October 28, 1912. U.S. Department of State, *Slavery*, 47.

304. Casement to Grey, August 24, 1912. Bodleian Library MSS Brit. Emp. S-22 G344a.

1908, the two governments agreed to recommence their frontier negotiations and withdraw their forces in the interim. Colombia had done so; Peru had not.[305] The Manaos Colombian consul traveled to Cotuhé to meet with the commission on its way out of the Putumayo, but it is unclear that he did so.[306]

When Casement learned in September that the consuls were indeed now traveling with Rey de Castro and Arana, he wrote the Foreign Office that the trip would certainly become a farce:

> In reality everything will be done to aid [Arana] in profitably exploiting the patient native race. The leopard does not change his spots; and the Aranas and Zumaetas have no more feeling for Indian humanity today than when they began their career of murder and pillage on the Putumayo.[307]

The two consuls were not taken in. Fuller later observed:

> Although Señor Rey de Castro is a high official of the Peruvian government, I do not consider the selection of a man of his reputation, for the duty of accompanying us, by any means a compliment to myself or to the United States government. His reputation is well known, and a matter of common talk in Iquitos.[308]

Fuller added:

> I thanked Señor Rey de Castro for his courtesy, but declined his assistance, stating that, from the advices I had received from the acting prefect at Iquitos, I understood his mission to be an investigation as to the conduct of their business by the local authorities, and hence not the same as mine, which was to report on commercial and labor conditions, those under which money being publicly collected for missionary purposes might be spent, and the conditions that might be met with by any American citizens who might elect to go there as missionaries. I also stated that it was beyond my province to sign formal acts with him regarding the internal affairs of Peru without explicit instructions from my government. I added that I expected and desired to travel quietly and independently, to see the people in their home life; that I had

---

305. Colombian Consul General, Manaos to the Consuls of Great Britain and the United States, Iquitos, September 2, 1912. Bodleian Library MSS Brit. Emp. S-22 G323.
306. Colombian Consul General, Cotuhé, to the British and American Consuls "Wherever they may be," September 26, 1912. Bodleian Library MSS Brit. Emp. S-22 G323.
307. Quoted in Goodman, *The Devil*, 196.
308. Consul Fuller to the Secretary of State. October 28, 1912. *Slavery*, 46.

complete equipment for so doing, and that neither assistance nor escort were necessary. When asked if I did not form one member of a joint commission with the British consul to investigate crimes committed in the Putumayo, I stated that I did not; that I might and might not travel in the interior with Mr. Michell; that I was not informed whether he had instructions to investigate possible criminal acts on the part of the British company operating there. When asked if I could give him a copy of my report, I stated that I could not do so, but that the Peruvian legation in Washington might, if they wished, apply to the government for it.[309]

Michell also complained that Rey de Castro acted throughout as if he was empowered to organize the consuls' trip and then expected that all three would author the final report.[310] The consuls' movements were scheduled and monitored constantly.[311] They were subjected to several delaying tactics and were required to witness staged Indian festivals and dance performances arranged along the way by Arana's staff. After one of the performances, "Señor Rey de Castro kept pointing to their fat and prosperous condition, and expiating on the happiness and obvious contentedness of the people." Michell further noted, "Whenever we tried to talk privately to the Indians an employé of the Company who knew the language would approach, and the people would immediately cease to be communicative."[312]

The entourage arrived at La Chorrera on August 17 and remained in the Putumayo until early October. Fuller commented in his report to Washington that, while the Indians reported former abuses, the conditions had seemingly improved:

> I doubt whether they know the difference between proper treatment at the hands of the whites and maltreatment, for the simple reason that the first idea of the white man they had was bad usage. In case of any trouble they would not be likely to appeal to the authorities. They would not understand how, and they would have no conception of government. The only way to protect them

---

309. Ibid., 47.
310. Michell, *Report*, 3–4.
311. In commenting on his hosts comportment in this regard, Michell observed,
> His [Rey de Castro's] anxiety not to lose sight of us was amusingly evident. Though totally unfitted for physical exercise, he followed us over fatiguing roads, through heat and storms, wherever we went, while Señor Arana, a heavy man, no longer young, and suffering acutely from sciatica, also accompanied us, uncomplaining but indefatigable (Ibid., 12).
312. Ibid., 4.

is to watch over them and their interests. All of the Indians we saw looked well fed and vigorous. The photographs which accompany the present dispatch, and which were taken by the writer, will give an idea of what they look like. Our interpreter, who was in the district at the time the atrocities were being practiced, says that there is no great difference between their appearance then and now. They are a small people and not over strong physically. We saw a considerable number in the various posts who bore the scars of old floggings. The Indians all knew who Arana was. They called him "Captain of the Peruvians" and evidently stood in great awe of him. He is much more to them than the whole Peruvian government.[313]

Then, too,

> As for the company, I believe that, having followed the policy of forcing everything out of the native labor that they could, they are now resting on their oars until the Indian population can recuperate and the rubber sources replenish themselves; but, believing (as a result of the way in which government has handled the pending prosecution) that they would be immune from interference, they would not hesitate a moment to repeat the past were it necessary to make a big showing to unload the property. In fact, it is hardly good policy for those in control to force the property while the company is in liquidation, but, rather, it is advisable to conserve the resources until after the settlement of the receivership.[314]
>
> Señor Tizon, at La Chorrera, I believe to be sincere and honest and trying to do the best he can. His ideas are good, and if allowed to work them out he should in time accomplish much for the good of the Indians. I am inclined to believe, however, that his authority at present is limited and that he will only be allowed a free hand with the reforms so long as they suit the business purposes of those in control of the company. If the company were recapitalized and the necessity for paying dividends on a heavy stock issue were to arise he would likely go.[315]

The Indian population at La Chorrera was down to two hundred individuals from a former high of two thousand. Consequently, "It is the

---

313. Consul Fuller to the Secretary of State, Iquitos. October 28, 1912. U.S. Department of State, *Slavery*, 51.
314. Ibid., 57.
315. Ibid., 58.

last section in which to expect maltreatment of the natives, for the simple reason that there are hardly any of them left to maltreat."[316] Meanwhile,

> While the two chief managers are men above the average [Tizón and Loayza], those in charge of the subsections (with a few exceptions) are very ordinary and, I believe, entirely capable of repeating the atrocities of the past if instructed to or offered inducements, such as commissions on rubber produced. In other words, the machinery is all there; and there is danger that the temptation to make a big showing preparatory to seeking new capital might bring a return to the old conditions. The sole value of the property lies in the labor. The product is inferior and, I believe, worked out to a considerable extent, and the only possible way to make a showing is to push the Indians.[317]

Fuller further noted,

> The strength of the [Peruvian] government's and the extent of their control is indicated by the fact that Consul General Señor Rey de Castro found it necessary, in order to secure information in regard to his mission, to rely entirely on the company. The fact is that this vast territory was handed over by the Peruvian government for a private business enterprise, at first Peruvian and later British, while the government made no effort to exercise sovereign rights or establish laws therein. The sole officials—two in number—were government officials in name only, being employees of the company. It is only now, after repeated exposure of the maltreatment of the natives, extending over a period of time of some eight or nine years, the government has even made a pretext of doing something. The pressure brought to bear in Lima has evidently borne some fruit, in that just before my British colleague and I left Iquitos, and apparently only after we had announced our intention of making the trip, the organization of a force of constabulary for the region was hastily undertaken. As men could not be secured in the day or two available, 25 of the soldiers stationed at Iquitos were drafted for the service. These men were selected on the day we left and sent up on the launch *Liberal* with us.[318]

For Fuller, the bottom line was,

> Say what you will, it is nothing more nor less than forced labor, whether it is secured and kept by the rifle or by a system of peon-

---

316. Ibid., 61.
317. Ibid., 59.
318. Ibid., 53.

age based on advances on merchandise. I believe that the Indians only work rubber in the fear of what might be done to them, based upon the experiences of the not far distant past. I am inclined to believe that "commissions" are still undertaken to get the Indians to work.[319] There is no doubt that the company is in a very bad way financially, and, as something will have to be done before long unless the price of rubber goes up, the temptation to abuse is imminent and strong. Considering the inaccessibility of the region (with consequent high cost of transportation for supplies and product), the unproductive nature of the soil (making food extremely scarce), and the very inferior quality of the rubber produced, it is hard to see how the enterprise can be made to pay without hard treatment of the Indians, forced labor to say the least.[320]

Meanwhile, in both London and New York, things were heating up. On August 4, not only did Canon Herbert Henson denonunce the Peruvian Amazon Company from the pulpit of Westminster Abbey, but that same day *The New York Times* published the story of Robert Isaacs, an elevator operator in the city, who claimed to have worked for Arana in the Putumayo and to have been a party to many brutal raids. He claimed, "I can go back to the Putumayo forests and show anybody heaps of rotting skeletons at the places where we surrounded and massacred the Indians."[321]

There was then an exchange between Henson and the attorneys of the Peruvian Amazon Company's directors, culminating in publication, on August 24, in *The Times* of London each side's acerbic letter to the other. By now the names of the members of the Board of Directors had been revealed, and Henson advocated that both they and Arana be prosecuted in a British court. The Peruvian Amazon Company attorneys demanded that his charges against the Company be retracted and the canon refused. The following day, the newspaper editorialized in favor of Henson.[322] In short, in Europe and the United States the scandal was spinning out of control—at least that of Julio César.

Unsurprisingly, in Peru the dynamic was somewhat different. The prestigious Lima newspaper *The West Coast Leader* (edited by an American) editorialized in agreement with the Lima correspondent of the *New York Herald*, who in turn quoted the Iquitos correspondent of *El Diario*, blaming the accusations of crimes in the Putumayo on "Colombian fron-

---

319. Ibid., 60.
320. Ibid., 6.
321. Quoted in Goodman, *The Devil*, 189.
322. Ibid., 205–209.

tier authorities who are friendly to Enrique Cortez, a Colombian banker whose partner is the notorious General Rafael Reyes, former President of Colombia. Cortez seeks the rubber consignment business seeking to destroy the Peruvian firm of Arana, who are the principal shareholders in the Peruvian Amazon Company."[323] It also accused Colombia of being sanctimonious, given that it was harboring criminals who had fled the Putumayo after being indicted by Judge Paredes. The editor noted further that any atrocities that transpired happened in the fastness of a vast Amazonian jungle that the Peruvian government found difficult to administer. This was particularly so after the *modus vivendi* agreements that removed authorities of both countries from the Putumayo, leaving it entirely in the hands of the rubber barons—thereby creating what Peruvian authorities referred to as "the protocol of the barbarization of the Putumayo."[324] President Leguía had tried to investigate but was deceived regarding conditions there even by his own authorities. Furthermore, the greatest criminals in the Putumayo were the Barbadians. Casement's report itself stated, "responsibility for these horrors falls exclusively on Great Britain, as the exports from those places, (the rubber districts) are shipped to British markets and carried by British vessels. Some of the employees are British subjects and the future of these regions depends upon British capital." The Paredes report was due to be published shortly in translation in the August issue of *Peru Today*. Until then it was important to withhold judgment.[325] For its part, the English-language publication *The Inca Chronicle* noted that it had published articles two years earlier on the abuses in the rubber districts and therefore welcomed Great Britain's involvement in the matter, since that would force the Peruvian government to investigate and institute reforms.[326]

Then, too, there was Arana's continuing political leverage in Peru. In early September, Senator Egoaguirre and his fellow senator Rojas from the Loreto jointly introduced a bill in the Peruvian Senate to create a commission to investigate who was slandering Peru and for what purpose. It also prohibited investigations by foreigners on the national territory. The measure was approved.

There then ensued an epic exchange in the Peruvian parliament. On September 12, 1912, Deputy Castillo, responding to *The Times*'s article on the Casement report, rose to his feet to demand a full investigation of the alleged atrocities in the Putumayo. According to the London newspaper,

---

323. *The West Coast Leader* (Lima, Peru), August 7, 1912, vol. 1, no. 32, 2.
324. Ibid.
325. Ibid.
326. *The Inca Chronicle*, August 18, 1912.

Peru did not deny the charges but seemed incapable of doing anything about them. Some 215 outstanding arrest warrants had gone largely unserved; several of the accused were walking about the streets of Iquitos, Lima, and Callao. National honor demanded that an investigatory commission be formed and that the criminals be punished (great applause).[327]

This prompted the deputy from the Lower Amazon district (which included the Putumayo) to rise in defense of the Casa Arana. Deputy Carlos de la Torre said that Castillo presumed guilt where there was none. The whole affair was a fabrication designed to extract money. He accused subaltern officials in Iquitos of trying to blackmail Arana. Then there was the Englishman [sic] who went to the Putumayo and lost his equipment there. Unless he was paid 10,000 pounds (later reduced to 5,000) he would initiate a smear campaign against the Company—"the one that we are now all aware of." Casement, too, had tried to extort money from the Casa Arana. As for the supposed flight of criminals, many were currently imprisoned in Iquitos and those that fled did so because an impetuous judge had arrived in the region and issued 150 [sic] baseless arrest warrants. While he was a bit shaky with his facts, the deputy (in Valcárcel's view) seemed to be simply parroting the line provided to him by Carlos Rey de Castro.[328]

Fellow deputy, Torres Balcázar, then stated:

> I have many relationships in Loreto where I have resided for a certain time, and from the communications that I have received I have become convinced that *all of the reputed crimes alleged to have happened in the Putumayo region are completely false* [emphasis in the original]. The Casa Arana is one of the most respectable [enterprises] in Loreto [and] principally its manager is a distinguished person and worthy of every consideration: Señor Julio Arana. Besides we should remember that for us the Casa Arana has earned very special merits, because with its own resources, with its own personnel, it has sustained and continues to sustain the Peruvian flag in those far-distant regions. I wish that my words appear in the record because, as I have said, I am profoundly convinced that we are dealing with blackmail here, as the honorable Deputy from the Lower Amazon has affirmed.[329]

Deputy Batta commented that nobody should believe that news of the scandal was limited to *The Times*. It had appeared in *Le Journal* and *Le*

---

327. *El Comercio*, September 13, 1912. Also Valcárcel, *El proceso*, 39–40.
328. Ibid.
329. Ibid., 41.

Matin in Paris as well. He was recently in New York City and saw a sign there in neon lights that proclaimed: "In Peru the life of a man is worth less than that of a mosquito." He believed that Great Britain was determined to remove the Putumayo from Peruvian control and was soliciting American support for the policy. Washington had dispatched its consul to the area. Castillo agreed to have Batta's remarks added to his, and the parliament then approved a resolution calling upon the government to appoint a Putumayo investigatory commission and provide a firm timetable for its investigation.[330]

On September 20, 1912, Peru's minister for external affairs (Germán Leguía y Martínez) responded. He noted that Paredes had already been there and found that most of the crimes antedated 1909 and there were few abuses any longer. Most of the Casa Arana employees were gone. Paredes himself had been working with Prezet in Washington to counter defamations of Peru. Furthermore, there had been recent new investigations on the ground by Carlos Rey de Castro, "the functionary closest to where the events were staged and whose wisdom, intelligence, and activity inspire the greatest absolute confidence." The Peruvian Amazon Company's liquidator, Julio César Arana, had also been sent to La Chorrera to ensure that the "cruelties" ceased.[331] Reference, of course, was to the August trip by Fuller and Michell to the Putumayo aboard the *Liberal* into which Rey de Castro and Arana intruded themselves in order to monitor the consuls' investigation.[332]

Leguía claimed to have heard back from his Manaos consul that the abuses were indeed over and that the minister anticipated receiving a full report shortly from Rey de Castro that would lay to rest all of the "malevolent propaganda" against Peru while bringing the scandal to a close. Finally, he had no need of appointing a new commission, since the one named the previous spring [which included Paredes] would soon begin its work.[333]

All of this was, however, a belated gesture at best, since, on September 27, 1912, the liberal politician, Guillermo Billinghurst, was sworn in as Peru's new president. Clearly, Arana's and Egoaguirre's political fortunes in Lima had taken a decided turn for the worse.[334] Furthermore, the Le-

---

330. *El Comercio*, September 13, 1912.
331. Reported in the September 30, 1912, issue of *La Prensa*.
332. Valcárcel, *El proceso*, 277.
333. *La Prensa*, September 30, 1912.
334. Not only was Billinghurst pro-worker, he was also of English descent, although there is no direct evidence that he was especially favorably disposed toward London. However, for Loretan regionalists, accustomed as many were to equating primal loyalty with descent and birthplace, Billinghurst's antecedents could not have been comforting. Billinghurst did not last long. When his programs were opposed by conservatives in Con-

guía article had been accompanied that same day by another in *La Prensa* written by Dora Mayer. She detailed the history of the previous two years and applauded the ongoing commitment of Great Britain to airing the Putumayo scandal. She denounced the appointment by Leguía of Egoaguirre to Peru's pending investigative commission. It simply undermined the commission's credibility. She also cautioned that, under the Monroe Doctrine, the United States seemed poised to seek control of the Amazon. She recommended that Peru cooperate closely with the Vatican as a counterweight to American designs. This same issue of the newspaper published a lengthy summary of Casement's Blue Book, but without any sort of editorial comment.[335]

On October 7, Michell was back in Iquitos and penning his report (received in London on December 5, 1912). On October 14, the British consul was given a packet of documents by Rey de Castro that the Peruvian had had translated for submission with Michell's observations to the Foreign Office. There were the requested statistics provided by Arana's section chiefs regarding rubber production in their jurisdictions. Some had expressed their consternation (along with refutations) of the articles in the London press of the alleged atrocities. Rey de Castro had written his own summary of the trip (to be sure, one which differed in substance, and certainly in spirit, from Michell's).

On October 17 the Colombian consul in Manaos wrote Michell saying it was common knowledge that the Peruvian Amazon Company was assisting its accused employees to escape, giving them money and Indians to help them reestablish themselves elsewhere. Abeldardo Aguero "is one of the most notorious of the assassins" and

> Seventy Indians, men, women and Children [*sic*] were given to him to work for him as a tenant of the Suarez firm at Chapacura on the river Tahuamano in Bolivia. These Indians have been kept as victims of the whip, the hatchet, and the revolver by Abelardo Aguero. More than thirty of the Indians have already died, or have been disposed of. Several of the women have been given as presents to his friends. Every four months he sends a young woman to his friend Augusto Jimenez at Cobiga, in Bolivia.[336]

---

gress he threatened to dissolve it. He attempted to arm the populace to fight the Army. However, in 1914, the then Colonel Oscar R. Benavides Larrea (Basque-surnamed) led a military coup that forced Billinghurst into exile in Chile, where he soon died. Benavides became a general and presided for eighteen months before handing over the reigns of government to José Pardo y Barreda.
335. Ibid.
336. Colombian Consul General, Manaos, to H. E. J. B. Michell, British Consul, Iquitos,

About this time, there was a curious telegram from Arana to a Lima newspaper, *La Crónica*, denying the allegation that he was attempting to sell Indians to the Manaos-Mamoré railroad construction project in Brazil.[337] Nevertheless, shortly thereafter, *La Prensa* reported on the harassment by officials in Iquitos of a Peruvian customs officer:

> The cause of the injustices against the victim Cherre is his knowledge that 120 families were sold by the *cauchero* Ruiz to the Casa Arana. These families were made up of Jivaro Indians, and once they were in possession of the Casa Arana it sold them to another company situated on the Madeira River, where there are few people due to the inclement climate.[338]

Upon being vindicated and reinstated by the Peruvian Supreme Court, Judge Valcárcel had resumed his duties in Iquitos. It was then that the local press launched attacks on him, including accusations of treason for having collaborated in London with *Truth* in the defamation of both the Casa Arana and Peru.[339] After Paredes wrote an editorial defending his fellow judge, he, too, became the object of renewed attacks in the rival newspapers. As in the past, such Iquitos articles were transmitted to Lima and appeared there in certain of its newspapers. Both judges were denounced by Rey de Castro as being key organizers of a shadowy Iquitos political faction, "La Cueva," that was opposed to both the Casa Arana and former Peruvian president Leguía's faction within the nation's ruling party.[340] The calumnious denunciations were repeated in the Peruvian parliament by Egoaguirre and his fellow Loreto senator.[341]

October 17, 1912. Bodleian Library MSS Brit. Emp. S-22 G323.
337. *La Crónica*, October 10, 1912.
338. *La Prensa*, December 3, 1912.
339. Valcárcel, *El proceso*, 316. Valcárcel rejected the charge (Ibid., 323).
340. Ibid., 317. Department of State, *Slavery in Peru*, 135.
341. Ibid., 271–73, 318. While some historians accept the allegation at face value, it can just as easily be regarded as one more obfuscatory tactic by Rey de Castro and the Arana forces. Chirif questions the very existence of *La Cueva*. He notes that none of the ostensible members of it ever claimed to be. Nor are there any known foundational or statutory documents regarding its existence, let alone a mission statement (unlike the "Loretan League" of the Loretan regionalists, which had all of the foregoing). Finally, Chirif finds it inconceivable that any group, let alone one of self-styled progressives, would call itself "The Cave"—with all of the term's association with troglodytes, wild beasts, and sinister darkness. Chirif, "Los informes," 79. It might also be noted that the ostensible "*La Cueva*" faction was opposed by *La Liga*, a formal organization of members of Iquitos's business and political elites. In 1913 (or the same year as Rey de Castro was publishing in Barcelona his defenses of Arana). Pilar García Jordan, "A propósito de redes sociales, económicas y políticas en el Iquitos de inicos del siglo xx," *Boletín Americanista* 56 (2006), 104.

Meanwhile, Arana was making certain that none of his station chiefs were apprehended. In Goodman's words:

> Andres O'Donnell managed to get himself safely to Panama even though he had been arrested. A legal quibble over the extradition [to Peru] papers had led the chief justice of Barbados to order O'Donnell's release from custody, and he promptly made his way to Bridgetown's wharf to get out on the next available ship. Armando Normand had escaped to Bolivia, was living in Cochabamba, in the center of the country, and was appealing to the Peruvian courts to quash his warrant. Victor Macedo, according to Consul General Lucien Jerome, was living openly in Lima; when the police went to the address they were given, there was no sign of the man, though he had been there until very recently. Jerome suspected that Macedo had been given a tip-off.[342]

And so it went in Aranalandia. Indeed,

> The machinery of the Casa Arana was so perfectly adjusted that it maintained paralyzed the judicial process for four years, during which the members of the judiciary employed the most absurd legal recourses [obfuscations]. In Iquitos, Julio César Arana was not only considered to be a patriot, a defender of Peruvian sovereignty against the pretensions of Colombia, and a civilizer of cannibal Indians, he also corrupted judges, politicians, mayors, commissioners, and functionaries. In Iquitos, the Casa Arana showered millions of *soles* on the Chamber of Commerce, the Municipality, the Departmental Council, the Beneficent Society.[343]

Many individuals in Iquitos, including officials, were beholden to Julio César. It was not uncommon for the Loreto's funding from Lima to be tardy. But it was not to worry, since a cash-strapped government employee could borrow a salary advance from the Casa Arana.[344]

---

342. Goodman, *The Devil*, 198.
343. Lagos, *Arana*, 144–45.
344. Indeed, we are provided one small insight into this pervasiveness by Pablo Zumaeta in his subsequent attack on the credibility of Judge Paredes. It seems that, when he was appointed judge, in 1907, Paredes asked the Peruvian Amazon Company [!] to send half of his salary to his invalid father. Zumaeta argued that, given the high cost of living in Iquitos, Judge Paredes could not possibly have survived there without accepting bribes. Pablo Zumaeta, *Las cuestiones del Putumayo: Memorial* (Barcelona: Imprenta Viuda de Luis Tasso, 1913), 6–7 [*Las cuestiones*]. The allegation raises another interesting point — namely that it was the Peruvian Amazon Company that was underwriting an Iquitos judge's salary. There is legitimate question regarding Zumaeta's authorship. While he signed the piece, it seems to have been written by Carlos Rey de Castro. More about that

Paredes himself summed up the situation as follows:

> The institutional life of Loreto has always been in the hands of the Casa Arana. This enterprise was all encompassing and dominating, strong and feared. Its millions made everyone genuflect, its influence was felt everywhere: the Chamber of Commerce, City government, the Departmental Board, the Beneficent Society, etc.—everywhere Mr. Pablo Zumaeta and his employees. The old political bosses of the Loreto helped the Casa Arana, who then through the support of this power were able to treat this land [the Department of Loreto], as they always have, as their feudal holding. The Casa Arana and its allies formed an oligarchy against which no one dared to raise his voice. *Since here there have never been political parties, even the Congressional representatives had to go along with the Casa Arana. Whoever was designated or supported by the Casa Arana was elected* [emphasis in the original]. Everyone served it submissively. And what is far more scandalous, every demand or request made to the government, no matter how unjust, always enjoyed the support of all these institutions under Arana's control, whose enterprise thereby exploited with impunity the prestige of appointed officials.[345]

On November 5, 1912, the leading merchants of Iquitos feted Arana at a banquet. Julio César assured the gathering that he had weathered the baseless attacks, and underscored the fact that he had been sent from London as the liquidator for the Peruvian Amazon Company as clear evidence that his good name in England remained intact. Furthermore, he had sacrificed much to bring civilization to the Putumayo:

> It is a region, watered by the river of that name, which has remained unexplored until a few years ago and which with part of its forests inhabited by cannibal natives has for a long time resisted every attempt at civilization. It was necessary to establish enterprises strong and powerful in capital and resources in order to achieve the domination over the tribes which were an obstacle to the march of progress. . . . I do not defend nor shall defend those who, calling themselves civilized, have perpetrated cruelties through degeneration or greed but I shall do all in my power that light should be shed and that the name of those men should be rehabilitated who acted in defence of the National territory, contributing to our sovereignty.

---

below. Valcárcel regularly underscores Zumaeta's obscenely close relationship with (and influence over) Iquitos's judiciary tribunal (for example, Valcárcel, *El proceso*, 255).
345. Cited in Ibid., 270.

> I do not know if perseverance in work and faith in the destiny of my Country, stimulated by our valuable and benevolent friendship, will lead me on to commercial success but in exchange I am certain that in all adverse vicissitudes of fate, I must preserve stainless the name left me by my fathers [parents] and which, with fortune or without, I wish to leave equally pure and clean to my children.[346]

American Consul Fuller reported on the event to Washington:

> As to public opinion in Iquitos, a large subscription dinner was given to J. C. Arana just before I left by the inner circle of the chamber of commerce. At this Consul General Rey de Castro and others made speeches lauding him and the company. Only one discordant note was heard. One of the speakers made the point that "throwing bouquets" was all very well, but that Peru and the whole civilized world were waiting to hear from Arana some word or proof to exonerate him from the charges under which he rests.[347]

British Consul Michell told London:

> The editor of the paper, Dr. Gamarra, got up at the table and said it was all very well to be presenting bouquets and so on, but it should be borne in mind that Arana's name was connected with very reprehensible practices; and it would be very convenient if that opportunity was taken by Don Julio to clear his name. This fell like a bombshell in the middle of the banquet, and Don Julio got up and said he had plenty to say in exoneration of himself, but this was not the occasion.[348]

Meanwhile, the next day, or November 6, S. Restrepo, head of the Colombian delegation in London sent a missive to the secretary of the House of Commons stating that his country had long sought negotiated arbitration of its border dispute with Peru. It had also on numerous occasions the treatment by Peruvians of Colombians in the Putumayo, as well as of the Indian population there. Restrepo concluded by welcoming any inquiry into Colombia's treatment of its own Indians.[349]

By mid-November 1912, Arana was in Manaos, where he would remain for the next three months. On November 30, 1912, the Peruvian ambassador in Washington forwarded to the U.S. secretary of state the published report authored by Paredes after his initial visit to the Putu-

---

346. Quoted in Goodman, *The Devil*, 201–202.
347. Valcárcel, *El proceso*, 70.
348. British Parliament, *Report*, 175.
349. S. Restrepo to Walter Legge, November 6, 1912. Bodleian Library MSS Brit. Emp. S-22 G335.

mayo. It applauds the efforts of Tizón and notes that the wholesale abuses prior to 1906, and those that continued afterwards, had all but ceased. The real culprits fled in the face of the impending arrival of both the Paredes mission and the two consuls. At the same time, Paredes was opposed to English intervention in the Putumayo—including a budding initiative to send Protestant missionaries there.[350] He further noted that the Peruvian Amazon Company was, after all, an English enterprise, and its directors were deserving of opprobrium. Regarding Arana and the subaltern administrators in Iquitos who were his minions, Paredes stated:

> There was a sort of tacit agreement to deny the facts, in spite of their being known with certainty. A species of false patriotism, foolish and mistaken, and a certain respect, based on servility and adulation, for the opulence of the Arana Co., caused everything to be kept under cover for a long time, even going so far as to deny absolutely the existence of the evil.[351]

Nevertheless, it could also be said:

> One of the reasons why the efforts of the [Peruvian] government were nullified was the exportation of the Barbados negroes by the English consul general, Sir Roger Casement, thus depriving the Peruvian courts of the important source of information, for there can be no doubt that their testimony would have thrown more light upon the case, illuminating dark places in the court proceedings, because the aid extended to these negroes—the real hyenas of the Putumayo—was the first step toward the breaking up of those bands of assassins.[352]

On December 10, 1912, Judge Valcárcel issued arrest warrants for Juan B. Vega[353] and Julio César Arana. In his own words, and at the age of thirty-two, he wrote, "Upon signing these orders, I had the awareness that my very life might be in danger, that my judicial career was cut short, and, worst of all, that my reputation would be sullied."[354]

---

350. Stanfield, *Red Rubber*, 170.
351. The Peruvian minister to the Secretary of State. Washington, D.C. November 30, 1912. U.S. Department of State, *Slavery*, 193.
352. Ibid., 194.
353. According to Consul Fuller,
    Señor Juan Vega, another implicated person, now a business man in Iquitos, left Iquitos very quietly and hurriedly by the mail steamer while I was away, going to Switzerland. It was stated that he left on account of private advices from Lima and that he went to Switzerland to avoid extradition. Consul Fuller to the Secretary of State. Iquitos, October 28, 1912. U.S. Department of State, *Slavery*, 56–57.
354. Valcárcel, *El proceso*, 273.

Three days later, an angry mob swarmed through Iquitos demanding Valcárcel's head. It seems that, despite a Peruvian law making it illegal to defame government officials, Iquitos authorities had looked the other way when an inflammatory circular was distributed in the city's streets. It stated:

> Today on the steamboat "Atahualpa" a coward flees, a miserable man with the title of judge has sentenced to prison many of your innocent sons, among them Julio C. Arana because he would not give him 20,000 *soles*. Will you consent to this? Will you allow the man who sold the Putumayo to the English for 20,000 pounds, he who robbed lands from your parents, leaving their unhappy children in poverty, to flee abroad to enjoy the fruit of his infamies? No, a thousand times no, the one responsible for the hecatomb of "La Pedrera," the man responsible for spilling of so much Peruvian blood, should be tried by you; he should lower his criminal head before the sovereign townspeople. Valcárcel shall not [be allowed to] go. The townspeople must make him render accounts, drag him through the streets, and purge him of the infamous crimes of *lèse patrie* that he has committed.[355]

So, Valcárcel joins the growing list of those who tried to blackmail Arana. Lima's *La Crónica* noted that Pablo Zumaeta had defamed both Paredes and Valcárcel. It declared:

> It is to be hoped that the government, jealous of its good name, will not wish to justify that which the previous one sustained as an excuse for its inaction, namely: that distance and difficulties in communicating had made it impossible to exert central authority in the Loreto Department, and that in these moments surely the government minister [in charge] will have ordered with full force that the [Loretan] authorities fulfill their obligations, by detaining the promoters of the scandal, while providing the most complete guarantees of the safety of the magistrates and ordering the immediate execution of the judicial order [of imprisonment], except in the case of those who can post a substantial bail. . . . If Señor Arana is innocent, as we hope, his detention . . . as well as that of Don Juan Vega, does not in any way damage them from a moral standpoint, given that their innocence should be easily proven.[356]

The newspaper concluded that, had Valcárcel been lynched, it would have confirmed to the world that the accused were indeed capable of the crimes in the Putumayo.

---

355. Ibid., 278–79.
356. *La Crónica,* December, 17, 1912.

There was additional drama provided by Lima's leftist newspaper, *La Acción Popular*. On December 12, 1912, it began publishing a series of articles by Captain Moya del Barco. He had recently commanded Peru's garrison in the Putumayo for seven months and was now accused of treason and espionage by his military commander, as well as by the civil and judicial authorities in Iquitos. The Peruvian captain had been imprisoned briefly, but was then released when it became evident that the charges against him were entirely spurious. It seems that he had allowed a Colombian Army officer to set foot briefly at La Chorrera. The man was treated politely, but his every attempt to garner intelligence had been thwarted.

Captain Moya del Barco argued that he had simply extended the minimum required of him under international rules of reciprocal military courtesy. In his view, the real reason for his legal difficulties and loss of his command was his unwillingness to turn a blind eye to the abuses in the Putumayo of his military superiors, Loretan officials and the Arana interests. He was an admirer of Judge Valcárcel for his honesty and courage. He also disclosed that the military officials in the Putumayo expected to be furnished with Indian women by the Peruvian Amazon Company and were then allowed to exchange them for others whenever they became bored with their concubine. Moya del Barco believed that the Indians would become excellent and loyal Peruvian citizens, prepared to defend their country, if better treated. He denounced the Barbadians for their crimes, but also believed that the directors of the Peruvian Amazon Company should be brought to Peru for trial. He further noted that those responsible included

> The general manager of the company as well as its Iquitos manager, the managers of La Chorrera and Encanto, and as the direct authors of many of these horrors the employees and section heads, in the majority Peruvians and Colombians; but as well Portuguese, Spaniards, Jamaicans and other nationalities who have direct or indirect participation in the crimes of the Putumayo. And, finally, those responsible through inertia for these crimes [include] all of the commissaries and justices of the peace of the Putumayo; as well as some of the authorities in Iquitos, such as the Justice Court, its fiscal and fiscal agents, who, having knowledge through public awareness of the criminal behavior of the aforementioned firm [PAC] in the Putumayo, have not taken any measures to secure punishment of the abuse, or even to avoid repetition of it in future.[357]

---

357. *La Acción Popular*, December 17, 1912.

Valcárcel had sent President Billinghurst an urgent telegram (on December 17) asking for assistance, and the response from Lima was a clear order to local officials to protect him. A chastened prefect appeared before the judge with the offer to detain anyone Valcárcel desired, while requesting that he put in a good word for him with the new administration (Valcárcel refused in the belief that the man had been entirely complicit in the plot against him).

On December 19, Julio Egoaguirre exclaimed in the Peruvian parliament that the rumor of the Casa Arana's implication in the anti-Valcárcel street manifestation was baseless, and that it had all been due to the judge's arbitrary closure of a popular Iquitos newspaper.[358] That same day, there appeared in Lima's *El Comercio* newspaper a letter from the "board of directors" of an Iquitos organization calling itself the Centro Loretano ("Loretan Center") that denounced both Paredes and Valcárcel as outsiders who belonged to the group known locally as La Cueva ("The Cave") whose sole purpose was to seize power in the town. The letter accused Paredes of having acquired *El Oriente* illegally and therefore deserving of imprisonment. The Paredes's report on the Putumayo was "a fantastic literary piece," and the man was both a pawn of the English and a member of London's Anti-Slavery Society. As for Valcárcel, in cowardly fashion he had waited until Arana had left for Manaos before issuing the warrant for his arrest, since the judge knew that Julio César would have easily debunked the accusations against him. In short, it was all a concerted campaign to discredit Arana, given the fact that he was being urged by his supporters to run for the Peruvian senate as the Loreto's most qualified candidate.[359]

Valcárcel was once again deathly ill and requested emergency sick leave. It was granted immediately. Lima ordered the prefect to provide the judge with transportation and medical assistance as far as Manaos (where he was to board a ship out of South America). In the event, the prefect stalled and eventually only provided inadequate transportation as far as the Brazilian border. Valcárcel later opined that the intention was for him to expire before reaching Manaos.[360] The judge was scheduled for medical treatment in Barbados, but after all of his legal battles he lacked the wherewithal to pay for it. It was then that his physician, (Basque-

---

358. On January 3, 1913, Julio César himself undermined that contention by stating in an interview that the manifestation was a reaction to the Valcárcel's warrant for Arana's arrest (Ibid., 274–76).
359. *El Comercio*, December 19, 1912.
360. Valcárcel, *El proceso*, 280–81.

surnamed) Dr. Arriola, leant him 70 pounds.[361] For the second time, Judge Valcárcel was leaving Iquitos "for his health" (never to return).

In Manaos, Arana, albeit under close surveillance by the British Foreign Office, remained impervious. From there, on December 30, he sent a telegram to Lima's *La Prensa* newspaper in which he noted that the Iquitos warrant against him proved that the Colombians had managed to convert even Peruvians into their agents. Julio César went on to note,

> I would proceed to Iquitos were not my presence indispensable in Europe to undo the machinations against me and to save the prestige of my Country. But I am confident that the high Tribunals of Peruvian Justice will annul the Order of Judge Valcárcel, who has abandoned Iquitos in a quasi-clandestine manner. I beg my Countrymen to suspend judgement until they know the defence I am about to make and which will produce a worldwide reaction in favour of Peru and leave my name unblemished.[362]

A short time later, the Iquitos High Court quashed the Vega and Arana arrest warrants for insufficient evidence.

Meanwhile, the Billinghurst administration was proceeding with its predecessors' initiative to form a new Peruvian investigative commission for the Putumayo. On December 10, 1912, Paredes wrote to London to the effect that his inclusion in it had been blocked by Senator Egoaguirre. However, he had now received his appointment and was scheduled to leave for the Putumayo within the next month. He noted, "It [the delay] has not been my fault, but that of the influences of the House of Arana with the government of Sr. Leguía."[363]

It was now that *El Comercio* published a lengthy article on the activities regarding the Putumayo of Roger Casement, the British Ministry of Foreign Affairs, the British parliament and the Anti-Slavery and Aborigines' Protection Society as possible threats to Peru's national sovereignty. At the same time, however, the newspaper reported receiving many letters from Peruvian ex-patriots in Europe, "manifesting to us the embarassments that these scandalous events were causing them, and underscoring the need for the Peruvian government to act, and in efficacious manner, to erase the bad impression that predominates in Europe today with respect to our culture and human sensitivities."[364]

---

361. Ibid., 320.
362. Quoted in Goodman, *The Devil*, 230.
363. December 10, 1912. Bodleian Library MSS Brit. Emp. S-22 G322.
364. *El Comercio*, December 11, 1912.

Part 3
The Hearings

The Select Committee

Unbeknown to Julio César, November 5, 1912, or the very day that he was being honored in Iquitos, British government agents had entered the London offices of the Peruvian Amazon Company with a search warrant issued by the speaker of the House of Commons. Its records were seized for examination by a committee constituted by fifteen members of Parliament. They were given a broad mandate that included the power to subpoena witnesses.[1]

On November 6, 1912, the Select Committee, as it was called, initiated its Hearings into the dealings of the Peruvian Amazon Company. While not continuous, the sessions lasted until April 30, 1913. The published proceedings[2] document 14,019 questions put to sworn witnesses and their answers. The enormous, double-columned work is 622 pages long. While the Select Committee's primary charge was to look into the behavior of the Company, and particularly its British Board of Directors, inevitably there would be considerable testimony about the atrocities in the Putumayo (who knew what and when?). The Select Committee's mandate also included consideration of the British government's oversight of abusive labor practices throughout its own empire and the wider world.

MP Charles Roberts was named chairman. He immediately sought out Roger Casement's advice regarding selection of key witnesses to be summoned. Casement himself would give extensive testimony on two occasions. While both men recognized the danger that Sir Roger could be perceived to be controlling or manipulating the inquiry, in fact they met regularly in private (frequently in one another's residence) over the sub-

---

1. Collier, *The River*, 222.
2. British Parliament, *Report*.

sequent months of Committee Hearings. Casement shared with Roberts unique documents and his own copy of Rey de Castro's edited version of the Robuchon text. The latter was annotated extensively regarding its omissions and inconsistencies. Indeed, Casement provided twelve pages of notes on the Spanish text that included extensive chapter summaries, cross references to points in his own Blue Book and suggested queries to be put to potential witnesses—particularly Arana.[3] Even when abroad during late 1912 and early 1913 (in Ireland, the Canary Islands, and South Africa), Casement monitored the Hearings and provided Roberts with detailed guidance.[4]

Nor can there be any doubt regarding Casement's position on Julio César. By now, he was completely convinced of Arana's culpability in the Putumayo atrocities. Most of the advice given to Roberts was designed to assist the Select Committee (and the wider world) to reach the same conclusion.[5] Like Hardenburg before him, Sir Roger used extremely disparaging language whenever referring to his adversary.

Initially, the Hearings went over well-trodden ground.[6] On November 13, Roger Casement was queried about the details and conclusions of his two reports. He stated unequivocally that he regarded Arana and Zumaeta to be "criminally responsible under Peruvian Law" for the 1908 "massacre" at La Unión and La Reserva of their Colombian settlers. The raid had to have been funded by the Peruvian Amazon Company, since up to eighty of its employees took part, and many were transported in its flagship *Liberal*.

Two of Pablo Zumaeta's brothers had been killed by Indians. Bartolomé Zumaeta had taken part in the raid against the Colombians and

---

3. Goodman, *The Devil*, 215, 218; Casement's notes to Roberts in Bodleian Library, MSS Brit. Emp. S-22 G344a. Casement commented in an aside, "The Robuchon book was clearly issued by the Aranas, through the Lima authorities, as a puff for the hopes for floating of a big company in Europe."
4. There are literally dozens of handwritten letters in this regard from Casement to Roberts in the Bodleian Library MSS Brit. Emp. S-22 G344c.
5. This was scarcely unprecedented. Hughes describes similar proceedings in 1837–1838 held by the British parliament to discuss the merits of the practice of transporting British convicts to Australia. As did Casement with Chairman Roberts, the chairman of the Molesworth Committee reported the behavior of transportation opponent Bishop Ullathorne who came by his house "and tried to coach me as to the best way of giving evidence." Robert Hughes, *The Fatal Shore: The Epic of Australia's Founding* (New York: Vintage Books, 1986), 495. Hughes notes further,
> The Molesworth Committee claimed to be an objective tribunal. It was in fact heavily biased show trial designed to present a catalog of antipodean horrors, conducted by Whigs against a system they were already planning to jettison. (Ibid., 493)

6. British Parliament, *Report*, 12–35.

was subsequently killed in a confrontation with the Indian chief Katenere. Casement refused to regard Bartolomé's death as "murder" since

> I believe he was rightly killed by the Indians, and the attempt to catch Katenere was not due solely to the fact that they killed Bartolomé Zumaeta, but to the fact that he was fighting for his own people. He was fighting for the Indians against the slavery being imposed upon them; he was a dangerous Indian.[7]

Casement believed that many of the Barbadians were complicit in the crimes (including murder) that they reported to him. Some had been offered bribes to lie to him—if not necessarily by the Company, then at least by some of its section chiefs acting on their own volition. When he returned to Iquitos, he offered to assist the local authorities to prosecute the guilty Barbadians, but only on condition that the "higher criminals" (i.e., PAC officials) be included as well. And nothing happened.[8]

Sir Roger testified that he had no doubt that Arana knew of the abuses in the Putumayo, and that he and Pablo Zumaeta kept their English directors in the dark. Neither man deserved to remain in authority. Since there were only two ostensible Peruvian officials in that region at the time of his visit, and both were beholden to the Company, he believed that the Peruvian government was kept pretty much in the dark. He gave the following overview of recent Putumayo history,

> Yes, I have a general impression—I might say almost a conviction—which I arrived at that the Aranas went as filibusters to the Putumayo, and ostensibly opened up trade relations with the Colombians, who were already settled there; then they expropriated the Colombians sometimes by a fair deal and purchase, and sometimes by, I think, unfair dealing and sometimes by armed attack. These various proceedings lasted through several years, and the Aranas were from time to time assisted by forces of the Peruvian government. I presume the Peruvian government would have looked upon it as repelling an attack made upon Peruvian territory by Colombia; but, as a matter of fact, in the specific cases I have in mind, the raids on the Colombians on the Caraparaná, no evidence was ever laid before me that the Colombians did meditate attacking, or did attack, the Peruvians. It was a raid by the Aranas in order to get more rubber, and to close the gates of exit for the Indians flying from the region.[9]

---

7. Ibid., 14.
8. Ibid., 16.
9. Ibid., 25.

Nevertheless, the articles in *Truth* did force a review of the matter. It seems that in November 1909, or in the midst of the budding scandal, the Peruvian Amazon Company's Board of Directors had resolved to themselves fund the expenses of someone who would be sent to South America to assume management of the Company. That initiative failed. Then, too, a letter was written to the president of Peru requesting that the allegations be investigated.

And, of course, the directors *did* constitute their own investigative commission that went out to Peru. Casement had traveled with it—not as a member but rather as a matter of convenience—charged by the Foreign Office to look into the plight of British citizens (the Barbadians), but neither into the plight of the Indians nor the broader operations of the Peruvian Amazon Company. Of course, he became well acquainted with the members of the Company's commission and held them in high regard. He believed that they had gone about their business with open minds and ultimately came to share his conclusions regarding the abusive system.[10] He also praised Juan Tizón, noting that three days after his (and the PAC commission's) arrival at La Chorrera, Tizón professed to be completely convinced of the veracity of the testimony of the Barbadians. He "practically placed himself in my hands."[11]

Casement noted that his introduction to the ill treatment of Indians in the Upper Amazon came from his reading, on his first journey to the Putumayo in 1910, of the Enock book, *The Andes and the Amazon*.[12] Asked if the evidence of abuse was blatant, Casement replied:

> A man of ordinary common sense could not have failed to notice it. For instance, before leaving La Chorrera on my way up we stopped at a place called Indostan, which was a food settlement to produce food for La Chorrera. There were about 40 Indian men and women employed in cultivating the ground under the direction of one of the agents of the company, a man called Zumaran [Basque-surnamed]. I heard groans, and I went into a house and found three Indian men with very high temperatures and fever. I found two little girls quite emaciated, and then a man appeared with two heavy chains upon him, one round his neck and another round his body, with two padlocks. I went back to the steamer and asked the Commissioners to come and see it. That was the first incident.[13]

---

10. Ibid., 33.
11. Ibid., 34.
12. Ibid., 121.
13. Ibid., 116.

There was also the matter of Carlos Rey de Castro. The final report of the Company's Commission was critical of the highly unusual Peruvian official's 4,600 pounds of debt on the Company's books. In Casement's view, the money was simply a bribe and he had confronted the Peruvian Amazon Company's manager about the matter,

> As soon as I saw it, I asked Mr. Gielgud: "How on earth can a Peruvian Consul-General be drawing money from a British firm? What is the meaning of it?" And he said Mr. Arana had explained that Mr. De Castro had rendered great services in the way of publicity I presume in that book [Robuchon's]. As a matter of fact, also he had rendered great services in inducing the Prefect at the time of the raid on the Colombians to allow the prisoners to go.[14]

And then, on December 11, 1912, Sir Roger was recalled before the Select Committee, and he characterized the state of the Indians in the following terms:

> The Putumayo Indians, the people of the Montana, the eastern section of Peru beyond the Andes, are very largely still wild Indians. There are numerous tribes, the majority of whom go quite naked. These people have absolutely no human rights. They are hunted and chased like wild animals. . . . They cannot own their own bodies or their wives and children. . . . If you shot a man in the streets of Iquitos there would be trouble but if you went 90 miles away, where there were wild or semi-wild Indians, there would be no trouble at all.[15]

Casement added:

> The raids into Colombia were to catch Indians who had run away. The Peruvian judge, Dr. Paredes, told me himself the following explicitly. I hesitated to recall it in a dispatch to the Foreign Office—it was so brutal, but one of the higher agents of the Company, Miguel Flores, had on one of these correrias after fugitive Indians surprised one Indian settlement in the forest with his men. The parent Indians had just time to fly, but not to take their children, and 18 small children were left in the houses. When the parents had got a little way off they waited, hoping the children would overtake them. The judge said that the parents, according to the statement, had seen what followed. Flores and his men took the children and dashed their brains out against the trees. That man [Flores] I found on the Putumayo. He was one of many; he had

14. Ibid., 32.
15. Ibid., 116.

been a chief of section. He was one of the men Commandant Polack had arrested and kept in irons, and whom Señor Julio Arana caused to be released.[16]

Sir Roger, with a flair for the dramatic, produced a collection of newspaper articles from the Peruvian press, some of which spoke of the general practice in the Amazonian lowlands of enslavement of the Indians. There were his own photographs of abused Indians. He then displayed a collection of miserable trade items that the gatherers were given in exchange for their rubber. He held up a rinky-dink, single-barreled shotgun for which he claimed an Indian would become indebted for two years. This prompted considerable discussion of the huge disparities in what the Company paid for its trade items versus the charge for them in the Putumayo, in addition to the value credited to the accounts of the Indians for their rubber versus its true worth in the world market.[17]

How many Indians died? The estimates (there were, of course, no actual censuses) varied considerably, and there were mitigating circum-

---

16. Ibid., 113. Arana would himself refer to two Peruvian raids across the Caquetá—one in 1908 and another in 1910 (Ibid., 617). It seems that by 1912 Gregorio Calderón was residing on the Upper Apaporís River. After selling out to Arana he had resettled as a *cauchero* on the Lower Apaporís (a distant tributary of the Caquetá and well within undisputed Colombian territory) taking with him many Huitotos. Calderón had to abandon his initial settlement after suffering no fewer than three attacks in twenty days by Carijonas Indians sent by the Peruvians (likely Arana) to kill his rubber gatherers and Huitotos (twenty-six Indian fatalities alone). Hamilton Rice, "The River Vaupés: Further Explorations in the North-West Amazon Basin," *The Geographical Journal* 44, no. 2 (1914), 148. We have another account of the fate of the *cauchero* named Agustin Ciceri. He was on the fringe of Aranalandia on the Yarí River. He had a total of forty Huitoto men and one hundred Carijonas gathering rubber for him. But then, in 1909, an armed force of the Casa Arana, under the command of José Fonseca, came to attack him while proclaiming that his Indians were theirs. They fled and actually ended up on the Apaporís River as well (Dominguez and Gomez, *Nación*, 199). Along the same lines (probably referring to later, but time unspecified) is the testimony in the account of a Huitoto who claimed to have been part of a group transported with many other Indians out of Aranalandia to the Yarí River by a Peruvian named Zumaeta. He established himself there and forced them to collect *balata*. There was some mention of his punitive measures, but he did not seem to be excessively brutal (by Putumayo standards anyway). He was eventually detained by the Casa Arana for having absconded with Indians. After his return to the Yarí, he became ill and died. During his absence, his indigenes were told to clean the path from Pedrera to Tarapacá. After Zumaeta's return, the narrator went back to him willingly to collect rubber. But Zumaeta became ill and died. Quoted in Pineda Camacho, *Holocausto*, 200–201. Finally, there is evidence that, in the mid-1920s, Julio César's brother, Remigio, was the head of a Casa Arana operation on the Caquetá inland a half-day's walk from the confluence of the Caquetá and Cuemaní. Gómez López, *Putumayo*, 130.
17. British Parliament, *Report*, 111–13.

stances. The Peruvian Amazon Company's prospectus spoke of a population of forty thousand Indians from thirty different tribes inhabiting the Company's Putumayo realm. The point it was making was that there was a considerable available labor force. Furthermore, a single *racional* (with a rifle to be sure) could supervise three or four hundred indigenes (i.e., management costs were minimal). Casement had estimated that over a decade the population of Indians had been reduced from forty thousand to about eight or ten thousand. Were there thirty thousand deaths?

On December 4, it was G. S. Paternoster's turn to testify. Before proceeding with publication, *Truth*'s editors had sought verification of Hardenburg's claims. Paternoster noted that he had checked out the few references that Walter provided (including a former schoolteacher and the editor of an American newspaper). Perkins sent a letter of support. He also contacted Colombian consul general, Francisco Becerra, and he handed over a notebook filled with the accusatory statements of expatriate Colombians familiar with the Putumayo. However, it was Hardenburg's altruistic commitment to justice and the fate of the Indians that convinced Paternoster of the American's sincerity. He had come to England at his own expense and was short of funds, yet Walter stated from the outset that he sought no payment. Were he a blackmailer, why, once in London, didn't he make some sort of demand upon the directors or management of the Peruvian Amazon Company? He never did so.[18]

Paternoster had the good fortune that the British consul in Iquitos, David Cazes, was home for a holiday and agreed to be interviewed. Cazes affirmed that the allegations were probably true, and also said that, while his lips were sealed regarding the matter, he had privately seen the report written in late 1907 by American Consul Eberhardt characterizing the

---

18. Ibid., 96. Paternoster would subsequently write,

> Mr. Hardenburg himself supplied the clearest possible evidence of the motives which animated him by explaining that he desired no recompense for the information he wished to supply. He placed the whole of the information which he had acquired at the disposal of the editor of *Truth*, in whose possession all the original statements of witnesses still remain... No one who talked with Mr. Hardenburg for five minutes could have doubted either his good faith or determination. Behind a quiet reserved manner there was revealed a man's burning hatred of oppression and a youthful zeal for the punishment of the oppressor. It was to stop the crimes and secure punishment for the criminals that he made his way to England. The United States have every reason to be proud of the part their young citizen has played in bringing to light the treatment received by the Putumayo Indians at the hands of the rubber lords of the Devil's Paradise, and it should be a pleasure to his countrymen, as it is to those who made his acquaintance in this country, to accord him the recognition for his services to humanity which is his due. G. Sydney Paternoster, *The Lords of the Devil's Paradise* (London: S. Paul and Co., 1913), 41–42.

circumstances of the Indians in the Putumayo as "slavery."[19] He admitted that, while there were many rumors of the Putumayo abuses in Iquitos, they were dismissed as the rantings of disgruntled former employees of the Casa Arana. Saldaña Rocca was said to be another vindictive individual whose campaign was launched out of revenge for Arana's refusal to make him a loan to set up a printing press.[20] In the event, Paternoster had treated the Peruvian Amazon Company and its Board of Directors circumspectly in his articles, simply because he found it impossible to believe that a British company would knowingly collude in the atrocities of the Putumayo.[21]

On December 17 and 18, David Cazes was the witness.[22] He had served as British consul in Iquitos from 1902 to 1911. He rejected the phrase "honorary"; rather, he had been "unsalaried." His real profession was that of merchant, owner of the Iquitos Trading Company. He had supplied goods to the Casa Arana and held Julio César in the highest regard. Cazes stated that Paternoster's comments regarding their exchange were largely inaccurate. He never claimed to have seen Eberhardt's report.[23] Then, throughout his testimony, Cazes was equivocatory—constantly underscoring the fact that either he could not be certain or that in answering he was relying upon memory rather than documentary evidence.

Cazes had indeed confronted Arana with the accusations of abuse of certain Barbadians in the Putumayo. But then, in November 1906, a

---

19. Ibid., 95.
20. Ibid., 158.
21. Ibid., 98.
22. While in his earlier testimony Casement had been critical of the failure of David Cazes to inform the Foreign Office of both the alleged slavery of Indians in the Putumayo and the specific complaints of the Barbadians, Sir Roger was now asking Roberts to go easy on the former Iquitos consul. Casement had given the chairman a letter that Cazes had received from Arana in July 1910 announcing the arrival of the PAC investigative commission, while underscoring that its mission was commercial (and not an investigation into past atrocities). For Casement, the fact that Cazes gave him the Arana letter was sufficient evidence that the Iquitos consul's allegiances were to his office, and not to Julio César. Casement therefore asked that the letter not be introduced into evidence—since to do so would be an unnecessary embarrassment of Cazes, on the one hand, and might underscore the close collaboration between Casement and Roberts, on the other. Sir Roger also seemed ambivalent regarding Cazes as a person. As a largely unpaid amateur diplomat, the man had clearly been over his head. Casement's recommendation to the Foreign Office that a true professional diplomat be named to the Iquitos post had cost Cazes his appointment. Yet Cazes had hosted Sir Roger in his home, and shared useful documentation and frank opinions with him regarding Iquitos politics, as well as Arana's control of them. Goodman, *The Devil*, 222–24.
23. British Parliament, *Report*, 155.

Barbadian in Arana's employ took refuge in the consulate, claiming that he and other West Indians had been ordered to go on Indian hunts in the forest. When Cazes informed Arana that his government took the allegations very seriously, Julio César promised to look into the matter, and eventually dismissed the accused perpetrator, Ramón Sánchez. However, Cazes did not go to the Putumayo to investigate (since he was unpaid, he could not afford the time). When asked if Mr. Sánchez had been tried for this behavior, Cazes replied,

> No; I tried to take the matter further when he arrived back in Iquitos; but in an interview I had with the Prefect he said they could not do anything without I would directly charge the man but I had nothing to substantiate my charges, nor would I take the responsibility for doing so.[24]

Cazes said that, since Arana was using British citizens to police his territory, he insisted that the Barbadians be brought back to Iquitos for repatriation. Arana denied that anything was wrong, but agreed that any Barbadian wishing to leave (even though under contract) could do so. Cazes could not remember the exact number. ("I am not certain whether it was 37, 47, or 27, but, anyhow, it was a fair number of them, and I had a very special interest in seeing that they were sent back to Barbados direct.") The Booth Steamship Company worked with him regarding their transportation.[25]

As early as 1901, Robuchon had told the (future) consul of the abuses of the indigenes of the Putumayo(!).[26] Confronted with his own letter to the Foreign Office in May 1905, Cazes admitted that he had informed London of the dire status of the Indians in the Putumayo. However, since they were not British citizens and there was as yet no London firm involved, the matter was beyond his official purview.[27] He later struggled with that point once the Putumayo was controlled by a London-based British company with a British directorate. However, some of his own employees had worked for Arana in the Putumayo and none had ever mentioned the atrocities.

If, indeed, he was aware of the Saldaña Rocca articles (they began appearing just about the time of the public stock offering of the Peruvian Amazon Rubber Company), why didn't Cazes inform the Foreign Office

---

24. Ibid., 157.
25. Ibid.
26. Ibid., 166. That would have been a year before Cazes's appointment as consul and before Robuchon had actually visited the region.
27. Ibid., 156.

of the allegations or contact the London firm for an explanation of them? The rather lame reply was that in Iquitos no one paid any attention to Saldaña Rocca, and that Arana later began the legal proceedings against their editor that caused him to flee Iquitos.[28]

Had not Germán Alarco sought him out after the *Truth* articles appeared to request a letter or statement from the English consul to the effect that the allegations were unsubstantiated and part of an attempt to blackmail the Company? Well, yes, but while he had refused, it was his belief that the Iquitos mayor was acting as a patriotic Peruvian, and not as a representative of the Company.[29] Not mentioned was the fact that Germán (along with his brother Abel) was a partner in something called the Arana y Alarco Company, a firm that imported goods into Iquitos (some of which were sold to Arana y Hermanos as well as to the Peruvian Amazon Company). However, it was also true that the Aranas had scaled back their mercantile businesses in Iquitos in order to concentrate their resources on developing the rubber operations in the Putumayo.[30]

And so forth. In short, Cazes was quite evidently too close to Arana to serve as a credible character witness.[31] He also admitted that he, too, had been hoodwinked by the Company. But given Arana's total control over access to the Putumayo, it was virtually impossible to secure independent reliable information regarding conditions there.[32]

Once Cazes finished his testimony, the Select Committee adjourned on December 18 for a Christmas break. Casement promised Roberts to turn over his remaining documentation and his private diary to the chairman, and then departed for a rest in the Canary Islands. Sir Roger was clearly worried that the Hearings might lead nowhere—indeed, that they might result in an excuse for the British Foreign Office to simply admonish the Peruvian government to undertake serious investigation and

---

28. Ibid., 158–59.
29. Ibid., 156, 166.
30. Ibid., 107, 146.
31. At one point, Cazes produced a letter of apology from Roger Casement regarding his negative testimony concerning the Iquitos consul. He regretted causing him discomfort and expressed his gratitude for the hospitality extended to him by Cazes during his stay in Iquitos, as well as the candor in answering his questions. Casement enclosed a letter from one of the Barbadians' repatriated by Cazes, writing of it, "As a certificate of character to a British consul—yourself—from a distressed British subject, the letter is probably unique." The letter actually regarded gratitude for springing the accused murderer from the Iquitos jail. The Casement exercise seemed to be more of a collegial gesture between two British consuls and was not particularly relevant to the matters under consideration. Ibid., 164.
32. Ibid., 157, 159.

reform while washing its own hands of the matter. In that event, it would be business as usual for the criminals and victims in the Putumayo. All of these concerns were put in a letter from Casement to Roberts dated January 3, 1913, and sent from Las Palmas. In it, Casement admitted to having overstepped the bounds of his assignment from Grey. In fact, the investigation,

> . . . had been practically a one-man affair—my own. I must tell you the truth—so that you may understand—and you will acquit me, I know, from anything like boasting. It was really my doing from the day Grey sent me off until the Report was published two year later.[33]

In December of 1912, Walter Hardenburg's book, *The Devil's Paradise*, was published in London. It began with an Introduction by Charles Reginald Enock reiterating his belief that the enslavement and mistreatment of forest Indians was a widespread practice in Peru.[34] On January

---

33. Quoted in Goodman, *The Devil*, 224.
34. Hardenburg and Enock, *The Putumayo*. The two men had never met before and were not true collaborators. The publisher had approached Enock requesting the introduction. The book contained only a portion of Hardenburg's original text—beginning with his journey down the Putumayo. Much of the work reproduces documentation from testimony Hardenburg recorded in Iquitos and various articles published in *La Felpa* and *La Sanción*, not to mention his own observations and opinions. It also reproduced Casement's Blue Book. On balance, the effect was much the same. Hardenburg entitled his manuscript "The Devil's Paradise: A Catalogue of Crime." It was dedicated to "Senor Benjamin Saldana Rocca, Dr. Dario A. Urmeneta and the other benevolent men, who have raised their voices in defence of the innocent and helpless victims of the gruesome PUTUMAYO and exposed to the horrified eyes of the civilized world the stupendous crimes daily and hourly committed there, this volume is gratefully dedicated by The Author." In his original conclusion Hardenburg states,
> In making these exposures, I have obeyed the dictates of my conscience and my own sense of outraged justice and now, that I have made them, and the civilized world is aware of what occurs in the vast and tragic selvas of the River Putumayo, I feel that, as an honest man, I have done my duty before God and before society and trust that others, who are in a position to do so, will take up the defence of these unfortunates and the prosecution and punishment of the human hyenas responsible for these crimes. (W. E. Hardenburg, "The Devil's Paradise, A Catalogue of Crime." Bodleian Library MSS Brit. Emp. S-22 G335, 164.)

Hardenburg now raised his hyperbole to new heights,
> As we have already shown, over the whole length and breadth of this vast region, reigns one perpetual, eternal and devilish carnival of crime: in short words are unable to convey any idea of this gruesome field of blood and crime and bleached skeletons, rotting under the falling leaves of the forest trees. It is a living hell. No wonder that the vegetation is so luxuriant here, for the soil has been deluged with the blood of so many innocent victims of the bestial greed and rapacity of these vile monsters that it should be the richest on

11, 1913, *The Nation* published a favorable review of it, one of many in the British press that were unequivocally devastating for Arana. *The Nation* review began:

> The greatest disaster that ever afflicted the human race has been the contact of Europeans with the natives of the Tropics. In the long record of mankind's misery and pain nothing else can for a moment compare with the cruelty, lust, and brutal greed by which the contact has everywhere been accompanied. Some generous-minded people have doubted whether man's noble qualities could ever have been evolved from apes and similar animals. If they knew anything of the conquest of native races by white men, of the slave-trade's history, or of the present condition of "niggers" and "Indians" under the control of Christian communities, they would rather doubt whether the breed of apes could ever have evolved anything so fiendish as European mankind has proven itself to be.[35]

It was now in early January that the Hearings resumed with the testimony of the current British consul in Iquitos, George Babington Michell. He stated that the rubber coming from the Putumayo presently was,

> ... consigned to Cecilio Hernandez at Iquitos. Cecilio Hernandez, I believe, is connected by marriage with the Arana family, and I understand these consignments are brought to Cecilio Hernandez under a claim by Mrs. Arana, who is said to be the principal creditor.[36]

Asked to recall his journey to the Putumayo with American consul Fuller, Michell underscored the constant interference of their movements by Julio César and Rey de Castro. He was favorably impressed by the sincerity of the section chiefs Tizón and Loayza. Told that Casement had a low opinion of Benito Lores, the present military commander in the Putumayo, and hoped that he would be replaced, Michell dissented. He thought that the man had integrity.[37] As for the Arana machine's grip on everything, Michell observed,

---

earth! Ibid.

There is another curious Hardenburg manuscript in the Bodleian Library—an unpublished short story written by him and entitled "The Flower of the Selva: Tale of the Upper Amazon." It regards the ruination of a forest maiden and her tribe by none other than José Fonseca. I reproduce it as appendix 3.

35. *The Nation*, January 11, 1913.
36. British Parliament, *Report*, 170.
37. Ibid., 173.

You must understand the Putumayo is practically their private property. We were like walking about in another man's garden, and were expected to admire the flowers.[38]

When asked to comment on what might happen should the Putumayo revert to the Casa Arana, Michell exclaimed, "God help the Indians!"[39] Michell noted that the prefect of the Loreto, as well as an "English trader" (without naming David Cazes), were present at the laudatory banquet given in Iquitos in Arana's honor in November 1912.[40] The consul then noted,

> I may say that the publication of Sir Roger Casement's report has had a very disastrous effect on the trade of Iquitos. The Alcalde, the President of the Chamber of Commerce, informed me that the publication has had the effect that the European firms, who are in the habit of giving credit to local firms, had completely closed their credit, even to persons who are engaged in other parts of the country, and in no way connected with the Putumayo.[41]

Michell concluded: "I believe everything is done at Iquitos, and I believe both Tizon and Loayza, however good men they may be, are more or less tools in the hands of Pablo Zumaeta."[42]

Louis Harding Barnes, chairman of the Company's investigative commission in the Putumayo, also testified. At his first interview, and before agreeing to join the commission, the possibility of his being hired as manager at La Chorrera was broached. On the trip to the Putumayo, PAC Secretary Gielgud expressed his skepticism that there was anything to investigate.[43] Barnes later characterized Gielgud as "easily hoodwinked."[44] Barnes's commission members wrote their report collectively and independently of Casement (without reading his). The two documents were, however, in substantial agreement. As for the charge that Hardenburg was a blackmailer, Barnes had no more than hearsay evidence and no firm opinion. His commission was never asked to investigate that allegation. As for the charge that Casement had also attempted to blackmail the Company, he found it to be "absolutely incredible."[45]

---

38. Ibid., 172.
39. Ibid., 174.
40. Ibid., 175. Earlier (Ibid., 173), Michell had opined that "an honorary Consul, a trading Consul, could well have a conflict of interest."
41. Ibid.
42. Ibid., 176.
43. Ibid., 84.
44. Ibid., 84–85.
45. Ibid., 76. Dr. Dickey subsequently stated in print that Casement was the most genere-

Barnes gave several interesting insights into the Company's operation. He noted that the Arana firm was by far the most important in Iquitos.⁴⁶ The two government officials at the Putumayo during the PAC commission's visit were both on the Company's payroll. It was Barnes' understanding that Commandant Polack had requested that the Peruvian Amazon Company's employees accompany his force during the "guerrilla frontier warfare" at La Unión. He believed that twenty-five employees participated.⁴⁷ There was no standard fee paid to the Indians for their rubber; rather each section chief was at liberty to set his own price.⁴⁸

The commission spent considerable time at La Chorrera, more than a month in all. Its manager, Victor Macedo, denied any wrongdoing or having knowledge of the atrocities. Toward the end of their stay, Macedo approached the commissioners with a request that they write a letter exonerating him, and they refused.⁴⁹ On every critical matter, the London directors were kept in the dark by Zumaeta and Arana. Several of Arana's relatives had Peruvian Amazon Company accounts on which they were overdrawn by "many thousands."⁵⁰ Both the Iquitos and Manaos offices of the Company regularly ignored the London headquarters' request for proper accounting with comprehensive and comprehensible balance sheets.⁵¹

Barnes now believed that several claims in the prospectus for the Peruvian Amazon Company's stock offering were erroneous. When he had tried to ignite a sample of Putumayo "coal," it would not burn. The only agriculture that he observed was of the subsistence variety to supply some of the food to the section stations. He saw little potential for much else. Despite the positive descriptions in the prospectus of other PAC properties, his opinion of Nanay "was most unfavorable," and he didn't see a whole lot of potential at Pébas either.⁵²

---

ous man he had ever known, constantly receiving requests for financial aid that he never refused:

> I reflected also that Casement could have gotten as much as £100,000 for whitewashing the wealthy Peruvian Amazon Company. (I myself was offered £500 if I would pen one letter simply stating that I as a company doctor had seen no mistreatment of the Indians). Instead he made a most careful factual and voluminous report that ruined the Company. Casement has lingered in my memory as singularly lacking in venality, a man to whom it would not occur ever to "sell out" to anyone. Mitchell, *Sir Roger*, 735.

46. Ibid., 84.
47. Ibid., 85.
48. Ibid., 90.
49. Ibid., 92.
50. Ibid., 78.
51. Ibid., 77–78.
52. Ibid., 91.

Barnes agreed with Casement that Arana was "the organizer of the criminal system," and did not believe that Julio César was the proper choice as liquidator of the Peruvian Amazon Company.[53] When told that, as recently as September 27, 1911, Chairman Gubbins was telling the shareholders that the majority of the abuses in the Putumayo had transpired before formation of the Peruvian Amazon Company, and that many improvements had been made, basing his contentions on the Commission's report, Barnes disagreed: "That is not taken from our report. I have no recollection of giving any dates of that kind." Nor did Barnes remember anything in the report that would have justified Chairman Gubbins, stating,

> And on the whole, the efforts of the Peruvian Amazon Company, Ltd., to improve the lot of the Indians in the Putumayo have met with a fair measure of success, though much still remains to be done before the primeval savage is converted into a civilized human being."[54]

Gielgud was summoned before the Select Committee on several occasions for extensive questioning. He noted that he had written a great deal of the PAC commission's report, but had not signed it (or been asked to). In point of fact, during the journey home he had remained in Manaos for a while to conduct Company business. He gave the other members a handwritten copy of his impressions, and they then finalized their report back in England. Pressed on the matter, Gielgud stated that he agreed in the main with the report.[55] He admitted that his earlier letters regarding his first visit to the Putumayo were given far too much weight by the PAC's Board of Directors, particularly since it reinforced their wishful thinking by dismissing the atrocities. Throughout his testimony, Gielgud emphasized that the letters were based on impressions and not actual investigation. He had been sent out solely to examine the books—not the labor practices or the conditions of the Indians. He now believed that care had been taken to hide the true reality from him.[56]

Gielgud was questioned closely and repeatedly concerning his understanding of the practices and valuations of the Peruvian Amazon Company. He assigned a total value of 592,828 pounds to the Putumayo asset on the Peruvian Amazon Company's books. That involved treating a 33,000-pound item, and another 11,000-pound one, differently than Zumaeta and Arana

---

53. Ibid., 80, 88.
54. Ibid., 90.
55. Ibid., 102, 105.
56. Ibid., 102–105, 126, 131.

wished. It had to do in part with some lingering obligation of J. C. Arana and Hermanos to Arana, Vega, and Cia. They still owed Juan B. Vega some money from the earlier transfer, and they wanted that obligation to be assumed by the Peruvian Amazon Company. They were successful in getting Gielgud to reclassify the entries in December 1909, although he insisted that it happened prior to his accepting his management position with the Company and was in no way connected to that offer.[57]

Gielgud had understood the more than 500,000 pounds spent on developing the Company to have been primarily the costs of buying out the Colombians, but also for the "conquest" (*conquistación*) of dangerous and hostile Indians, not to mention expenses incurred defending the Peruvian Amazon Company's (and Peru's) territory against Colombia's pretensions and PAC Indians against would-be Colombian usurpers. There was considerable discussion over whether the Spanish word *conquistar* and the English one "conquer" meant the same thing. Arana's side would argue that the Spanish term meant "attract," and referred to the practice of trying to induce the Indians to take trade goods in return for collecting rubber (to work off their debt).[58] Gielgud described the system as follows:

> The general nature of the procedure, as I was given to understand it, was that a party of men would go into an outlying part of the forest; they would build themselves a bit of a house; they then would endeavor to get into touch with the natives; they would take a lot of trade goods with them which they would show to the natives and endeavor to get the natives to believe they could not do without them and so induce them to work.[59]

The issue was exacerbated when the Peruvian Amazon Company was unable to produce critical Putumayo account books for the telling period of 1907 and 1908 (or the first two years of PAC ownership in the area).[60] There was also the critical matter of whether the Indians were treated as property. There had been an internal difference of opinion over whether they should be counted as worth 10 pounds each. Macedo argued that they represented 30 pounds per head when referring to the Morelia section, and Arana concurred because the Indians there were rebelling and the costs of their *conquistación* were higher than elsewhere.[61]

---

57. Ibid., 379–89, 415–16.
58. For example, Ibid., 397–408 and elsewhere, 419.
59. Ibid., 397.
60. Ibid., 300–301, 310.
61. Director Read testified in response to a query from the chairman of the Select Committee that it was fine to calculate one's assets in terms of laborers rather than acres

In Gielgud's view, it simply made sense to calculate assets in terms of numbers of Indian workers because the land itself without workers was valueless and the worth of the fragile infrastructure was infinitesimal.[62] Indians sometimes fled into the forest (and even across the Caquetá) and had to be pursued or ransomed from the Colombians.[63] Then, too, in one internal PAC memo Tizón had referred to the many accomplishments of "the *Conquistadores* of the Putumayo"—meaning the whites that had taken over the area.[64]

When Gielgud opined that Arana remained the best man to serve as liquidator, his interrogator asked if he now believed Julio César to be a transformed individual:

> No, I do not think he is a converted man, but there has been a large amount of publicity given to the matter, and, personally, I am inclined to think, I will not say a continuance of the old regime, but the continuance of the management of that district in the hands of a man who has, at any rate, felt the force of public opinion and may to that extent be considered as holding himself responsible, is less likely to be harmful to the Indians than the retirement of that company from the Putumayo and the subsequent entry of a large number of individual traders.[65]

At one juncture, there was a flap over missing account books for the critical years 1909 and 1910 that would presumably shed light on how the section chiefs were compensated—not to mention for what (such as *conquistación*). Gielgud had to admit that Arana, Marcial Zumaeta, and Abel Abarco had taken away much documentation from the London office once the Peruvian Amazon Company entered liquidation—presumably to be used for that process in South America. Gielgud had little idea of what was in the several tin boxes.[66] However, the missing account books were eventually produced.

Part of Gielgud's salary had not been paid, and he was therefore one of the Peruvian Amazon Company's creditors. Nevertheless, the Peruvian Amazon Company had cash flow, since more rubber was

---

thereby eliciting the question: "It would be perfectly right valuation for an American slave owner in the old cotton days to value his estate at so much per slave, would it not?" The flummoxed reply was: "I do not agree. That is a question I cannot answer." Ibid., 308.
62. Ibid., 397–98, 419–21.
63. Ibid., 403, 409.
64. Ibid., 421.
65. Ibid., 130.
66. Ibid., 373–74.

arriving at present from the Putumayo than at any time in the Company's history.⁶⁷

Board Chairman Gubbins was called before the Select Committee. In general, he agreed with the Casement report. Asked about his reservations he noted that there were a few internal discrepancies. Regarding the number of Indian deaths, for instance, you could not simply subtract current population estimates from past ones in order to fix an absolute number of fatalities (i.e., thirty thousand).⁶⁸ This dramatic reduction did not necessarily imply purposeful genocide. Many of the fatalities could be attributed to the introduction of European diseases into the area regarding which the natives possessed little immunity, and the indirect effects of fatigue from rubber gathering and transporting. Then, too, if the abuse became excessive the Indians might escape the control of their persecutors by simply fleeing deeper into the forest or migrating to another region. There was the internecine fighting among the tribes themselves, as many were implacable enemies. There was also the fact that some of the area's Indians, the numerous Huitotos in particular, were easily domesticated, while others (notably the Bora) resisted attempts to dominate them. Regarding the latter, there was a constant state of hostility with the whites that produced fatalities on both sides.⁶⁹

Still and all, Gubbins did not dispute the allegation that horrible atrocities had been committed by PAC employees.⁷⁰ When pressed about whether he was aware of the raids to capture and punish Indians, a possible violation of English law against slavery, Gubbins admitted that he had never inquired into the matter, while adding, "The subjection of Indians by commercial companies is the condition prevailing in the whole of the Amazon Valley."⁷¹

The *Truth* articles had come out shortly before Arana arrived in England from Manaos. Gubbins took the initiative on his own to call on

---

67. Ibid., 397.
68. Casement had estimated that the indigene population had declined from about forty thousand to ten thousand.
69. Paredes had also characterized the Bora as the most noble and fierce of all the tribes. Due to their mistreatment by whites, they were particularly prone to flee ever deeper into the forest. Three (of many) of the Bora "captains" or leaders had once cooperated with the Peruvian Amazon Company, but, at the time of Paredes's visit, only one continued to do so. There was said to be a Bora group that had never been contacted. The Bora occupied the largest and most rubber-endowed territory in Aranalandia, so, despite their resistance, they were the constant objects of the whites' conquest agenda. Chirif, "Los informes," 100–101.
70. British Parliament, *Report*, 185–86.
71. Quoted in Ibid., 236.

David Cazes, and had been assured by the Iquitos British consul that the allegations were likely untrue and that Arana was a good man. Throughout the autumn of 1909, Gubbins never doubted Arana's version of the facts. Nevertheless, as far as he could recall, he had never seen the Saldaña Rocca articles, even though Arana ordered copies sent to the London office. While the chairman had to admit that Julio César (and Abel Alarco) regularly kept information from him and the other directors, he still trusted Arana.[72] Indeed, while, "Very strange to say, Mr. Arana makes a most favourable impression on everyone he comes in contact with."[73]

Clearly, the Board had known that the two reports—one from its own Commission and the other by Roger Casement—were going to be devastating for the Peruvian Amazon Company and Peru alike. It sent a letter to Pablo Zumaeta on January 11, 1911, and another on February 3, stating the directors' disbelief that Pablo could have possibly been ignorant of the atrocities, and blaming his lax management for the fact that the British Foreign Office might well publish Casement's report. In point of fact, Arana himself had expressed reservations to Gubbins about both Pablo Zumaeta and his brother Lizardo and felt they should be retired once they had made some money.[74]

There was one red herring in the midst of the Hearings. A circular letter, supposedly sent from Iquitos to the section heads of El Encanto and La Chorrera, had ostensibly just turned up in the files in the Peruvian Amazon Company's London office. Signed by Pablo Zumaeta and dated December 31, 1907, it stated:

> Dear Sirs—The Board of Directors of the company are extremely anxious that the orders given by our predecessor firm, regarding the good treatment which should be received from the officials by the Indians working in the rubber extraction, shall be strictly maintained in force; both from the necessity of caring for them, since semi-civilisation and adaptation to work of each individual means for the company capital invested; and because this will ex-

---

72. Ibid., 435–36.
73. Ibid., 190. Collier describes their situation as follows:
> ... an incredible picture emerged—of the British directors' strange servitude and the almost hypnotic power Arana exercised over them. In London, the Peruvians had conducted much of the company's correspondence in private, routing letters to Pablo and Lizardo which his associates never saw. Sometimes, at a Board meeting, he would read extracts from these letters, but they were never permitted to examine them more closely—nor were copies placed in the company's files. Only as a managing director and then chairman had Gubbins achieved the status of a desk. Until then he and the others had studied documents or signed checks at a blank table. (Collier, *The River*, 237–38)
74. Ibid., 213.

ercise influence as a means of encouragement to attract the tribes who are still outside control, which is advisable in its interests. The Company is determined to keep the professional medical men and assistants who, for some time past have been engaged in both departments, and will always give preferential attention to any orders for medicines which may be given at the request of such doctors; not only such medicines as may serve for curing the patients, but such as may in their opinion be necessary to prevent the outbreak of any epidemic and for the destruction of centres of infection, etc. It must be borne in mind that every one of the individuals working for the company in those zones means to it a capital invested which it in no way wishes to lose, or to allow to fail to give a reasonable yield; for this reason it will consider as an undue waste of its capital any disablement for work which may be brought among the Indians by reason of ill-treatment or neglect on the part of the officials in whose immediate charge they are, and will proceed without hesitation to discharge any guilty parties and to demand by legal means that the authorities of the country shall punish everything involving misuse of force or any offence against the life and health of the Indians. The Company knows that in the case of the almost uncivilised people whom it places under the supervision of its civilised officials it will always be necessary to use more strictness in order to maintain discipline in the work, but it in no case desires that such strictness should be carried to the extreme, or that it should go outside the limits of justice and humanity."[75]

Wow! Unfortunately, Gubbins could not recall ever seeing this. The surprise document essentially blunts most of the accusations against the Peruvian Amazon Company and its key employees even before incorporation in Great Britain. It was supposedly circulated widely, yet to date no copy had ever been volunteered by anyone. It coincided with the height of Saldaña Rocca's campaign against Arana and would certainly have been provided to the Iquitos public as past and present evidence of humane PAC policy regarding its Indians. One would have assumed that, were it genuine, it would have been produced early on by Arana to his Board of Directors as proof positive that Hardenburg's charges were spurious. In short, it seems to have been one more self-serving, if ham-handed, exercise in fabricating evidence after the fact. It was never taken seriously or even referenced again during the London Hearings, or in Arana's subsequent documents and publications in his own defense.

---

75. British Parliament, *Report*, 238.

Gubbins then dropped his own little bombshell. On December 10, 1912, he had received a telegram from Arana stating that he planned to return to London in February to testify before the Select Committee. As a foreign national, he was not required to do so, but wished to clear his name. However, it would take Julio César some time to arrange his affairs.[76]

Arana had not liked the idea of forming a PAC commission because of his belief that the Peruvian government might take umbrage at such English intrusion on its sovereignty. For the same reason, Julio César had insisted that Casement's mission be cleared by Lima. Gubbins and Read were delighted by Grey's choice of Sir Roger as his investigator.[77] The Peruvian government acquiesced almost immediately—Arana then dropped his objection and leant full cooperation. Tizón had been very open and helpful to the visitors.[78] It had been Gubbins's idea (and not Julio César's) that the Company's commission should focus solely on the present and future, with an eye toward reform, rather than upon past abuses (particularly those antedating formation of the Peruvian Amazon Company).

As chairman, Gubbins felt he was issued, "responsibility without power."[79] He, and the other English directors, had hoped to improve their influence, but had no effective control and, "were trying to do quietly what we could not do violently."[80] Frequently, Arana corresponded directly with South America and kept information from the Board. The financial reporting from the South American offices proved tardy and dilatory. Gubbins had no idea why Zapata was paid 7,000 pounds, and only a vague one regarding Rey de Castro's services to the Company.[81]

Gubbins further testified that the Board had been totally unaware of Eleonora's senior note for 60,000 pounds against the Company. He stated that, "Arana never pressed us for payment of her money, and this mortgage was sprung on us behind our backs without any intimation whatsoever."[82] It was Pablo Zumaeta's doing and ". . . that is the main cause why I expressed my great distrust of him to Sir Roger Casement."[83] Gubbins stated that he and Arana argued about it. Zumaeta was urging Julio César to use Eleonora's mortgage to take back control and eliminate the English influence over the Arana's business affairs.[84]

---

76. Ibid., 197, 238.
77. Ibid., 219–20.
78. Ibid., 212–214.
79. Ibid., 213.
80. Ibid., 216.
81. Ibid., 218–19.
82. Ibid., 198.
83. Ibid.
84. Ibid., 199.

Nevertheless, Chairman Gubbins felt Arana was continuing to act in the best interests of the British creditors and shareholders[85] and Director Read concurred.[86] While Gubbins agreed that it might appear illogical, he and the other directors unanimously believed that Arana was the most suitable man to serve as liquidator.[87] To appoint an Englishman would raise Peruvian hackles and suspicions. He would be ineffectual. Then, too, Arana controlled 83 percent of the stock. He could simply make a motion, vote his shares, and appoint himself. Gubbins's lengthy experience in Peru had convinced him that the only effective liquidator would be someone who had connections and understood thoroughly the lay of the land. Who better than Julio César?[88] While Arana might be under indictment in Iquitos, given his popularity and influence there, Gubbins doubted that the warrant would ever be served.[89]

Former Board member Medina (senior) subsequently best summed up the directors' thinking in his written brief to the Select Committee,

> It may seem incredible to the Committee that Mr. J. C. Arana should not have been cognizant of all these atrocities, but I believe that he was only in that region, for short stays, two or three times, and when one remembers how easily other people have been deceived, it seems to me only reasonable to suppose that Mr. Arana may have been deceived in the same way by trusting what was told to him by his chief agents, who, to all outward appearance, were respectable men. With my knowledge of, and acquaintance of Mr. Arana, I am fully convinced that he is innocent of any knowledge of, or complicity in, those awful deeds, and considering the large amount of money I have lost and all the trouble and worry the matter has caused me, I am not likely to be biassed [*sic*] in his favour.[90]

He then noted that, by accepting the role of liquidator, Arana had actually sacrificed his own best interests:

> Mrs. Arana was the principal Peruvian creditor, so that the temptation for him was great to keep away from the liquidatorship, to press judicially for Mrs. Arana's claim and to get that large amount of money and, in the process, with his local influence, to try and

---

85. Ibid., 199, 432.
86. Ibid., 325–26.
87. Ibid., 205.
88. Ibid., 194–98.
89. Ibid., 217.
90. Ibid., 620–21.

possess again (as his friends seem to have suggested before) of the best part of the business. Many a man would have taken this course. On the contrary, Mr. Arana accepted the post and has not pressed his wife's claim; he shouldered an immense responsibility knowing the odium that he incurs, the opposition and the danger. Is it possible to act in this way if not inspired by the motives that he has set forth, *viz.*, the good of the creditors and of the shareholders andthe ultimate success of a business to which his name is attached? And not only is he the best liquidator for the creditors and shareholders but also for the aborigines themselves. They will be well treated because they are under a man who will not be hoodwinked in future, who knows his responsibility to English law and that it is only by reform and good administration that he can save the huge interest he has at stake and retrieve his name. I now see that he has come to England to defend himself which, by itself, is a presumption of his innocence.[91]

In the opinion of Director Lister-Kaye, after reading Casement's report: "it was impossible to think that he [Arana] was ignorant" of the "criminal system," but he probably did not know all the intimate details. It should be remembered that the impression of Arana was by no means altogether bad. Even now, Lister-Kaye thought that Whiffen and Hardenburg were blackmailers, and the latter a forger as well.[92] Lister-Kaye concurred that Arana had been the best choice as liquidator. Also, Julio César was undoubtedly chastened by events and would initiate more reforms in the Putumayo.[93]

Director Read also still held Arana in high regard. He was a humane administrator and "an extremely straightforward man."[94] Like the other directors, Arana had been deceived by "the same wretched employees."[95] Read was quite sobered by the revelations and felt (in retrospect) that *Truth* had done a great service in exposing the abuses. He was now thoroughly fed up with the Peruvian Amazon Company, and wanted nothing more to do with rubber. He noted that, in 1909, or when things were going well, he tried on three occasions to resign from the Board. However, once the scandal broke he would not have dreamed of doing so. It would have been cowardly.[96]

---

91. Ibid., 621.
92. Ibid., 257.
93. Ibid., 288–89.
94. Ibid., 352, 368.
95. Ibid., 372.
96. Ibid., 372–73.

Gubbins had met and cooperated with Roger Casement on more than one occasion. He testified that Casement held out hope that he had influential friends who might invest up to 100,000 pounds in the Peruvian Amazon Company. That could have proven to be its salvation, as well as the means with which the Board might have assumed real control of the enterprise. However, that capital infusion failed to eventuate once the PAC chairman provided Casement with candid information regarding the Peruvian Amazon Company's severe financial distress.[97]

Gubbins believed that the situation on the Putumayo had improved immensely. If Casement's number of 30,000 fatalities over a decade was accurate, there once had been an average of 2,500 deaths annually. While still too many, the previous year of 1912 there were only 25.[98] Gubbins was now working directly with an empowered Tizón to effect reforms. They were communicating about the need for better roads and a project to introduce mules into the Putumayo to alleviate the Indians' burden of transporting rubber on their backs.[99]

There was corroborating testimony provided by an Englishman, Henry Samuel Parr. He was twenty-one years old when he answered a PAC employment ad in London and ended up running the Company's store at La Chorrera beginning in November 1909. He was there when the Board mandated that the contracts of European employees be cancelled or renegotiated. He could have left or stayed. The original contract was for a salary of 100 pounds annually, plus full room and board, as well as transportation to and from Europe. His pay was now raised immediately to 10 pounds monthly and was to be increased soon to 15 pounds per month. Should he stay for three years he would receive a 100 pounds bonus. None of his contracts had provisions for a "commission" on collected rubber. He was later made section chief at Último Retiro in August 1911.[100] When he arrived there, floggings had ceased out of fear of the pending Casement investigation—the former stocks had been used as building material for a staircase. Parr had met all of the men accused of atrocities by Casement and found them to be gentlemen. However, he admired Casement, too, and felt that his investigation had been necessary.[101]

Parr had replaced a man named Miranda. While not certain, he suspected that Miranda (who ran the section for only eight months) fled

---

97. Ibid., 198–200.
98. Ibid., 197.
99. Ibid., 213.
100. Ibid., 336, 342–3.
101. Ibid., 344.

over fear of being accused of having committed former crimes earlier in a remoter section. Parr saw scarred Indians (about half the men had been flogged at one time) and noted that many used to flee regularly and were pursued by search parties of four or five men. When they brought back fugitives, they were given special bonuses. He heard that Macedo once sent a man named Jiménez in pursuit of thirty Indians who had absconded across the Caquetá. Jiménez returned with ten or twelve. Parr never noticed scars on women or children. The Indians were entirely docile and gave him no trouble.

Once Tizón (whom Parr greatly admired) had taken over in the Putumayo, treatment of the Indians improved notably. When the Englishman arrived in Último Retiro, there were slightly fewer than two hundred Indians on his roll. That number had increased to 209 by the time he left in October 1912, since some had come out of the forest to work of their own volition. The Indians were not "guarded" any longer. They were free to run away, and were not pursued even if they left owing rubber for trade goods already advanced. He had heard that in former times rifles were used to shoot natives, but no longer. Everyone still carried one for sport and protection against jaguars (some Indians had been eaten by them and Parr saw scars on others from attacks). Rubber production was going up. Tizón mandated that no Indian be required to carry more than 30 kilos of rubber (Casement had reported loads of sixty and more), and mules were on the way.

Parr was now back in England after finishing his three-year contract with the Peruvian Amazon Company and was mulling over his future. He had a job offer from Arana to return to the Putumayo and was considering it.[102]

On January 29, 1913, a British author and publisher, Norman Thomson, wrote to Chairman Roberts offering to testify before the Select Committee. He had attended every session save one, and was in possession of considerable documentation from Colombian sources. He enclosed partial galleys of his forthcoming "Red Book" on the Putumayo atrocities. He noted, "I have not mentioned [in the Red Book] that Vega—a Colombian traitor and criminal of the firm Arana, Vega Cia—enabled Arana to secure control of the region, and that this Vega is, I am informed, a relative of the Cortes Bank people [the Medinas]." He claimed to have a sworn affidavit from a Colombian witness to two Arana bribes of Zapata at La Chorrera for 8,000 pounds and 5,000 pounds respectively. Thomson was allowed to appear briefly, but

---

102. Ibid., 337–47.

it seems that his testimony was largely dismissed because of his obvious Colombian ties and sympathies.[103]

While the Select Committee still awaited Arana's appearance, on February 7, 2013, U.S. Secretary of State, P. C. Knox, sent an extensive report entitled *Slavery in Peru* to the Congressional House of Representatives, accompanied by a letter of transmittal signed by President William H. Taft. The battle was now joined on both sides of the Atlantic. Indeed, the U.S. government's documentation included Casement's Blue Book and an abbreviated version of the Paredes investigation. It also published the reports sent to their Department of State by consuls Eberhardt and Fuller—and these would inform some of the questions put to Arana during his testimony before the Select Committee.

On February 20, 1913, Julio César sent a letter to his supporters in Iquitos who were vetting in the local newspapers the possibility that he might run for a seat in the Peruvian Senate. He was refusing the honor because

> . . . I have concluded that the seats in the National Congress should be occupied by men with solid administrative and political preparation; those capable of truly contributing to the solution of the gravest problems in our institutional life; and not by people like myself, without any qualifications to invoke before their fellow citizens other than that of assiduous dedication to work and a determined willingness to defend the sacred rights of the country. It seems to me that to abandon my efforts as a businessman, exchanging them for those of a legislator, would produce a dislocation that would be damaging to the public interest . . .[104]

Arana went on to note that he had become the target of, "an unprecedented defamation campaign" and was fighting for his very survival. It would be irresponsible for him to become distracted by public duties in Lima. He also noted that he did not wish to provide more ammunition for anti-Peruvian propaganda to the foreign interests arrayed against him.

---

103. Ibid., 442–43. See also Norman Thomson to Chairman Roberts, January 29, 1913. Bodleian Library MSS Brit. Emp. S-22 G344a. Thomson's father had been the former director of the Jamaica Botanic Gardens and a pioneer of rubber plantations (he had supervised planting of sixty thousand trees in Colombia in 1882). The British Foreign Office had published two of his papers on rubber (1903 and 1904), and about that time the elder Thomson had tried to interest foreign investors in rubber production in the "Colombian Putumayo."

104. Valcárcel, *El proceso*, 370.

His election to the Senate at this time might be misconstrued (by international opinion) as a disrespectful challenge to the Peruvian systems of justice and public administration.[105]

On February 26, 1913, Arana arrived in Lisbon on his way to England. Interviewed by a reporter, he noted that the accusations against him were the doings of his rivals in the rubber business. While he no longer doubted that there had been ill treatment of natives in remote districts, he had never been to them. In a different Lisbon interview, Julio César stated that he looked forward to testifying before the Select Committee, since

> . . . having paid a visit of inspection to the region [Putumayo] between August and October last at the same time as the Consuls of Britain and the United States and representative of the Peruvian government visited the region. . . . I think my conduct and innocence will be properly judged there on this occasion, my case having already been rigorously investigated by my own country, where I have been entirely freed from responsibility by the higher tribunals.[106]

On March 2, Arana reached England and was met by many journalists. *The Observer* published his photograph and commented:

> He is a stoutly built man of forty-five, about five feet nine inches in height. His short-pointed beard is growing somewhat grey, but, as one would expect from a man of Spanish race, he is naturally dark. In manner he is simple and kindly, and quite free from affectation. That he is a good businessman is abundantly clear, but he is at the same time most charitable. He assists not only his friends who need help, but some of those who are not his friends.[107]

Julio César brought with him many documents that he believed would exonerate him completely from any responsibility for abuses in the Putumayo. He was accompanied by Marcial Zumaeta, who served as his translator. Marcial emphasized that he and Julio César had personally visited the Putumayo recently and found the Indians there to be satisfied and in good health. They had photographed the unscarred backs of many to demonstrate to the world that the talk of atrocities in the Putumayo was exaggerated "romanticizing." Were it all true, Julio César would simply

---

105. Ibid., 371.
106. Quoted in Goodman, *The Devil*, 231.
107. Quoted in Ibid.

be a murderer. He had returned to England to tell the entire truth, as he knew it, and to defend his honor.[108]

Meanwhile, Roger Casement had written Chairman Roberts from South Africa on March 4, 1913, to the effect that

> Julio Arana is bringing home [to England] "faked" statements. He has probably bribed John Brown to say things agreeable—and, of course, Rey de Castro he has in his pocket. Rey de Castro, by the way (I don't know if I told you) is at Lima (altho' their consul friend!) on a charge of swindling a lady! Jerome wrote me this some time back. He got with the ladies' [sic] confidence and then relieved her of her property—so she says—anyhow the case is still before the courts.[109]

Before testifying before the Select Committee, Arana had another pending matter—he had to appear before Justice Swinfen Eady of the Chancery Division of the High Court of Justice on the Strand. Some PAC shareholders (with the assistance of the Anti-Slavery and Aborigines' Protection Society) were demanding dissolution of the Peruvian Amazon Company and removal of Julio César as its liquidator. The verdict of the court case was read on March 19. Justice Eady had used Casement's reports as his main source of information.[110] In his ruling, the justice noted that the rubber had been gathered in "an atrocious manner," and that,

> If Arana did not know he ought to have known. Having regard to this, and to his position as vendor, he is the last person in the world to whom the winding up of the company should be entrusted.[111]

The Anti-Slavery and Aborigines Society then published a booklet with Justice Eady's entire scathing denunciation of the Peruvian Amazon

---

108. *Daily Mirror*, March 3, 1913; *Morning Advertiser*, March 3, 1913.
109. Roger Casement to Chairman Roberts, March 4, 1913. Bodleian Library MSS Brit. Emp. S-22 G341. Nor, does it seem, that the legal difficulties of key members of the Arana team were limited to Rey de Castro. A few days later Lima's *La Prensa* newspaper published a letter from Germán Alarco denouncing the fact that in a telegraphic exchange between the Pro-Indígena Society and a Señor Bustamante of Huarás, his brother, Abel, was referred to as the "Macabre Assassin of the Putumayo." It seems that Abel Alarco was under police guard in Huarás after being accused of shooting Bustamante's wife three times. Germán had his doubts about the veracity of the charge, but, in any event, wanted to set the record straight. His brother had never set foot in the Putumayo—residing as he did in Manaos for years before assuming his duties in London (*La Prensa*, March 9, 1913).
110. Goodman, *The Devil*, 233.
111. Quoted in Collier, *The River*, 253; *Daily Telegraph*, March 20, 1913.

Company, its Board of Directors, and Pablo Zumaeta and Julio César Arana as well.[112]

In the event, the justice's verdict mandated that the Peruvian Amazon Company be dissolved. However, in reality, it would have little practical effect. Julio César remained in control of the business on the ground and would continue in future to refer to it as the Peruvian Amazon Company whenever it suited his purposes.

While in London, Arana learned that Eleonora was gravely ill in Switzerland, her condition in part exacerbated by her stress over the scandal. He postponed his scheduled March 26 appearance before the Select Committee in order to rush to her side.

Nor was the news out of South America from Lizardo and Pablo encouraging. The price of Amazonian rubber had remained static in both 1911 and 1912, but would be halved in 1913. That year marked the precipitous collapse of South America's rubber boom (now totally eclipsed by plantation production in Asia). The economies of both Manaos and Iquitos were in free-fall.

On April 8, 1913, Arana appeared before the Select Committee. Committee Chairman Charles Roberts opened the questioning by stating that their charge was to determine the responsibility of the directors of the Peruvian Amazon Company as British citizens, not Arana's responsibility. Nevertheless, he noted that grave accusations had already been leveled against Julio César and that he had a right to a fair and impartial hearing of his responses to them. As a Peruvian, Arana was under no obligation to self-incriminate himself in testimony before an English court (or this Committee), and he was free to refuse to answer any question. However, his testimony was being taken under oath, so his answers could subject him to the same obligations and penalties as if given in a court of law. Chairman Roberts then approved the request of Arana's attorney, Douglas Hogg, to inform and advise his client through a court-appointed interpreter (one Dr. Mascarenhas) and then serve as Arana's spokesperson in communicating the response to each query.[113]

Arana sought to introduce the several documents in his possession, and Roberts ignored the request (shades of Casement's letter to the effect that they were "faked").[114] In the event, despite months of exhaustive

---

112. Goodman, *The Devil*, 233.
113. British Parliament, *Report*, 458–59.
114. There is a collection of handwritten letters in the Bodleian archives that may have been part of this material (Bodleian Library, MSS Brit. Emp. S-22 G335.). They basically depict the Peruvian Amazon Company in a favorable light, but seem contradicted by overwhelming evidence to the contrary on the points in question. For instance, there is

(and exhausting) questioning of the directors of the Peruvian Amazon Company, during his testimony Arana would scarcely be asked anything by his interrogators regarding the internal workings of his London Board.

The chairman himself conducted the first day's questioning. From the outset, Roberts sought to ascertain when and how much Julio César knew. Did he accept the charges of the Putumayo atrocities in the reports of Casement and Paredes? The reply was that, while they contained certain exaggerations, Arana now believed them to be fundamentally accurate.[115] However, it was only by reading them that he had learned of the abuses. His personal visits to the Putumayo were for but a few days, such as in 1901 and 1903, and then only for the purpose of dealing with Colombians with whom he had pending business matters.[116] The abuses in question dated from the time that the Colombians were in control of the Putumayo; not on his watch.[117]

Roberts brought up the Robuchon book edited by Rey de Castro. Apparently, in his original notes the French explorer had stated:

> The Indians care nothing for the preservation of their rubber trees, and rather desire their destruction. Eager to recover their lost liberty and their independence of former days, they think that the whites who have come into their dominions in quest of this

---

a letter ostensibly written by Perkins on March 20, 1908, when about to depart from El Encanto that reads as follows:

To whom it may concern—

I state the following without compulsion or pressure of any kind. That early in January of this year I was in the house of a Colombian named David Serrano with whom I had some business connections. That the Peruvian steamer Liberal & a government launch landed at that point & Sr. M. Loayaza [sic] told me that as my life was in danger there from the Colombians to come with them & that he would pay the amount Serrano was indebted to me & aid me in recovering my baggage & that of my companion which was in the woods about four hours walk from that place. The country being under military control I came down to the house named Encanto & instead of going directly to Iquitos as did my companion, W. E. Hardenburg, I remained at Encanto in hopes of recovering the baggage, a thing which I found to be impossible. I wish to especially state that while on the steamer & during my long wait at Encanto I was treated with the greatest kindness by Mr. Loayza, the chief & all his subordinates, as was also my companion so far as I have any knowledge. Faithfully stated W.H. Perkins

This statement may well have been written by Perkins, but as a requirement of his liberation from El Encanto. If so, one can judge its sincerity by comparing it with the lengthy excerpt from Perkins's subsequent damning testimonial sent to Great Britain during the Select Committee Hearings.

115. British Parliament, *Report*, 460, 477–78, 486.
116. Ibid., 466.
117. Ibid., 473.

valuable plant will go away when it has disappeared. With this idea, they regard with favour the disappearance of the rubber trees, which have been the cause of their reduction to slavery.[118]

Rey de Castro had omitted this paragraph, as well as Robuchon's reference to the Colombians as having been the original white settlers of the Putumayo (an inconvenient fact in the ongoing struggle between Peru and Colombia over its sovereignty). Arana noted that the editor (Rey de Castro) did not know English and therefore may have failed to understand that passage. When told by Roberts that he was translating from Robuchon's French,[119] Julio César gave a rather obfuscated reply. Robuchon's notes existed in English, French, and Spanish. The Spanish version had gone to the Peruvian government and was the basis of the Rey de Castro publication. Arana had only received the French and English ones. He was uncertain why the passage was missing in one or more of the versions but insisted that he had never seen it before.[120]

And what of the claims of the enslavement of the Putumayo Indians that were made as early as 1907 in the writings of American Consul Eberhardt and certain Iquitos Catholic priests? Arana's response was that he was never informed of such allegations directly by either (and he was well acquainted with the priests in question). Had he been, he would surely have investigated the charges.[121] He also noted that, in 1908, the Peruvian government constituted an investigative commission [Prefect Zapata's] that Julio César accompanied during its visit to the Putumayo. While he would also contend that the real purpose of the mission was to establish a stronger Peruvian military presence in the Putumayo, on their arrival, certain PAC employees were in detention for the raid at La Reserva. Arana claimed that Loayza had remained behind at La Unión, so he was not a detainee. He had heard that Bartolomé Zumaeta was at La Reserva, but was uncertain. He did claim that Bartolomé was not among the detainees.[122] The Zapata commissioners failed to witness cur-

---

118. Ibid., 461.
119. Casement had insisted that Roberts ask about the omission of the passage in the original French text, in which the French explorer wrote of the Indians' desire to be rid of the Casa Arana. A handwritten copy of the French original had been among the documents seized by British authorities at the headquarters of the Peruvian Amazon Company (Goodman, *The Devil*, 235).
120. British Parliament, *Report*, 461, 464–65, 490.
121. Ibid., 461. As we shall see later, these were no doubt Augustinian missionaries appointed by the Vatican and not by the Loreto's generally corrupt secular religious hierarchy.
122. Ibid., 490–91. Casement was told that Bartolomé had fled to the Abisinia district on learning that Polack had been ordered to detain the leaders of the La Reserva attack in

rent abuses of either Indians or Barbadians, or to uncover any historical evidence of such, and were told by the section heads that the allegations against them were spurious.[123]

When asked why he seemed to always be the last to know about conditions in his own domain, yet both Casement and Paredes were quick to produce detailed reports of the abuses, Julio César responded that Casement had the advantages of his experience in the Congo (he knew what to look for) and his knowledge of English (so he could interview the Barbadians). Arana himself possessed neither. Compared to Arana, Paredes had spent a considerable amount of time in the Putumayo and also had the advantage of knowledge of the allegations (including access to the Casement report) of specific abuses and the names of alleged perpetrators that he intended to target.[124]

Arana was quite equivocal when questioned about Saldaña Rocca. On the one hand, he was dismissive and claimed that no one in Iquitos took the man seriously, while, on the other, he suggested that the articles against him stemmed from his refusal to advance the publisher funds for his printing business. At times, Julio César's testimony was contradictory regarding when and how he learned of Saldaña Rocca's campaign against him.[125]

When pressed regarding his reasons for founding the Peruvian Amazon Company, Arana noted that it was simply to raise capital. Chairman Roberts produced a document by Julio César that stated: "The constitution of a society in England with a capital in London, and in which the Arana Brothers will hold the greater part of the shares, will contribute without doubt in the most efficacious way to the recognition of the property of Peru in the Putumayo." Arana replied that "The region of the Putumayo is a matter of large discussion at the present moment, and to have a Company there with English and Peruvian capital would greatly facilitate matters in connection with the questions of frontiers."[126]

There were extraordinary glimpses into an Arana rubber gathering operation on the Acre River—on both sides of Peru's border with Bolivia. It seems that a certain man now possibly residing in London, (Basque-surnamed) Norzagaray, either had worked there as the general manager[127] or been sent out by the Peruvian Amazon Company's Board of Directors

---

anticipation of the prefect's arrival. Mitchell, *Sir Roger*, 643.
123. Ibid., 461–64.
124. Ibid., 466–67.
125. Ibid., 462–64.
126. Ibid., 467.
127. Ibid., 362.

(in November 1909) to make recommendations for improving the financial results of the operations. Norzagaray wrote,

> We can assert that the industry is better organized in the zone which belongs to Bolivia, and that the product is obtained there under more economical conditions. However, this organization is unfortunately founded on a regime of exploitation which must be properly considered as inhuman, and consequently worthy of censure."[128]

Arana claimed that those gatherers were all whites and were paid well. As a consequence, most renewed their contracts with the Company. Chairman Roberts, however, produced a letter from Abel Alarco referring to the use of Indian labor on the Acre.[129] The report on the Acre River was written in May of 1910. Lizardo Arana was in charge of that operation by then.

Nor were the business activities limited to the Putumayo and Acre, since, when the Peruvian Amazon Company was on its last legs financially, Abel Alarco was ordered to travel to the Upper Purús River in Brazil to collect money there owed to the Company.[130] Furthermore, the Peruvian Amazon Company owned large estates near Iquitos at Nanay and Pébas.[131]

When asked why he had founded the Peruvian Amazon Company, Arana answered:

> Because I had other businesses on various rivers, the Acre River and other rivers, which was even more important than my business in the Putumayo, and for that reason, as the business was increasing very greatly, I wanted to turn it into an English Company and with the object of getting more capital, and for the various purchases I was making, I naturally required more capital.[132]

With this observation, Julio César opened a window on a vast world that remains (to my knowledge) totally uninvestigated by scholars. We have all been focused upon the Putumayo scandal to the exclusion of Arana's extensive activities elsewhere. There are small glimpses of them in the Norzagaray connection.[133]

---

128. Ibid., 428.
129. Ibid., 495.
130. Ibid., 109.
131. Ibid., 91. In the case of the latter, in 1903, a missionary priest had filed a complaint with the Peruvian government alleging that the Arana Brothers abused the Indians there and seized their women and children. Ibid., 167.
132. Ibid., 489.
133. We could also mention the fact that Arana continued to send his steamships up other

Then there was the matter of Carlos Rey de Castro. Asked if he was on Arana's payroll, the answer was in the affirmative. The Manaos consul had represented Arana in both Lima and Rio de Janeiro. In the former, his assignment was to secure clear land titles to its Putumayo holdings for the Casa Arana. In Rio, he pursued the Arana y Hermanos claims for rebates of certain contested fees paid in the past to Brazil. Rey de Castro had received extraordinary payments in addition to a monthly stipend of 50 pounds.[134]

The give and take was quite frustrating to the questioners. Arana was proving to be an elusive target, and one very adept at prevarication. At one point, a frustrated committeeman, Swift MacNeill, complained to Chairman Roberts; "This witness never answers a question Yes or No, and the first thing I submit is to tell him to answer Yes or No, and make his explanations afterwards."[135] Nevertheless, it seems fair to say that Arana's rhetorical exercise failed to convince. *The Manchester Guardian* noted,

> The much-talked-of Julius Caesar Arana came before the Putumayo Committee today—a heavy, yellow-faced man, with black-grey hair and beard, who gave his evidence in rapidly spoken conversational asides to the interpreter beside him, emphasizing it from time to time by a quick gesture of his fat short-fingered hands. . . . Mr. Arana hardly impressed one as a witness. At this time of day it is really too late to minimize the atrocities or deny, as he did, the accuracy of the consul's reports.[136]

At the conclusion of the first day's Hearings, Hogg informed the Select Committee that Eleonora's health had taken a turn for the worse, and Julio César was anxious to rejoin her in Switzerland.[137] Chairman Roberts promised to try and conclude Arana's testimony the following day. According to Collier, much of the British press reported that Arana had acquitted himself surprisingly well during his first day of testimony.[138] Then, too, Julio César

> . . . left the witness-chair this afternoon on his own showing a much wronged man, accused of bribery, cruelty, or connivance in cruelty, and participation in a system of unspeakable savagery,

---

Amazonian rivers. In 1905, he offered both passenger and cargo shipping on the Purús and Acre rivers, directed from Manaos, and, in 1906, the *Liberal* was making trips up the Ucayali in between voyages from Iquitos to the Putumayo. Valdez Lozano, *El verdadero*, 49–50.
134. British Parliament, *Report*, 472.
135. Ibid., 475.
136. *Manchester Guardian*, April 9, 1913.
137. British Parliament, *Report*, 473.
138. Collier, *The River*, 251–53.

while all the while his relations to the natives were those of Robinson Crusoe to his man Friday—of a paternal agent of civilisation towards a friendly and contented race.[139]

Day two was dominated by a continued rhetorical debate over the translated English meaning of two Spanish words in Peruvian Amazon Company records—*conquistar* and *correría*. There were significant Peruvian Amazon Company budgetary items referring to both—particularly for the purchase of firearms. The implication was that the weapons were purchased to "conquer" the Indians and that those who failed to become willing workers (or who subsequently absconded) were "chased" down (*correría*) and captured, tortured or killed as an example to the others. Julio César contended that *conquistar* referred to beguilement—a campaign to win over the Indians with trade goods.[140] The firearms were likely the key, since everyone carried a weapon in the dangerous jungle for self-defense against jaguars and for hunting. *Correría* was equally benign and referred simply to searching for illiterate Indians who owed money to the Company for trade goods (in return for rubber) who had simply gone back to the forest without paying off their debts.[141] From their persistent questioning regarding these meanings, as well as the evidence of Spanish semanticists, it is fair to conclude that the Select Committee remained skeptical about Arana's linguistic interpretation.

---

139. Quoted in Goodman, *The Devil*, 236–37.
140. While Arana failed to convince the Select Committee of this interpretation, in fairness to him Joaquín Rocha claimed that during his 1903 visit to the Putumayo (or before Arana was in total control of it):
> Whenever one encounters a tribe of savages whether known already or not, or which has had no dealings with whites before, it is said that the individual who is successful in entering into business dealings with this tribe and who gets it to extract rubber and to plant fields and build a house in which he can live in their midst, has conquered them. Of these Indians who have now entered into the great and common task of civilized people, it is possible to consider them to now be incorporated into civilization. In effect, many of them learn to speak Spanish; when a priest passes through these places they receive the sacrament of baptism; at times that of matrimony as well and they take a Christian name with which to be known by the whites, and a number of these Indians leave their settlements and acquire insensibly new ideas and knowledge and some of the needs of civilized people. In exchange for the rubber that they extract and for their work in the gardens and as oarsmen, they receive from their conqueror linen cloths, work tools and mirrors, beads, needles and other trinkets. Rocha, *Memorándum*, 102.

There is no mention whatsoever of intimidation by their "conqueror."
141. British Parliament, *Report*, 474–76. In his analysis of Arana's testimony before the Select Committee, Colombian anthropologist Pineda Camacho considers this debate under the subheading "Tower of Babel." Roberto Pineda Camacho, "El comercio infame: El Parlamento Británico y la casa cauchera peruana (Casa Arana)," *Boletín de Historia y Antigüedades* 89, no. 817 (April–May–June, 2002), 386–90.

When the subject shifted to his accusers, Arana first asserted that Hardenburg was a blackmailer and forger. He claimed to be in possession of the bank draft for 830 pounds that Hardenburg had presented for collection in Brazil, maintaining that the amount had been altered chemically from an original 10-pound amount. Nevertheless, when his interrogator asked repeatedly whether Arana was accusing Hardenburg of forgery, Julio César wavered. Eventually, he admitted that he could not be certain that the American had crafted the actual forgery (sensing possible vulnerability to a charge of libel), but continued to claim that he had presented the falsified document for collection.[142] Queried about Whiffen (disguised in the Hearings as Mr. X until he insisted on testifying in person to clear his name), Arana claimed that the captain had asked for 1,000 pounds to produce a favorable report on the Putumayo for the British Foreign Office. Julio César would repeat his accusations against the two men in his subsequent testimony.[143]

There was also the strategic use of English (and lack thereof). When it suited his purpose, Arana cited the Casement report, contending that he had had a few parts of it translated for him. At other times, it was his lack of English that justified his ignorance of specific allegations against him or his Company. At one point, when confronted with the contention that Normand was paid a commission on the amount of rubber delivered, Arana declared that he now understood that to be the case, but he had only learned about that payment system by reading Casement. While he agreed that a commission system could easily lead to abuse, he had since abolished it.[144] In any event, in his view, it made no sense for the Company to torture or exterminate the scarce labor supply upon which its success was predicated.[145]

---

142. Ibid., 480–82.
143. Ibid., 492, 500.
144. Unfortunately, there was a letter from Alarco to Zumaeta, dated June 17, 1909, that stated,

> On examining the Putumayo accounts we find that the system adopted over there of granting to the heads of sections a bonus on the amount of rubber brought in by the Indians and of allowing the higher officials of the Putumayo a certain percentage of the profits shown in the balance-sheets thwarts its own object, as we attribute the bad quality of the rubber to the desire on the part of heads of sections to obtain the greatest quantity possible without regard to the quality of the same, the quantity being what they derive their profit from. We likewise consider it a mistake to allow the Putumayo officials a percentage of the profits, as the said officials will try to swell out the balance-sheets in order to thus get a bigger profit, so that this bonus system is productive not of emulation sought after, but of which is quite the contrary, viz., the demoralization of the officials. (Ibid., 311)

145. Ibid., 473.

Committeeman Swift MacNeill was the day's last interrogator, and he chose to revisit Arana's accusations against Hardenburg. Asked if he now believed the American to be a forger, Julio César failed to do so explicitly, but added, "I have to repeat what I said before, that the suspicion falls on two persons, on the buyer and on the endorser." Swift MacNeill replied, "Turn round and see Hardenburg before you. Do you not know him?" And Arana responded, "Yes, I am very pleased to see him here, because we can enter into other details."[146] The day's proceedings ended on that note.

Arana would testify for a third and final day. Throughout his testimony, Julio César had made the distinction between the civilized and uncivilized parts of the Amazon. The former had settled *racionales* who were part of a market system that, in the case of rubber gathering, was based upon debt peonage. While the system might appear punitive, the *racionales* agreed to short-term (one to three year) contracts that they regularly renewed voluntarily. The uncivilized regions were inhabited by illiterate, innumerate savages who (at least in the case of the Putumayo) were cannibals to boot. These jungle-dwellers paid no taxes, recognized no national sovereign, and remained pagans. It was a part of the Casa Arana's mission in the Putumayo to tame and civilize the Indians for their own benefit. The idea was to provide structure (government) and bring them into the debt peonage system with trade goods. However, one's risk was considerable, since the outstanding debts were harder to enforce against people who could simply melt away into the forest. Arana acknowledged that, in the 1907 prospectus for the Peruvian Amazon Company's stock offering, there was the statement: "There is an abundance of labour, the Indians being naturally submissive, and eight or ten civilized men can control 300 or 400 of them."[147]

When asked if he had personally recruited the Peruvian Amazon Company's directors, Julio César replied no—rather it was Abel Alarco (by then in London to serve as general manager of the proposed Company) who had done so. Asked if Alarco was not subsequently "kicked out" of the Company, Arana conceded that he was "retired" because he could not get along with the directors. Pressed again on whether the real purpose of forming the British company was to gain protection of its assets in the sovereignty struggle between Peru and Colombia, Arana conceded that there was that advantage, but after twenty or thirty years of activity in the Amazon he was seeking to retire from the business.[148] The subject of

---

146. Ibid., 487.
147. Ibid., 489.
148. Ibid.

health, and particularly his painful chronic sciatica, came up periodically throughout his testimony.

Asked whether his brother-in-law, Bartolomé Zumaeta, was present during the assault upon the property of David Serrano, Arana stated that he wasn't certain because he himself was not there and "In the Putumayo there are several Zumaetas."[149]

Regarding his activities on the Acre River, it seems that those scattered properties had appeared as assets of the Peruvian Amazon Company and were sold for 30,000 pounds during 1909 and 1910. Arana claimed that he had received instructions while in Manaos from the London manager (presumably Alarco) to sell the Acre holdings, as the Peruvian Amazon Company was running short of capital. It was unclear whether directors were made a part of that important decision.[150]

Pressed again by Committeeman Douglas Hall as to whether he believed that atrocities had been committed in the Putumayo, Arana admitted that, after seeing the Casement and Paredes reports, he did. However, he continued to insist that they happened in the past and that there were exaggerations—such as the allegation that Indians were lined up in a row as targets for displays of marksmanship or burned alive. Julio César insisted that he had not personally been able to verify the accuracy of such reports. And then there was the following exchange,

> Hall—Do I understand that you have taken no steps of your own accord to find out whether they are true or not?
>
> Arana—If these persons who are denounced are no longer in the region how is it possible to prove it?
>
> Hall—That is not an answer to my question. Have you taken any steps?
>
> Arana—What I have occupied myself in, is to find out whether they still commit these crimes.
>
> Hall—That is not an answer to my question. I repeat my question—Have you taken any steps to find out whether the statements contained in Sir Roger Casement's report are true or not? I want an answer—yes or no.
>
> The Interpreter—The reply is exactly the same as before. I cannot get him to say yes or no.
>
> Hall—Do you decline to answer that question?

---

149. Ibid., 490.
150. Ibid., 491.

Arana—It is between yes and no, as the [Peruvian] authorities are there [in the Putumayo] to prove it.

Hall—But you have not yourself taken any steps?

Arana—I have taken steps, but I have not been able to prove anything.[151]

There were also some interesting exchanges regarding the treatment of the Indians as property. At one point, Arana noted that there were disputes with Colombian rubber collectors in the Putumayo because they were "stealing" Indians from the Peruvian Rubber Company.[152] Nevertheless, there was also the notice sent out to its overseers in the Putumayo by the Casa Arana in 1904:

> Iquitos, 18th December, 1904
> Dear Sir, for reasons which are in the knowledge of all, the traffic of minors under any form or pretext whatsoever is rigorously prohibited by the laws of this country, and under our ordinances any person contravening this provision and the regulation of our establishments in Putumayo shall be dismissed, and the natives shall be sent back at their cost, but the expense of the journey here and the return and the care necessitated by having them here have to be charged to them.[153]

Arana's attorney, Douglas Hogg, was the last to query his client. He got Julio César to state that he had come before the Select Committee willingly to answer any of its concerns. Regarding the question of who forged the Muriedas document, Arana maintained that the Committee members would have to reach their own conclusions after examining the evidence. Finally, Hogg asked Arana pointedly when he first learned of the atrocities, and Julio César replied, "The only information I got is from the publications in "Truth" after which time the [Zapata] commission was sent out to make enquiry."[154]

Five days later, or on April 15, it was Hardenburg's turn to testify; Arana was not present. Walter had come to London at the request of the Anti-Slavery and Aborigines Protection Society. His book, *The Devil's Paradise*, had been published to considerable critical acclaim in late 1912, and there was likely little that he could add to that story. However, the

---

151. Ibid., 493.
152. Ibid., 494.
153. Ibid., 497.
154. Ibid., 500.

Peruvian Amazon Company's key directors had all expressed their belief that Hardenburg was a blackmailer and forger. The Society wanted him to come to England to clear his name and, when informed that he lacked the funds to do so, it raised them through a solicitation. He brought with him an affidavit from Walt Perkins (now resident in Lovington, New Mexico) that supported every detail of Hardenburg's account, while vouching for his good character.[155] Nevertheless, Lagos believes that Hardenburg had a financial stake in confronting his accusers. Later in 1913 there was to be a second edition of *The Devil's Paradise*. Would the publisher have proceeded had the allegation that Hardenburg was a forger gone unchallenged?[156]

Then, too, Hardenburg was at pains to clear his name. On April 10, 1913, he wrote Chairman Roberts stating that he was aware that Arana had retracted the allegation before the Select Committee, but what about the directors of the Peruvian Amazon Company? In his testimony, Sir John Lister-Kaye had stated his belief that Hardenburg was a blackmailer. Walter sought vindication because, he said, "I am quite prepared to go into any British court and establish my innocence, but I am extremely anxious to give no pretext for a trial in South America where justice moves very slowly, but where I should probably get prompt trial and no justice."[157]

On the stand, Hardenburg noted that he had learned in September 1909, or about the time that the first article appeared in *Truth*, that the Peruvian government had agreed to reimburse him and Perkins 500 pounds for their lost possessions. Hardenburg did not believe that the Peruvian government's decision to pay the compensation was related to the London scandal, rather he opined that it was probably because in the Putumayo he had been shot at and detained by a Peruvian military mission.[158]

The American testified that, while in Iquitos, he had tried to keep his delving into the Peruvian Amazon Company a secret, but it was quite possible that Arana learned of it. Certain Colombian friends had put him onto former Company employees in the Putumayo who might be willing to talk. They, in turn, provided the names of others. However, some people were too frightened to go on record with their stories. Others could very well have informed the Arana people of the investigation. Hardenburg

---

155. Goodman, *The Devil*, 243.
156. Lagos, *Arana*, 337–38.
157. W. E. Hardenburg to Charles Roberts. April 10, 1913. Bodleian Library S-22 MSS Brit. Emp. G342.
158. British Parliament, *Report*, 503–504, 512.

maintained that the abuses in the Putumayo were general knowledge in Iquitos. In his words,

> I cannot say that it was so commonly talked about that you would hear it on every corner or anything like that, but I think the majority of the people knew of the main facts of the exploitation of those Indians up there, and how they were being forcibly compelled to work and being flogged. Whether they knew of the burnings I do not know, but I think many of them did.[159]

He added that Germán Alarco (Abel's brother) was mayor of Iquitos during his stay, and that official was later quoted in the press while on a trip to London after the allegations had been made in the *Truth* that "these things had been long known in Iquitos."[160]

Was Hardenburg aware that a letter was sent on June 6, 1909, from the Iquitos office of the Company to the London one. The one that stated:

> We have knowledge that a person called Hardenburg is traveling towards England, who has been preparing a business of blackmail against you, as it is said. We mention this to you for your guidance."[161]

Hardenburg was ignorant of that letter and denied its allegation. He also rejected the charge of having forged the Muriedas bank draft and claimed to have never met with a Mr. Guarnier in Manaos. He did note that he had a letter of introduction from a Colombian in Iquitos to a well-known Colombian in Manaos named Espinosa who would likely give him additional information on the Putumayo abuses. It was Espinosa who convinced the Colombian consul in that city to accompany Hardenburg to the bank and vouch for his identity, a prerequisite for cashing the draft that Muriedas had endorsed over to the American.[162]

Hardenburg was questioned closely about his Colombian sympathies and admitted that he had been treated well by many Colombian friends and acquaintances before, during and after his travels in the Putumayo. And, of course, he had just testified about the help he received in Manaos from Espinosa and the Colombian consul. Hardenburg felt that, while there was much rivalry, the better classes of people in both countries were reasonably tolerant of one another. He then noted, "I cannot give many cases, but you meet fair men you know, men like Dr. Ego Aguirre [*sic*], I do not remember him saying anything derogatory [about Colombians]."

---

159. Ibid., 508.
160. Ibid., 511.
161. Ibid., 504.
162. Ibid., 510.

When he was then reminded that it was Egoaguirre who had called him a blackmailer in print, Hardenburg could only comment "I am surprised he did that."[163]

That same afternoon, it would be Captain Whiffen's turn on the witness stand.[164] There ensued an analysis of a copy of the patched together (by Arana) letter in which Whiffen purportedly asked for payment in return for saying that he had seen no "irregularities" in the Putumayo. He now testified that he thought that phrase referred to the business of illegal rubber exports by Putumayo PAC employees to Brazil. Whiffen acknowledged that almost all of the writing was his, albeit reflective of the alcohol consumption, except for a couple of key words that significantly altered meaning. There was also some question as to whether certain words had been moved around during the reconstruction of the document.[165] In any event, Whiffen denied in the strongest of terms having ever offered to sell his reports: "I swear that it is absolutely impossible that it ever passed through my mind to offer to suppress any report for a sum of money."[166] He could not even remember clearly having written the letter before destroying it.

After learning that he had been denounced as a blackmailer at the Peruvian Amazon Company's December 31, 1908, annual Board meeting, late the next spring Whiffen had filed a suit for damages against Arana—but by then the Peruvian was back in South America, so the papers were never served. Whiffen was aware that Arana had subsequently defamed him in the Peruvian press—including accusations that

---

163. Ibid., 513. Not exactly. In point of fact, Hardenburg had known about Egoaguirre's accusation since his earlier stay in London. In a letter to R. A. Bennett at the *Truth*'s offices, dated February 11, 1913, Walter seemed to include Egoaguirre among the Arana forces trying to discredit him and Whiffen. It is in this letter that Hardenburg declares that his tight finances precluded his coming to London to testify before the House Select Committee, but he was more than willing to provide a notarized affidavit should it be deemed useful. Clearly, Bennett communicated the situation to the Anti-Slavery and Aborigines Protection Society, and the decision was made to fund Hardenburg's expenses. W. E. Hardenburg to R. A. Bennett. February 11, 1913. Bodleian Library, MSS Brit. Emp. S-22 G323. Hardenburg, while not an academic and hence less sensitive to the "rules," was clearly capable of plagiarism. In 1910, he had published an ethnographic article (W. E. Hardenburg, "The Indians of the Putumayo, Upper Amazon," *Man* 10, 1910, 134–38), that contains considerable material that was clearly lifted without attribution out of Fuentes, *Loreto* 2, 114–21.
164. British Parliament, *Report*, 516–28.
165. Michael Taussig (*Shamanism*, 118) reports the forgery as self-evident, adding, "It was grotesque. It was banal." In point of fact, neither Arana's nor Whiffen's testimony established such definitive certainty.
166. British Parliament, *Report*, 520–24.

the captain himself had "improper relations" with the Indians and was habitually drunk throughout his stay in the Putumayo. Whiffen denied both allegations.[167]

Once Arana was finished with his verbal testimony, he was free to leave England immediately—which he soon did. However, he left behind his last word in the form of a "Paper" that he handed over to the Select Committee, expounding his views regarding the true history of the events and some of their protagonists. It was in English and was prepared by his secretary, Marcial Zumaeta, after translating the Spanish responses by the deponent. It is a carefully measured exercise that underscores Julio César's disagreement with several points brought out in Casement's Blue Book, as well as earlier testimony by him and others before the Select Committee. He was also handing over supporting documentation in the form of Peruvian newspaper articles and copies of relevant contracts and correspondence (presumably the material that Chairman Roberts had refused to admit during the Hearings). Arana was, of course, under no obligation to prepare such a statement at all, any more than he had been to return to London to testify before the Select Committee.

The text was notarized on April 4, 1913, and begins with Julio César's personal oath regarding the veracity of what follows. He first recounts a brief history of his involvement in the Upper Amazon, emphasizing his limited personal experience in the Putumayo. He had visited it on but four occasions between 1899 and 1910. His first two visits were for a few days and limited to negotiating financial matters with his Colombian partners and debtors. Neither then nor subsequently, when accompanying Zapata in 1908 and the British and American consuls in 1910, had he ever witnessed evidence of abuse, let alone an actual atrocity. He had not invented the labor system, but rather inherited it by acquiring the Colombians' holdings. He rejected, categorically, Casement's contention that he obtained some of them through coercion and violence—offering to provide written evidence of the purchase of every one.

Casement had mischaracterized Julio César's visit with Zapata to the Putumayo. He had accompanied the prefect and Consul Rey de Castro at the request of the Peruvian government and as a part of the installation

---

167. Ibid., 522–24. Casement noted,
   Also a Dr. [Dickey] at Encanto for 14 months who saw a lot of Whiffen says the story the J.C.A. gang circulated of his immorality &c. &c. is an absolute lie. He says Whiffen acted all the time on the Putumayo like a gentleman & the whole thing was a put up game to blacken his character when they found he knew too much. (Mitchell, *Sir Roger*, 682)

of a two-hundred-man military force in the area. Despite the accusation by several witnesses, he denied having paid Prefect Zapata any amount to procure the release of prisoners. Commandant Polack had never detained PAC employees, rather Prefect Zapata had ordered that several Colombian ones captured at La Unión be brought to La Chorrera—several of whom died en route of disease or from their wounds.

As for the articles in *La Felpa* and *La Sanción*, Saldaña Rocca wrote them and then demanded money not to publish. Whiffen and Hardenburg did the same, offering to sell Arana their texts and silence. Hardenburg was not only a blackmailer, he was a forger as well. Arana had intended to sue him after the *Truth* articles, but was dissuaded from doing so by his London legal counsel. His relationship with Rey de Castro was entirely honorable and professional—Peru's Manaos consul was an attorney who was hired to pursue land titles to the Putumayo in Lima and in Rio de Janeiro the repayment of money owed by Brazilians to the Peruvian Amazon Company. British Consul Michell's statement that the rubber from the Putumayo was being consigned to his wife's account in Iquitos was simply untrue.

He denied, categorically, that the Indians had ever been subjected with force. And so forth.[168]

Two key contentions are points 40 and 41:

> 40. I deny that it was impossible for anyone who visited the Putumayo to be ignorant of the atrocities. These appear for the most part to have taken place in the remotest sections which I did not visit, and as the Committee have learned from the evidence of other witnesses, it was quite possible for anyone who had not an illuminated mind, such as Sir Roger Casement had, to visit the principal stations [El Encanto and La Chorrera] without becoming aware of any irregularities. 41. I was absolutely ignorant of orders having been given for raids, to be carried out or of rewards having been promised in connection therewith. It was a term of the articles of partnership of J. C. Arana and Hermanos that 10 per cent of the profits were reserved for distribution amongst the employees who distinguished themselves by their good conduct and aptitude for work.[169]

The day following Whiffen's testimony, it was Arana's attorney, Douglas Hogg, center stage arguing his client's case before the Commit-

---

168. Ibid., 612–18.
169. Ibid., 616.

tee.[170] He began by underscoring that most of the "evidence" of atrocities in the Putumayo was essentially rumor and hearsay. Still, by this time, Hogg was prepared to accept that horrific abuses against the Indians were indeed committed. Nevertheless, with the exception of Captain Whiffen, none of Arana's accusers ever told him about their knowledge of the abuses. Those who chose not to included American Consul Eberhardt, the Iquitos priests and Hardenburg. Hogg planned to return to Whiffen's statements at the end of his presentation.[171]

So, what did (and could) Arana know? When the charges appeared in *La Felpa* and *La Sanción*, Julio César's overseers in the Putumayo were outraged and sued Saldaña Rocca for defamation. When, in April 1908, Arana himself visited the Putumayo to investigate, his managers unanimously denied having committed any atrocities. Arana failed to witness any or see any evidence that there had been some in the past. Of course, none of that was surprising, since those accused had the biggest stake in concealing their crimes from their employer. Consequently, it was only after the articles in *Truth*, and the subsequent extensive reports of Casement and Paredes, that the allegations gained credibility. Since then, Arana had cooperated with every initiative to investigate them—and then to rectify the situation in the Putumayo. He had even come to Great Britain on his own initiative and volition to defend his good name.[172]

Hogg also asked the Select Committee members to suspend judgment regarding the system of debt peonage—advancing goods to be repaid in bartered rubber—a system equated to "slavery" in some of the testimony. Similarly, the *correrías* were simply another form of the former, but now applied to illiterate Indians who could avoid their debts by fading away into the forest. One should remember that the system was legal and the debts were enforceable under Peruvian law. In purchasing his properties in the Putumayo from Colombians who possessed no land titles, in reality Arana was buying the modest infrastructure and the business operation. The latter included indebtedness on a particular company's books—much of it deriving from goods advanced to Indians on credit. It was in this sense that Arana came to "own" Indians. Undoubtedly, the system could and did lead to abuses on occasion, however, abuse was not inherent in it. Hogg opined,

---

170. Ibid., 528–37.
171. Ibid., 530–31. Hogg also mentioned that he was not privy to the captain's actual testimony (given but the day before when he was absent from the Hearings) and knew of its content only through what he had read in the newspapers.
172. Ibid., 530–32, 537.

Nobody could suggest that the horrors which one has read of in the Blue Book, and horrors which, I suppose, never were equaled in the worst days of the Congo, nobody can suggest that those horrors were perpetrated merely with the idea of getting as much rubber as possible; they were the wanton cruelty of men to whom cruelty must have become almost a mania. When one reads of the burning alive, the shooting in wanton sport, the using of Indians as targets, and all the abominable cruelties and outrages perpetrated, no one can look upon that as efforts to see how much rubber could be got. These things were done by men who had lost all sense of propriety, and were guilty of these outrages merely to gratify their lust for horrors of that kind.[173]

To suggest that Arana knew of such behavior and did nothing about it defied logic. Who had a greater long-term financial interest in the stability and well being of his labor force? Why would he knowingly alienate or liquidate it?

Hogg recognized that Whiffen was the one witness who claimed to have informed Arana of the atrocities early on, or in the spring of 1909. He then did what he could to undermine the captain's credibility. There was discrepancy in the dates furnished by him and Mr. Cazes as to their meeting in Iquitos (was it March as Cazes claimed or April as did Whiffen?). Then, too, Whiffen declared that he could remember nothing about writing the famous letter demanding payment for his silence, while subsequently providing minute detail as to what transpired that midnight between him and Arana in the Motor Club. It would simply have to be up to the Select Committee members whom they chose to believe.[174]

Hogg certainly gave a thought-provoking and spirited defense of his client. Unfortunately, it contained a fundamental error of omission. The notion that the barbarities of the Putumayo were the work of debased, even insane criminals, ignored the fact that many of the accused were related to Arana. His own half brother, Martín, stood accused of torture, and his deceased brother-in-law, Bartolomé Zumaeta, was characterized as a sadistic and perverted monster. Arana himself had remarked in his testimony that there were many Zumaetas in the Putumayo—presumably his in-laws. Pablo Zumaeta and Lizardo Arana were now running the Iquitos office. Pablo's brother-in-law, Abel Alarco, had managed the London one. He and his brother, Germán, mayor of Iquitos, were principals in

---

173. Ibid., 530.
174. Ibid., 534–37.

the firm Arana, Alarco and Cia. Germán Alarco, while visiting England during the *Truth* scandal, had requested of English Consul Cazes a letter declaring that the Putumayo allegations were false, but was then quoted, almost disingenuously and dismissively, in an English newspaper as saying the practices were common knowledge in Iquitos. If Augustinian priests in the city denounced the treatment of Indians in the Putumayo and demanded that the Peruvian government initiate reforms, it was all news to Arana. In short, it absolutely defies anyone's imagination that Julio César, micro-manager *par excellence*, was kept totally in the dark by his entire tight network of consanguineal and affinal kinsmen forming the inner core of the Casa Arana and its management of Aranalandia, not to mention Julio César's toadies in the municipal, departmental, and national governments and judiciaries.

There was also the issue of the articles in *La Sanción* and *La Felpa*. Arana testified before the Select Committee that he had learned of them only belatedly after copies were sent to the Peruvian Amazon Company's London office about a year after their publication. It is unthinkable that he was ignorant of the several months of public weekly attacks upon him in Iquitos, including cartoons that went so far as to lampoon Julio César as a haughty exterminator of Indians. Then, too, there was the fact that Saldaña Rocca published in installments the entire text of his own legal brief that accused Arana of personal responsibility for crimes committed in the Putumayo. This, of course, rubbed an extremely touchy issue, since his claim of personal ignorance of the atrocities was now Arana's prime defense.

There is the matter of the untimely death (at fifty-two years of age) of Saldaña Rocca in Cerro de Pasco, felled by the bullet of a Chilean assassin on April 17, 1912. After leaving Iquitos, he had founded three crusading newspapers in Lima and two in Cerro de Pasco, all of which failed. He had spent five years moving from one city to another in futile flight from "an invisible persecution" that followed him.[175] From the standpoint of the Arana forces, this demise *was* certainly convenient. Recall that almost two years earlier (June 28, 1910) Hardenburg had made the obvious recommendation to the Anti-Slavery Society that the former Iquitos newspaper editor should be at least interviewed, if not summoned to London to actually testify.[176] By April of 1912, the prospect of a public airing of the Putumayo scandal was probable. Clearly, any testimony by Saldaña Rocca, even more than that of Hardenburg or Whiffen, posed a major threat to the Arana forces. Might they have had a hand in his murder?

---

175. http://diariodeiqt.wordpress.com/2009/04/15/benjamin-saldana-en-el-recuerdo/.
176. Walter Hardenburg to Travers Buxton. June 28, 1910. Bodlein Library MSS Brit. Emp. S-22 G322 (file 2 of 2).

There was indirect support of such a sinister thesis in a certain side play during Arana's testimony before the Select Committee. Seated silently in the gallery were the brother and the sister-in-law (A. Kate Rance) of C. H. Rance, an English public accountant who had worked for Arana in Manaos from June 1911 to June of 1912. Casement had met with him and advised him to resign his position. He did so and embarked from the city on a ship for Great Britain. He died on board and was buried at sea. The cause of death was listed as "delirium tremens." The man was not a heavy drinker and no one who knew him believed this to be possible. On March 31, 1913, the bereaved sister-in-law sent a letter to Chairman Roberts stating that it was her belief (and that of the family) that Rance was the victim of foul play designed from keeping him from testifying before the Select Committee.[177]

It was clear from the final days of the Committee's Hearings that the capacity of Great Britain, the United States, and the two acting in concert was fraught with its own problems and limitations. Two witnesses, Edward Morel and Harry Johnston, an old African consular service hand with some South American experience, testified regarding the behavior of British firms in Africa and the current state of affairs in the Congo. Both observed that there was presently a strong and ongoing British consular presence in the Belgian Congo that had all but eliminated the former abuses there. However, while it might be desirable to install a similar (British) watchdog presence in the Putumayo and the Upper Amazon in general, there were serious impediments. After all, under the 1885 Berlin agreement regarding the Congo, Great Britain was accorded certain oversight treaty rights (and obligations) there. The same was not true in any South American venue, and excessive British investigative activity on the continent could very well be resented by any and all of its sovereign states. Nevertheless, according to Johnston, the abuse of indigenes was prevalent throughout the Amazon, and the present disclosures of the atrocities in the Putumayo highlighted the need for serious reform. Johnston stated:

> I think it is perhaps one of those useful flashlights which are suddenly turned on to a dark corner, but I know it has been going on in many other directions, and it will probably go on until some native races are exterminated, if something is not done to fix responsibility.[178]

---

177. A. Kate Rance to Chairman Roberts. March 31, 1913. Bodleian Library MSS Brit. Emp. S-22, G-345–48.
178. British Parliament, *Report*, 550.

There were also certain North American and broader international sensibilities in play. The United States would likely oppose British actions in Latin America that smacked of economic or political penetration. The Americans would have to be consulted regarding any future system of foreign safeguards against atrocities being repeated in the Putumayo. However, that brought up another delicate matter, namely the degree to which a joint effort by Great Britain and the United States might be viewed (and opposed) by the rest of the world as an initiative designed to impose growing Anglo-American hegemony upon the southern continent.[179]

In this regard, the predominant concern was over the blatant imperialist ambitions of Kaiser Wilhelm. Indeed, early twentieth-century German machinations in Africa, Asia, and Latin America presaged the outbreak of World War I in 1914. In 1903, the American and German navies had narrowly avoided a battle off the Venezuelan coast. American president Theodore Roosevelt was convinced that German machinations and aspirations in Latin America, and particularly Brazil, would provide the first real test of the Monroe Doctrine.

In actuality, there was a longstanding German presence in South America. We have noted that Germans predominated in the ranks of the Jesuit missionaries that sought to evangelize broad swaths of the Upper Amazon. By the end of the nineteenth century, Germans were prominent in the lists of the commercial elite of both Manaos and Iquitos. There is evidence of their considerable presence in rural Colombia as well.[180] German (and French) interests were the most predominant foreign influences in Loretan merchant firms and, as a consequence, both countries maintained consuls in Moyobamba.[181] Then, too, there was extensive German immigrant settlement in southern Brazil. Kaiser Wilhelm openly sought to promote and enhance it. In short, Brazil had become a cornerstone of Wilhelm's imperialist policy—one to be pursued through the stimulation of German emigration and continued penetration of German commercial interests there, without excluding the possibility of eventual gunboat diplomacy as well.[182]

After months of Hearings (thirty-six sessions with interrogation of twenty-seven witnesses), the Select Committee was ready to publish its findings. In the main, it found that the evidence of the atrocities was

---

179. Ibid., 538–57 and elsewhere.
180. Dominguez and Gómez, *La economía*, 43.
181. Santos-Granero and Barclay, *Tamed Fontiers*, 57–58.
182. John C. G. Röhlm, *Wilhelm II: Into the Abyss of War and Exile, 1900–1941* (Cambridge: Cambridge University Press, 2014), 213–18.

overwhelming and that they both predated and postdated formation of the Peruvian Amazon Company. The combination of Saldaña Rocca's and Hardenburg's testimonials (published respectively in *La Felpa*, *La Sanción*, and *Truth*), the Casement and Paredes reports, as well as that of the Peruvian Amazon Company's own investigative Commission—all of which were in substantial agreement—simply outweighed Arana's denials, evidence, and self-serving prevarication.[183] The Select Committee members agreed with Justice Swinfen Eady that Arana, "was the last man who should have been appointed as liquidator."[184]

Regarding the ostensible savagery of the Indians, the Select Committee observed:

> Much stress is laid by the apologists for the Arana firm upon the traces of this sort of ritual cannibalism. The Committee thinks that the above extracts give a fair account of the character of the hapless Indians, whose abominable and inhuman oppression is a black stain upon civilization.[185]

Special condemnation was reserved for the Peruvian Amazon Company's English directors. The Select Committee concluded:

> It appears from documentary evidence before the Committee that it was always intended to keep the ultimate control in Peruvian hands. The objects of forming the British Company were said to be partly the want of further working capital, partly a wish on the part of Señor J. C. Arana to retire, and in part a desire to gain the backing of a British interest as an offset to a syndicate formed in the State of Maine to develop by the aid of the Colombians the rubber districts of the Caqueta and the Putumayo. It may be added incidentally that the contemplated formation of the company had its influence on the atrocities. It has been suggested that the Arana firm had no conceivable motive for driving away or exterminating the Indians. It is obvious, however, that for the flotation of the company it was indispensable to show an increasing output of rubber in the three years before the date of the issue of the prospectus, as was in fact shown. There is no doubt that during that period there was severe pressure to expand the output of rubber by terrorizing the Indians. . . . The British directors therefore undertook the direction of a business of which they were by their own account most imperfectly informed, and in which from the

---

183. British Parliament, *Report*, iv–v.
184. Ibid., xix.
185. British Parliament, *Report*, xxxi.

outset they might be outvoted and in which on their own statements they were dominated by Señor Arana.[186]

While they were certainly ignorant, they could not be absolved "from the charge of culpable negligence as to the labour conditions that prevailed under the Company." Then, too,

> No care at all was taken, and the employees were, in fact, a gang of ruffians and murderers, who shot apparently from sheer lust of blood, or burnt, tortured, and violated in a spirit of wanton devilry. . . . The directors then assumed positions to which are inseparably attached responsibilities they failed to discharge, and in the opinion of your Committee their conduct on this head is deserving of severe censure. They should not lightly have exposed to risk the good name of England.[187]

The PAC directors stubbornly refused to consider any compromising evidence and accepted, without serious question, Arana's representations. They then tarried, and even obstructed, formation of their own investigative Commission, while some of the worst crimes were perpetrated, and more Indians died.[188]

Concerning the accusations against Hardenburg and Whitten, the conclusions were:

> Mr. Hardenberg [sic] came back of his own accord from Canada expressly to deny the charges against him. The Committee went through the documents and heard his explanations. His story appears to be satisfactory, and in fact the dossier of documents compiled against him does not preclude his explanation as now given. In view of the fact that he had been compensated for personal ill-treatment by the Peruvian government, the allegations of Señor Arana were certainly not enough to absolve the directors from the duty of immediate inquiry. As to the charge of blackmail brought by Señor Arana against Captain Whiffen, your Committee refer to the

---

186. Ibid., vii.
187. Ibid., xvi.
188. Ibid., xvii–xviii. To be sure, in March of 1914, a bill was introduced in the House of Commons to extend Great Britain's Slave Trade Act to other forms of coerced labor, but then died after a first reading. In July of that same year, Lord Lytton, successor to Charles Roberts as chairman of the Select Committee, introduced his own bill into the House of Lords that would make the director of any British company culpable in matters of enslavement. Ignorance of the situation on the ground would no longer be an acceptable defense. The definition of "slave" was to be extended to anyone held in service against his free will. This initiative likewise failed after a first reading. Goodman, *The Devil*, 248–49.

Minutes of Evidence for the two versions of this incident. Captain Whiffen appeared at his own request before the Committee to deny on oath the charge brought against him and to give his explanation. On his own account of the events, he was entrapped under circumstances which he himself describes, into writing a document which on its face value was incriminating. The production of that document to the directors afforded them ground for discounting his testimony. But the directors were too ready to seize upon anything which would discredit hostile witnesses and appear to have made no attempt to communicate directly with either Mr. Hardenberg [sic] or Captain Whiffen in order to satisfy themselves as to the value of these gentlemen's assertions.[189]

Then, there was Julio César:

After carefully weighing all the evidence placed before them your Committee is convinced that Señor Araña [sic], together with some or all of the partners in the vendor firm, must have had criminal knowledge of the atrocities perpetrated by his agents and employees in the Putumayo. But they are also satisfied that he did not communicate his knowledge of the atrocities to the British directors before the "Truth" revelations. Subsequently he denied that the atrocities had occurred and repudiated all knowledge of them. He maintained his denial of all knowledge both before the High Court in the shareholders' actions and before your Committee.[190]

In contemplating the Hearings themselves, we might keep in mind Terry Eagleton's admonition,

. . . nobody who has read a government communiqué can be in the least surprised that truth is no longer in fashion. Gross deception, whitewash, cover-up and lying through one's teeth: these are no longer sporadic, regrettable necessities of our form of life but permanently and structurally essential to it. In such conditions, the true facts—concealed, suppressed, distorted—can be in themselves politically explosive; and those who have developed the nervous tic of placing such vulgar terms as 'truth' and 'fact' in fastidiously distancing scare quotes should be careful to avoid a certain collusion between their own high-toned theoretical gestures and the most banal, routine political strategies of the capitalist power-structure. The beginning of the good life is to try as far as possible to see

---

189. Ibid., xviii.
190. Ibid., xxxiv.

the situation as it really is. It is unwise to assume that ambiguity, indeterminacy, undecidability are always subversive strikes against an arrogantly monological certitude; on the contrary, they are the stock-in-trade of many a juridical enquiry and official investigation.[191]

---

191. Terry Eagleton, *The Ideology of the Aesthetic* (Cambridge: Basil Blackwell, 1990), 379–80.

Part 4
The Aftermath

Survivors and Their Fates

After his last day of testimony, Julio César stopped attending the London proceedings. Nor, as a foreign national, was he vulnerable to any kind of British criminal charges emanating from them. So, he set about winding up his affairs and left England never to return.

While it was true that the U.S. government had published its damning *Slavery in Peru* report in February 1913, or at the height of the Hearings by the Select Committee, by the following summer the Americans were equivocating. Newly appointed British ambassador to Washington, Cecil Spring Rice, was informed that the Americans felt that further investigation and reform in the Putumayo should be carried out by the Peruvian government. Continued pressure from Great Britain and the United States risked aggravating a Peruvian nationalistic backlash and could easily become counterproductive. Then, in September 1913, the Americans closed their Iquitos consulate and reassigned Stuart Fuller to Durban, South Africa.[1] For its part, the British Foreign Office maintained its consulate in the Loreto, but with the instruction not to intervene in the Putumayo or intrude in the Peruvian government's initiatives. After all, rubber extraction there was no longer in the hands of a British company and was therefore no longer of direct British concern.

Before leaving England, Hardenburg gave a series of public lectures, culminating on April 23, 1913, at the annual meeting of the Anti-Slavery and Aborigines Protection Society. Many prominent persons were in attendance, and the American praised the Society's Putumayo work in, "stripping off from the limbs of the slaves the corroding chains of a slavery

---

1. Goodman, *The Devil*, 263–64.

of the most detestable kind known in modern times."[2] Both Collier[3] and Paternoster had depicted Hardenburg as an idealistic all-American boy with the pragmatic Christian values of his Methodist upbringing. Walter was denounced by Julio César himself as a thief, forger, blackmailer, and Colombian agent. Apparently unbeknownst to either his admirers or the Arana inner circle were Hardenburg's socialistic politics. After his move to Canada, he had written a work defending socialism.[4] In May, "Comrade" Hardenburg published an article, "The White Man's Burden," in the new Canadian socialist journal *The New Review*, in which he denounced the treatment by whites of the world's indigenous peoples in the name of profit. To wit,

> Among the most tragic of the many horrors that have cursed the human race under the baneful rule of Capital, nothing else can for a moment compare with the cruelty, lust and greed that have everywhere accompanied the contact of Europeans with the natives of the Tropics. It seems that under the goad of capitalist ambition, men who have had the advantage of modern education and the "moral benefits" of our Christian civilization, will readily throw off their humanity and yield to this insidious and demoralizing influence, which transforms them into veritable devouring beasts. And once the teeth of these human hyenas penetrate into the quivering flesh of their victims, it is almost impossible to make them loose their hold.[5]

---

2. Quoted in Ibid., 243–44.
3. Collier, while an embellisher, was a thorough researcher. It seems inconceivable that he was unaware of Hardenburg's many publications in Canadian socialist outlets. However, given the anti-Communist sentiment in the United States at the time of publication of *The River That God Forgot*, it is possible that Collier chose not to even invoke, let alone sympathize with, Hardenburg's political beliefs.
4. W. E. Hardenburg, *What Is Socialism? A Short Study of Its Aims and Claims* (Vancouver: Dominion Executive Committee [Socialist Party of Canada], circa 1912). Hardenburg adroitly highlights the poem "To England's Man" by Percy Bysshe Shelley as being the precursor to Karl Marx's famous admonition "workers of the world unite." Ibid., 7. Written about 1912, Hardenburg's work predicted that the proletariat of the planet were about to unite into an international movement that would sweep all before it and erect a democratic "Social Commonwealth" devoid of private property. Like so many other socialist writers of his day, Hardenburg failed to perceive that the world was poised on the precipice of the Great War in which workers, serving under their national flags, would slaughter one another by the millions. After his 1913–1914 contributions to *The New Review*, it seems that Walter put down his pen until publishing, in 1922, a book on mosquito control (W. E. Hardenburg, *Mosquito Eradication* [New York: McGraw-Hill Book Company, Inc., 1922]). It was apparently his last publication before his death in 1942.
5. W. E. Hardenburg, "The White Man's Burden," *The New Review* 1 (May 1913), 495.

Then, in a series of articles from June, 1913, through January of 1914, Hardenburg published his views regarding the Putumayo in *The New Review*. In the lead one, he denounced the main villain,

> This individual was no other than Julio César Arana, to-day a multi-millionaire, a "captain of industry," a "gentleman"—and the designer and chief beneficiary of the system that has resulted in the ravishing, torture and murder of some 40,000 human beings![6]

By October 1913, Julio César Arana was back in Iquitos with his family.[7] Despite the ongoing international scandal, Julio César was once again on the safe turf that he not only dominated, but where he was the object of considerable genuine admiration. He was clearly in charge of the assets of the Peruvian Amazon Company—never mind the complaints of his critics and the ruling of an English judge. In theory, Julio César was removed as the liquidator of the Peruvian Amazon Company by judicial order on March 19, 1913, at the insistence of the shareholders. However,

> Shortly after the making of the Winding-up Order, J. C. Arana attended by request upon the Official Receiver, when he was requested to give possession of the assets of the Company in South America. Whilst expressing willingness to assist in the liquidation, Arana asserted that there were local creditors in Iquitos whose claims had arisen under his personal guarantee and who relied upon him to look after them, that in Manaos the business had been carried on under the name of his firm and that local creditors there also held him responsible for their debts. His intention was to see all local creditors paid. He also indicated, that in the event of any attempt to take possession of South American property by the Official Receiver it would be necessary for him (Arana) to act in relation to the security held by Mrs. Arana as well as on behalf of the local unsecured creditors.[8]

It was never clear if the Winding-up Order could be enforced in South America, and Arana did use Eleonora's mortgage to exert total control.[9] From a Loretan point of view, this likely went unnoticed, since

---

6. W. E. Hardenburg, "Story of the Putumayo Atrocities 1," *The New Review* 1 (June 1913), 632.
7. Lagos, *Arana*, 354.
8. W. J. Warley, Senior Assistant Official Receiver. March 17, 1914. Bodleian Library. MSS Brit. Emp. S-22 G335, 20.
9. The Peruvian Amazon Company was not dissolved legally until 1920 by Julio César. So, for a number of years, Arana remained in charge of what technically remained an English enterprise (García Jordán, "La misión," 271). Even after its dissolution, he con-

the Peruvian Amazon Company was always and by everyone referred to simply as the "Casa de Arana,"[10] according to Lagos,

> Julio César Arana... carried on with his business activities, traveling to Manaos and imposing his sovereignty in the Putumayo. The rubber man continued to put into practice the law of the jungle in this disputed territory, [that was] devoid of judicial, police and military authorities.[11]

This seems to be a hyperbolic statement. Clearly, Arana was somewhat chastened by the international scandal and clearly had his domestic enemies as well. Then, too, there would soon be the English missionaries whose very presence portended renewed scandal should there be new abuses. Some reform had been in the air even before the flight of many section heads and the subsequent humanitarian changes instituted by Tizón. Writing in the autumn of 1910, Loreto prefect Francisco Ayalza Paz y Soldán estimated the rubber production of La Chorrera's stations to be 250 metric tons annually and that of El Encanto's sections 150 metric tons. The quality of the product in the latter was superior to that of the former. However, 100,000 rubber trees had been planted recently in the former with "imported seeds." This was promising, since the density of rubber trees in plantation settings was ten to twenty times greater than occurs in the wild.[12] Miguel Loayza stated:

> With a view to the future of the rubber industry, two years ago plantations of rubber began to be made, there being at present writing [September 12, 1912] more than 50,000 trees of over 1 year and 12,000 of lesser age distributed in the sections and here [El Encanto]. By the time that these young trees may be tapped a remunerative form of rubber, easy for the company and for the native rubber worker, will thus have been established.[13]

During this new phase of his Putumayo operations, Arana exported his rubber in partnership with the Iquitos firm of his friend and relative

---

tinued to use the designation whenever it suited his purposes—until at least 1933 (Bákula Patiño, "Introducción," 93).
10. Valcárcel, *El proceso*, 276.
11. Lagos, *Arana*, 355.
12. Paz y Soldán, *Mi país*, 164–65.
13. M. S. Loaysa to Señor C. Rey de Castro. Encanto. September 27, 1912. United States Department of State, *Slavery*, 81. By 1906, the Peruvian government was well aware that the plantation system of Southeast Asia, combined with the depradation of rubber trees by Amazonian rubber gatherers, threatened the industry's future. It decided at that time to dedicate about 10 percent of the country's tax on rubber exports to the replanting of rubber trees. Flores Marín, *La explotación*, 55.

by marriage, Cecilio Hernández. Prior to this new arrangement, Hernández had ranked as the thirteenth largest rubber exporter in Iquitos; now, in control of the Putumayo's production, he was the first.[14] Meanwhile, Julio César remained fully in charge of the situation on both the Putumayo and in Iquitos. He and Zumaeta quickly satisfied other partners who had shares in the Putumayo enterprise and then, on July 1, 1914, in a transaction registered in Manaos, Julio César bought out the interest of his brother-in-law. In 1915, Pablo Zumaeta was president of the Iquitos Chamber of Commerce and, a few years later (1919), he became the city's mayor. Despite the shift of the rubber boom to Asia, Arana was adept at finding a niche market—his latex production on the Putumayo remained steady throughout the World War One.[15] There is also some evidence that Julio César may have maintained holdings in the wider Upper Amazon. In late 1913, the *London Star* reported that,

> In 1909–1910 the Peruvian Amazon Company sold its properties in the Acre district, but it is alleged by the Anti-Slavery Society that some of the former employees of the Aranas are still at their hellish work in this new area.[16]

Over the short term, the scandal festered unabated. In July, 1913, or a month after the Select Committee's report was published, the Anti-Slavery and Aborigines Protection Society constituted its own watchdog subcommittee to press for actions by the House of Commons—chaired by none other than Charles Roberts and composed of many members of the original House Select Committee. In November, a petition to that effect was circulated and signed by many prominent Englishmen—including four lords. It called for closer regulation of British companies operating abroad, an elaboration of antislavery treaties with foreign countries and assignment of consular officials whose purpose would be to monitor labor conditions in the more inaccessible parts of the world. The results were forwarded to British Prime Minister and Lord of the Treasury H. M. Asquith.[17] Yet nothing really happened.

The London Hearings and international scandal did, however, trigger an immediate triangulated public rhetorical exchange. G. Sidney Paternoster published a popularized, book-length summary of his articles in *Truth* together with Casement's Blue Book.[18] Publication of the Case-

---

14. Stanfield, *Red Rubber*, 169.
15. Ibid., 180.
16. *London Star*, November 11, 1913.
17. T. F. Buxton and Charles Roberts to H. M. Asquith. November 26, 1913. Bodleian Library MSS Brit. Emp. S-22, G335.
18. Paternoster, *The Lords*.

ment reports, the American government's *Slavery in Peru*, the detailed proceedings of the House of Commons' Special Committee, the report by Britain's Iquitos consul Michell of his trip with American consul Fuller to the controversial area, and the running commentary in the London press all constituted an unrelenting and exhaustive public damnation of Arana and the Peruvian Amazon Company.

Unsurprisingly, the Colombians smelled blood in the water, not to mention opportunity on land. The London proceedings were a huge black eye for Peru and the Peruvian Amazon Company. The Colombians immediately began issuing a series of publications that denounced the Peruvian atrocities, while asserting Colombia's superior historical claim to the contested Putumayo/Caquetá. In April 1913, or practically coterminous with conclusion of the London Hearings, British writer and publisher Norman Thomson released his work *The Putumayo Red Book*.[19] It was a masterful text designed to guide the English reader through the rhetorical maze surrounding the border dispute, while providing a succinct summary of the salient points in Casement's report, the *Truth* articles and the protracted London Hearings.

On April 8, or the very same day that Arana began his testimony before the Select Committee in London, Colombian consul general in Manaos, José Torralbo, wrote Julio César a letter in the February 19 edition of *El Heraldo* complaining of his treatment of the Indians. Torralbo then informed Arana sardonically (i.e., as if Arana were ignorant of the matter), that his "fellow countryman," Juan Olegario Vargas, had just sold (March 29) about five Huitoto Indians at the mouth of the Purus River. Torralbo asked rhetorically, "How much will the Peruvian Amazon Co. gain by this little transaction?" He was sending a copy of his letter to both the Select Committee and London's Anti-Slavery Society. Torralbo concluded by stating:

> I hope you have succeeded in the intention you expressed here when you sailed [from Manaos to England to testify], of "ruining the English Consul, author of the 'Blue Book,' & causing the

---

19. Norman Thomson, *The Putumayo Red Book* (London: N. Thomson & Co., 1913). He had shown the work to London's Colombian Legation and they were delighted with its point of view. In Thomson's own words, "I am doing the Republic [Colombia] a service. I have asked for no funds, and as a matter of fact when making arrangements for publication insisted that in the circumstances could not accept financial assistance. But the Colombian Chargé d'Affairs has on his own responsibility arranged for distribution of copies. We see, however, no prospect of recovering our out-of-pocket expenditures either directly or indirectly. . . ." Norman Thomson to Chairman Roberts, March 13, 1913. Bodleian Library MSS Brit. Emp. S-22 G344a.

downfall of the Cabinet of that powerful country." That would be one more success for the enslaving power of gold, even when it represents blood, ruin, robbery, arson, in short, devastation.[20]

Then, too, immediately after the Select Committee's Hearings were concluded, there began a six-month exchange in the pages of *The Times* of London. On April 29, 1913, there appeared, in its monthly *South American Supplement*, a favorable review of a recent article in the *Boletín* of the Colombian Ministry of Foreign Affairs that expounded the legitimacy of that country's claims to the Putumayo, but also emphasized its historical and present willingness to negotiate peacefully with Peru. *The Times* critic demurred from opining about the legalistic merits of the arguments, but applauded Colombia's stance by noting, "It will be a great thing, not only for the natives of the Amazon Valley, but for Latin-America at large, if the Colombian and Peruvian governments would substitute for warlike threats and provocative actions a rivalry in good works for the benefit of those stricken regions."[21]

On May 27, 1913, the *Supplement* reviewed the Thomson *Red Book*:

> The book itself, which is admirably written and arranged, is in part a restatement of the horrible scandal which is now so notorious; it breaks new ground, however, in its clear *exposé* of the diplomatic issues involved, a knowledge of which is essential to all those—and they are many—who have made up their minds that this monstrous reign of cruelty and injustice shall cease. . . . Mr. Thomson . . . draws largely on Colombian sources, notably the works of Dr. Olarte Camacho, and he may be accused by Peruvian sympathizers—if in this matter there are any left—of Colombian bias.[22]

On May 27, the *Supplement* published a letter from S. Restrepo, chargé d'affaires in Colombia's London Legation, defending Colombia's rights and history in the Putumayo:

> All the evidence which the Select Committee has taken—all the evidence available in fact, with the natural exception of the allegations of Arana in defence of himself and of his company—goes to prove that, as long as the Putumayo remained in the undisputed possession of its rightful owner, Colombia, the aboriginal tribes

---

20. José Torralbo to Julio César Arana. April 8, 1913. Bodleian Library MSS Brit. Emp. S-22 G332.
21. Cited in Norman Thomson, *The Putumayo Red Book*, 2nd edition (London: N. Thomson, 1914), xl [*The Putumayo*].
22. May 27, 1913, *The Times, South American Supplement*.

enjoyed safety and peace, and that if any attempts were made to bring them within the pale of civilisation these were confined to the influence of the missionaries and the discipline of the peaceful settler unacquainted with the "higher" and intensive methods of Reduction—the gun, the whip, and the machete—introduced by Arana and his associates.[23]

Restrepo also raised an objection. While he welcomed the light shed upon the Putumayo atrocities by Casement's report and the London Hearings, he believed that Colombia was slandered in both without any redress—such as the opportunity to rebut. After all, there had been considerable testimony by Arana's directors and others that accused Colombians of establishing the atrociously abusive rubber gathering system in the Putumayo in the first place (although, as we have seen, some of them did). Then, too, both London's insistence that Peru investigate the allegations and the request to Lima by the United States for retribution for Hardenburg and Perkins could be interpreted as tacit international recognition of Peruvian sovereignty in the Putumayo.[24]

The June *Supplement* published letters from Thomson restating the Colombian position, as well as from Eduardo Lembcke making the Peruvians' case and expressing a willingness to resolve the differences peacefully. Thomson noted that he had received a letter of gratitude from the Colombian government for his *Putumayo Red Book*, as well as assurance that the president of Colombia favored negotiations. Thomson suggested that Great Britain and the United States arbitrate. Furthermore, he asserted that the collection of rubber in the Putumayo should be suspended and that an International Board be established to oversee labor practices in the entire Upper Amazon. *The Times* approved all of these recommendations.[25]

Enter Carlos Larrabure y Correa and Carlos Rey de Castro. The former was now based in Paris as director of the Peruvian government's Information Office in Europe. He wrote a lengthy rebuttal (dated June 19, 1913) to S. Restrepo's May letter that was then published in the June *Times Supplement* as a paid advertisement of the Peruvian government. It also quickly appeared in Spanish.[26] It consisted of three chapters, the

---

23. S. Restrepo, "Colombia and the Putumayo Question," *The Times*, May 27, 1913. The language in this letter is that of a native English speaker and corresponds a lot to Norman Thomson's literary style (possible shades of the ghost writing by Carlos Rey de Castro).
24. British Parliament, *Report*, 206.
25. Thomson, *The Putumayo*, xliii–xlvi.
26. Carlos Larrabure y Correa, *Peru y Colombia en el Putumayo* (Barcelona: Imprenta Viuda de Luis Tasso, 1913).

first of which ("The Land Titles") reviews the convoluted history of the border dispute and seems most certainly to derive from the author's vast experience regarding that subject. However, the second chapter ("The Occupation") and third and final one ("The Humanitarian Argument") both reflect a major shift in focus and literary style. They muster Casa Arana insider information and conclude with a defense of both Julio César and the Peruvian Amazon Company. They invoke issues that Carlos Rey de Castro would soon dwell upon in his own signed works and even incorporate similar orthographic errors (e.g., "Cassement" for "Casement"). By this time, Carlos Rey de Castro was also residing in Paris, and it is obvious that he actually penned the last two chapters in the Larrabure y Correa work.

The Colombian government was quick to respond. It commissioned Thomson to write a rebuttal that was published as the so-called "White Book" on the Putumayo.[27] It is a most detailed, point-by-point refutation of the Larrabure y Correa text. Then, the September *Supplement* published comments by Peruvian president Billinghurst that his government's attempts to clean up the situation in the Putumayo were hampered by the region's clouded sovereignty, and so it was in his country's interest to resolve that dispute with Colombia. Thomson also sent a letter stating that arbitration was likely and imminent. So, *The Times* declared its involvement to be over.

On November 7, 1913, there was formal denunciation of Arana in the Colombian parliament. It contained several inaccuracies, including that the Peruvian Amazon Company had been liquidated. (Arana had been appointed liquidator, but the Peruvian Amazon Company survived for several more years, albeit under Julio César's complete control.) It also spoke of the atrocities committed by Armando "Hernand" (instead of Normand). It cited testimony from the London Hearings and praised Thompson's Putumayo *Red Book* as "moving and sadly humane." It referred to Arana's Putumayo operations as Dante's Hell.[28] In January 1914, Thomson published a second edition of *The Putumayo Red Book* that included a lengthy new introduction outlining much of the previous 1913 interchange between the parties.[29] There was also a section on Thomson's proposed "International Board," now represented as an all but done deal.

Then, on January 27, 1914, the Colombian minister of foreign affairs published a circular calling for protection of Indian tribes and the sup-

---

27. Norman Thomson, *Colombia and Peru in the Putumayo Territory* (London: N. Thomson & Co., 1913).
28. Villegas and Yunis, *Sucesos*, 156–57.
29. Thomson, *The Putumayo*, xxix–lviii.

pression of their enslavement. The matter was to be on the agenda of the forthcoming Pan-American Conference, and the Colombian Legation in London requested the support of the Anti-Slavery and Aborigines Protection Society.[30]

Joseph F. Woodroffe weighed in with his own book, *The Upper Reaches of the Amazon*, based upon his years of experience in Brazil, Bolivia, and Peru.[31] He had worked as an accountant for Arana in Manaos and then for several months in the Putumayo. While he was there his considerable debts, due to the collapse of rubber prices during his stint as a merchant on a tributary of the Ucayali river, were sold to creditors in Iquitos. *He had become an indentured servant:*

> I was now a victim of peonage; from this day on my life was a living hell, and it must not be wondered at that for a while I lost all hope of ever becoming a decent member of society again, and must frankly admit that I adopted, without thought or contemplation, some of the many habits of my companions, each peculiar to the country, and only upon the arrival of the English Commission, sent to investigate into the atrocities, was I able to make a really serious attempt to get out of the rut into which I had so unfortunately fallen.[32]

Casement and the PAC commissioners had urged Woodroffe to leave.

In a heartfelt letter to Reverend Harris, written in December 1913, Woodroffe attributed that influence and advice with saving his dignity and soul. While he expected to be vilified like Hardenburg and Whiffen before him, he was determined to tell his story, "and no slanderous or libelous statements will prevent me from doing what I consider to be my duty to my Maker."[33] Woodroffe simply affirmed the beatings of the Indians, and likened their circumstances to slavery. He did not dwell on the lurid details that spiced the Hardenburg, Casement, and Paredes accounts. He did make one original contribution to the list of abuses by adding the behavior of Peruvian military authorities:

> It was a common occurrence for an officer to neglect his duties in order to visit El Encanto or La Chorrera to exchange his woman

---

30. February 27, 1914. Pedro M. Carreño to Travers Buxton. Bodleian Library, MSS Brit. Emp. S-22 G332.
31. Joseph F. Woodroffe, *The Upper Reaches of the Amazon: Types of Brazilian Rubber Gatherers* (London: Methuen & Co., 1914) [*The Upper Reaches*].
32. Ibid., 141.
33. J. F. Woodroffe to Reverend Harris, December 3, 1913. Bodleian Library MSS Brit. Emp. S-22 G332.

on some frivolous pretext, and his desires were always met, perhaps to the extent of taking some child barely in her teens, and in a semi-savage state, from her parents into the hands of a man thoroughly brutalized by vice, and who, after ruining the girl both morally and physically, returns her to the agents of the company.[34]

Colombian poet Cornelio Hispano, who was also resident in Paris at this time, weighed in with an impassioned defense of his country's claim to the Putumayo that included pointed denunciation of "Julio Arana, the unfortunately celebrated personage whose name will forever remain united to the most execrable crimes of the present epoch. . . ."[35]

Hispano further opined that,

> If the universal press and especially the British government had not addressed the matter, by this time Julio Arana and his Agents would be dancing on the mutilated cadavers of the 10,000 Huitotos who still roam as ghosts through their forebears' destroyed jungles.[36]

About the same time, Colombian writer Demetrio Salamanca T. wrote a two-volume work, *La Amazonía Colombiana: Estudio geográfico, histórico y jurídico en defensa del derecho territorial de Colombia*, defending Colombia's claims to the Putumayo/Caquetá.[37] Publication of volume one was delayed until 1916 and then volume two, which was to have had most of the material regarding the Putumayo scandal, was destroyed by the Colombian government because of perceived inaccuracies!

The Arana forces counterattacked through a remarkable series of publications that appeared in 1913 and early 1914. The effort was clearly spearheaded by Carlos Rey de Castro—who, depending on one's point of view, was either Arana's tribune or occluder. He would sign three of the polemical works and was probably the ghostwriter of the others.[38] These included two by Pablo Zumaeta[39] and one by Julio César Arana, in addition to his likely authorship of the last two chapters in the Larrabure y Correa book. The "Zumaeta" works were in response to negative articles

---

34. Woodroffe, *The Upper Reaches*, 138.
35. Hispano, *De París al Amazonas*, 253.
36. Ibid., 260.
37. Salamanca T., *La Amazonía*.
38. Chirif concluded as much after effecting a content analysis of the literary style, vocabulary, and peculiar construction of paragraphs. Alberto Chirif, "Presentación" in Carlos Rey de Castro, Carlos Larrabure y Correa, Pablo Zumaeta and Julio César Arana, *La defensa de los caucheros* (Iquitos: CETA, 2005), 58–59.
39. Zumaeta, *Las cuestiones;* and Pablo Zumaeta, *Las cuestiones del Putumayo: Segundo memorial* (Barcelona: Imprenta Viuda de Luis Tasso, 1913).

in Paredes's Iquitos newspaper *El Oriente*. The first memorial was actually completed on December 27, 1912, and mounts an attack upon judges Paredes and Valcárcel. The second, dated in Iquitos on February 27, 1913 (or while the London Hearings were still in progress), gives a detailed rebuttal of many points in the Paredes report and defends the Iquitos judiciary's exonerations for lack of evidence of both himself and Julio César.

The tone of the Rey de Castro pieces was smarmy toward Arana's critics and oleaginous in his defense. The man's own questionable moral standards notwithstanding, Rey de Castro was both clever and crafty. While clearly outgunned, and operating from the low ground in terms of the court of international public opinion, Rey de Castro put up a spirited fight on behalf of Julio César Arana and Pablo Zumaeta.

Also in rhetorical play during the Putumayo scandal was the Spanish black legend among Anglophones versus the "anti-gringoism" of their Latin critics. Hardenburg, Enock, Casement, Michell, and Collier all invoked conquistadors and/or inquisitors at some point in their discourses as embodiments of Latin authoritarianism and close-minded cruelty. In crafting his defense, Arana relied upon his confidante and minion, Carlos Rey de Castro.

As the former Peruvian consul in Manaos, he certainly understood how to play upon Peruvian nationalistic sentiment. He was also adept at underscoring the hypocrisy of Europeans and North Americans who consumed rubber, but without questioning how it was produced. He found great irony in the moralizing of Great Britain and the United States—given their own records of enslavement and genocide of the world's indigenous peoples. Had not both of them participated in the transatlantic slave trade that resulted in creation of the New World's slave-based economy?

Then, too, hadn't Great Britain created Australia as a penal colony in which convicts were treated with incredible cruelty, earning it the sobriquet of "Hell on Earth"?[40] During the nineteenth century, several indigenous tribes were exterminated in what was called "Manifest Destiny," the juggernaut of white expansion across the Australian and North American continents.[41] Hughes reports that the Aborigines were sometimes hunted for sport, there being "no more harm in shooting a native, than is shoot-

---

40. Hughes, *The Fatal Shore*, and elsewhere. For imprisoned Irish political dissidents, Australia was Great Britain's "Siberia" (Ibid., 181), and for all of its convicts a precursor of the Gulag.
41. Ibid., 120, 272–78, 414–24; Benjamin Madley, *An American Genocide: The United States and the California Indian Catastrophe* (New Haven & London: Yale University Press, 2016).

ing a dog," given that "it was preposterous to suppose they had souls."[42] Murphy reports the local campaign in Shasta for raising bounty money to be paid for Indian scalps, whereas the Northern California Indian Affairs superintendent informed Washington in 1961 that the Indians were "being hunted down like wild beasts and killed."[43]

Nor were such Anglo abuses matters of the distant past. If the London proceedings were essentially a British undertaking, in close collaboration with the Americans and within the same decade of the alleged atrocities in the Putumayo, had not Great Britain invented[44] the concentration camp to inter Boers during its South African campaign and the United States killed about fifty thousand Filipinos while subjecting their homeland to American rule?[45] Boers and Filipinos continued to be restive under their Anglo overlords at the time of the London Hearings. Lynchings of Blacks throughout the American South with near judicial impunity continued and would do so for decades longer (producing thousands of victims between between the Civil War and World War II).[46] Nor were the other sanctimonious European commentators in this narrative guiltless. During the decade of Arana's alleged Putumayo atrocities, had not Germany committed genocide of the Herrero and Nama tribes in Namibia,[47] and weren't the French still operating the world's quintessential diabolical prison system on "Devil's Island"?[48]

Then, too, there was the legacy of recent American imperialism in Latin America itself. The United States, victorious in the Spanish-Amercian War, imposed onerous conditions on Cuba as the price for its "independence" and has never relinquished Puerto Rico. There was also Theodore Roosevelt's blatant support of ostensible rebels that severed a chunk off of Colombia in order to facilitate construction of the Panama Canal.

---

42. Hughes, *The Fatal Shore*, 277.
43. Madley, *American Genocide*, 295–96.
44. This is historically inaccurate in that Spanish general Valeriano Weyler rounded up rural Cubans who might be supporting rebels during the insurrection of the 1890s and concentrated them into villages where they could be more easily controlled. Tom Gjelten, *Bacardi and the Long Fight for Cuba* (New York: Viking, 2008), 66.
45. Robert D. Ramsey III, *Savage Wars of Peace: Case Studies of Pacification in the Philippines 1900–1902* (Fort Leavenworth, KS: CSI Press, 2007), 103.
46. Bryan Stevenson, "A Presumption of Guilt," *The New York Review of Books* 64, no. 12, 8–10.
47. J. B. Gewald, "The Herero and Nama Genocides, 1904–1908," *Encyclopedia of Genocide and Crimes Against Humanity* (New York: Macmillan Reference, 2004).
48. W. E. Allison-Booth, *Hell's Outpost: The True Story of Devil's Island by a Man Who Exiled Himself There* (New York: Minton, Balch & Company, 1931). That prison was not closed until 1953.

The Americans first tried to negotiate a deal with Bogotá, but, when that failed, they fomented Panama's independence movement from a nation that was all but prostrate from its recent War of a Thousand Days.

July 9, 1913, Julio César penned an open letter (see appendix 4) from Manaos in which he denounced the lengthy "campaign of defamation" against him—including the attempts to arrest him in Iquitos and the pronouncements of the Select Committee. Every one of his critics had their own ulterior motives, and all were pawns of Colombia. Attached to Arana's letter was the Spanish version of the Larrabure y Correa answer in *The Times Supplement* to S. Restrepo, and the two memorials supposedly written by Pablo Zumaeta. Arana promised to produce his own pamphlet soon, but that it would take time. It was released but a few weeks later[49] and included a front note, dated (in Barcelona) August 15, 1913, by Marcial Zumaeta.[50] The "Arana" text was largely self-serving clarifications of, and elaborations upon, his responses during the London Hearings.

It might be noted that Julio César signed "his" open letter in Manaos in July, a scant four months after concluding his testimony before the Select Committee. Three attached documents had all been published in Barcelona, the city where Rey de Castro just happened to be during this period—as would be the Arana treatise a month later. The letter's unusual vocabulary, its misspelling of "Cassement" that occurred across all six of the Barcelona publications and the fact that it was open (i.e., to no one in particular—scarcely Arana's *modus operandi*), all suggest that Rey de Castro probably wrote the letter and sent it to Manaos from Europe for Arana's signature.

The first Carlos Rey de Castro text, signed with his own name, is dated September 6, 1913 (in Paris).[51] Then, on November 28, 1913, he finished a second work with the identical title while in Barcelona,[52] denouncing

---

49. Julio C. Arana, *Las cuestiones del Putumayo*: *Folleto No. 3* (Barcelona: Imprenta Viuda de Luis Tasso, 1913) [*Las cuestiones*].

50. He was Eleonora's younger brother residing in London where he worked as Julio César's personal secretary. The young man was apparently reasonably fluent in English and was the likely translator of the Arana team's position in letters to editors of English newspapers. His translations were no match for the eloquence and sophistication of native speaker Norman Thomson.

51. Carlos Rey de Castro, *Los escándalos del Putumayo: Carta abierta dirigida a Mr. Geo B. Mitchell, Consúl de S. M. B.* (Barcelona: Imprenta Viuda de Luis Tasso, 1913) [*Los escándalos*]. The "open letter" to "Mitchell [sic]" is a rebuttal of the consul's report to his government after his investigative trip to the Putumayo in the company of American consul Stuart Fuller.

52. Carlos Rey de Castro, *Los escándalos en el Putumayo: Carta al director del Daily News & Leader, de Londres. — Nuevos artículos alarmistas. — Plano de la zona sindicada. — Inglaterra en*

two "alarmist" articles that had appeared a few days earlier in a London newspaper, the *Daily News & Leader*. That year, he authored another work signed in Paris and likewise published by his Barcelona publisher.[53]

In February 1914, Rey de Castro published his account of the *pobladores* (settlers) of the Putumayo.[54] Most of the text regards the known history and ethnography of its indigenes. However, the final section recounts the historical evidence supporting Peru's claim to the Putumayo. Then, basing his argument upon the earlier Carlos Larrabure y Correa text, Rey de Castro contends that the original settlers of the Putumayo came from the Central Andes, and that the area was once under Inca domination (*ergo* Peru had a pre-historical claim to the Putumayo). He then summarized the Arana side of the argument by noting that the "evidence" had demonstrated beyond a doubt that rubber-gathering in the Putumayo was not based upon a regimen of violence and terror, and that, whatever past crimes might have been committed, they were exceptional cases resulting from the conditions in that remotest of areas (and would likely have transpired under similar circumstances irrespective of the nominal sovereignty of any other country). But the main weakness in the allegations of the accusers regarded their questionable witnesses, particularly Casement's, who were,

> a) Negroes from Barbados [the racism is intentional and is repeated elsewhere in the Arana defense statements][55] who convict

---

crisis (Barcelona: Impenta Viuda de Luis Tasso, 1913).
53. Carlos Rey de Castro, *Antagonismos económicos, protección y librecomercio: Tratado de comercio entre Perú y el Brasil* (Barcelona: Viuda de Luis Tasso, 1913). In it he argued that a freedom of commerce treaty with Brazil would foment economic development in the Loreto and Brazilian borderlands, provide relief by removing tariffs and lowering the costs of goods for the *caucheros* (besieged as they now were by competition from Southeast Asian rubber), and afford Peru the opportunity to perfect the existing free navigation treaty with Brazil (Ibid., 153–54). Rey de Castro's benefactor, Julio César Arana, would obviously be benefitted by all of these outcomes. It should be added that it was never entirely clear how Arana's rubber left South America. There was concern that some of his overseers in the Putumayo might be stealing and smuggling some of the product into nearby Brazil. Captain Whiffen testified before the Select Committee that he believed that Julio César was querying him regarding that possibility. As for the rubber that remained on the Company's books, it was clearly to Arana's advantage to transport it first to Iquitos, pay the export duty there and then transship it aboard Booth Steamship Line vessels—given their exemption from Brazilian tariffs. Peru's export duty on rubber was 2.5 percent versus Brazil's 23 percent—more than ten times greater (Bákula Patiño, "Introducción," 87).
54. Carlos Rey de Castro, *Los pobladores del Putumayo: Origen—nacionalidad* (Barcelona: Imprenta Viuda de Luis Tasso, 1914), 47–50.
55. Nevertheless, the Select Committee concluded:

themselves as having figured in criminal cases as authors and accomplices; b) illiterate Indians, whose recollections are all mixed up with fantastic narrations, inspired by the desire to revenge ancestral offenses, fruit of their ancient conflicts or persecutions, and c) a small number of white employees, without freedom of action, some of whom [Colombians] were interested in harming Peru, and all induced, by the passions and feuds fomented by life in the jungle, to alter the truth.[56]

Rey de Castro then denounced Great Britain and the United States for siding with Colombia to gain commercial advantages in the Putumayo/Caquetá for their industries there. He particularly noted that the United States was negotiating the terms of compensation to Colombia for orchestrating the secession of Panama in order to build the Panama Canal[57] and was also supporting Colombia's territorial claim in the Putumayo—as part of the broader attempt to normalize relations between those two countries.[58]

All of these Spanish-language texts were printed by the same Barcelona publishing house (1913–1914 must have been a banner time for Luis Tasso's widow!)[59] and were sprinkled with photographs of healthy and robust Indians. In any event, the entire collection was therefore (despite some of the subtitles) obviously intended more for the South American public than any English-speaking audience. There was also a campaign in the region's media. An example of the kinds of "unsolicited" testimonials appearing in the Lima press was the letter of Antonio Menacho, sent from Iquitos to the *El Comercio* newspaper:

> The indictments found at Iquitos in the so-called Putumayo case have made a deep impression among us and we cannot understand

---

The Barbadians were recruited in 1904–5 for the Putumayo at a time when there was no emigration law in Barbados. They were ill-treated and were coerced into ill-treating others. A large batch of them was sent back through the help of the British Consul at Iquitos in 1905, and the emigration from Barbados to the Putumayo was stopped. Subsequently, in 1907, a Barbadian emigration law was passed. When batches of labourers are being recruited under contract, proper emigration officers and provisions for repatriation are required. (British Parliament, *Report*, xxi)

The Barbadians in the Putumayo were paid 5 pounds monthly (British Parliament, *Select*, 376) and keep. Their transportation to and from Barbados was also covered.

56. Rey de Castro, *Los pobladores*, 51.
57. The United States would eventually issue a formal apology and give 25 million dollars to Colombia.
58. Ibid., 53.
59. Lagos notes that thousands of copies of these Rey de Castro works were printed, mainly for distribution in Lima. Lagos, *Arana*, 345.

how a group of brave Peruvians may thus be prosecuted; men who have for the last twelve years maintained the territorial integrity of Peru in that distant portion of the Fatherland. If the employees of Araña [sic] Brothers are to be punished for having defended their lives, and they are to be hounded for having created a substantial income for the government by subduing those ferocious infidels, who are also cannibals, what punishment do the latter deserve for their murders and their banquets on the flesh of those unfortunate Peruvians? Please note the following and most explicit account: Noé Montalban of Moyobamba was eaten by the Okanias [sic] Indians in July 1902. The Boras Indians, in 1903, killed and ate [Basque-surnamed] Theobaldo Aguirre, son of Col. Aguirre, chief of Police of Lima ... and Victor Puelles of Ancachs. In that same year, and by the same Indians, the following persons were cannibal food: Ernest Zumaeta, brother-in-law of Mr. Julio C. Araña [sic], Isaías Izquierdo, his wife and two sons, a Mr. Menendez of Amazonas, and Manuel Rosenberg of Huaraz. Twelve employees of Araña [sic] Brothers were eaten in 1904 by the ferocious Andokes, among them Sergeant Rodrigues of Lima, who had belonged to the troops stationed at Loreto, [Basque-surnamed] Enrique Esquelaga of Lima, José Rangel of Iquitos, and Mariano Hernández. José de Almeidas, employed by the Araña [sic] Brothers, was killed in 1905 and his body found ready for a feast. About the middle of the same year Enoch Richards of Lima was killed by a rifle shot, as he appeared in his doorway; his companions buried him and left the place in order to report the occurrence. Before they could return the Indians dug up the body and ate it. It was in 1905, also, that Crisóstomo, a former police inspector in Iquitos, was assassinated while bathing. In May 1910 Bartolomé Zumaeta, another brother-in-law of Mr. Julio C. Araña [sic], was killed, but not eaten, due to the bravery of Mariano Cubas, a discharged soldier, who watched over his body, rifle in hand, until help came. . . .[60]

The struggle for the public's attention entailed visual imagery as well. Chaumeil underscores that the cover of the Hardenburg and Enock book is a photo of Indians chained together. There was also the photo of a near-skeletal dead Indian woman fixing the viewer with an accusatory stare. It was reproduced numerous times and, according to Chaumeil,

> Worthy of attention is the fact that the caption varies according to the context of the publication. At times it is an Indian woman

---

60. Paternoster, *The Lords,* 306–308.

condemned to die of hunger in the Alto Putumayo (Hardenburg 1912), and at others of a Huitoto or Bora slave agonizing on the Yubineto River, the fault being that of the Peruvians of the Casa Arana—according to the firm's detractors—or of the Colombians according to Arana and his acolytes (*Variedades*, 1912). One observes in this the whole dimension of the veritable "war of images" among the protagonists of this confrontation, given that both parties employed the same means of diffusion.[61]

We have already noted the extent to which Rey de Castro insisted upon photographing healthy and content Indians during the visits by the American and British consuls to the Putumayo—even contriving festive events for that purpose. The Casa Arana's Brazilian photographer during that jouney, Silvino Santos, would eventually marry Julio César's ward. Arana financed a three-month stay in Paris for Santos, so that he could learn cinema-making at the Pathé and Lumière Brothers studio.

During late 1913, Silvino spent two months filming the Indians in the Putumayo to provide visual proof that they were being treated well. He dispatched the negatives to Lima by sea in order to have positive copies made, but they were destroyed when the ship was torpedoed by a German U-boat in 1914.[62] Santos returned to the Putumayo in 1916 for a second shoot that resulted in the film *Indios Huitotos do Rio Putumayo*. Arana paid to have the film produced and distributed in 1917. It was shown in several South American venues over the next several years. One commentator, writing in 1928, described the "latent" message of the film as being, "the great leap toward progress and civilization that the cannibal and anthropophagi savages of these regions had made."[63]

On balance, it seems fair to say that Julio César Arana (and Carlos Rey de Castro) were relatively successful in manipulating their public image in Iquitos and Lima, but were hugely outgunned in the international arena. The combined weight of the publications by Hardenburg, Casement and Paredes upon both British and American public opinion were uniformly anti-Arana. As we shall see, even the Vatican would both demand and then orchestrate reforms in the Putumayo.

Nor did Rey de Castro's publications go entirely uncontested on Peruvian turf. Judge Valcárcel had barely survived his journey from Iquitos to Manaos, but, by March 1913 (or while the Select Committee was still

---

61. Jean-Pierre Chaumeil, "Guerra de Imágenes en el Putumayo," in Alberto Chirif (ed.), *Cien Años después del caucho: Cambio y permanencias en las relaciones con los pueblos indígenas* (Lima: Tarea Asociación Gráfica Educativa, 2009), 48.
62. Goodman, *The Devil*, 252; Chaumeil, "Guerra," 53.
63. Quoted in Chaumeil, "Guerra," 56.

in session), he was recovered and in Panama. It was there that an incensed Valcárcel began writing his own damning rejoinder, replete with minute detail regarding the atrocities committed in the Putumayo, as well as the Arana machine's persecution of its critics (himself, Saldaña Rocca, Paredes, Hardenburg, Whiffen, Casement). At the outset of his book, Valcárcel states:

> So, I am going to say the entire truth in this unfortunate matter, the naked truth, without euphemisms or reticence. I know that I am exposing myself to everything. I know that my enemies will be innumerable. I expect anything: but nothing can now stop me because I have the profound conviction that I am doing a service to my country and to [the cause of] justice. . . .[64]

Valcárcel's text incorporated most of the Paredes report's key points regarding the atrocities, as well as the Arana machine's political manipulations of them. There had been interminable delays and obfuscation in the prosecution of those responsible for the crimes. As earlier noted, in Iquitos the local court had ruled that, since there were now 255 outstanding indictments over the Putumayo allegations, before any of the accused could be incarcerated all would have to be. This was impossible, of course, given that many of them were no longer even in Peru.[65]

There were several curious cases. Despite having been indicted, Luis Alcorta and López Zumaeta were still employed in the Putumayo as overseers. When Elías Martinengui opted to flee the country, he left the Putumayo on the *Liberal* accompanied by his four favorite Huitoto women.[66] One of the accused, Vicente Macedo, was detained briefly in Lima. The ex-Prefect of Loreto opposed extradition of Macedo to Iquitos and he had been released by order of President Leguía's secretary. The succes-

---

64. Valcárcel, *El proceso*, ii. The dedication read as follows:
  The ex-president of the Republic of Peru, Don Guillermo E. Billinghurst, who did so much to improve the lot of the indigenes of said Republic, and to the Anti-Slavery and Aboriginal Protection Society of London to whose generous and well-intentioned actions are owed the lives of 10,000 aborigines of the Putumayo saved from assassination, I dedicate these pages of horror.

  Valcárcel finished his text sometime near the end of 1913, and in his conclusion, he refers to the publication by Rey de Castro (*Los escándalos del Putumayo*) that was published in November of 1913 (Ibid., 309). The Valcárcel work was not actually published until 1915, and appeared in a limited edition of one thousand copies. Interestingly, the judge clearly feared further foul play and possible persecution, since he insisted upon signing personally every copy of the book as a guarantee that its text was authentic (i.e., unadulterated).
65. Ibid., 20–21, 316.
66. Ibid., 243.

sor prefect of Loreto, (Basque-surnamed) Julio Aguirre, claimed there was no one of that name in detention. Valcárcel believed that someone had simply destroyed the relevant documents.[67]

The judge was critical of the attitude that equated Peruvian patriotism with silence and a cover up of the atrocities. He argued that, after the exposé, it was only by confronting the horrors and thereby extirpating them would it be possible to reinstate Peru's international standing. Therefore, conducting an impeccable investigation, followed by fair Hearings and just punishment of the guilty, was the only true Peruvian patriotism.[68] He was disdainful of the supposed patriotism of Julio César Arana; after all, was he not responsible for replacing a Peruvian company in the Putumayo with an English one simply because it suited his personal interests.[69]

Valcárcel was particularly incensed over the collusion of Iquitos's judiciary with the Casa Arana:

> The first 60 to 80 pages of the indictment presented in 1907 are a painful recital, that arouses one's indignation on account of the slowness and red tape in preparing the same, and the dilatory and evasive manner of reaching conclusions. The court issued a writ declaring itself incompetent to act, to which I referred in my first report, basing the same upon a treaty of temporary neutrality which had been signed with Colombia regarding this region; and in harmony with the attitude of his superiors, the Judge of First Instance, whose name I withhold, proceeding with audacity bordering on insolence, went even farther and issued that astonishing decree which will mark, without doubt, a new epoch in the annals of administrative justice; pigeonholed indefinitely."[70]

As well as with this:

> ... from 1908 until 1912, [Egoaguirre] wielded a decisive influence in the Peruvian government in everything referring to the Department of Loreto; and the Commissaries of the Putumayo and all of the public functionaries of Iquitos were named through Egoaguirre's orders; as a consequence for the President of this republic, the Putumayo continued to be pure invention, until the final months of 1911.[71]

---

67. Ibid., 24, 350.
68. Ibid., ii–iii.
69. Ibid., 303.
70. Ibid., 314–15.
71. Ibid., 271. It also became clear that, while a senator, Egoaguirre continued to work as Julio César's personal attorney and was on the Company payroll (Ibid., 274).

Nevertheless, under the Billinghurst administration, there seemed to be some attempt from Lima to implement justice in the Putumayo. By May 1913, Armando Normand had been detained in Bolivia, and the British Foreign Office believed that he would be returned to Peru to face charges.[72]

Then, too, Paredes wrote to Buxton with an update on developments in Iquitos. Judge Herrera had reissued about two hundred warrants, although the names of neither Julio César Arana nor Pablo Zumaeta were among them. Nevertheless, according to Paredes,

> Now that the second stage of the action has been reached, the development of the question will go apace. After all, we must congratulate ourselves that the plenary procedure should have been reached. This is so much gained. You will have read the plenary pamphlets which Zumaeta has published against Dr. Valcárcel, the judge who succeeded me, and against me. They are the least that could be expected from a man guilty of such stupendous crimes. I have written a book in my own defense; but it falls within the range of the government's authority, it has forbidden me to publish it without its permission. But I am determined to publish it at whatever cost [he never did]. Dr. Valcárcel has written a book about the Putumayo, and I know it to be in press. I can assure you that, in all these affairs, my proceedings have been straight, & that I fear nothing, calumny least of all. I feel very proud and highly honoured at being a member of this society [the Anti-Slavery and Aborigines Protection Society].[73]

However, according to papal apostolic delegate in Perú, Angelo Scapardini, there was now in Lima, at the highest levels, the belief that the Putumayo scandal had been orchestrated from London to discredit

---

72. W. Langley to The Reverend J. H. Harris. May 24, 1913. Bodleian Library MSS Brit. Emp. S-22 G332. By December 1912, or while the Hearings in London were still is session, Armando Normand appeared at the Peruvian consulate in Cochabamba, Bolivia, requesting that it forward to the Peruvian minister of justice a letter in which he defends himself against the Putumayo charges. *El Ferrocarril de Cochabamba*, December 24, 1912. By late 1913, Normand was in Lima's Guadalupe prison destined to be sent to Iquitos to stand trial. There is a published interview with him in which he recounts his life. (see appendix 5). During the next year, Bolivia would detain Abelardo Aguero and Augusto Jiménez (working for the Suárez Brothers on the Beni River) and turn them over to Peruvian authorities. They also reported that Victor Macedo had fled their country. Pedro Suarez, Bolivian Legation in London, to Travors Buxton. September 9, 1914. Bodleian Library, MSS Brit. Emp. S-22 G332.

73. Rómulo Paredes to Travers Buxton. May 9, 1914. Bodleian Library MSS Brit. Emp. S-22 G332.

both Peru and the Casa Arana to drive down the value of the Peruvian Amazon Company in order to force its sale to British interests at a huge discount.[74] Scapardini also noted that President Leguía was himself reputed to have an ownership interest in the enterprise, as did many other prominent Peruvian politicians.[75]

In any event, by early 1914, both the Peruvian and Colombian governments were on record with statements by their supreme leaders favoring a negotiated settlement of their longstanding territorial dispute. Thomson was in a self-congratulatory ecstasy for having fostered an International Board to redress the humanitarian catastrophe in the entire Upper Amazon. It would likely become a reality within the year. Nevertheless, neither Peru nor Colombia committed to serious bilateral negotiations under the world's watchful eye. Rather, each country took baby steps toward strengthening its presence on the ground in the Putumayo. In anticipation of the International Board, Colombia prepared a proposal seeking permission from it to place one or two boats on the Putumayo pending ultimate resolution of the conflict.[76] Peru erected a wireless telegraph tower at La Chorrera to bring that outpost into instant communication with Iquitos (and Lima). The prefect of Loreto ordered that a trail be blazed from the Putumayo to the Napo River—Julio César had his eye on that area as well.[77]

Arana's nemesis, Sir Roger Casement, was embarked upon his own fateful and fatal adventure. As British consul in Santos and Rio, he had expressed his veiled admiration of Brazil's Germans. In his 1905–1906 report on German trade activity in Santos, he noted, "In paper and cement Germany has practically superceded British and all other exporters to this part of Brazil, and she is steadily entering into competition in a variety of other articles once largely supplied from the United Kingdom."[78] Casement pronounced the Germans to be potentially the country's most talented foreign entrepreneurs ("Germany in Brazil—the teuton on the Amazon—would work more amazing things than the English in India").[79]

By 1909, Casement was sending surreptitious donations from Brazil to Irish nationalist causes. He wrote Morel at this time that war between Great Britain and Germany seemed unavoidable, ascribing the

---

74. García Jordán, "La misión," 257, 259.
75. Ibid., 262. As we have seen, Jerome shared that belief. However, García Jordán underscored that the allegation has never been proven.
76. Thomson, *The Putumayo*, lx.
77. Stanfield, *Red Rubber*, 179.
78. Quoted in Angus Mitchell, *Roger Casement* (Dublin: The O'Brien Press, 2013), 132.
79. Mitchell, *Sir Roger*, 362.

major blame for it on the British.[80] In sum, "He was not the only Irish Germanophile. Many looked to Germany for support in their struggle for Home Rule."[81] Casement subsequently informed Morel that the Irish cause would be his life's work once the Putumayo Select Committee Hearings were over.

As his Irish nationalism became personal obsession, Sir Roger increasingly regarded Germany to be the necessary European counterweight to British imperialism. In 1911, he wrote a piece entitled "The Keeper of the Seas" that was later included in a collection of his published (previously mainly anonymous) anti-British essays, completed on September 1, 1914—or literally within a few days of the outbreak of World War I.[82] Roger denounced the Entente Cordiale, signed in April 1904 between France and Great Britain, that was designed to divide much of Africa and Asia into their respective spheres of imperial expansion (for example, Great Britain in Egypt and France in Morocco), while, at the same time, thwarting an ascendant Germany's initiatives in Europe and on the wider world stage. According to Casement,

> It can be summed up in one phrase. German expansion is not to be tolerated. It can only be a threat to or attained at the expense of British interests. Those interests being world-wide, with the seas for their raiment nay, with the earth for their footstool—it follows that wherever Germany may turn for an outlet she is met by a British challenge.[83]

In 1905, Germany tested the Entente Cordiale when Kaiser Wilhelm went to Tangiers and supported its sultan's sovereignty. Tensions escalated as Germany and France mobilized troops and threatened to go to war over the issue. What has been called the "First Moroccan Crisis" was resolved at the Algeciras conference the following year, attended by all of the European powers and the United States. Germany proposed a compromise that was supported by only Austria-Hungary; the rest recognized French colonial hegemony (but within an ostensibly still "independent" Morocco).

Then, in 1911, after France sent a large expeditionary force into the Moroccan interior in defense of its minion, the sultan of Fez, Germany (by then engaged in an arms race and naval buildup with Great Britain)

---

80. Mitchell, *Roger Casement*, 142–43.
81. Bryant, *Roger Casement*, 200.
82. Sir Roger Casement, "The Keeper of the Seas," in Sir Roger Casement, *The Crime against Europe: A Possible Outcome of the War of 1914* (Philadelphia: The Celtic Press, 1915), 5–8.
83. Ibid., 5.

decided to test the British-French alliance once again. A German gunboat was sent to Agadir under the pretext of protecting German commercial interests in the area. Great Britain supported France and the international tension triggered a major collapse of the German currency and stock market. Wilhelm had little choice but to withdraw from Morocco, but the coalitions and battle lines of World War One were clearly established.

For Casement, the real enemy, and a supremely arrogant one at that, was now Great Britain. British imperial hegemony was the result more of "unrivalled position in the lap of the Atlantic, barring the seaways and closing the tideways of Central and North-eastern Europe," rather than to any particular English genius. Furthermore, Ireland was the key:

> This highly favoured maritime position depends, however, upon an unnamed factor, the unchallenged possession and use of which by England has been the true foundation of her imperial greatness. Without Ireland there would be to-day no British Empire. The vital importance of Ireland to England is understood, but never proclaimed by every British statesman. To subdue that western and ocean-closing island and to exploit its resources, its people and, above all its position, to the sole advantage of the eastern island, has been the set aim of every English government from the days of Henry VIII onwards.[84]

On his was back from visiting his brother in South Africa in 1912, Casement visited Germany and was favorably impressed. In June of 1913, he was in Connemara to see for himself the devastation of a typhus epidemic there. The poverty and disease that he witnessed prompted him to label it the "Irish Putumayo," and its inhabitants the "white Indians."[85] On August first, he retired from the Foreign Office. He believed that war between Germany and Great Britain was immanent, and would provide the opportunity for Irish independence. He immediately began writing and speaking in favor of a Free Ireland, and became one of the founders of the Irish Volunteers—conceived as a counterweight to the pro-British Ulster Volunteers. In April 1914, Germany landed two thousand rifles and two million rounds of ammunition in Ireland for the Irish Volunteers.[86]

In July 1914, Casement traveled to the United States to solicit Irish-American support for the campaign. The next month World War I broke out, and, in October, Sir Roger was in Berlin soliciting German support for the Irish nationalists. He volunteered to recruit an Irish Brigade

---

84. Ibid.
85. Goodman, *The Devil*, 252.
86. Ibid., 253.

among the growing population of British prisoners of war that would be trained by the Germans and landed in Ireland, along with a large shipment of arms, in support of the pending Easter Uprising. The Germans demurred at the last minute and Casement tried to head off the suicidal uprising by his fellow nationalists, but to no avail. So, he rushed to Ireland to be with his comrades, and, on April 21, 1916, he was captured ashore by the British after barely surviving a shipwreck off the Irish coast.

In July, while awaiting his trial in London's Brixton Prison, Casement received a telegram from Julio César Arana, sent from Manaos, that read:

> On my arrival here am informed you will be tried for High Treason. Want of time unables me to write you being obliged to wire you asking you to be fully confessing before the human tribunal your guilts only known by Divine Justice regarding your dealings in the Putumayo business. . . . Inventing deeds and influencing Barbadians to confirm unconsciously facts never happened, invented by Saldana [sic] and thief Hardenburg. . . . I hold some Barbadians declaration denying all you obliged them to declare pressing upon them as English Consul and frightening them in the King's name with prison if refusing to sign your own work and statements. You influenced the Judges in the Putumayo affair who by your ill influence confirmed your own statements. You tried by all means to be a humanizer in order to obtain titles fortune, not caring for the consequences of your calumnies and defamation against Peru and myself doing me enormous damage. I pardon you, but it is necessary that you should be just and declare now fully and truly all the true facts that nobody knows better than yourself.[87]

Despite an outpouring of appeals for clemency from prominent British and international celebrities, Sir Roger was condemned to death. On July 29, 1916, the U.S. Senate passed a resolution favoring commutation of Casement's sentence. For some reason, the White House dallied for three days before forwarding it to London. It arrived shortly after Sir Roger Casement ascended the gallows on August 3.[88]

Persons like George Bernard Shaw, Arthur Conan Doyle, and W. B. Yeats (who penned a poem in Casement's honor) had written in support of the defense. After the execution, W. B. DuBois called the deceased a patriot and martyr. An exception was Joseph Conrad, Casement's confidante in the Congo, who did all that he could to distance himself from the controversy. Clearly, within British government circles, the treason of

---

87. Quoted in Ibid., 256.
88. Bryant, *Roger Casement*, 296.

a former colleague and collaboration with the enemy in an ongoing war was simply unforgiveable.

However, the death sentences of some of the other leaders of the Easter Rebellion were commuted to life imprisonment, so Casement's determined enemies could not be certain of his execution. Just as the campaign for clemency was heating up, the British government claimed to have found Roger's diary (later labeled the "Black Diary") that recorded his ostensible secret life as a homosexual and pedophile. While, at the time, details were leaked to the press, for decades all access to the original texts was denied for reasons of "national security." Not surprisingly, this has led to controversy regarding the very authenticity of the manuscripts—one that continues at present.[89]

According to Mitchell,

> Casement was tolerant to most faiths, including Moslems, Hindus and Buddhists, and also respected magico-religions of West Africa and shamanic rituals of the Amerindians. There is also an element of agnosticism in some of what he writes. What he vehemently despised was religious intolerance and sectarian hatred. Among his final words, in an extraordinary gesture of reconciliation, he wrote: 'My goodwill to all men; to those who have taken my life equally to those who have tried to save it—all are my brethren now.'[90]

---

89. See "The Diaries Controversy" in Angus Mitchell, editor, *The Amazon Journal of Roger Casement* (London: Anaconda Editions, 1997), 15–56 [*The Amazon Journal*]. Mitchell clearly concluded that Casement had been framed. In another publication he reproduced, as an appendix, the remembrances of Dr. Dickey. He and Casement were on board ship together and Roger was reading a letter that he had received from Conan Doyle. He told Dickey that Conan Doyle had written, or was writing, a book called *The Lost World* based on information supplied to him by Casement,

> According to Sir Roger, Doyle had enquired if there were, along the Amazon but separated from it by considerable distance, a high plateau. He had responded that there was and described its main features. Doyle had used it in his story. (Mitchell, *Sir Roger*, 737)

The next day, the subject came up again and Casement told him that he had given Doyle information about sexual depravity among Congo natives and the novelist had inquired if he had similar stories from the Amazon. Casement now asked Dickey, if, as a medical doctor, he knew of any. Dickey understood that his fellow physician, Conan Doyle, was interested in the possibility that what civilized persons take to be depravity might be regarded among "savage peoples' to be normal behavior. Dickey began recounting so many cases that Casement excused himself. "He returned from his stateroom with a notebook bound in black and with yellowish lines. For at least the next half hour, maybe three quarters, I recited instance upon instance of sexual perversion among the Indians." Ibid.,738.

90. Ibid., 319. Dickey reported that he had seen Casement weep while observing Indians

Two months later British foreign secretary Grey received a letter from Arana's London attorney asking if Casement had replied to Arana's telegram. He had not.⁹¹

After all of the *Sturm und Drang* surrounding the Putumayo scandals, the world had moved on. Indeed, it plunged right over the cliff into World War I. The precipitous fall from grace of the "traitor" Sir Roger Casement in British public opinion assumed its own life. The first minister of the British Legation in Lima noted that "the intervention of His Majesty's government in the further investigation of charges that were originally formulated by a man whose name is now the subject of universal reprobation" was unwelcome, and advised that the Putumayo no longer be mentioned in Great Britain's diplomatic encounters with Peru.⁹² The plight of a few Huitotos and Boras in some backwater simply disappeared off the international radar screen.

Rómulo Paredes, perhaps somewhat amazingly, managed to hang on until 1919 in Iquitos as a newspaper publisher critical of Arana. In 1918, he penned a poem called "A Samarem" (To Samarem), a mythical Indian chief. It decries Peruvian white society, and particularly the *caucheros*, imploring Samarem to descend the river flowing from his wilderness utopia in order to conquer and then rule (in the interest of racial harmony and national glory) over the perfidious and crass society established by whites in Peru (see appendix 6). The poem caused local and national outrage, and even embroiled Paredes in litigation.⁹³ He was eventually forced to leave Iquitos, possibly in fear for his life,⁹⁴ and resettled in Chiclayo. He continued there as a literary figure and influential newspaper pundit until his death in 1961.

Enter God

As early as the autumn of 1908, Hardenburg had written a letter to the Anti-Slavery and Aborigines Protection Society that ended: "In conclusion, I would say that the whole region of the upper Amazon, and espe-

---

being boarded for sale as slaves, exclaiming "it's cruel! I can't stand it." Dickey therefore opined, "I can imagine that he suffered no greater torment to his last living hour than when he collected evidence of the atrocities in the Belgian Congo and in the Putumayo, neglecting no detail of torture and butchery." Ibid., 736.
91. Goodman, *The Devil*, 259.
92. Ibid., 264.
93. Chirif, "Los informes," 80–81.
94. Ricardo Vírhuez Villafane, "Rómulo Paredes: Canto Samarem," http://4.bp.blogspot.com/_IR08RZyyyUo/R41H6sFmYSI/AAAAAAAAAlw/gztRyllkdNU/s1600-h/SAMAREM.jpg.

cially the Putumayo district, offers a vast and promising field for missionary work."[95] In his Blue Book report, Casement had recommended that the Peruvian government be required to encourage and fund missions in the Putumayo. After all, he had found that well-intentioned Christian missionaries were about the only counterweight to King Leopold's policies in the Congo.

Such a seemingly simple recommendation invoked the extremely complex history of Christianity in the New World over the previous four centuries. By the time of the American independence movements, the Church (including its religious orders) had acquired vast estates throughout the region and exacted tithes on many private holdings as well. In effect, the Catholic Church was the largest single landowner in Latin America when Simón de Bolívar began his campaign.

Under Spanish rule, throughout the Hispanic Americas most of the region's clergy had originated in Spain—the result of longstanding collaboration between the Vatican and the Spanish government. While rebel victories on the battlefield led directly to replacement of Spain's civil administration in most of the region (excepting Cuba and Puerto Rico), the new continental governments were forced to deal with an existing Catholic hierarchical structure with continuing strong European roots.

Nevertheless, throughout the nineteenth century, the considerable secular power of the Catholic Church in southern Europe would itself come under serious assault. The unification of Italy (1860) by liberal secular forces reduced the Vatican State's holdings in Central Italy to its present enclave within the city of Rome.[96] In Spain, the process of *desamortización* (disentailment from mortmain), initiated by Godoy in 1797 and completed by Madoz in 1855, had stripped the Church of many of its landholdings and traditional tithes in that country. There would be similar echoes throughout Latin America—the details varying by country—as the new independent republics struggled to forge their political infrastructure and national identity. The history of this historical process in Colombia and Peru is of particular interest to our study.

In 1818, Colombia expelled its religious orders, but then, in 1821, Colombia's first Republican Assembly sought to establish diplomatic relations with the Vatican. However, the latter dithered, clearly concerned that it would infuriate Spain. So the pope put off naming any Colombian

---

95. Walter Hardenburg to Travers Buxton, October 9, 1908. Bodleian Library MSS Brit Emp S-22 G322 (file 2 of2).
96. This led to a half-century standoff between Church and state that lasted until the famous Concordat signed by the pope and Benito Mussolini in 1929.

bishops until 1827 and failed to recognize Colombian sovereignty until 1835. This caused the Colombian Congress to prorogue the agreement known as the *patronato*. It called for government control over the granting and defining of ecclesiastical jurisdictions, the financing of new church institutions and closure of existing ones, management and investment of church tithes, and the nominating of parish priests to their bishops and of bishops to the Vatican.

For the next couple of decades, there was tense mutual tolerance between Church and state, and, in 1844, the Jesuits were allowed to return to Colombia during the term of President Alcántara Herrán (1841–1845). However, serious trouble began when liberals and conservatives emerged as the two contenders (and remain so to the present) for power in Colombian politics. The former were viewed as anti-clerical progressives and the latter as the defenders of the Church. In effect, Colombia began experiencing a series of religious wars between liberals and conservatives—albeit the allegiances and alliances of individuals were not always fixed.

A prime example of such shifting was the career of General Tomás de Mosquera y Arboleda (1798–1878). He would be Colombia's president on four occasions. He had fought against the Spaniards with Bolívar and suffered a severe wound to his face that impaired his speech permanently. He was elected president in 1845 of what was then known to be the Republic of New Granada as the candidate for what would become the Conservative Party. Nevertheless, during his first term he worked for the separation of Church and state—indeed, most of his policies were closer to the liberals' agenda, a posture that alienated his former supporters. In 1848, there was a congressional debate regarding the "foreignness" of the Jesuit missionaries that culminated in legislation requiring that any missionary be a Colombian citizen. After leaving office and succeeded by a liberal, in 1849, de Mosquera moved to New York City, where he would fail in business.

Meanwhile, in 1850, the liberals again expelled the Jesuits and incurred Vatican condemnation for its agenda favoring universal suffrage, freedom of speech, freedom of the press, freedom of conscience, etc. Colombian priests railed against the government, and the conservatives initiated a series of armed rebellions. The liberals expelled three high church dignitaries, including the archbishop of Bogotá, Manuel José de Mosquera y Arboleda (brother of the former president), and, in 1853, declared the separation of Church and state. Nevertheless, it could be said that under the liberals,

> ...in 1850 the land-grabbers took a decisive step by legalizing the splitting-up of the *resguardos* [Indian reserves]. What followed from this was sheer pillage, which resulted in the proliferation of minute scraps of land granted to the Indians, and the transformation of millions of peasants of mixed blood—who now [1984] form half of the Colombian people—into a rural sub-proletariat.[97]

Missionaries, who would civilize and thereby subordinate the Indians, were viewed as a critical phalanx in implementing the new social and economic order.

This is illustrated well by the missive sent by the provincial Legislature of Casanare in 1854 to the archbishop of Bogotá demanding the immediate dispatch of "zealous, active, worthy and faithful" missionaries to the area who would gain for Christ and society "thousands of wild beasts [Indians] more fearsome than the jaguar and the crocodile." In the event that the archbishop refused, the legislature was disposed to bring in "Protestant missionaries or those of any other religion."[98]

In 1854, Tomás de Mosquera returned to Colombia and led the successful "artisans' revolution" against the dictator José María Melo. He was then the presidential candidate for the Liberal Party in the election of 1857, but lost to the conservative Mariano Ospina Rodríguez.

The following year, the Granadine Federation was created (which included Panama and parts of what would ultimately become northeastern Brazil). It divided greater Colombia into federal states, and Tomás de Mosquera became the governor of the Federal State of Cauca,[99] whence he opposed President Ospina. In 1860, Governor de Mosquera declared the independence of Cauca and was joined by the states of Santander and Tolima; they proclaimed him their governor as well. He led his troops to victory in the resulting civil war. In 1861, he was made interim president of the new United States of Colombia. Under Ospina, in 1858, the Jesuits had been allowed back into the country, but now they were expelled once again. All religious orders were banned. President de Mosquera also moved quickly to nationalize all Church property held in mortmain, declaring that the land should be given to those who worked it.[100]

Tomás de Mosquera was elected to a new term for 1863–1864 (his third as president of what is today is called "Colombia"). Then, in 1863,

---

97. Bonilla, *Servants*, 57.
98. Ángel O. A. R. Martínez Cuesta, "San Ezequiel Moreno, Misionero en Filipinas y Colombia," *Thesaurus* 52, no. 1,2,3 (1997), 487–88.
99. His natal region and the one that recurs throughout this text in its several manifestations as Popayán, Pasto, and the Putumayo/Caquetá.
100. Bonilla, *Servants*, 50–52.

President/General de Mosquera led his army against that of Ecuador in the War of the Cauca, which was triggered by Ecuador's conservative President Gabriel García Moreno's decision to empower the Catholic Church, opposed by the country's liberals. President de Mosquera provided the latter with aid. It was now that he decided to recreate the former Gran Colombia and summoned García Moreno to a meeting to negotiate the details of incorporating Ecuador into it. García Moreno failed to show and war broke out. Ecuador invaded the Cauca, but was defeated soundly by the Colombians at the Battle of Cuaspad. The conflict was resolved by an armistice that recognized the status quo ante between the two countries. That same year of 1863, Colombia formulated a new constitution that guaranteed freedom of religion, thereby terminating the Catholic Church's monopoly and banning its intervention in political affairs. This infuriated the Colombian clergy.

When his term expired in 1864, de Mosquera went to Paris as his country's ambassador. He returned to Colombia, and, in May of 1866, ran successfully for a fourth term as president. His continuing tensions with the Catholic Church and with some liberals led him to abolish the regular session of the Congress in 1867. This dictatorial move provoked severe criticism from Pope Pius IX, and de Mosquera was deposed in a coup d'etat in November of 1867. By then, many of his fellow liberals had acquired disentailed former Church properties, given originally to the poor, and feared losing them once the corrupt politicans who had facilitated the expropriation were brought before President de Mosquera's justice and therefore supported the conservatives' coup.

After the coup, de Mosquera lived in exile for three years in Lima, returning to Colombia in 1871 for an unsuccessful run at the presidency. He was then elected as president of Cauca State, a post that he held until 1873. In 1874, Colombia again had a leftist government that sought to protect the Indians. It prohibited the use of violence against them and tried to protect their remaining lands. However, it was a time of fiscal crisis and political disturbance, so such measures were lightly enforced. In 1878, de Mosquera died on his farm in Puracé, close to his birthplace in Popayán, the city where he was buried. His legacy was vilified and condemned by the conservatives, who depicted him as the prime persecutor of religion.

There was now considerable strife, but the conservatives and the Catholic Church prevailed. In 1885, the Jesuits were allowed to return, and the following year a war-weary country forged a new constitution that would remain in effect for more than a century (or until 1991). Its first words were, "In the name of God, supreme source of all authority. . . ."

According to Bonilla, the Catholic Church now had total control over the country. There was immediate legislation requiring the secular authorities to respect Canon Law. Colombia would be subject to the Concordat—an official Catholic country. The Church was to be compensated for its lost lands and buildings and was allowed to acquire new ones. There was to be an annual subsidy for the Colombian Catholic Church, in the amount of 100,000 pesos, to be distributed in consultation with the apostolic delegate. Initially, it was decided to allocate 25,000 pesos to the missions. The Church would serve as the official guardian of the academic curriculum from primary school through higher education. There was also a Convention of the Missions that ensured that only Catholic orders could carry out missionary activity in the country and would do so with financial support from the state.[101]

Then, in 1890, there was new legislation, Natives' Law 89, that divided Colombia's Indians into those who were already "domesticated" and living on reserves and the remaining nomads. The former were allowed to practice their traditional form of government and lands within their *resguardos*, although the state was empowered to administer them. Then there were the indigenes still roaming the forests. They were placed under the aegis of missionaries. Three years later, new legislation exempted such Indians from Colombian laws and conferred on the missionaries the power to exercise civil, penal, and judicial authority over their charges. In 1902, these powers were renewed and enhanced. Public lands could now be given to the mission orders to underpin and finance their work. The state subsidy of the missions was tripled to 75,000 pesos. Education in the missions was placed directly under the control of the missionaries, and they were empowered, in consultation with the apostolic delegate, to appoint the civil authorities within their mission precincts. Under these measures, three-quarters of the country was now effectively mission territory.

Assessing the ecclesiastical history since Independence, with respect to Peru, Turvasi wrote:

> It is a known fact that towards the end of the colonial period the clergy of Peru fell into a most degraded and demoralized condition, owing largely to the benefices given to Spanish priests. They accepted them as stepping stones for promotion to better positions in Spain or Mexico. As a result, the higher clergy took very little interest in their subordinates, who generally came from the lower class. Like any other branch of Spanish administrative circles,

---

101. Ibid., 53–55.

promotion was closed to the colonial born. The war of independence and a long series of revolutionary wars sustained this state of corruption in the Church. Bishops were chosen by the government, which retained the right of nomination to a vacant see. In choosing, the government was guided more by the political pliability of the bishops than by any regard for their piety or religious zeal. The bishops' stipends were apt to fluctuate if they did not respond to the exigencies of the party in power.[102]

The same could be pretty much said for Colombia. The situation there by the mid-nineteenth century could be summed up as follows:

> ... the vast concentration of landed property in clerical hands, though denounced a century earlier, still continued, and the poor use made of such large tracts of land greatly hampered the economic and social development of the country. This, combined with the greed for land of the rich Creole families, inevitably worsened the situation of the growing and impoverished mass of Indians and mestizos, thus bringing into being a new rural sub-proletariat.[103]

Nevertheless, even in countries like Peru and Colombia, where Catholicism persisted as the official religion, the Catholic Church's secular powers would erode. Peru underwent its own disentailment of church properties and privileges. Legislators, ranging from confirmed anti-clerics to moderate Catholics, and convinced that reform was essential, unilaterally reconfigured the status of Catholicism within their country. During the nineteenth century there was considerable dialogue and a degree of cooperation between the government and the Vatican regarding missionizing of the Amazon. This was played out against the backdrop of both colonial history and contemporary realities. The missions had been neglected by Peruvian bishops and were in a deplorable state.[104] It seems clear that at least some elements within a chastened Catholic hierarchy were anxious to demonstrate that the Church could be a progressive, modernizing force.

As early as 1802, civil and ecclesiastical authority in the future Loreto was upgraded when the Spanish government created both a governor-

---

102. Turvasi, *Giovanni Genocchi*, 84.
103. Bonilla, *Servants*, 52.
104. Turvasi, *Giovanni Genocchi*, 84. Turvasi cites the work of Marcel Monnier, *Des Andes au Para* (Paris: Chez Plon, 1890) as documenting the corruption of both the Peruvian and Ecuadorian churches.

ship and general commandancy of Maynas[105] (under the Viceroyalty of Peru) and a coterminous Maynas diocese (within the jurisdiction of the archbishop of Lima). This was designed to stop the relentless expansion of Portugal into the Upper Amazon. The newly constituted Maynas Region, headquartered in Chachapoyas, consisted of present-day Peru's Loreto and Amazonas departments, as well as parts of contemporary Bolivia, Ecuador, Colombia, and Brazil.[106] After independence, a Peruvian reduced version persisted as the Subprefecture of Maynas that encompassed all of the country's Amazonian Oriente.

In 1830, the sub-prefect of the Peruvian Maynas issued an order to the effect that all indigenes working involuntarily for whites were henceforth released from such service.[107] Nevertheless, given that the Indians were indolent and had high mortality from their diet of poisonous reptiles and forest fruits[!], administrators were required to send *peones* to the capital to work for the regional government and merchants—whoever asked for them—but they were to receive fair compensation.[108]

Meanwhile, Peru created its Departamento de Amazonas in 1832, and its administration was mandated to cooperate with the missionaries in the reduction of savage Indians. García Jordán notes that this was the first attempt in Latin America to ally church and state in a common progressive and patriotic civilizing mission. It was mandated that the missionaries in the new department would receive an annual subsidy of 2,000 pesos for tools and farm implements for development of their settlements.[109]

It is clear that the civil administration of the Subprefecture of Maynas had its own problems keeping officials on the ground. In 1835, there was an order from the sub-prefect of Maynas that the governors and other civil appointees could not leave their posts without official permission. It seems that many would assume their duties for a few months to launch private initiatives and put supervisory staff in place, sometimes transferring whole groups of Indians from one locale to another, before absenting themselves. In 1838, the Subprefecture of Maynas ordered that administrators, judges, residents, and missionaries should collaborate in converting the Indians, but without abusing

---

105. Written earlier in this text as Mainas, when referencing Golob.
106. Bákula Patiño, "Introducción," 20–21.
107. Ibid., 69.
108. Jesús Victor San Román, *Perfiles históricos de la Amazonía peruana* (N.p.: Centro de Estudios Teológicos de la Amazonía; Centro Amazónico de Antropología Práctica; Instituto de Investigaciones de la Amazonía Peruana, 1994), 118–19.
109. García Jordán, "El oriente," 228–29.

them.[110] Nevertheless, in 1850, it could still be said that the indigenes were the ones being targeted by officials for obligatory labor—and without compensation.[111]

The initial history of the Maynas diocese proved rocky, when, in 1821, its first bishop, Hipólito Sánchez Rangel, a Spanish national and monarchist, abandoned his post and fled. There were several attempts to replace him that failed. In 1836, (Basque-surnamed) José María Arriaga accepted the post of bishop of Chapapoyas (with jurisdiction over much of Maynas). That same year, the college at Ocopa for the formation of missionaries was reopened under the auspices of Spanish Franciscan friars. In 1840, Arriaga proposed creation of *La Obra de la Propagación de la Fé*, an initiative to evangelize the Peruvian Amazon. It would be

> ... eminently patriotic, given that the civilizing of those regions implies inevitably the broadening of our national life. With the Missions we would have not only faithful sons of Religion, but also citizen defenders of our Country, untiring workers who in the social realm would take active part in our national enhancement.[112]

Furthermore, Arriaga proposed funding the OPFé from private donations. Neither the financially strapped new republic nor the existing residents of the region would be expected to fund the new missions. He also believed that the missionaries, in most cases, should be Peruvians rather than Spaniards. Given that the Oriente was a vast undeveloped region with ill-defined boundaries, yet regarded by all Peruvian nationalists as a major key to the country's destiny, it is scarcely surprising that the bishop's plan was welcomed, even by the anti-clericals within the government.[113] In the event, the joint initiative was stillborn for lack of funding and personnel. However, its legacy was to define the region's uncivilized indigenes as both the obstacle to economic development and its potential key component as the future work force.

---

110. Ibid., 229.
111. San Román, *Perfiles*, 119.
112. Quoted in Pilar García Jordán, "Las misiones orientales peruanas: instrumento de pacificación, control y tutela indígena (1840–1915)," *Canadian Journal of Latin American and Caribbbean Studies/Revue canadienne des études latino-americaines et caraïbes* 13, no. 25 (1988), 90.
113. There was a backlash in 1849 when some parliamentarians questioned the wisdom of bringing European missionaries (mainly Spanish) into Peru. A crisis was averted with the compromise that all foreign missionaries would be required to swear an oath of loyalty to Peruvian law and respect for the country's civil and ecclesiastical authorities. García Jordán, "El oriente," 231.

At the same time, there were initiatives to foment immigration of foreign agriculturalists (as another way of colonizing the vast tropical forest). In 1832, the Department of Amazonas decreed that immigrants would be accorded title of up to forty *fanegas* (about sixty acres) of land, varying according to their capacity to cultivate it. They were also to be given 30 pesos on arrival. In 1852, 50,000 pesos were appropriated for this purpose—half to go to Europeans and the other half to immigrants from other parts of the world. In 1853, under (Basque-surnamed) President José Rufino Echenique (1851–1855), the Council of State approved 100,000 pesos to incentivize immigration in the Loreto. The settlers were also to receive free river transport to their holding and tools and seed to develop it. There was a plan to recruit 13,000 German immigrants for the Oriente. In 1872, a Sociedad de Inmigración Europea (European Immigration Society) was founded. In 1893, new legislation authorized payment of the passage of immigrants from their place of origin, tools and seeds, two hectares of free land and perpetual title to any additional land purchased from the state for 10 soles per hectare and freedom from taxation for five years.[114] Generally speaking, the initiatives failed to meet expectations, but by century's end there were small French, German, Italian, and Chinese populations in the Oriente.[115]

Peruvians in other areas were encouraged to resettle in the Oriente. In 1845, there was legislation to guarantee them title to all of the land that they were cultivating, as well as freedom from taxation for twenty years.[116] Of course, all of this colonization of the Oriente by immigrants and Peruvians transpired on "vacant land," which is to say Indian territory.

In 1857, the new Provincia Litoral de Loreto (encompassing our study area) was created with its capital in Moyobamba, converted to full departmental status in 1866.[117]

In the 1870s, there appeared a number of articles advocating cooperation between the Church and state in civilizing the Indians, including one by a missionary priest, Francisco Sagols, published, in 1875, in the *Boletín de la Sociedad Geográfica de Lima*. He argued that the establishment of missions would incorporate both the Indians and the resources of their immense territories into the nation, thereby making Peru the world's leading republic. Hyperbole aside, Sagols also asked, "isn't it an affront, a contradiction, to see still [in Peru] in the nineteenth century so many wandering, ferocious, can-

---

114. San Román, *Perfiles*, 123–26.
115. García Jordán, "El oriente," 233.
116. San Román, *Perfiles*, 128.
117. Ibid., 116–17.

nibalistic bestial people?"[118] As the future historian of the Franciscan missions, (Basque-surnamed) Bernardino Izaguirre, would say, it was necessary "to go in pursuit of the wandering Indian, with the only object to convert him into a civilized man, a Christian, and also a Peruvian."[119]

This renewed activity in the Amazon coincided nicely with the Lima government's interest, after the loss of a part of southern Peru to Chile in the disastrous War of the Pacific, to develop (not to mention define) its tropical territory. The initiative, as earlier, was framed in the rhetoric of outright "colonization." By reducing Indians into settlements, and thereby converting them from nomadic forest dwellers into Christian agriculturalists and useful citizens, Peruvian missionaries (working in consort with the national government) could create a whole new Amazonian economy and society. At the same time, they would be flying the Peruvian flag—thereby asserting sovereignty over contested territory.[120] By the 1890s, the missionaries were viewed as critical collaborators in the creation of the "New Peru," a role that persisted formally until the the end of Leguía's second term in 1930.[121]

However, the plan faced formidable obstacles. During the 1870s, there was a serious internal dispute among the country's Franciscans. The head of their missions accused those in charge of the Franciscan College in Lima of refusing to produce missionaries.[122] Given the recent liberal reforms, the Peruvian Church had lost some of its resources, and the Peruvian government possessed little wealth at this time.[123] So, it was far from clear who would (or could) fund the new missionary initiative. Nor was there any lack of doubters. In 1887, influential writer and academic, Carlos Lissón, argued that development of the Loreto depended entirely upon establishing an infrastructure there that would attract intending family-farm agriculturalists. Settling them along the navigable rivers was the best prospect for the region. He had no use for either prominent capitalists or missionaries:

---

118. Quoted in García Jordán, *Cruz y arado*, 150.
119. Bernardino Izaguirre, *Historia de las misiones frnaciscanas y narración de los progresos de la geografía en el Oriente del Perú, 1619–1921* (Lima: Taller Tipografía de la Penitenciaría, 1922–1929), vol. 12, 191. Izaguirre, born in the Bizkaian town of Mendata on June 28, 1870, was the prior of the Franciscan *Descalzos* in Lima for many years. He compiled a massive fourteen-volume history of the order's Amazonian missions from the sixteenth century to his day. Ibid., vol. 1, 7–8.
120. García Jordán, *Cruz y arado*, 163; García Jordán, "El oriente," 224–25.
121. Ibid. We shall consider how, in some regards, the collaboration continues to the present-day, despite Peru becoming a "secular" nation.
122. García Jordán, *Cruz y arado*, 151.
123. Ibid., 155, 161.

Nothing of entrepreneurs, nor of the ancient missions' system. In its time it produced very venerable priests, martyrs and wise men that spilled their blood and served the sciences well with their grand discoveries; but their hardships and efforts have been sterile. . . . The evangelical word finds no echo in the savages. They only respect force; for that reason we wish there be the soldier by the worker's side; because the rifle and the plow together are the symbol of peace and of labor.[124]

These sentiments were echoed in 1890 by (Basque-surnamed) Manuel Patiño Zamudio, who declared in a lecture to Lima's Geographical Society that "civilization does not spread in the jungles simply through the words of the missionaries, because the savage Indians are, in general, untamable, traitorous, and rebellious regarding civilizing influence."[125]

Lissón was skeptical that Peru could attract much international immigration.[126] It was simply too distant and isolated from Europe, and was devoid of intrinsically attractive expanses for intending agriculturalists, such as the vast plains of Venezuela and the pampas. He believed that the settlers in the Loreto would have to come primarily from other parts of the Peruvian nation.[127] Nevertheless, by this time, within the rhetoric of an evolving Peruvian immigration policy, the preferred newcomer was the Anglo-Saxon—prized for his reputed seriousness and work ethic.[128] Oth-

---

124. Carlos Lissón, *Breves apuntes sobre la sociología del Perú en 1886* (Lima: Imp. y Librerías de Benito Gil, 1887), 61 [*Breves apuntes*].
125. Quoted in García Jordán, *Cruz y arado*, 201.
126. Nearly half a century earlier, or in 1849, President Ramón Castilla had promoted and passed an immigration law designed to attract foreigners to the Oriente. Among other incentives, the newcomers were to be exempted from Peruvian taxes for ten years. Alberto Chirif, "Ocupación territorial de la Amazonia y marginación de la población nativa," *América Indígena*, 35, no. 2 (1975), 271.
127. Lissón, *Breves apuntes*, 88–96. Two decades later, Loreto prefect Francisco Alayza y Paz Soldán would dissent in a report to Lima. He urged that true agriculturalists be recruited from Spain and Italy with incentives. As for Peruvians,

To send people from the Coast for this purpose does not produce good results, in addition to the scant existing population in said zone, only the most pernicious elements and those inept for agricultural work would come, those who would be very quick to return to Iquitos without providing it. The dwellers of the Sierra are equally unsuitable given that they suffer greatly from the tropical diseases and climate. In should be kept in mind that it is indispensable to bring *true farmers* and not city dwellers and persons habituated to industries other than agriculture, energy or moral fibre. (Alayza Paz y Soldán, *Mi país*, 161)

128. García Jordán, *Cruz y arado*, 182. In 1888, H. Guillaume, Consul-General for Peru in Southampton, published a book designed to attract British immigrants for Peru. H. Guillaume, *The Amazon Provinces of Peru as a Field for European Emigration* (London: Wyman and Sons, 1888).

er Europeans, and particularly Spaniards (easily assimilated given their knowledge of the national language), were felt to be superior to Peruvian nationals; most of the missionaries in the religious orders were Spanish nationals. Yet, Peru and Spain had been recent battlefield adversaries in the Chincha Islands War (1864–1866).

As for the British, they had financed the Chileans in the War of the Pacific. Nevertheless, they had extensive investments in Peru and held much of its war debt. This was used periodically to extract economic concessions. In 1891, Peru granted one and a half million hectares of its tropical montane to the British Peruvian Corporation. It was supposed to develop infrastructure and attract European immigrants and Peruvian settlers to a vast agricultural development. The project was a complete failure and became the object of considerable scandal.[129]

And, of course, whether based upon civilizing Indians, attracting immigrants,[130] or both, such colonizing plans had their natural opponents. Rubber collecting in Peru was expanding exponentially at this time.[131] That industry was entirely dependent upon indentured workers—Indian and non-Indian alike. In the case of the former, the exploitation in many areas (not just the Putumayo) crossed over the line into outright slavery—reinforced and enforced by extreme coercion (including beatings, torture, and

---

129. Ibid. Lissón regarded Great Britain to be Europe's supreme power in 1886. So, despite some of its pushy dealings with Peru, it had to be respected. For him, Germany and France were superior European partners (the French had financed much of Peru's railroad infrastructure). As for the United States, he noted "There is no American brotherhood." The North Americans were interested solely in securing profitable investments within the Peruvian economy that they would exploit from afar as absentee landlords (Lissón, *Breves apuntes*, 79–80).

130. Indeed, all of the nations of Amazonia sought to attract European immigrants to the vast wilderness. In 1906 and 1907, British explorer Percy Harrison Fawcett was commissioned to map the disputed border between Peru and Bolivia in the Acre region. His diaries are replete with encounters with Austrian, English, Swiss, French, and German immigrants along his way. See Lt. Col. P. H. Fawcett, *Exploration Fawcett: Arranged from His Manuscripts, Letters, Log-books, and Records by Brian Fawcett* (London; Hutchinson & Co., 1953), 38, 49, 50, 55, 56, 57, 66, 84. He reported visiting the headquarters in La Esperanza of the major Bolivian rubber barons, the Suarez Brothers:
> Here we found several British mechanics in the service of the firm, well paid to look after the launches. The clerks, Germans to a man, were openly hostile to them. (Ibid., 91)

The explorer was aware of the budding Putumayo scandal and reported similar atrocities against the indigenes in Bolivia. He recounted *correrías* there to capture Indians for sale as slaves and likened the *enganche* system to near slavery as well. (Ibid., 55–58).

131. Arana entered the Putumayo in 1886, or the same year described by Lissón in his sociological analysis of the country. In his chapter on developing the Loreto, he fails to even mention rubber as one of its tropical resources! (Lissón, *Breves apuntes*, 60–64). In short, Peru was a late arrival to the Amazon's rubber boom.

murder). The rubber interests needed to control dispersed Indians who could then collect rubber throughout the forest, not concentrate them into mission settlements where they would while away their time attending school and church services.

There was also ambivalence among the Peruvian clergy regarding the missionary friars. Peru's own clergymen had long ignored the opportunity (or responsibility) to proselytize in the forests. In some cases, the secular clergymen were aligned closely with local business and political leaders. In short, viewed from the perspective of Iquitos, the whole religious reform was the imposition from the outside (distant Lima and Rome) upon local life in the Loreto.

By the last decade of the nineteenth century, as the plans for colonizing the Peruvian Amazon foundered, a debate broke out between those who believed that simply establishing a white presence in traditional Indian territory counted as "civilizing," an interpretation that obviously favored the *caucheros* and their supporters in places like the Loreto, and those who wanted to transform the region through agricultural settlements inhabited by converted Indians, migrant Peruvians, and European immigrants.[132] In 1896, the latter were successful in reviving the OPFé with the full support of the country's bishops. The private initiative was headed by the daughter of Peru's liberal president, Pierola (he was named its honorary leader), and received a small subsidy from the Peruvian Congress.

In 1900, the bulletin of the OPFé stated:

> It is high time that the government should make its moderating action felt on the shores of the Amazon, Marañon, and other rivers. There are traders without conscience who carry cruelty to the point of carrying off wives and children from the homes of the unhappy savages. They enslave them after taking advantage of their unpaid services and corrupting their morals. One can understand that such procedure discredits the civilized life and the religion which nominally these soulless beings profess, thus driving the natives from the centers of population, and exposing honorable explorers, the missionaries, and even the authorities themselves, to the vengeance of the Indians.[133]

Meanwhile, Pope Leo XIII manifested strong interest in his church's missionary activities worldwide. In an 1892 encyclical commemorating the accomplishments of Christopher Columbus, he underscored the

---

132. Ibid., 185.
133. Quoted in Ibid., 27.

Church's continuing obligation to proselytize in the Americas until every soul right down to Patagonia had been converted to Christianity. Then, in 1894, he sent a letter to his Peruvian bishops ordering them to evangelize their jungles.[134] The situation in Peru was indeed dire. The Franciscans were charged with the forest missions, and, at this time, there were only a total of eight in the country's Amazon region, all far from the Putumayo. Five friars and four lay brothers ministered to a mission population of slightly under eight thousand persons,[135] in a world in which the indigenes numbered in the hundreds of thousands.

It was now that the Vatican, through its proselytizing arm of the Sacred Congregation for the Propagation of the Faith, stepped forward with its own plan for sponsoring renewed missionary activity in the Peruvian Amazon. It proposed dividing the vast area into three Apostolic Prefectures. In 1897, the Peruvian Congress passed the necessary enabling legislation, and, in 1898, President Piérola signed it into law. So, on February 5, 1900, the Vatican created the new pre fectures, including that of the Apostolic Prefecture of San León de Amazonas—which included much of the Putumayo region. However, Along the Putumayo River it was impossible to establish any mission owing to the abuses of the rubber traders against the Indians whom they mistreated and murdered for frivolous reasons, seizing their women and children.[136]

The new prefecture overlapped the existing Diocese of Chachapoyas, to whose bishop the secular clergy of Iquitos reported. Its key secular priest, Father Correa, was complacent and complicit with the rubber barons—turning a blind eye to the plight of the Amazonian Indians. In 1901, The Sacred Congregation for the Propagation of the Faith assigned the Augustinian order to Iquitos to head up the prefecture, while serving as missionaries throughout most of the Loreto. The Augustinians (along with the Dominicans) possessed resources and excess personnel at this time, having been forced to abandon the Philippines after those islands were lost to the United States in the recent Spanish American War (1898). In December 1900, four Augustinian friars and one lay brother (all Spaniards) arrived in Iquitos to take over the new Apostolic Prefecture of San León de Amazonas, headed by Father Paolino Díaz.

Almost immediately, Father Díaz was complaining about his treatment by the local authorities. Rather than cooperate with him, they put obstacles in the way of the Augustinians at every turn. He did manage to

---

134. Lissón, *Breves apuntes*, 161.
135. García Jordán, *Cruz y arado*, 201–203.
136. Turvasi, *Giovanni Genocchi*, 73.

establish a new mission at the rubber station Nazareth on the Marañon River, and assigned Father Bernardo Calle to it. Then, in 1902, the prefect of the Loreto asked the Augustinians to establish a mission on the upper Napo River, an area where Peru and Ecuador were contesting sovereignty (as a way of demonstrating the Church's recognition of Peru's territorial claim). Yet, at the same time, Portillo was dissuading Díaz from extending his purview to the Putumayo, because the area was contested by Peru and Colombia. In late November of 1903, the Peruvian government transported a large contingent of troops to the Igaraparaná River, and Father Díaz was very desirous of going to the Putumayo himself; he referred to it as "the empire of Satan."[137]

Father Díaz was devastated by the desolate state of the former missions along the Amazon and Marañón; most abandoned and with their Indians now "dispersed among the rivers of Brazil or Bolivia, and . . . reduced to the most degrading and frightful slavery." It had been "all destroyed in a short space of time by the greed of soulless men calling themselves 'the standard-bearers of civilization and progress.'"[138] Then, on June 2, 1904, driven to desperation by their white overlords, the Indians attacked Nazareth. Father Calle and a secular priest, Muñoz, were among those killed.[139] The following year, Díaz reported Indians fleeing from rubber stations into the forest, and then, in May of 1907, he traveled to Rome to report on conditions in his jurisdiction. His conclusion was that it was "Impossible to civilize the natives as long as the rubber traders continued their slave raids and other outrages which the pen hesitated to describe."[140]

While the last thing that the nervous *caucheros* wanted were critical friars competing for the attention of the indigenes and disposed to come

---

137. Ibid., 30. Shades of "The Devil's Paradise."
138. Ibid., 30–31.
139. Ten years later, the Augustinian's apostolic prefect in Iquitos, Fr. Miguel San Román, noted that the Indians and missionaries at Nazareth had developed mutual trust and a close working relationship when the area was flooded with opportunistic rubber gatherers who usurped the territory, reconcentrating in the Indians' "savage hearts their mortal hatred that sooner or later had to explode with all of the terrible ferocity of which a provoked tiger is capable." Indians, armed ironically by the *caucheros*, swept the area, killing every white that they found and burning their property. A band of Indians arrived at the mission and offered to protect it against the marauders. But they were "false," and once inside the precincts fell upon the missionaries. Father Calle was mortally wounded and lay dying when a sympathetic Indian finished him off to end his suffering. Isacio Rodríguez Rodríguez, OSA and Jesús Álvarez Fernández, OSA (eds.), *Monumenta histórico-augustiniana de Iquitos: Volumen Tercero 1910–1915* (Valladolid: Centro de Estudios Teológicos de la Amazonía [CETA], 2001), 494–95.
140. Turvasi, *Giovanni Genocchi*, 31.

out in their defense,[141] the Augustinians initially focused their main efforts upon Iquitos itself. In 1903, they inaugurated the College of Saint Augustine. It would become the city's leading educational institution (and continues to function to the present day). That same year the Peruvian government noted that the office of the National Treasury at Iquitos was to give 25 pounds sterling monthly to the Apostolic Prefect of San León de Amazonas for support of the missions. The following year, a decree signed by (Basque-surnamed) Eguiguren [minister of justice and religion], in Lima, ordered the Iquitos authorities to dedicate a site in the town for construction of a mission headquarters, as well as a monthly stipend of 5 pounds sterling towards the costs of said project.[142]

On April 13, 1907, a Peruvian governmental decree assigned Iquitos to the bishop of Chachapoyas. On August 27 of that year, Augustinian Father Prat filed a report that was most critical of the situation in Aranalandia. There were "massacres and slave raids everywhere." Then, on September 8, 1907, Prat noted:

> An intelligent and zealous and energetic leader, who can be no other than a Peruvian bishop, must reside there [the Loreto]. If he is not a bishop he will not be heard by the government and the people. If he is not Peruvian worse still, although the missionaries under him can be of any nation. . . ."[143]

Meanwhile, Father Díaz was losing hope. In September of 1908, he reported that he had ceased to visit the Putumayo because of the outright dangers to missionaries there posed by the resistance of the Casa Arana to any outside interference in its affairs.[144] Then, in November 1908, Díaz expressed his desire to access the Putumayo from his mission on the Napo River, noting that he was encountering resistance (surely from the Casa Arana). On October 4, 1909, Father Alemany, the apostolic prefect of San Francisco de Ucayali, sent his annual report to the Congregation of the Propagation of the Faith in which he denounced

> . . . the horrible practice of going out in war parties to hunt down unarmed native forest dwellers. The inevitable consequences of this practice are: the death of the native heathens, captivity for the

---

141. Lagos, *Arana*, 112.
142. Turvasi, *Giovanni Genocchi*, 26–27.
143. Ibid., 86. It should be noted that Prat's comments coincided with the appearance of Saldaña Rocca's newspapers. While he did not mention them, it is possible, even probable, that they influenced his reporting. One or both of these reports were likely the reference made during Arana's testimony before the Select Committee of which he claimed ignorance.
144. García Jordán, *Cruz y arado*, 234.

women and children, their exile forever along with unspeakable suffering; in a word, the bloody business that should be wiped out for many reasons. Some businessmen along these rivers get control of tracts in the middle of nowhere and in a few years come out rich, some even very wealthy. They are able to do this through the work of thirty, forty or more families of native heathens whom they have reduced to a state of degraded slavery. They also buy them and then sell them off separately with the painful and pitiable separations and departures this entails. Only a few days ago I witnessed how the buying and selling of two native heathen girls was carried out for the ridiculous price of forty shekels. We witnessed the weeping and lamentation of the girls as they were violently separated from their friends. A Jewish man bought them. Where did he take them? As can be seen by anyone, this moral perversion, along with the criminal commerce, places serious obstacles in the way of the missionaries' zeal and the conversion of the heathens.[145]

Furthermore, the limited funding for his missions from the Peruvian government had been suppressed (likely through Arana's influence or that of his minions). This, in turn, prompted a petition from the OPFé, on October 11, 1909, to the Peruvian Congress for restoration of the funding. By then, the first Hardenburg article had been published in London. The petition stated:

> It is not a question of distant colonies, neither of a simple sentiment of human dignity. It is on national soil that these offenses are committed (how often have the Apostolic Prefects asked for alms to ransom our Indians, and to prevent their being sold as slaves in Brazil?), and these unhappy Indians have no protection other than forty-one missionaries [in all Peru] assisted and defended by a Society of Ladies.[146]

This initiative was unsuccessful.

In 1903, Pope Pius X had succeeded Leo XIII and inherited the former's mantle regarding the Catholic Church's missionary initiatives. For the first years of his papacy, Pius X was fixated upon his antimodernism campaign within the ranks of the clergy, but then the Putumayo scandal erupted in late 1908 and throughout 1909. So it was, on June 6, 1910, that the pope appointed the able Giovanni Genocchi as his roving investigator of the conditions of the Indians in South

---

145. Turvasi, *Giovanni Genocchi*, 1–2.
146. Ibid., 32.

America, and particularly the status of the Catholic missions established there to minister to them. Genocchi was to recommend possible reforms. While this was clearly a response to the contemporary scandal, it reflected Vatican concerns over the past and future ones as well. Where was the Church when the earlier atrocities were being committed? Was it culpable—if only by virtue of its silence and indifference? And now, there was the danger that Protestant missionaries might be the ones to spearhead any religious reforming initiative.[147]

Then, in October 1910, Father Alemany reiterated his accusations of abuses of the Indians by the rubber traders—noting that it was customary to buy and sell indigenous latex collectors openly, there were regular raids into Indian territory to burn down houses and then

> seize the defenseless women and children, kill the old, and fight the men who struggle to defend themselves, their families and the land they don't want to leave. But the men are either slaughtered or dragged off to the place where they will work at extracting rubber. They will do this for many years under the most severe vigilance and rigorous discipline. Often the whole course of their natural life will be spent collecting, distributing, carrying and transporting the aforementioned commodity from place to place.[148]

Meanwhile, immediately after his return to England from his first visit to the Putumayo, Casement began advocating creation of a Christian mission in the Putumayo. As early as March 2, 1911, he mentioned in a letter to Louis Mallet at the Foreign Office that, "it would have to be a Catholic mission for the Lima Govt. recognizes only the one church."[149] Sir Roger believed that his native Ireland would be supportive of the idea, and proposed that three missionaries be sent out from there.[150] Then, on March 17 and March 21, Casement forwarded his two reports to Sir Edward Grey confirming the accuracy of Hardenburg's allegations, adding the results of his own interviews and research, while appending the recommendation that the British Foreign Office officially advocate creation of a Catholic mission in the Putumayo.

On June 15, 1911, Monsignor Manuel Bidwell, chancellor of the Catholic Archdiocese of Westminster, at the bequest of the British For-

---

147. García Jordán, "La misión," 259–60. Scapardini, informed Cardinal Merry del Val at the Vatican Secretariat of State that the Peruvian government's primary reason for constituting the Paredes investigative commission was to mollify the Vatican (Ibid., 260).
148. Turvasi, *Giovanni Genocchi*, 2–3.
149. Mitchell, *Sir Roger*, 114.
150. Turvasi, *Giovanni Genocchi*, 11–12.

eign Office informed the Holy See about the Casement report. Bidwell commented that

> The consular officer sent out to the Putumayo expressed it as his personal opinion that much good would result from the establishment of a Roman Catholic Mission, with headquarters in Iquitos or some other convenient centre, which could from time to time send representatives to inquire into the welfare of the natives in the outlying districts and generally watch over their interest, and from inquiries which he made of the Peruvian authorities in Loreto, he understood them to look with favour on such an idea.[151]

In late May of 1911, the Vatican's secretary of state, Cardinal Merry del Val, sent a letter to the apostolic delegates of Peru, Bolivia,[152] Colombia, Costa Rica, and Mexico, as well as the papal internuncios of Chile and Argentina, informing them of the pope's concern over the circumstances of indigenous peoples throughout the Americas. They were to investigate and report on the numbers and activities of the Catholic missionaries in each of the countries.[153] A few weeks later, the Holy See received the contents of Casement's report, and thereby learned that the situation in the Putumayo was worse than that reported by Father Alemany for the Ucayali. Then, on June 15, Monsignor Bidwell informed Rome that Casement was recommending establishment of a Roman Catholic mission, headquartered in Iquitos or some other convenient center, to look out for the welfare of the Indians. It was Bidwell's understanding that the civil authorities of the Loreto favored the plan.[154] On July 19, Casement wrote to Gerald Spicer:

> As to a mission I saw Monsignor Bidwell yesterday & had a long talk with him. He says Cardinal Mercy del Valle [*sic*] is very much in earnest indeed—and that the Vatican will do all it can in the matter. He agrees with me that it is much better to try & get a mission directed from home instead of one solely dependent upon Lima or the local ecclesiastical authority of Peru. The latter will be brought into line he asserts definitely—that opposition

---

151. Ibid., 5.
152. Genocchi noted that he heard a fightful Bolivian anecdote. There were frequent military efforts to interdict Indian attacks upon travelers to Santa Cruz, but they failed. It was then that the city experienced an outbreak of smallpox. For sanitary purposes, the garments of the victims were burned. But it was then suggested that their blankets be left in the forest where they would surely be confiscated by the Indians. This was done and it seems that four or five tribes were exterminated in this fashion. Ibid., 90.
153. Ibid., 3.
154. Ibid., 5.

on their part to a mission having headquarters elsewhere than in Peru would be met and overcome by the Vatican—that this can be counted on. The chief difficulty is one of funds.[155]

That same day he wrote to his supporter, William Cadbury, that for the Putumayo mission to happen, the funds had to be raised in England. Peru would promise support if asked, but was impecunious and could therefore not be relied upon over the long run:

> The Archbishop's man I saw yesterday (Monsignor Bidwell) is very keen. He thinks if an endowment of £10,000 can be raised at home the Vatican will move firmly and energetically in the matter & will secure us against any ecclesiastical opposition in Peru. The FO will secure us the open support and official approval of the Lima govt. who may give a subsidy of possibly £500 a year. With such an income assured it, a Catholic Mission could be sent to the Putumayo composed of high-souled, devoted men and we should have really done something of the highest value for those unhappy people—and possibly, too, have broken the back of the evil over a wider area than the Putumayo forest alone. . . . the Duchess [of Hamilton] is writing to the Duke of Norfolk about the Mission, but I learn privately that English Catholics are "broke" and won't subscribe anything—I have more hope of Irish Catholics & hope of Englishmen of no particular church—or rather of all churches but one faith.[156]

The duchess of Hamilton then wrote Casement, "My husband wants me to say he will gladly contribute £100 towards the funds you are trying to raise, but indeed would quadruple the amount if instead of peace you were getting up a punitive expedition to extirpate those vipers off the face of the earth."[157]

In October 1911, Merry del Val ordered the Vatican's chargé d'affairs in Lima, Monsignor Quattrocchi, to inform the Peruvian foreign minister, Porras, of the, "deplorable facts, invoking provisions to suppress abuses no less contrary to religion than deplorable to dignity and the prestige of Peru."[158] Porras welcomed the Vatican's concern, but noted that the problem was in large measure due to the indeterminate status of a territory claimed by Peru, Colombia, and Brazil. He thereby fired the opening shot in what would be several attempts by his government to gain Vati-

---

155. Mitchell, *Sir Roger*, 495.
156. Ibid., 497–98.
157. Ibid., 507.
158. Turvasi, *Giovanni Genocchi*, 3.

can recognition of Peruvian jurisdiction in the Putumayo. Quattrocchi was already aware of the contents of the Paredes report (submitted to the Lima government but a few weeks earlier) and informed Merry del Val that it had identified three hundred persons who should be arrested and tried for their crimes, yet not even ten had been detained as yet. He added that the three hundred were but a small fraction of those that, "continue their trade in human flesh, undisturbed and triumphant." He added:

> In a festival given by the members of an English rubber company, one of the events on the program consisted of target shooting with fifty natives as the targets. These natives, running up the hill, were shot to death by the revelers, who were drunk with whisky and champagne. As a matter of fact, I could cite horrors of this kind, or worse, by the thousands.[159]

If, in March 1907, Father Correa and his fellow secular priests in Iquitos were still obedient to the prefecture, by the time that papal envoy Genocchi had reached Iquitos, in late 1911, they refused to recognize the prefecture's authority: "Indeed, as Peruvians they boast of the government's support, and seek to keep alive the population's hatred of the foreign intruders, the poor Augustinian missionaries, who are all Spaniards."[160]

Shortly thereafter, American Consul Fuller informed Washington:

> As to the proposed establishment of missions in the district [the Putumayo], the company representative and Señor Arana state that they would not mind missions of Peruvians, but they are noncommittal as to what their attitude would be toward missions of foreign nationality. It is easier to understand this when one bears in mind the fact that the authorities of the State church of Iquitos do not favor the establishment of missions in the Putumayo, fearing that they will not be allowed a free hand. In other words, the company is willing, if they must have missionaries, to have those whom they can keep under their thumb.[161]

---

159. Ibid., 4.
160. Ibid., 83. For many dozens of letters and reports authored by Augustianians in Iquitos documenting their travails in the city during the first decade of the twentieth century, see Isacio Rodríguez Rodríguez, OSA and Jesús Álvarez Fernández, OSA (eds.), *Monumenta historico-augustianian de Iquitos. Volumen Segundo 1903–1909* (Valladolid: Centro de Estudios Teológicos de la Amazonía [CETA], 2001).
161. Consul Fuller to the Secretary of State, Iquitos, October 28, 1912. United States Department of State, *Slavery*, 62.

While Genocchi's purview ultimately embraced several South American countries, the Putumayo scandal was clearly on the front burner. The British Foreign Office advised its Lima representative, Lucien Jerome, to render all possible assistance to Genocchi as he passed through the city before going to the Putumayo, underscoring that it was this papal emissary's prime assignment to determine the feasibility of establishing a Catholic mission there.[162] According to Turvasi, "Jerome, who was a fervent Catholic, had put so much energy into anti-slavery propaganda that he made himself rather troublesome to his government."[163] During Genocchi's stay in Lima (from September 12 until October 10), he worked closely with Jerome analyzing the current status and needs of the Peruvian missions, and particularly those of the Loreto. They were all woefully understaffed and underfunded. This close working relationship between the British Legation and the Apostolic Delegation would prove both durable and influential in the formulation of the future mission in the Putumayo.[164] On October 6, 1911, the apostolic delegate sent his first report to Merry del Val in which,

> Genocchi related to the Holy See his observations concerning the state of the missions, calling special attention to certain deficiencies. In the Urubamba the Dominicans seemed to work well, although some said that they accommodated themselves somewhat to the abuses of the rubber traders. However, the Apostolic Prefecture had few means because they were totally separated from their religious Province in Peru. If the Prefecture were united in some way with the wealthy Province, it could expand and prosper. In the Ucayali, the *Descalzos*, exclusively Spanish and very revered in Peru, were sacrificing themselves. The recent suppression of their mission schools by order of the Superior General had greatly affected and disheartened these zealous Franciscans. They might take courage again if some appropriate satisfaction were obtained from Rome. To this end, Genocchi considered writing about the *Descalzos* to the Congregation of the Propagation of the Faith and to the Congregation of Religion because they were doing a great deal of good among the abandoned populations of the mountain area and among the Indians. The Jesuits of Lima were willing to accept missions in those territories, but they first wanted to be recognized by the government as were all other religious congregations of Peru. This problem did

---

162. Turvasi, *Giovanni Genocchi*, 12, 35.
163. Ibid., 35.
164. Ibid., 53–56.

not seem insurmountable to Genocchi. The Redemptorists were doing immense good by traveling around the country preaching missions. However, Genocchi pointed out: "Many isolated pastors are taken to drunkenness and lust, acting as implacable collectors of matrimonial and baptismal fees, and making themselves hateful to those half-savage populations."[165]

Before departing for the Loreto, Genocchi informed Jerome that the pope himself was keenly interested, "in taking some practical steps soon to give help to those poor Indians of the Putumayo above all, and afterwards to those of other provinces."[166]

It was now that Genocchi departed Lima for Panama, Belem, and the eventual trip up the Amazon to Iquitos. He obviously thought that his mail would be monitored by whomever, since he left behind in Lima a code with the apostolic delegate whereby he could communicate the details of both his reception by the Augustinians and the condition of his personal health. During his weeks of research in the Peruvian capital, he had already formed strong opinions regarding the situation. While en route to Iquitos, he informed the Vatican:

> Yesterday I departed for Panama, where I shall arrive on the sixteenth. I shall cross the Isthmus by train and after eight days will enter the Atlantic to go to Brazil and more precisely to Belém do Pará, near the mouth of the Amazon River. There I shall embark again and proceed up the river for more than 2,000 miles, in order to reach Iquitos and beyond, where the rubber traders have committed, and to this day continue to commit, the most incredible cruelty, including murdering the Indians by the thousands, with the result that some tribes have been exterminated. It is there that the Church must accomplish its work of setting up good mission centers which will be a check on the inhuman rubber traders as well as the health of the surviving Indians.[167]

Genocchi also expressed his doubts regarding the Peruvian government:

> I have spoken with the [Peruvian] Minister of Foreign Affairs, and he gave me letters of introduction. I trust more in the intro-

---

165. Ibid., 59. This portrait of the complexity and the semi-autonomy of each of the religious orders vis-à-vis one another (and even the Vatican), not to mention the many issues surrounding the country's secular priests and their bishops, makes any reference to the "Catholic position" in Peru simplistic.
166. Ibid., 57.
167. Ibid., 64.

duction of the English government, which concerns itself with my mission as if it were its own and sends instructions to its consuls and representatives to support me.[168]

The Vatican certainly harbored less concern regarding Colombia, and Genocchi did not visit that country on his fact-finding tour. He obviously could have done so easily as part of this itinerary from Callao to Iquitos. In Colombia, there was already a vibrant Capuchin[169] missionary presence in the Upper Putumayo, based in the Sibundoy Valley and dating from 1893, when the bishop of Pasto[170] first invited that Catalan religious order into his jurisdiction and, by 1896, two friars and a lay brother had established a residence in Mocoa in the Alto Putumayo, followed, in 1899, by another in Sibundoy. By 1900, there were eight Capuchins evangelizing the region.[171] They found a large Sibundoy Indian settlement in place. The Indians were conversant with Catholicism from their contacts with missionaries in earlier centuries and understood some Spanish. There were white former quinine and rubber gatherers who had settled in the valley and coopted land for livestock raising, as well as newly arrived poor whites that had been welcomed by the Sibundoy as refugees after being displaced by a recent earthquake (1899).

---

168. Ibid.
169. The Capuchins were founded in 1525 by Mateo de Bascio as an offshoot of the Franciscans. They were cloistered and devoted to caring for the infirm. By the early seventeenth century, they were establishing overseas missions, including in Nueva Granada (Colombia). Expelled from Colombia in the mid-nineteenth century, along with other missionary orders, in 1873, the Capuchins expelled from Guatemala and El Salvador by liberal governments were invited to come to Ecuador (they were expelled once again from there in 1896). In any event, the new Concordat of 1887 between Colombia and the Holy See permitted Capuchins to return (Bonilla, *Servants*, 64; Kuan Bahamón, "'La Misión," 39–45). The Concordat established Catholicism as Colombia's official religion, while providing the Church with subsidies. On December 16, 1890, Congress authorized, "the government, in accord with the ecclesiastical authority, to proceed to reduce to civilized life the savage tribes that inhabit Colombian territory bathed by the Putumayo, Caquetá and Amazon rivers and their tributaries." (Ibid., 75)
170. This future Catholic saint Ezequiél Moreno y Díaz (native of La Rioja, Spain) became an Augustinian Recollect after attending seminary in Monteagudo, Navarra. He served for many years as a missionary in the Philippines before taking charge of the facility at Monteagudo. In 1893, he assumed the bishopric of Pasto. He would share the Capuchins animus against liberals, having been formed in an ultramontane and Carlist environment (Ibid., 58–59). He became an active opponent of the Colombian Liberal Party and urged conservatives and all Catholics "to defend their religion with Remingtons and machetes," proferring absolution should they have to do so. (See https://www.lablaa.org/blaavirtual/revistas/credencial/octubre1993/octubre2.htm.)
171. Ibid., 45.

The Capuchins launched several expeditions to the Middle Putumayo, including the future Aranalandia. However, none of them resulted in permanent residences (missions).[172] In point of fact, for the first three decades, the Capuchins restricted their activity to the Upper Putumayo, launched from Sibundoy.

The Capuchins in the Sibundoy initially found themselves in the midst of a bitter struggle. The whites had been usurping Indian property and/or renting it at ridiculous rates and introducing livestock where there was no fencing to keep it from destroying Indian crops. The whites were totally imbued with the longstanding colonialist mentality that viewed indigenes as inferior and their territory up for grabs. For their part, many of the Indians had fled the settlement, but were fighting back. There was considerable episodic violence between the adversaries. The friars initially sided with the Indians, causing the white settlers to complain to the bishopric in Pasto.

It was then that the Capuchins tried to resolve the tension by convincing the Indians to set aside land for an exclusively white settlement, which they did in 1902. Nevertheless, the whites resisted being displaced from existing holdings and the struggle continued. So, both the Sibundoy Indians and whites pleaded their case in Pasto with civilian and religious authorities, before the former prevailed. However, their main village in the Sibundoy Valley now contained white missionaries and there was an all-white settlement up the road. The friars founded a school, but the Indians refused to send their children to it.[173]

In 1904, the Vatican declared the region to be the Caquetá Apostolic Prefecture, and, in 1905, named the Catalonian Friar Fidel de Monclar, head of the Capuchin mission in Sibundoy, as its first apostolic prefect. He was an extraordinarily determined and intelligent man who appointed the equally strong-willed Fray Estanislao De las Corts as his assistant. According to Bonilla, "The character of the two crusaders was matched by their brusque military manner of speech."[174] Friar Monclar felt it to be his primary charge to civilize (i.e., Christianize) the Indians of the Putumayo/Caquetá, a mission that was coterminous in his mind with economic development of the region (in white terms, of course). Faced with the challenge of defining as either savage or civilized the indigenous inhabitants of the Sibundoy, the seat of his mission, Monclar stated:

---

172. Kuan Bahamón, "La Misión," 47–48.
173. Ibid., 66–73.
174. Ibid., 74.

I do not believe that it would occur to anyone to count the Indians of the Caquetá and Putumayo among the civilized, given that even though the Mission has accomplished a lot in this regard there remains much to be done: a savage race is not civilized in a few years, it requires various generations to pass before they abandon their repugnant customs and absurd traditions and give up their innate laziness to devote themselves to work and modest industries. Their dress, language, instincts, superstitions, aversion to live in settlements, and a thousand other details convince anyone who visits these places that their Indians are not yet civilized. The same General Don José Diago, Special Commissary of the Putumayo, who, because of his enmity against the Missionaries, has worked to have this Ministry [Interior] issue a dictamen against the Mission, has said various times in my presence the following words, "The Indians nauseate me, I cannot stand to be in their presence, their savagery repels me." Such is the manner of thinking and speaking of the many who are not imbued with the spirit of charity with respect to these poor Indians. They should not cause repugnance in anyone, since they are our brothers, and with patience and time we can bring them up to the level of the Indians who inhabit part of the Provinces of Pasto, Túquerres and Obando. Those Indians indeed can be counted among the civilized, due to the constant work among them of the Church over many generations, and to the constant contact with whites over the course of enough years, they now constitute peoples in which the individuals have forgotten the ridiculous traditions and superstitions of their ancestors. And they practice the Christian religion, with relative purity, speak no other language than Spanish, cultivate the earth with regular meticulousness, interact intimately with whites, conducting business with them and dedicating themselves to modest industries. These Indians are identified with the customs of the civilized and have adopted the majority of their habits. This is so much the case that the Indians of this Mission, including those of Santiago who have communicated with this Ministry, label as whites the Indians, inhabitants of the Department of Nariño, because they wear shoes and dress like civilized persons.[175]

Given the recent Colombian legislation regarding the Indians and missionaries, it was clearly in Monclar's interest to have the Indians of the Sibundoy classified as "uncivilized." The missionaries would thereby

---

175. From a document in the Colombian National Archives quoted by Dominguez and Gómez, *Nación,* 43.

be in complete control of both religious and civil affairs within their precinct. Those who criticized the friars were denounced as liberals desirous of separating Church from state. According to Fray Luis de Pupiales in 1905, "Liberalism is the father of all sins because this sin is the worst of all errors . . . we say this to the liberals, wolves dressed in sheep's clothing."[176]

Despite the sporadic incursions of Capuchins from Colombia into the Caraparaná River area between 1900 and 1906, or when the area still had a number of Colombian settlers, this missionary activity remained concentrated in the river's upper regions.[177] Most of the Sibundoy Valley was treated by the Capuchins as *baldío*, or unoccupied commons, and, in 1911, new legislation stated:

> . . . the part of of the Sibundoy Valley that is baldía will be distributed as follows: to each of the existing towns of Santiago, San Andrés, Sibundoy and San Francisco and to that of Sucre, 300 hectares; and to the [future] benefit of each of these, 100 hectares; to public instruction of each of them, 100 hectares; to the church of each, 100 hectares; in each of the towns of Santiago, San Andrés and Sibundoy, for demonstration plots operated by the Marist brothers, 50 hectares; in the town of Sibundoy, to support the establishment and to sustain a special college for forming missionaries, 100 hectares; and for the colonists or cultivators, the number of hectares to which they are entitled according to the law.[178]

All of this was carved out of what until then had been tribal territory. Furthermore, other legislation assigned to the Indians only two hectares per household while colonists were allotted four and allowed to "buy" from 50 to 100 hectares. The Indians were prevented from alienating their holding. In short, the indigenes were now marginated to an extreme in their own ancestral homeland.[179]

The missionaries tried to force the Indians into nucleated settlements dominated by a church. This went against the grain of Sibundoy culture, and the Indians resisted.[180] Furthermore, the Capuchins railed and campaigned against the indigenes' casual sexual relations (they were not particularly prone to marry), unwillingness to wear white clothing, and alcoholism. To exacerbate matters, the Capuchins decided to raise livestock themselves on prime lands over which they managed to finagle jurisdic-

---

176. Ibid.
177. Restrepo López, *El Putumayo*, 20.
178. Cited in Kuan Bahamón, "La Misión," 121.
179. Ibid., 121–22.
180. Bonilla, *Servants*, 80–81.

tion by various means. The Indians began to kill mission livestock, which, of course, enraged Monclar and De las Corts. Many Indians began fleeing into the forest to escape missionary control. Given the harsh conditions they encountered, this led to an excess of deaths over births, degenerative illness, famine, and suicide (around four hundred in 1908 alone).[181]

Indian resistance proved short-lived—they had little choice but to knuckle under to the friars' authority. Montclar instituted what Bonilla called "a real theocratric dictatorship."[182] He appointed the police and nominated all candidates among the indigenes for their own council. He imposed curfews, had to approve any indigene meetings and instituted a system whereby neighbors reported to him on one another's morality. He instituted a system of fines, beatings and incarceration for delicts. The indigenes were also required to provide the missions with uncompensated corvee labor. Monclar also had the power to redistribute land within the valley irrespective of indigenous arrangements and traditions.[183]

During the Bardados to Belem leg of his journey to Iquitos, Genocchi happened to share his vessel with Peruvian soldiers from the Loreto who were returning home after receiving medical treatment for illnesses contracted in their skirmishes against "Bolivians in the Coquetá" [surely it was Colombians in the Caquetá]. From them, Genocchi learned firsthand that

> The exploitation of the Indians, reduced to slavery, is horribly barbaric, as we know, but it makes little impression on either the Peruvians or their neighboring peoples. Coercion and the whip are necessary with the Indians, else nothing can be obtained from them: this is the principle. To buy them, sell them or catch them as you can, is regarded as no greater fault than the smuggling of goods in Europe.[184]

On November 13, 1911, Monsignor Manuel Bidwell forwarded Genocchi's views to the British Foreign Office:

> The information I have just received from Cardinal Merry del Val with reference to the proposed Putumayo Mission is as follows:
>
> (1) If it is thought best that that missionaries should be English or Irish, this will not give rise to difficulties with the local Ecclesiastical authorities. You will remember that Sir Roger Casement suggested that Irish missionaries should be sent out, as he thought it would be easier to raise funds in Ireland if this could be done.

---

181. Ibid., 82–83.
182. Ibid., 86.
183. Ibid., 86–87.
184. Turvasi, *Giovanni Genocchi*, 65.

> (2) The missionaries will be able to support themselves if they can count on a minimum income, from other than local sources, of 400 pounds sterling a year.[185]

Genocchi arrived in Iquitos on December 23, 1911, and, unsurprisingly, ultimately made the Putumayo the cornerstone of his damning report to the pope. His notes for it included excerpts from Saldaña Rocca's *La Felpa* and *La Sanción* newspapers.[186] Genocchi was privy to the as yet unpublished Casement report and made copious notes from it. He wrote down such accounts as that of the Barbadian Frederick Bishop, who had told Sir Roger about Martinengui, a man with twenty or thirty Indian concubines, who, on learning that one had syphilis, had her suspended, flogged, and her genitals branded. Adolphus Gibbs had reported the case of an Indian placed in stocks who escaped after becoming so thin that he was able to slip his hands out of his chains. One Arana section chief, Augusto Jiménez, sent an eighteen-year-old Indian lad after the escapee, and he was soon recaptured. Jiménez then ordered the boy to cut off the man's head.[187] Then, too, Genocchi reported:

> Father Alemany tells me about the traffic in human flesh carried on by Antonio Vasana, a Spaniard, a fugitive from Ceuta: hundreds and thousands of women and children. This criminal keeps going further inland into the upper Ucayali and eludes all searches. Most corrupt authority! . . . Among other things, Father Alemany impressed upon me that, while traveling with his confrere Father Cornejo four of five years ago, he was on a steamboat full of Indian girls from eight to sixteen years of age. They were all sold to a Jewish man and probably brought to Europe. The priest could do nothing to prevent it.[188]

Furthermore,

> The Campas Indians of Ubriqui River were dwelling peacefully in their houses when suddenly there fell upon them men sent on a slave raid by one of the traders of the upper Ucayali who lives near Unini. These, without warning, attacked the innocent Campas, seizing those whom they could, killing many men so that few escaped their cruelties. Even up to now the number of their victims is not known. It is certain that many were found in a state of de-

---

185. Ibid., 61.
186. Ibid., 78–79.
187. Ibid., 37.
188. Ibid., 71–72.

composition. All of the houses of the Ubiriqui were burnt. These deeds had exasperated the Indians. Father Alemany warned: "If no effective remedy is applied, later on we shall not be safe even in the mission village, nor shall we be able to continue to win over and civilize the natives who dwell in our forests."[189]

It was now, or on February 11, 1912, that Genocchi departed for Europe, arriving in Rome on April 4. En route, March 6, 1912, he wrote to Cardinal Merry del Val, recommending that an Indian mission be established post haste at La Chorrera. He concluded:

> Crimes against the poor Indians continue. They are kidnapped, sold and murdered. It is true, the situation is not as bad as before, but it still exists to a small extent. . . . There are very recent facts which occurred during my stay in Iquitos, for which I have the best proof. I have manifested some to the English Consul who was unaware of them.[190]

Consequently, it was natural that several rubber traders and employees viewed missionaries among the Indians as smoke in their eyes: "A short time ago a Peruvian told Father L. Alvarez that the missionaries ought to be treated as criminals because, by instructing the Indians, they deprive us of our beasts of burden." Genocchi, who recorded the testimony of Father Alvarez, commented: "This man only translated in clear-cut language the ideas which are dominant in Peru's uncultivated regions. Unfortunately, many of the government's employees are of the same opinion, but they fear the press and England."[191] Genocchi noted that certain territories should be allocated to the missions and declared off limits to anyone else.[192] His visit to the Loreto had also convinced him that Iquitos needed its own bishop, and that he should be Peruvian.[193] Furthermore,

> What demands greater attention is some serious provision for Iquitos, where the few Augustinians are very much disliked. They do and can do almost nothing. The Prefect Apostolic for the mission, so-called de Amazonas, Father Paolino Díaz, is now in Europe, and will certainly come to Rome. I have heard he wants to leave his prefecture, and to remove his religious because the hostility of the people and of the government render them impotent. Msgr.

---

189. Ibid., 72.
190. Ibid., 79.
191. Ibid., 80.
192. Ibid.
193. García Jordán, "La Misión," 261–62.

> Quattrocchi is of the opinion that he should not be dissuaded too strongly from his purpose because that mission really needs another kind of worker, not so crushed and depressed.[194]

Upon his arrival in Rome, Genocchi had an immediate and lengthy audience with Pius X and then went to work detailing his findings and recommendations. There seems little doubt that he had extraordinary papal access, and exerted considerable influence upon the Vatican's thinking.[195]

Meanwhile, Sir Roger was grappling with his own challenges on the religious front. The February 1912 issue of the *Annals of the Propagation of the Faith*, the journal of the Franciscans' English Province, noted that four friars, led by Father Leo Sambrook, had been selected to go to the Putumayo and that the fundraising in England for the effort had garnered 10,000 dollars to date—including donations from Anglican clergymen like the Dean of Petersborough, Archdeacon Potter, and Canon Hensley Henson. There were other non-Catholic contributors as well, such as the Duke of Hamilton and Lord Rothschild.[196]

By March of 1912, Casement had shared his report with Randal Davidson, the Anglican Archbishop of Canterbury, in an effort to enlist his support for a Catholic mission in the Putumayo. While Davidson agreed that it was "one of the blackest stories of cruelty that I have ever read," he questioned whether the mission had to be Catholic, and whether the Anglican Church could officially support a Catholic undertaking. Casement responded to the archbishop in a letter on March 12, 1912:

> I have lived so much abroad, and so much among savages, that I fear I have come to regard white men as a whole as Christians as a whole, and not sufficiently to realize the distinctions that exist at home, and separate them into separate schools of thought. In what I felt to be an appeal to a common pity, and a common compassion that animated all kindly civilized men, I was, I fear, underrating the influences that separate Christian Churches, and perhaps revealing myself as something of a heathen.[197]

Casement had organized the Putumayo Mission Fund and its fundraising campaign was to be launched in April—timed to coincide with publication of the Blue Book, as well as an event designed to honor Edmund Morel. To Casement it seemed obvious that the mission would

---

194. Turvasi, *Giovanni Genocchi*, 60.
195. Ibid., 117–18.
196. *Annals of the Propagation of the Faith*, 75, no. 500 (February 1912), 255 [*Annals*].
197. Turvasi, *Giovanni Genocchi*, 105–106.

have to be Roman Catholic (Article Four of the Peruvian Constitution declared the country officially Catholic), and the Putumayo Mission Fund had sent out a circular to that effect.[198] Nevertheless, there was ambivalence within English circles over supporting a Roman Catholic initiative, fueled in part by the ambitions of Protestant missionary groups to at least be included. Despite his prestigious committee—one that included Sir Arthur Conan Doyle, Lord Rothschild, and future home secretary of the British government, William Joyson-Hicks—the effort to meet the Fund's 15,000-pound goal foundered.

Towards the end of May, Bidwell informed Genocchi of the struggling Casement effort to raise support for the Putumayo mission. Genocchi then traveled to London determined to advance the initiative. On June 19, he met with Bidwell, Casement and Percy Browne (the banker in charge of the Putumayo Mission Fund's fundraising) to discuss strategy, particularly in light of the public unwillingness of Protestant missionary societies to accept their exclusion from consideration.

Genocchi knew that the pope was about to issue an encyclical on the lamentable state of affairs in the Latin American missions. He now advised the Vatican to hold off releasing it in order to allow the British Foreign Office to publish Casement's Blue Book as well as its own conclusion that only a Catholic mission was feasible under the circumstances. Hopefully, the Protestant missionaries would accept their fate. Meanwhile, Genocchi also recommended that the missionaries be Irish Franciscans as recommended by Casement. Giovanni and Sir Roger, having met for the first time, were now allies in their common cause.[199]

Two days later, Genocchi left London for Ireland. He carried a letter of introduction from the Rome-based head of the Franciscan order to the Irish Franciscans, but it was "not imperative." The Irish Province demurred, citing its own limited personnel. Genocchi had the capacity of

---

198. June 15, 1912. Bodleian Library MSS Brit. Emp. S-22 G346.

199. While secretly baptized a Catholic by his mother, Casement was raised a Protestant. By this time, however, and probably reflecting his growing Irish independentist sympathies as well, Casement wrote in a letter to friends: "I am more Catholic than anything else." Quoted in Turvasi, *Giovanni Genocchi*, 115. Nevertheless, he wrote to his dear friend and "Irish mentor" Alice Stopford Green,

> I send back the *Catholic Bulletin* having read the criticism which does not much appeal to me. I hate an "official" view—and this great official church with its preposterous claims to be the beginning and end of all life is a mental and moral stumbling block. Life is more beautiful than death—and the world we live in and should work for more lovely than all the plains of heaven. There can be no heaven if we don't find it and make it here and I wont [sic] barter this sphere of duty for a hundred spheres and praying wheels elsewhere.

(Mitchell, *Sir Roger*, 472)

obtaining a direct order from Cardinal Merry del Val to the Irish Franciscans to collaborate with him, but he decided instead to turn to the English province. Its father provincial, Peter Hickey, agreed to take on the Putumayo mission, and several of his friars immediately volunteered for it.[200] Shortly thereafter, Pius X wired Father Hickey with gratitude, while conferring his blessing upon the English mission. The designated missionaries began studying both Spanish and tropical medicine in anticipation of their new responsibilities.[201]

Then, on July 6, 1912, Pius X issued the papal encyclical *Lacrimabili Statu Indorum* (The Pitiful Condition of the Indians). On July 26, it was sent to representatives of the Vatican in several Latin American countries, along with the directive that it be distributed to the bishops, ordering them to address the "deplorable" state of their Indians.[202] In the event, the timing antedated publication of Casement's Blue Book (on July 13, 1912), but had the advantage of underscoring that the Vatican had embarked on its own comprehensive and longstanding initiative, and was not merely reacting to a British scandal.[203] In his encyclical, Pius X underscored the need "to deliver the Indians from the slavery of Satan and of wicked men." He declared guilty of a grave crime,

> Whosoever shall dare or presume to reduce the said Indians to slavery, to sell them, to buy them, to exchange or give them, to separate them from their wives and children, to deprive them of goods, and chattels, to transport or send them to other places, or in any way whatsoever to rob them of freedom and hold them in slavery; or to give counsel, help, favor, and work on any pretext of colour to them that do these things, or to preach or teach that it is lawful, or to cooperate therewith in any way whatever.[204]

So, just as Casement fed Select-Hearings Chairman Roberts with his questions and agenda, one can only imagine that Gennochi (either directly or indirectly) penned such words for the pope.[205]

---

200. Ibid., 108.
201. Ibid., 117, 133.
202. Ibid., 109.
203. Ibid., 111.
204. Ibid., 119.
205. Turvasi had examined the actual handwritten version of the encyclical and found the handwriting belonging to someone other than Genocchi. However, it seems that Genocchi was well aware of the content of the encyclical before its publication. Furthermore, he had prepared extensive notes on the history of papal pronouncements regarding slavery, as well as western and Christian efforts to abolish it beginning in the thirteenth century down to the present. It is unclear whether this was background research for the

On July 12, Casement wrote to Gerald Spicer that the Peruvian Amazon Company was doomed and its assets would soon be under Arana's control unless a major outside investor, like Andrew Carnegie, could be found at once. The letter ended on a realistic note by stating:

> I think the best thing to hope for is a strong Catholic Mission that could be got together, and against which no plausible argument could be sustained. A Protestant Mission is out of the question and even if it could get into the Putumayo (which it couldn't) it would be of much less service than a good Jesuit or Franciscan Mission. There is a body in Peru now—"the barefooted Friars" they are termed colloquially—who could, later on, be used if we caught enough support at home (of which I am not hopeless) to guarantee the start.[206]

There was, however, a problem. The budding scandal did produce in London a competing initiative to send Protestant missionaries to the Putumayo. After all, had not the Catholic Church had more than ample time to prove itself up to the challenge, and failed? Genocchi's mission was itself tacit affirmation of that failure. On July 13, he informed Rome:

> Yesterday I spoke with one of the heads of the foreign Protestant missions, whom I had already met in Rome. I hope to persuade him to oppose the attempt to establish Protestant missions in the Putumayo. The Anglicans are already persuaded. The Nonconformists are still resisting.[207]

On July 15, *The Times* of London published a comprehensive summary of Casement's Blue Book. It called upon all Christians to support a Catholic mission in the Putumayo, noting that, "The existing system cries aloud to Heaven." The following day, the newspaper published the appeal prepared by Genocchi, Browne, and Casement at their June 19 strategy session:

> While there are doubtless many people in this country who would wish to entrust any remedial undertaking to a Protestant body, it must be borne in mind that, according to the Peruvian Constitu-

---

encyclical, or for some other purpose. Ibid., 100–104. It might also be noted that Genocchi presented a paper, as early as 1907, on Christianity and slavery over eighteen centuries at the second Italian Anti-slavery Convention held in Rome on December 3–4, 1907. Ibid., 147–52.

206. Mitchell, *Sir Roger*, 455–56. It might be noted that, in 1907, the request of England's Baptist Missionary Society to found a mission in the Peruvian Amazon had been denied by Peru's government on constitutional grounds. Turvasi, *Giovanni Genocchi*, 114.

207. Ibid., 110.

tion, work of this kind could only be permitted if entrusted to the Roman Catholic Church. It is therefore suggested that a Roman Catholic mission should be sent to the Putumayo, far away though it is, and difficult as any work carried on under such conditions must be. For years to come the operations of these missionaries must consist, less of abstract religious propaganda, than of human fellowship inspired by compassion and the desire to uplift and benefit materially. For this large sums will be required, both initially and in the direction of providing an annual income, but in view of the expenditure the Church is itself prepared to make, a sum of 15,000 pounds will ensure the definitive establishment of a Christian mission on the Putumayo. We therefore appeal for this sum, not only to members of the Roman Catholic Church, but to all those whose hearts may in any way have been touched by the recital of one of the most terrible tragedies which has resulted from the commercialism of our time.[208]

On August 5, *The Times* published a sermon delivered by Hensley Henson, Canon of Westminster Abbey, which declared:

Unfortunately, many men, instead of employing money to make friends of the poor, and to propitiate heaven justifying the words of Jesus Christ, who calls wealth unrighteous mammon, i.e., often fruit of injustice and incentive to injustice. A worthy example is the horrendous story of rubber collectors in the Amazon region, especially in the area where the Putumayo passes through. Even elsewhere, in South America, in Africa, in Mexico atrocities are committed, but, as far as we know, they are not as many nor as diabolic as they are in the Putumayo. . . . Sir Roger Casement, who has fully earned the right to direct us, holds that the establishment of a Christian mission in the Putumayo district would be organized and carried out by the Roman Catholic Church. This opinion has been publicly endorsed by Sir Edward Grey. I hope that many English Churchmen will send contributions to the fund, which has been opened by the Duke of Norfolk and others, in order to raise without delay the sum (15,000 pounds) which is said to be required. It has been officially announced that the projected mission is to be entrusted to English Franciscans, an arrangement which will undoubtedly give satisfaction to this country. This is no time, when the Indians are perishing, to debate the merits of Churches and to inflame the mind with the recollections of eccle-

---

208. Ibid., 111–12.

siastical differences and conflicts. For my part, I prefer to recall the glorious achievements of Roman Catholic missionaries in the past and in the present. I refuse to see in them any other character than that of fellow-Christians called to an urgent and difficult work. I rejoice to aid their effort, and I pray God to bless it.[209]

Nevertheless, an August 8, 1912, letter notes that a meeting was held in a London hotel to discuss the proposal by the Evangelical Union of South America and the South American Mission Society to establish a Protestant mission in the Putumayo.[210] To Sir Roger's chagrin, the schism over Catholic versus Protestant missionaries cost him the support of some of his Protestant friends—most notably that of his close African ally, the Reverend John Harris.[211]

Throughout the late summer and into the fall of 1912, the "Catholic initiative" proceeded apace, but now encountered new difficulties. There was the delicate internal issue of assignment to the Franciscans mission territory that had been (since 1900) under Augustinian aegis.[212] Then, too, if initially the Peruvian government had been willing to support a mission staffed by foreign missionaries of any nationality, as long as they reported directly the Holy See, there was now the suspicion that the Vatican was colluding with the British government to embarrass Peru in order to support Colombian claims to the Putumayo. For Lima, the near simultaneous timing of the papal encyclical and publication of the Blue Book could scarcely be coincidental. So Peru's minister for foreign affairs informed Msgr. Scapardini that "the Peruvian government would not consent to grant facilities for, or permit access to the Putumayo of a mission even if sent by the Pope, should it be composed of British subjects."[213] There was the added fear that any incident of violence or crime reported to London by English missionaries might become a pretext for direct British intrusion into Peruvian affairs, or even an invasion of the country.[214]

On September 13, 1912, the Peruvian government's Vatican chargé d'affaires, (Basque-surnamed) Goyeneche, contacted Monsignor Eugenio Pacelli, the pro-secretary of the Congregation of Extraordinary Ecclesiastical Affairs (and the future Pope Pius XII), rejecting English missionaries in favor of Spanish ones:

---

209. Ibid., 122.
210. August 8, 1912. Bodleian Library MSS Brit. Emp. S-22 G322 (file 1 of 2).
211. Mitchell, *Sir Roger*, 115.
212. García Jordán, "La Misión," 262.
213. Turvasi, *Giovanni Gennochi*, 126–27.
214. García Jordán, "La Misión," 262, 265.

The Spaniards know the language of the country, have already proven themselves and have other missionaries in Peru. Peru recognizes the very lofty aims of the Holy See and indeed is grateful to it for the benefits of evangelization, but it cannot say the same for England which has imperialistic aims. Anticipating, therefore, difficulties which could arise, the Minister prefers that the Missionaries not be English. For example, it could happen that a missionary is killed by the savages. England could use this as a pretext to intervene, accusing the Peruvian government of not knowing how to protect the security of English subjects. Today the Monroe Doctrine is in force all over America; but who knows? Things could change. Thus, the Minister, though he recognizes the disinterested and very lofty aims of the Holy See, cannot fail to consider this purely religious aspect alongside the political element as well.[215]

When informed of this by Pacelli, Genocchi responded that the use of English missionaries in the enterprise would reduce rather than enhance the likelihood of English intervention in Peruvian affairs, since it would provide greater credibility to both the English public and its government (Edward Grey's conviction as well). Nor, given the recent history of the abuse of Spanish Augustinians in Iquitos, was Genocchi likely to favor the introduction of more Spaniards into the Loreto. He was also leery of its civil authorities, and felt that Prefect Alayza y Paz Soldán was not altogether trustworthy.[216] The English missionaries would remain under the Holy See and would be ordered to maintain political neutrality in all political matters. The Vatican would likewise go to conscious lengths not to recognize the authority of either Peru or Colombia in the Putumayo as long as it remained contested territory. By this time, or on September 27, the Colombian government had expressed its concern to the apostolic delegate in Bogotá that establishment of the Putumayo mission through the Prefecture of San León de Amazonas was at least tacit recognition of Peruvian authority in the region.

On October 2, 1912, Cardinal Merry del Val informed the Peruvian government and the apostolic prefect of San León de Amazonas of the pope's decision to establish a mission in the Putumayo. However, the boundaries of its jurisdiction were left ill defined on purpose, so as not to draw a firm line between Peru and Colombia.[217]

Peru remained silent for most of the month, so, on October 28, Merry del Val sent a telegram to Apostolic Delegate Scapardini in Lima:

---

215. Turvasi, *Giovanni Genocchi*, 126.
216. Ibid., 135.
217. Ibid., 128–29.

Response is urgently required regarding dispositions of the Peruvian government towards Catholic mission Putumayo. Missionaries' departure would be next November 12. Must have certainty that government would give warm reception establishing missions in designated territory. If government should obstruct mission, Holy See would be obliged to publish its failure, declining responsibility. *Odium* would fall on Peruvian government, arousing new ferment in the Anglo-Saxon world and would furnish pretext of greater Protestant intervention against Peru's national interest.[218]

But then, on October 29, Percy Browne received a letter from the British Foreign Office stating that Peru was willing "to give definite support" to the English missionaries and would provide them transportation from Iquitos to the Putumayo in a government vessel. Consequently, the missionaries, and Genocchi himself, left Lisbon for Iquitos on November 28. The Booth Company provided them free first class passage and a donation of 250 pounds, along with the assurance of furture collaboration. Pius X had made a personal monetary donation to the effort.[219]

However, the challenges were far from over. By the time that the Franciscans, five in all counting a lay brother skilled in carpentry and agriculture, left in November of 1912 for South America, the Putumayo Mission Fund had raised less than 3,000 pounds. Casement would later note that the Putumayo mission had been financed by a single anonymous donor.[220] While the party was in transit, Monsignor Scapardini telegraphed Merry del Val with Peru's request to establish the diocese of Iquitos immediately, with authority over the old and new missions of the Loreto. It all had to be done before the Peruvian Congress adjourned on December 10. Merry del Val replied that such decisions could not be made so precipitously, and that in no event could the Putumayo mission be placed under the control of any future diocese. Rather, it would remain under the aegis of the Holy See through the Sacred Congregation for the Propagation of the Faith. To do otherwise would be to take sides in the territorial dispute between Peru and Colombia.[221] Furthermore, creation of the diocese would institute indirect governmental control over the missions, given that there was an agreement between Lima and the Vatican that accorded

---

218. Ibid., 130.
219. Ibid., 135–36.
220. Mitchell, *Sir Roger*, 115. The donor was George Pauling who made a fortune building railroads around the world, but particularly in South Africa where he worked closely with Cecil John Rhodes in the development of Rhodesia. Turvasi, *Giovanni Genocchi*, 133.
221. García Jordán, *Cruz y arado*, 238.

the former the right to nominate candidates to fill vacant bishoprics within the country's secular church hierarchy.

Peru's president now insisted, if there was to be any hope of selling Peruvians on the plan to introduce English missionaries into the Putumayo, "we must find a formula which would entitle the government to declare to the Nation that the region is Peruvian not only in the political sense but also in the religious sense."[222] It was all a ploy to confer Vatican recognition upon Peruvian claims to the Putumayo.

Merry del Val again threatened to abort the program (which had received international acclaim) while placing the blame upon Peru. It would be the Peruvian government that blinked. In mid-December, it resolved to send Minister Goyeneche to Rome to negotiate the future diocese. Then, on December 27, Peru's apostolic delegate met with the country's president and minister for foreign affairs, and he was assured that the Peruvian government would protect the persons and belongings of the English missionaries, as well as facilitate their undertaking. While they still thought that it was all about English money and interests, they did not wish to offend the Holy See.[223]

It was then (December 28) that a telegram was received in Rome from the apostolic delegate stating that a hostile reception awaited the missionaries in Iquitos. There had already been a premature demonstration by "anti-clerical, free thinkers" on December 5 (they erroneously believed that the missionaries were arriving that day).[224] In the event, the new prefect, Juan José Calle, traveled to Manaos to meet and escort the missionaries to Iquitos. Genocchi believed that his detractors in Iquitos, who regarded him as a rash intruder, had hired a cutthroat to murder him upon his arrival. His boat was delayed by a "providential" storm that enabled government agents to uncover the plot and arrest the would-be assassins.[225] During their journey together, Genocchi and Calle forged a

---

222. Turvasi, *Giovanni Genocchi*, 136–37.
223. Ibid., 137.
224. Ibid., 138. This would seem to be a subterfuge. No one other than Arana (and his supporters) seemed capable of organizing such manifestations in Iquitos. The leader of these was one Nicanor Seavedra, the town's mayor in 1911. "Free thinkers" simply did not get elected to such high posts in Arana's town. This was likely one more attempt by Loretan regionalists designed to undercut the influence of both foreign missionaries and the Vatican in the region, and should be understood against the backdrop of the government's simultaneous efforts to establish local religious control through creation of an Iquitos-based diocese with jurisdiction over the missions throughout the Loreto (including Aranalandia).
225. Ibid., 138–39. There is again room for skepticism, since this was not Arana's *modus operandi*. He did not try to murder the likes of Hardenburg or Whiffen, let alone Valcár-

close relationship, and the prefect promised to protect and support the missionaries. He agreed to provide a good government vessel for transportation of the missionaries to the Putumayo. Calle was quite pleased that the mission would include (Basque-surnamed) Friar Olano, a Peruvian missionary. He commented, "A Peruvian priest would be very desirable in the Putumayo with the English. His presence would much more easily remove any political bias."[226]

Once in Iquitos, Genocchi did not believe it necessary to travel with the Franciscan missionaries to La Chorrera. While the mission had largely been his creation, he resolved to return to Europe; he would continue to support the Putumayo mission from afar.

The English missionaries left Iquitos on January 21, 1913, and arrived at La Chorrera on February 3 without incident. By February 10, they had opened their first school and planned to start another at El Encanto.[227] Merry del Val wrote to Scapardini on February 15, 1913,

> The present English missionaries at the right time could be successively replaced by Spanish language ones, when the new mission, that has been erected for purely evangelical reasons and absolutely none of a political character, is solidly established, is devolving its assignment efficaciously and has as well satisfied all of the concerns of public opinion, particularly English, regarding [possible commission of] crimes against the Indians of the Putumayo.[228]

The police commissioner welcomed the missionaries warmly and pledged them his support. Father Sambrook was then escorted throughout the region by Tizón. Everywhere, the indigenes seemed to the Franciscan to be healthy and in good spirits. Not only had they welcomed the missionaries, their chief had prepared a Huitito-Spanish dictionary

---

cel or Saldaña Rocca (at least while the latter was in Iquitos). The Arana *modus operandi* was more to discredit the foreign critics and run the Peruvian ones out of town. To murder the papal envoy would have been right over the top.
226. Ibid., 139.
227. Testifying before the House Select Committee in February of 1913, Henry Lex Gielgud noted that in the Putumayo,
> As regards the steps that were taken with respect to stopping the abuses, personally I believe that they have been fairly effective. I do not say for a moment the conditions are absolutely perfect. I do not suppose they are. But I do think a very great advance will be continued particularly in view of the fact that owing largely to Sir Roger Casement, a mission of Franciscan Friars has been sent out, a mission for which, personally, I have been doing a good deal. (British Parliament, *Report*, 386)

228. García Jordán, "La Misión," 265.

to help the friars with their work.[229] Then, on May 15, Father Sambrook wrote the Putumayo Mission Fund Committee, "As far as I know at present coercion for rubber is happily a thing of the past."[230] Father Sambrook remained optimistic throughout the first year and somewhat beyond that his work was proceeding well, and with the full blessing of the Casa Arana and its key employees.[231]

Of course, in positing this pollyanish view, Sambrook lacked much real experience. The Augustinians, supported by the OPFé, continued to found and staff missions in the Loreto.[232] Nevertheless, the more sobered Augustinian apostolic delegate in Iquitos, Father Miguel San Román, O.S.A., wrote on May 8, 1914: "that which is impossible can never be realized; while there are *caucheros* there will be no missions; this we can take to be indisputable."[233] He then recounted the disastrous experience with Jericho (Jericó) Mission at Pebas. It was founded in 1910 and seemed to have bright prospects. But then in December 1911,

> There arrived a day, a fateful day for God's cause. Ten civilized *caucheros*, well-armed, went into the forest effecting a correría (Indian hunt) in search of Indians that pertained to the Mission, and finding them they brought them to the banks of the river, having burned their houses, stolen their bananas, thrown their beds, cutlasses and knives (their only tools and belongings) into the water, leaving them with their charming and suffering bodies. Once on the riverbank they forced the women and their children into a canoe and sent it ahead down river, and they tied up the men and put them in a raft (a boat made of sticks that floats in the water), and they set out for who knows where and without having any idea of where their women and children were ending up. Despite being bound they were guarded by various men who would come to pay dearly for their actions. That night, while their captors were sleeping peacefully, the Yaguas [the Indians in question] stayed awake and one of them managed to untie himself and then untie his companions, and they all without second thoughts took revenge murdering their conductors. Because of these events the

---

229. Turvasi, *Giovanni Genocchi*, 140–42.
230. Ibid., 141.
231. Ibid., 140–41, 143.
232. Like the Franciscans, they struggled financially both due to the nosedive of the economy in the Peruvian Oriente with the collapse of the rubber boom, and then World War I's disruption of the world economy. At the end of 1914, the OPFé chapter in Lyon. France, sent 4,000 francs to the Augustinians in Iquitos in support of their missionary activity. Rodríguez Rodríguez and Álvarez Fernández, *Monumenta*, 521.
233. Ibid., 495–96.

Indians flee from contact with the whites who pursue them for having killed their pursuers.²³⁴

The indigene women and children simply disappeared and were never heard from again. Jericho Mission was done for, and it soon fell into ruin. In an earlier Augustinian report, it was stated that the attackers were from the "sadly famous" Putumayo, and that some of them were detained two months later for their actions and jailed in Iquitos. However, they were released from custody within six months.²³⁵

There are two comments to be made. Its "Pebas estate" was one of the assets listed in the Peruvian Amazon Company's stock offering. Then, too, there is Father San Román's depiction of the *correría* as an "Indian hunt" (*caza de indios*). Not for him were any of the earlier hairsplits in the London Hearings when Arana tried to argue that the Spanish word meant things like "attracting" or "indebting" the indigenes, rather than running them to ground.

Meanwhile, as early as October 10, 1912, meetings sponsored by the Protestant Evangelical Union of South America were held in London to launch its own Putumayo Mission Fund. It sought to raise 10,000 pounds for the initial effort, and, by November 12, it had almost half of the money in cash contributions and pledges. It was preparing to send Elliott T. Glenny, former medical missionary at Cusco, and Frederick C. Glass, who had twenty years experience as an engineer and missionary in Brazil, to find a suitable location in the Putumayo for the Protestant mission.²³⁶ Consequently, in December 1912, and again in May 1913, the Protestant Evangelical Union of South America sent out expeditions of its own. According to Mitchell,

> Like the antagonism whipped up between English and Norwegian explorers over the race to the Antarctic, the missions to the Putumayo became a peculiar contest between contending missionary groups and showed how far the Catholic and Protestant churches in England had drifted from any common Christian vision. A number of figures who had been prepared to help Casement with the Morel Testimonial and other appeals gradually began to distance themselves from Casement's insistence that the Peruvian mission had to be Catholic.²³⁷

---

234. Ibid., 496.
235. Ibid., 201–202.
236. Edward I. Reed to Charles H. Roberts. November 12, 1912. Bodleian Library, MSS Brit. Emp., S-22 G335.
237. Mitchell, *Sir Roger*, 466.

On March 7, 1913, Colombia's president, Carlos Eugenio Restrepo, issued a decree creating the Misión Comisionaria del Putumayo designed to enhance Catholic missions on the river—although the effort would be directed from Mocoa and focus upon the Upper Putumayo. Article Five states that the purpose was to "Attend with greatest interest to the civilizing of the inhabitants of the Territory, procuring to reduce the roving Indians to fixed settlements, and to accustom them, by gentle means, to obedience and submission to the laws."[238] This simply enhanced Monclar's influence, of course, albeit in Colombia but not Peru.

On August 30, Montclar wrote to Buxton that he stood ready to collaborate in the civilizing of the thousands of indigenes in the Lower Caquetá and Putumayo. He noted that his missionaries had founded twenty-five schools and taught one thousand two hundred Indians in the Upper Putumayo, but were unable to work in the lower stretches due to its contested sovereignty between Peru and Colombia, as well as, "the many cruelties that have been committed by the *caucheros*." He requested any support from the London society and offered to it the unconditional collaboration of his Capuchin missionaries.[239] Given Monclar's previous and subsequent strategies, one can regard the appeal for funds to be genuine, the offer to collaborate less so.

The Protestants persisted for a while. In a letter on January 16, 1915, there is mention that the Evangelical Union of South America sent an expedition across the Andes toward the Putumayo to contact indigenes "who, we understand, had fled in large numbers from the Arana territories and taken refuge in the forests around Caqueta."[240] There is an extract from a letter by Dr. E. T. Glenny, a participant, outlining its strategy. It claimed that the Colombian government was willing to help them, but the Peruvians were opposed. The plan was to enter the Igaraparaná region from the Lower Caquetá via the Cahuinarí River. While still in the Caquetá, he came across twenty Colombians who had about sixty Indians in their employ. All had fled Arana, and there was now practically no rubber being gathered in the eastern Putumayo—the indigene population of its only remaining rubber station numbering only a dozen. They found the area to be an uninhabited wilderness with many abandoned *malocas*. Consequently, "the establishment of any meaningful missionary work whatever in that region was an absolute impossibility." Glenny predicted that

---

238. *Misiones católicas del Putumayo. Documentos oficiales relativos a esta Comisionaria* (Bogotá: Imprenta Nacional, 1913), 6.
239. Father Fidel de Montclar to Sir Thomas Powell Buxton. August 30, 1913. Bodleian Library MSS Brit. Emp. S-22 G332.
240. January 16, 1915. Bodleian Library MSS Brit. Emp. S-22 G323.

both Iquitos and Manaos were doomed, but would be no great loss since, "Morally, the places leave Sodom and Gomorrah a hopeless second."[241]

The Evangelical Union dispatched a follow-up expedition to the Upper Putumayo, but found the Capuchins entrenched there "with definite political purposes."[242] So any Protestant missionary activity in the region seemed out of the question. There is no evidence that British Protestant missionaries ever managed to penetrate any part of the Putumayo/Caquetá.

On February 17, 1915, there was a letter from the oversight Committee of the Putumayo Mission Fund (unsigned, but likely written by Travors Buxton) to Reverend Stuart McNair of the Evangelical Union of South America, stating:

> I have now had a letter from the Secretary of the Roman Catholic Mission, to whom I wrote by direction of our Committee. He concurs with the view that the activities of the Arana Company have been reduced owing to the fall in the price of rubber, but thinks that the atrocities had practically, if not entirely, ceased before this Reduction, owing to the more humane regime of M. Tizon, who was in control when his mission arrived. He does not know of any great Reduction of the number of natives in the district effected by the Casement report and does not admit that the people are scattered and on the move. He states that their mission has two permanent stations from which they have been able to reach the greater portion of the districts where the Arana Company were at work, with satisfactory results.[243]

By 1916, the normally diplomatic Sambrook was beginning to express his reservations regarding his Putumayo missions. Little could be accomplished without the cooperation of the Casa Arana, yet it was increasingly evident that the needs and goals of the business and the missionaries were incompatible. The rubber enterprise needed workers scattered lightly and broadly throughout the forest and dedicated to their task. The missionaries wanted settlements with schools, churches, and civilization—social, cultural, political, and religious education and indoctrination. Sambrook also complained that civil and judicial authority was vested in the rubber enterprise, and respected its will. He also noted that execrable abuses and physical punishment continued unabated and alienated the Indians against all

---

241. Extracts of Letter from Dr. R. T. Glenny. Bodleian Library, MSS Brit. Emp. S-22 G323.
242. January 16, 1915. Bodleian Library MSS Brit. Emp. S-22 G323.
243. To Reverend Stuart McNeil, February 18, 1915. Bodleian Library MSS Brit. Emp. S-22 G332.

whites—including the missionaries. In sum, in the Putumayo everything was subordinated to "immediate commercial interest, and the civilizing of the native is completely ignored." As a consequence, although the former abuses and punishments of the indigenes had been somewhat worse, the English Catholic mission had not made a substantial difference.[244] Even now, a document from the Prefecture of Loreto stated that the Indians

> Find themselves beneath the dominion of the Peruvian *caucheros* who can count upon the support of the government and the armed forces of said country. In Yubineto they have established a military encampment and constantly launch incursions into the Putumayo to capture Indians by force and take them to the determined places of their rubber concessions.[245]

All of this concurs with oral histories collected in the 1980s from elderly members of the several Putumayo tribes. They reported how their ancestors fled en masse from Arana's forces. They were pursued, captured, and punished. Some were killed as an example to others. Then, ironically a few days after the execution of Casement in London, there was a major uprising by Huitotos focused in the Alto Cahuanarí that resulted in the capture and brief occupation of Atenas station. One of the Franciscan missionaries working in the Putumayo reported to the British consul that the whites had reaped "a little of what they had sown." A month later, Father Sambrook reported that a Peruvian expeditionary force of thirty troops, armed with machine guns, had retaliated—recapturing Arana's Atenas station, burning the indigenes' houses to the ground and rounding up those who remained alive.[246]

The situation was actually considerably more complicated, as reported by Colombian anthropologist Pineda Camacho after he analyzed several indigenous oral histories of the events: It seems that the central figure was the Huitoto *capitán* named Yarocamena, who dominated five *malocas* in the Atenas area. His eldest son was either killed outright by an oppressive white overseer (who whipped his Indian charges for failing to bring in sufficient *caucho*) or drowned while fleeing his wrath. The death of his son prompted Yarocamena to exact revenge. He fortified his *malocas* and dug escape tunnels for their residents should they come under siege. During the next year, he stockpiled rifles and ammunition, and then tried, unsuccessfully, to convince other *capitanes* to join him in rebellion. Meanwhile,

---

244. Summarized and quoted in García Jordán, "La Misión," 266–67. García Jordán, *Cruz y arado*, 239.
245. Quoted in Pineda Camacho, *Holocausto*, 190.
246. Goodman, *The Devil*, 264–65.

some of his men captured and occupied the Atenas station. This prompted the Peruvian military to send in two thousand troops from Iquitos.[247] The final denouement was the siege upon a *maloca* with walls fortified by rubber balls that absorbed the soldiers' bullets, including those fired from machine guns. The resisters were defeated only after the military successfully lobbed a burning missile onto the roof of the besieged *maloca* that then burned to the ground. Yarocena and several of his senior men escaped, while most of the women and children perished. He came to recognize that the rebellion had been a mistake and ordered his subjects to resume gathering rubber for the overseers. The rebellion did have the effect of improving the treatment of the Indian, if only to avoid such future conflicts.[248]

Meanwhile, Monclar's activities did not, of course, go entirely unnoticed by Colombian authorities. His budget requests for road-building alone were certain to prompt political debate in a cash-strapped country. Some Bogotá politicians even demanded that the Capuchin mission be investigated. So it was that, in 1913, in the wake of the hostilities with Peru, it was resolved to separate jurisdictionally the *resguardo* territory of the Putumayo from the Department of Caquetá. The new Putumayo territory was placed under the authority of General Joaquín Escadón. He wanted to bring its semi-autonomous theocratic regime under greater national control (for example, empowering civil courts over the ecclesiastical one), so he and Friar Monclar were immediately at cross-purposes. It was no contest. Before it was over, Monclar managed to have the general dismissed by civilian authorities and excommunicated by the Colombian Catholic Church. There could no longer be any doubt who ruled in the Alto Putumayo.[249]

It was Montclar's goal to build a road to the Putumayo and thereby link the remote region to the Colombian nation. For many years, albeit with hiatuses, he would control a state-appropriated budget with which he first finished the section between Pasto and Mocoa and then extended it downriver toward Puerto Asís. Thanks to these developments, the Upper Putumayo would be opened to additional white settlement and evolution of the economy from rubber gathering and exploitation of other forest products to the raising of cattle, tobacco, cacao, and bananas.[250]

There then began a remarkable three decades of missionary activity that became the phalanx of economic development as well—spearheaded by Montclar:

---

247. Surely fewer!
248. Pineda Camacho, *Holocausto*, 155–58.
249. Bonilla, *Servants*, 115–18.
250. Stanfield, *Red Rubber*, 186–87; 190–91.

He championed the foreign-born missionaries as Colombian patriots, true Christians, and resolute conquerors of the last-known land on Earth. He likened the Capuchins to sappers, missionary servants who could bring light to the eternally savage while improving business opportunities, and perhaps even turning the Peruvian flank by pushing down the Putumayo and then taking the Napo. Montclar's political and cultural rhetoric, combined with his personal lobbying of state and church officials, made the Capuchins seem critical to the pursuit of progress, sovereignty, and justice.[251]

In 1918, Bogotá dispatched Dr. Thomas Marquéz as its fiscal inspector to the Putumayo (*Visitador Fiscal del Putumayo*) to look into both Father Monclar's road project and the general state of his missions. Given his incursions into public affairs, the Capuchin priest had many critics. However, Marquéz quickly became a strong proponent of Montclar's activities and vision. By this time, the Capuchins were fully ensconced in the Upper Putumayo, and Monclar realized the importance of asserting a right to free navigation on the river if Colombia was to develop proper commerce with Brazil. He decided to dispatch one of his friars down the Putumayo to Manaos, where he would purchase trade goods and rent a vessel to deliver them to Puerto Asís. Marquéz agreed to accompany the priest—Gaspar de Pinell—on what would become a nine-month canoe journey (March to November 1918).[252]

In April, the travelers were ascending the Caraparaná on their way to El Encanto (hoping to establish a precedent). They passed about ten Colombian families (all of whom had a few Huitoto workers) who recounted their difficulties in former years with the Casa Arana. Now, they simply eked out a living by farming and hunting and fishing for game that they sold to Arana's passing vessels. Otherwise, they were left alone.

Miguel Loayza welcomed the travelers graciously at El Encanto. In addition to his impressive headquarters, there was a small Peruvian military garrison of about twenty-five men under the command of a Captain Urdiales. There was also a Telefuncke (telegraph) station for sending messages to the outside. Arana maintained a small vessel, the *Callao*, at El Encanto crewed by Huitoto Indians. It collected rubber throughout the area for transfer to the *Liberal* on its tri-monthly visit.[253]

Loayza facilitated their request to visit the English Franciscan mission in his district (several hours away on foot). He provided them with a

---

251. Ibid., 188.
252. Gaspar de Pinell, *Un viaje por el Putumayo y el Amazonas. Ensayo de navegación* (Bogotá: Imprenta Nacional, 1924) [*Un viaje*].
253. Ibid., 35.

white guide and four Indian porters. When they arrived at San Antonio, they were greeted by the startled Fathers Burne and Fitzpatrick, and their lay brother assistant, Edwin O'Donell. They were told that, initially, the missionaries established schools here and at El Encanto. However, they had received little support from either Arana or the Peruvian government. There was no encouragement on their part of sending the Indian children to school. When World War I broke out, the missionaries' regular supplies (funded in England and delivered by the Booth Line) dried up. They had been reduced to celebrating an occasional baptism and mass twice weekly. While the Arana firm did not help them (regarding them as foreign spies), it did not dare to hinder their work openly out of fear of another international scandal. Furthermore, soon after their arrival, all Arana employees were put on salary to avoid the accusations of abusing the Indians, as happened under the former commissions' system. Pinell was told by the Franciscans that there had been no corporal punishment for several years, and the Colombians claimed that the Indians were now receiving more for their rubber.[254]

There were still Indian issues. The Franciscans described the previous year's rebellion that lasted for weeks and ended only through the duplicity of certain informers inside the fortified compound that ultimately led to the torching of the building. The Boras remained all but unconquered and dangerous. No one dared go near their territory unarmed. Then, too, the elderly Indians had given up their cannibalism, but only out of fear of white retribution. The missionaries estimated that there were about 2,300 Indians remaining on the Caraparaná and 6,200 on the Igaraparaná. In 1916, their superior [Sambrook] and another Franciscan priest returned to Europe for health reasons, and they believed that it would soon be their turn to leave.[255]

Back at El Encanto, Loayza went out of his way to praise the Franciscans while pumping the two travelers on the purpose (particularly regarding the Putumayo/Caquetá) of their trip, "so we were careful to respond with much circumspection so as not to compromise or contradict ourselves."[256] They felt that they were fundamentally unwelcome, and, after securing a letter of safe passage from Captain Urdiales to Peruvian authorities at the Brazilian frontier, they departed without visiting the

---

254. In the event, these statements appear overly optimistic. It should be noted that the friars had but a tiny window on the Putumayo and other observers reported that the round-ups, with their exemplary corporal punishment of absconders, continued in at least the more remote areas of Aranalandia (Stanfield, *Red Rubber*, 198).
255. Pinell, *Un viaje*, 37–41.
256. Ibid., 43.

Igaraparaná. Pinell baptized some Colombians during the descent to the Putumayo and estimated that the entire Colombian population remaining in the area (including some Arana employees) amounted to thirty-five to forty individuals. On the Putumayo, they encountered some nervous Ecuadorians who feared that they were from the Casa Arana. At the border, the letter from Captain Urdiales was critical in assuaging the suspicions of the lieutenant in charge. He asked pointedly if any of their Indian crew were natives of the Igaraparaná or Caraparaná.[257] Clearly, the Peruvian officials were protecting Arana's interests as much as policing their national territory.

In September 1918, Antonio Pastrana, Colombian commissary in the Caquetá, reported that Las Delicias had suffered an attack from four Peruvians, supported by a small army composed of fifty well-armed Indians, who imprisoned four persons and appropriated a shipment of rubber and supplies. The Casa Arana was responsible for the attack.[258] By now, "acting prefect of Loreto W. Pinillos Rossell suggested that the small garrisons along the Ecuadorian and Colombian borders be reinforced to counter Capuchin and Colombian programs of colonization and road building."[259]

Meanwhile, in Manaos, Pinell and Marquéz set about purchasing their cargo and negotiating for a vessel to transport it to Puerto Asís. Pinell noted:

> Shortly after having arrived at the capital of Amazonas, Señor Don Julio Arana, principal owner of the Caraparaná and Igaraparaná regions, came to visit and told us that he had been informed by radio from El Encanto of our trip; he placed himself entirely at our disposal, visited us frequently enough, always appearing to be much interested in the successful outcome of our undertaking, speaking of the advantages that free navigation of the Putumayo would afford to his business. Without abusing the rules of gentility and courtesy, we were very careful not to be taken in by the influence and offers of said gentleman; and anyone who knows the history and actual state of affairs of the Putumayo will easily understand the reason for our behavior.[260]

They were ultimately able to contract the vessel *Yaquirana* and begin their return to Colombia. Once in Peruvian waters, however, they were approached by the *Callao* while ascending the Putumayo, now commanded by the region's military supervisor, Captain Manuel Curiel. There was

---

257. Ibid., 43–46.
258. Lagos, *Arana*, 355.
259. Stanfield, *Red Rubber*, 195.
260. Pinell, *Un viaje*, 85.

a contingent of somewhere between twelve and fifteen soldiers on the *Callao*. Curiel ordered the *Yaquirana* to El Encanto over the protests of its captain (it seemed a pointless detour from his intended direct ascent to Puerto Asís). At El Encanto, they were greeted effusively by Miguel Loayza, who assisted them with telegrams to Iquitos over the next five days requesting permission for the Brazilian vessel to proceed to Colombia—thereby establishing the right of free navigation on the Putumayo. Pinell's conclusion was that,

> We are completely convinced that the *black hand* that stirred up the matter was the Casa Arana and the Consejo Curiel. The one in whom we noted much perplexity was Señor Loaisa a consummately intelligent and astute man, who for more than twenty years has directed the affairs of said firm on the Caraparaná. He witnessed and took part in many of the events in those places while the rubber businesses were passing from Colombian to Peruvian hands. He was there when the commissions sent by foreign governments visited those regions regarding the great international scandal produced by disclosure of the crimes of the Putumayo. He had to work things out with many Peruvian government employees, some of whom had been his sworn enemies to the point of even striking him on occasion. There do not lack those who regard him as the biggest Putumayo criminal or the moral authority behind the greatest crimes; but the fact remains that he has always remained unharmed by all of these storms, without any criminal act by him ever having been proven to date.[261]

Pinell further noted that, had they been ignorant of the history, they might easily have been taken in by the charming Loayza. In short, it was all an exercise in good-cop/bad-cop with Loayza pretending to be their advocate while Curiel argued that, without the approval from the authorities in Iquitos, he was powerless. In the event, permission to proceed was denied to the *Yaquirana*, although the two travelers could return to Colombia in a canoe that Loayza was now offering them. To accept was to abandon their mission, so they declined and returned to Manaos.[262]

It was there that Arana sought them out again at the Capuchin convent. He offered to try and arrange for them to meet the *Liberal* in October, travel on it to El Encanto, and then complete their journey to Colombia by canoe. Arana claimed that the Peruvian authorities were

---

261. Ibid., 93–94.
262. To establish the precedent of free navigation on the Putumayo they needed to traverse its Peruvian section in a modern vessel.

prohibiting them from traveling by launch from Manaos to Puerto Asís. It also became evident that Arana was privy to the letters sent earlier by Father Montclar to the bishop of Manaos, soliciting his help in opening the Putumayo to Colombian commerce and free navigation. It seems that the Brazilian prelate was the neighbor and close friend of Julio César. He had consulted Arana before crafting his reply to Monclar.[263]

Marquéz decided to return to Bogotá via Pará while Pinell accepted Arana's offer of a letter of safe passage. The friar traveled to the rendezvous point and transferred to the *Liberal* along with his (Colombian) Indian crew. He recounted a sharp exchange that ensued after he mentioned to his Peruvian fellow passengers that the colors of the rainbow they all happened to be contemplating reminded him of the Colombian flag. The Peruvians all concurred that such a flag would never fly over the Putumayo as long as there was a single Peruvian capable of bearing arms.[264]

Pinell ascended by canoe from El Encanto to Puerto Asís. His mission had been unsuccessful insofar as it failed to either deliver its cargo or set a precedent for free navigation of the Putumayo, but it did underscore the issues. Brazil protested formally, and the Colombian Congress, by unanimous vote, censured the Peruvian actions. The Colombian government filed a demand with its Peruvian counterpart for compensation of the missionaries for the loss of their cargo (much of which was perishable and had spoiled during the delays in Manaos).[265]

According to Restrepo, Pinnell and Márquez founded a mission for Huitoto Indians at what is today the town of Güepí on the Peruvian bank of the Putumayo River.[266] By 1926, it had two priests and a resident college with 125 students—almost all Huitotos.[267]

In October of 1918, the English Franciscans closed their mission at San Antonio and returned to Europe. The Middle Putumayo, and Aranalandia in particular, were now entirely devoid of missionaries—Catholic or Protestant. One historian, writing in 1927, noted that the number of conversions by the Franciscans was "insignificant": the vast majority of the Indians of the Middle Putumayo remained "unbelievers."[268] A Franciscan from the Ucayali, Father Luis Olano, had visited the area that year

---

263. Ibid., 107.
264. Ibid., 144–45.
265. Pinell, *Un viaje*, 54–55.
266. Ibid., 17. These were Huitotos that had fled Aranalandia. At the time of Pinnell's visit, they were actively engaged in transporting salt to a port on the Napo River along a jungle trail that linked it together with the Putumayo and Caquetá.
267. Restrepo López, *El Putumayo*, 21.
268. P. Senen, and F. Tejedor, *Breve reseña histórica de la mission agustiniana de San León del Amazonas, Perú* (El Escorial: Imprenta del Real Monasterio de el Escorial, 1927), 101.

and reported that missionary activity there was impossible given the attitude and hegemony of the Casa Arana. The new pontifical representative in Lima, Lorenzo Lauri, informed the Vatican that nothing was likely to improve, given that the Peruvian Amazon Company had,

> ... formed with those inhabitants of this extremely remote district a slave colony, and that the government ignored absolutely the Company's affairs, given that almost all of Peru's politicians were either officers or shareholders of it.[269]

Nevertheless, Peru's administration was pleased by the departure of the "foreign missionaries." Olano was appointed the overseer of the Putumayo missions, but refused to assume his post. Nor were the Augustinians anxious to become involved. Both missionary orders viewed such an assignment to be a fool's errand. Meanwhile, the area was placed under the jurisdiction of the Prefecture of San León in the early 1920s, headed by a Spanish Agustinian, Sotero Redondo, who, at the insistence of the Vatican, was forced to take out Peruvian citizenship. That gesture notwithstanding, several secular priests were now under Redondo's aegis, which triggered a lengthy dispute. The wider Peruvian Catholic hierarchy agreed to assist the restored missions as the best means of evangelizing and civilizing the Loreto's indigenes. The Peruvian government was likewise pleased and provided the Prefecture of San León with a boat, accorded free passage to missionaries throughout the Oriente on its own vessels and granted free access to telegraphic service to all religious personnel.[270] However, subsequent reports by the missionaries continued to document the impossibility of working effectively in the Putumayo, given the Company's policies. "All-powerful" Julio César Arana received a visit from Redondo to discuss ways of improving their collaboration—and, as usual, nothing came of it.[271]

Nor did much seem to have changed elsewhere in the Oriente. Writing in 1929, the apostolic vicar of Ucayali, (Basque-surnamed) Monsignor Francisco Irazola, denounced the continued practice of rounding up Indians in the Tambo and Uinini river regions. The *correrías* were effected mainly by other indigenes under the command of whites. The raids were mainly to capture youths and children who were then sold for an average of twenty Peruvian pounds each. There is also the exchange of goods with whites in the forest stations in exchange for Indian children. In the

---

269. García Jordán, *Cruz y arado*, 240.
270. Ibid., 220–21.
271. Ibid., 242–43.

Tambo and Unini areas, in particular, there are many enslaved children and their owners are not punished by the authorities because the latter are "participants in the business."[272]

Adios Aranalandia

Pinell had some interesting observations regarding the state of Arana's business in 1918. Julio César had recently commissioned an optimistic film on the Putumayo as part of a strategy to either sell the property or float another public offering of shares in it. It also seems that a Colombian named Norzagaray, likely the same man mentioned in the House Select Hearings of 1912–1913, had pioneered new techniques for more efficient rubber gathering and processing on his Brazilian holdings, and Arana sent observers there to learn about them.[273] They worked and probably saved Arana's business over the short term. However, Julio César now had a nearly two-year unsold stockpile of rubber and was hard pressed to support the several thousand Indians gathering it for him. In Pinell's view, Julio César was likely pursuing a hard line over hegemony of the Putumayo to force the Colombian government's hand into making him an offer for Aranalandia. The Capuchin missionary thought that Julio César would be a willing seller who would believe the transaction to be honorable resolution of the years of conflict between the two countries.[274]

Some things never changed. In the 1925–1926 report on the Putumayo/Caquetá authored by Fray De las Corts it was stated:

> One time (*the caucheros*) ascended as far as the Eguaicurú, also called Elvira, tributary of the Yarí. They managed to capture an Indian and, with the objective of the discovering the whereabouts of the others, they bound his hands and feet and put him on a spit with fire beneath it. The tormented Indian resisted revealing the refuge of his brothers. The Peruvians took him half dead to La Chorrera, where they continued torturing him until he revealed the location of the fugitives. On that occasion, they ([the Peruvians] transported from the Yarí by the Puerto Huitoto path a great number of Indians that they embarked in four huge canoes. This occurred in 1912.[275]

---

272. From a letter quoted in San Román, *Perfiles*, 156.
273. Norzagaray was apparently an experienced innovator, since one of the charges of the Commission sent out by the Peruvian Amazon Company to investigate its South American business was to "Report on Norzagaray machine." MEMORANDUM, 2.
274. Ibid., 107–109.
275. Cited in Kuan Bahamón, "La Misión," 134.

A missionary report in the Colombian National Archive notes that, in February 1919, the Colombian Manuel Quintana was traveling up the Caquetá with canoes loaded with his rubber and a number of Huitotos who had fled from Aranalandia that he was harboring. A raiding party of some thirty armed men from the Casa Arana intercepted them and killed and wounded several of the Indians. Quintana and his rubber were transported to the headquarters of the Casa Arana. He was relieved of his *caucho* and equipment. It was only after much protest that he was able to secure the liberty of the survivors.[276]

In 1919, Augusto B. Leguía led a military coup that returned him to the Peruvian presidency. His supporter, Julio Egoaguirre, reentered politics as a senator from the Loreto.[277] By 1920, Iquitos was in desperate straits. Its share of Peruvian exports had fallen to 1.08 percent from 17.65 percent in 1910.[278] Surprisingly, President Leguía had initiated negotiations over the territorial dispute, and the Colombians were making gains in the continuing talks. In 1920, the Peruvian Amazon Company was finally formally dissolved and the Casa Arana was again in legal possession of the Putumayo (albeit nebulous, given that Julio César still lacked formal recognition of his land rights by the Peruvian government).

Then, in late 1920, while he was in New York on business, Arana received a summons from the Iquitos Chamber of Commerce to return immediately to help it confront the economic crisis. So, in December 1920, he was back and working with Egoaguirre on a bill that would remove all taxes on rubber as a stimulus of the troubled industry. He was then prevailed on to run for the vacant second Senate seat from the Loreto in the June of 1921 election. According to one [Peruvian] biography of Julio César, published on its eve:

> To conclude, we would say that the constitution and flourishing of the great enterprises of the Putumayo are due exclusively to [Arana's] perserverant energy and work, and his personal fortune and that of his wife, earned over more than forty years, has been employed entirely in the exploitation of this rich eastern region, destined in the future to be an emporium of wealth, not only because of the great variety of vegetative resources that it contains, but also due to the mineral deposits extended throughout all of the zone, especially the evidence of petroleum, iron, coal, etc.,

---

276. Cited in Gómez López, *Putumayo*, 237.
277. By 1922, he headed the Ministry of Justice, Religion, Education, and Welfare, and shortly thereafter he succeeded Germán Leguía y Martínez as president of the Council of Ministries.
278. Stanfield, *Red Rubber*, 196.

these latter being in plain sight, as well as of other classes of minerals. On the other hand, the climate of this region is among the most healthy, as its Indians knew not the many diseases that exist now, and lived without the necessity of doctors or medicines, until the whites introduced many of the illnesses known in the civilized world. I present here, in broadstroke, the biography of this illustrious and patriotic citizen, this grand professor of energy who is named don Julio C. Arana. He, who situated himself for forty years in the jungle and settlements of the Amazon that were his prison, surrounded by immense rivers and at times completely out of touch with people like himself, he went with his trade goods and personnel, or the forced laborers who accompanied him from one place to another, at times hit or miss, with the certainty that neither before nor behind them was there any protection should they fail. With a simple canoe or *piragua*, at times, to make long journeys in rivers spiked with snags, with fearsome waterfalls and populated by monsters, without more yearning than their love of work nor greater ideal than the patriotism that guided them, with the object of sustaining our sovereignty in the region. The dense forest presented them with what seemed an unpassable barrier; yet they penetrated into all of its gloom, opening the way with axe or machete, passing by the trunks, limbs and lianas that seemed brave hosts of beings charged with resisting them. Suffering the incarnated assault of the mosquitos and other innumerable pests that populate this environment. Illnesses germinated in their worn-out bodies. Fevers, sores, ulcers, beriberi and homesickness were their habitual companions. They remained without the most basic items that are indispensable for life, and in these conditions they felled the forests, building dwellings and fomenting agriculture and raising livestock, establishing great fields of sugar, rice, corn, beans, vegetables, yucca, etc. And since then Señor Arana has introduced there the first class cows and sheep that today provide the fresh meat for the region.[279]

Arana won the Senate seat easily and moved to Lima. Then, on August 5, there was an anti-centralist revolt in the Loreto against the Peruvian government, fueled particularly by the fear that the Putumayo might be turned over to the Colombian adversary. The rebels issued their own currency and managed to hold out in Iquitos for six months. But, the following March, the city was reoccupied by loyalist forces, and the rebels fled up the Napo River into Ecuador. The revolt persisted a bit longer in

---

279. Pinell, *Excursión*, 198–99.

some areas, including Arana's Caraparaná, but it soon fizzled out.[280] It is not altogether clear what role Arana played in these events.

Shortly after Julio César assumed his senate seat, the Leguía government accorded him formal Peruvian title to 5,774,000 hectares of land between the Caquetá and Putumayo rivers.[281] In 1922, he received Peruvian governmental assurance that he would be compensated should he lose his lands to Colombia.[282] By this time, it could be said that,

> Peruvians, in particular the Casa Arana, are working all of the territory between the Putumayo and Caquetá, and of the Tagua and Caucayá downsteam, in other words the southeast; it is certain that they exploit the left bank of the Caquetá; they have under them more than 30,000 Indians . . . The steamship *Liberal* and various launches leave every three months from the Caraparaná loaded with balata . . . In the Lower Caquetá there are no Colombians nor anyone else who can report on anything.[283]

As a senator, Julio César continued to protect the regional interests of the Loreto. On August 18, 1923, he made public the radiogram from Iquitos mayor Pablo Zumaeta:

> By agreement of the Council I direct myself to you to call your attention to the grave situation that Iquitos is going through. There is a virulent epidemic, with alarming mortality, that accompanies the pitiful general misery, the lack of employment, the consequences of devaluation of [our rubber] production, the shortage and high costs in every respect of the subsistence of life in general.[284]

Meanwhile, in 1924, three books were published in Colombia that fanned the flames of its nationalism, while underscoring the importance of concluding successfully the Salomón-Lozano treaty negotiations. Father Pinell's work, *Un viaje por el Putumayo y el Amazonas. Ensayo de navegación*, recounted his nine-months-long canoe journey in the company of Dr. Thomás Marquéz. The friar's companion was the former secretary of (Basque-surnamed) Rafael Uribe Uribe, a recently assassinated (1914) Colombian minister and associate of Rafael Reyes.

---

280. Stanfield, *Red Rubber*, 196–97.
281. There was, however, a problem. Under Peruvian law anyone could claim up to 30,000 hectares of vacant" land. At its discretion, the Congress could approve larger grants. Since this was never the case with Aranalandia, Julio César's concession was ultimately void under his country's own legislation. Dominguez and Gómez, *Nación*, 184.
282. Lagos, *Arana*, 356–57.
283. Quoted in Pineda Camacho, *Holocausto*, 190.
284. Quoted in Lagos, *Arana*, 359.

Marquéz, who was twenty-nine years old at the time of this venture, had a degree in international law, and was a brilliant journalist and a future roving inspector of several of Colombia's South American embassies. He would ultimately become the publisher of Bogotá prestigious *La Nación* newspaper. In short, Marquéz brought his personal celebrity to bear on the journey.

It was in 1924 as well, that Ricardo Gómez A. published his book, *La guarida de los asesinos: Relato histórico de los crímenes del Putumayo* (The Murderers' Hide-out: Historical Account of the Crimes of the Putumayo). There is a frontispiece photograph of José Francisco Gómez, the author's uncle, that bears the caption, "Distinguished writer, son of Pasto, cowardly murdered on the Putumayo River the night of September 29, 1905." Ricardo was born into a liberal family that suffered greatly in Colombia's War of a Thousand Days. The boy was orphaned when his father was killed in the fighting . His uncle, José Francisco, who was trained as a printer and then co-published the newspaper *El Eco Liberal* before becoming a journalist himself, was arrested and imprisoned in Popayán. He escaped and fled into Ecuador, where he was named Secretary of the Governance of the Oriente (the area contested by the three Andean countries). He resigned that post and took both his nephew and his own family down the Napo to Iquitos and then to the Putumayo, where he hoped to become a *cauchero*.[285] While working at La Florida, José Francisco had a confrontation with the Casa Arana's overseer, Juan Allende. Gómez declared it his intention to go into town and blow the whistle on all of the abuses that he had witnessed. Allende allegedly dispatched a messenger overland via the Napo to Iquitos to give the alarm. And again, allegedly, Julio César Arana himself gave the order to stop Gómez at any cost.[286]

It was then that Gómez departed La Florida by boat with his wife, two sons, and two daughters, as well as four Huitoto Indians as oarsmen. The night of September 29, the head Indian shot José Francisco, and his eldest son, Artemio, as they slept. His wife, fellow Pasto-native (Basque-surnamed) Mercedes Eraso, was able to flee with her surviving children. According to Ricardo, her subsequent attempts to get Colombian President Reyes to look into the matter were ignored.[287] It was also said that when Arana learned that the Indian who had executed the order to kill Gómez was being pursued by the authorities in Iquitos

---

285. Ricardo Gómez A., *La guarida de los asesinos: El relato histórico de los crímenes del Putumayo* (Pasto: Imp. "La Cosmopolita," 1933), (first published in 1924), 187 [*La guarrida*].
286. Ibid., 172–76.
287. Ibid., 177.

for questioning, he ordered the man killed as well by members of his own tribe.[288]

Ricardo Gómez A. arrived in the Putumayo as a fifteen year old and spent five years there working at several stations owned by Arana. He claimed to have been interviewed by Saldaña Rocca while still an adolescent, providing the Iquitos journalist with his own testimony of the atrocities. He also recounts confronting the infamous and fear some Victor Macedo after the assassination of his uncle, with the accusation that Allende had been behind the murders of the man and his son.

Ricardo Gómez A.'s account was lent an air of authority by a prologue written by Sergio Elías Ortiz, publisher of the Colombian journal *Boletín de Estudios Históricos*. Subsequent editions carried an appendix of twenty-five pages or so of laudatory reviews of the book's first edition by noted Colombian intellectuals and newspapers. The 1933 edition, the fourth printing, included a substantial introductory essay by noted historian Arcesio Aragón entitled *La fuerza del derecho* ("The Force of Law") and dated October 1932.[289] It laid out the history of Colombia's "legitimate claim" to the Putumayo/Caquetá region. Aragón stated that, in 1905, Rafael Larrañaga sold his inheritance for a "plate of lentils" after his father had died (with symptoms of having been poisoned). Arana then took over the properties of all the other Colombians. Some he had killed by the Indians and others were burned out. Still others were taken to the Iquitos prison (known as "the business office of the Casa Arana"), where they were told that they could sign over their property for the sum on offer or simply rot in jail. Then, there were the conclusions of the Casement and the PAC commission reports regarding Arana's treatment of the natives, descriptions that would have embarrassed Attila the Hun or Genghis Khan.[290]

Ricardo Gómez A. claimed to have witnessed many of the atrocities that he describes—having worked under Martinengui (whom he describes as a "man-wolf")[291] and Miguel Flórez (fat and physically repulsive), the enforcer sent to punish rebellious Indians and to pursue those who fled.[292] He also described Fonseca's alleged crimes, but, never having worked with him, admitted they were based upon hearsay. The author was clearly aware of the Saldaña Rocca articles, Hardenburg's book, and Casement's Blue Book (he does not mention the

---

288. Gómez López (ed.), *Putumayo: La Vorágine*, 449.
289. He had just published an historical opus magnus, the massive book *Popayán*, in 1930.
290. Gómez A., *La guarrida*, 43–46.
291. Ibid., 86.
292. Ibid., 92.

Paredes report, but had to know of it). It is my impression that many of his accounts of the crimes were embellishments of ones provided by these diverse sources. One can appreciate the flavor of his writing in his treatment of Armando Normand, several of whose acts he claimed to have witnessed. Accordingly,

> It was Normand, an individual who displayed titles of nobility and who had fine external appearances; a great lover of literature and as such disposed to send articles frequently to newspapers in London and Lima; but with an impure soul feeding upon criminal concupiscence. Cowardly hypocritical like none other, who was incapable of expressing feelings, much less to assume the arrogance peculiar to brave men. Murderer yes; he finished off in cold blood and without concern over the means or kindly consideration toward the unfortunate victims that fell into his clutches.[293]

It was all a morality play and admonition. Given Colombia's glorious history in defeating the Peruvians in past battles, where were the brave Colombians who were needed today to stand up to the crimes and arbitrary usurpations by the Peruvians in the Putumayo? Gómez A. was openly supportive of the Salomón-Lozano proceedings.[294]

Then, again in 1924, Colombia dispatched a governmental investigative team to its Putumayo/Caquetá region.[295] One member of it was the parliamentarian and Colombian poet José Eustasio Rivera. Upon his return to Bogotá, he published some newspaper articles on their findings and, in particular, the danger posed to Colombian sovereignty over the farflung region by the lack of proper infrastructure (particularly roads), efficient administration, adequate defense forces, and sufficient awareness of the dangers posed by the influence of

---

293. Ibid., 82–83.
294. Ibid., 135–36, 221.
295. It should be noted that a prime source of the situation in the region was the annually report sent by the Capuchins to both the Vatican nuncio and the national government in Bogotá. For example, in 1922 the report stated,
> It is commonly used expression among the *caucheros* to "conquer Indians," which means subject anindividual, family, farm or tribe, and oblige them to collect rubber in detriment to their freedom. Once conquered, they are handed from one owner to another, although now they are not usually transferred from place to place. Civil action has not yet managed to regulate the rubber trade, nor has its influence over the *caucheros* been felt in these regions.... We know that in the part of the Lower [Middle] Putumayo occupied by a very well known commercial house, horrors are committed against the Indians. A short time ago a missionary reported [hearing from] eyewitnesses that said house (Arana) obliges a great number of Indians to do forced labor, under threat of terrifying punishments. Kuan Bahamón, "La Misión," 134.

the Casa Arana over a portion of the territory.[296] While there is some confusion regarding the point, it is doubtful that Rivera ever visited the Middle Putumayo.[297] However, he wrote a novel about *caucheros* entitled *La vorágine* ("The Whirlpool" or "The Maelstrom"), that first appeared in late 1924 and became an instant bestseller.

The novel is about the search of Clemente Silva, a white Colombian and lower-class father, for his indentured son—who was reputed to be gathering rubber for the Casa Arana in the Middle Putumayo. The father makes his way to El Encanto and says the following:

> Well so: I followed the tracks of Lucianito [his son] toward the Putumayo. It was in Sibundoy that they told me that a pallid young man, in short pants, who couldn't have been more than twelve years old and whose only baggage was some clothes tied up in a rag, had gone downriver with some men. They didn't say who he was, nor where he came from, but his companions predicated vociferously that they were going in search of the rubber stations of Larrañaga, that heartless man from Pasto, partner of Arana and other Peruvians who, in the Amazon Basin, had enslaved thirty thousand Indians. . . .[298]

A carnival festival was in progress when Clemente arrived in Aranalandia. The revelers were quite drunk and included Indians from several tribes and whites from Colombia, Venezuela, Peru, and Brazil, as well as blacks from the Antilles. They clamored for alcohol, women, and such goods as cans of tuna fish. There was music everywhere. Then, suddenly, a Company overseer fired his Winchester to silence the crowd:

> "Rubber gatherers," he exclaimed, "you know of the munificence of the new owner. Mr. Arana has formed a company that is the proprietor of the rubber trees of La Chorrera and El Encanto. You must work, you must be submissive, you must obey! There is nothing left in the storehouse to give you. Those of you who were unable to get clothing will have to be patient. Those of you who are asking for women should be apprised that forty are coming on the next boats, listen well, forty, to be distributed from time to time among the workers who distinguish themselves. Besides there will soon depart an expedition to conquer the Andoque tribes and it

---

296. Eduardo Neale-Silva, *Horizonte humano: Vida de José Eustasio Rivera* (Madison: University of Wisconsin Press, 1960), 285. [*Horizonte*].
297. Ibid., 276–77.
298. José Eustasio Rivera, *La vorágine* (Caracas: Biblioteca Ayacucho, 1993), 115 [*La vorágine*].

is charged with collecting the women wherever they are found. So, give me your attention: any Indian who has a wife or daughter should present her to this headquarters so it can be determined what to do with her."[299]

The desperate father then met with Miguel Loayza to ask what it would take to free his son? Loayza looked Lucianito up in his account book, and noted that he owed two thousand soles to the Company. It would have to be paid off. At the same time, Loayza was making an exception reluctantly since, "We are not in the business of selling personnel. To the contrary: the enterprise is looking for people."[300] The novel then moves on to other action and protagonists[301] in other venues.

One might also mention (Rivera's precursor) the famous Brazilian novelist Euclides da Cunha (1866–1909) who toured the Brazilian Amazon in 1905 as part of a joint Brazilian-Peruvian Commission sent to the Purús River region to gather information prior to agreement of the two countries' common border. Da Cunha, author of the seminal Brazilian novel *Os sertões* (Rebellion in the Backlands), published a collection of essays including two about the *seringeiros* (or humble *caucheros*—the non-indigenous gatherers). The work was entitled *À margem da história* and was intended as prelude to his real work—an Amazonian novel to be called *Lost Paradise*. Unfortunately, he died prematurely in 1909 before he could write that work.[302]

By 1925, international shipping no longer ascended the Amazon to Iquitos, which meant that Peruvian rubber now had to be transboarded in Manaos after paying Brazilian duties. This, of course, raised the transportation costs to the producers. It also was an incentive for them to export their rubber illegally, the contraband diminishing the export taxes paid to the Peruvian government.

Arana was able to get the export tax on rubber leaving Iquitos removed altogether.[303] He also fought effectively to secure funds for com-

---

299. Ibid., 116.
300. Ibid., 118.
301. A prominent one is the Basque-surnamed Don Gabriel Zubieta. According to Rivera's biographer, the author met a man with this name in his travels and decided to use it, albeit not the personage, in his novel. Neale-Silva, *Horizonte*, 304.
302. Lucía Sá, Introduction. Voicing Brazilian Imperialism: Euclides da Cunha and the Amazon, in Euclides da Cunha, *The Amazon: Land without History* (Oxford: Oxford University Press, 2006).
303. Sensitive to the possible charge that this was self-serving, he noted that, since 1920, all of the rubber producers had ceased operations, given that the price of the product failed to cover their costs. He claimed to now be nothing more than one more of the Loreto's struggling rubber men. For years, he had been out of his former export and import

pletion of the city's civil hospital and its National College. However, Julio César was not simply a Loreto regionalist. He came to be viewed as a progressive legislator on the national stage—one concerned with improving the education and health systems. He was instrumental in the decision to provide incentives for establishment of air service between Lima and Iquitos.[304] On January 8, 1924, Lima's *La Crónica* newspaper editorialized:

> The representative from Loreto, Mister Arana, formulated yesterday in his Chamber two requests of transcendental importance. Reference first is to a reform of the demarcation of the national territory, and the other regards the necessity of implementing a general census [of the inhabitants] of the republic. With solid arguments, free from truculence and base rhetoric, the senator from Loreto made everyone see that the country once again needs to launch reform as quickly as possible. Besides, [Arana] requests that the geographical demarcation be effected by the Geographic Society.[305]

Meanwhile, Arana was far from convinced that the Putumayo might one day be Colombian. On August 31, 1923, he sent a letter to all of Peru's ministries, and particularly that of Foreign Affairs. It denounced the fact that the Marconi Wireless Company, charged with operating Peru's postal and telegraphic services, had listed El Encanto's radiotelegraphic station as a Peruvian installation within Colombian territory.[306] Then, on January 5, 1924, Arana presented a plan to the Peruvian Senate that would define unilaterally the country's boundary with Colombia in the Loreto, including creation of a province of Bajo Amazonas with a district whose eastern border would be the west shore of the Caquetá (thereby placing the Caraparaná and Igaraparaná entirely in Peru). It was debated and passed by the Senate's Commission of Territorial Demarcation, which also voted to solicit input from the Ministry of Foreign Affairs.[307]

Presumably, that was never forthcoming, since negotiations between the two countries were proceeding apace. President Leguía could not have been amused by the earlier Loreto rebellion, and he had reason to be concerned over international repercussions of an endless Loretan boundary

---

businesses. So, despite the accusations of his critics that he was being self-interested, he believed that he was advocating a measure that was critical to the general well being of the Loreto. British Parliament, *Report*, 361.
304. Ibid., 356.
305. Quoted in Ibid., 359.
306. Pinell, *Excursión*, 200.
307. Ibid., 202–204.

dispute, particularly after the Putumayo scandal. Then, too, there were the new realities of post-World-War-I international relations. While a battered France retained its African and Asian colonies, Germany's aggressive foreign policy would now be on hold until the 1930s. The European continent's economy was completely shattered by the conflict, and its casulaties numbered in the tens of millions. Great Britain and the United States were the relatively unscathed victors in World War I. So, the last thing that a president of Peru would court at this time was Anglo-American displeasure.

Indeed, between February 27, 1924 and April 29 of that year, Senator Egoaguirre ironically served as acting minister of foreign affairs during the absence of the incumbent, Alberto Salomón Osorio (who was embroiled in the bilateral negotiation with Colombia). Paralleling the peaceful discussions between the two countries, both tried to reinforce their physical presence in the Putumayo/Caquetá. Colombia sought to institute its own steamship service on the Putumayo, but effective river trade continued to be a Peruvian monopoly, its steamships making port regularly in Puerto Asís on both legal and contraband runs. In 1924, the appointed Colombian commissary for the region even drew up plans to expel Peruvians (and particularly the Casa Arana) from the Putumayo/Caquetá.[308]

By 1925, the Salomón-Lozano Treaty was being debated by Peruvians. Many in the Loreto regarded it to be a treasonous sell out of their interests. The agreement would prove to be the beginning of the end for Arana's Putumayo empire. It accorded Colombia the east bank of the Putumayo (including both El Encanto and La Chorrera, or about 60 percent of Arana's "holdings"). He immediately offered to sell his interest there to Bogotá for 2,000,000 pounds.[309] The Salomón-Lozano Treaty also gave Colombia a section along the Brazilian border that came to be known as the Leticia Trapezoid and guaranteed freedom of navigation on the Putumayo to both countries' vessels.

The terms of the proposed agreement would not become public until 1926, when they were released by Brazil, irritated as it was over creation of the Leticia Trapezoid without having been consulted.[310] Arana now published a ninety-three-page document directed to the Peruvian parliament and public denouncing the treaty. The work begins with the statement that he is providing a compilation of information that was being systematically withheld by the Peruvian government from the newspapers. In

---

308. Stanfield, *Red Rubber*, 198–99.
309. Ibid., 203.
310. Lagos, *Arana*, 360–61.

effect, according to Julio César, the Leguía government was trying to limit debate as it rushed the treaty through parliament for approval. The work is a compendium of papers, reports and former newspaper articles, as well as a detailed chronology of negotiations and treaties between Peru and its neighbors regarding common borders in the Putumayo/Caqueta, all demonstrating that Aranalandia belonged to Peru. Included are a number of protests of Salomón-Lozano signed by Loreto representatives in parliament and dozens of Iquitos residents.[311] It was vintage Carlos Rey de Castro, although I have been unable to determine his whereabouts (or even if he was still alive) at the time.

Roger Rumrrill claims to have seen a handwritten forty-four-page document supposedly authored by Julio César at about the same time, in which Arana notes that, at the urging of President Pardo, he spent about 1.7 million soles acquiring Colombian-owned properties to secure his Putumayo domain as Peruvian. He claimed to have constructed there 604 kilometers of rail lines and 175 kilometers of roads to link the Putumayo with its tributaries. He leant the vessels *Liberal* and *Cosmopolita* to the state in the interest of national defense. Between 1901 and 1920, he had shipped 103,960,000 Peruvian libras worth of product through Iquitos customs, paying export tariffs, as well as an additional duty of 200,000 libras on the goods he imported for his Putumayo enterprises.[312]

On December 20, 1927, the treaty came before the Peruvian Congress, and Arana vehemently opposed it:

> I vote against [it] because it provides no compensations, we are giving up the best of our Amazonian frontier, with the towns of Leticia, Loreto, Loretoyacu, Huata Yacu, Santa Sofía, Victoria, as well as La Chorrera and El Encanto, along with the wireless towers of Leticia and El Encanto, where there are commissaries, schools, defense installations, port toll stations, all without receiving any compensation.[313]

Nevertheless, it was approved by a vote of 107 to 7. Julio César and Julio Egoaguirre cast two of the handful of dissenting ones.

It should be noted that, after losing the vote, Arana, Egoaguirre, and two other Loretan representatives sent a letter to the mayor of Iquitos urging acceptance of the new reality. The law had to be obeyed, since noncompliance would produce disorder, "and Loretans through bitter expe-

---

311. Julio César Arana, *El protocol Salomón-Lozano, o el pacto de límites con Colombia: al Congreso Nacional* (Lima: La Tradición, 1927). For the petitions see pp.29–44.
312. Rumrrill, "El sueño," 8.
313. Quoted in Lagos, *Arana*, 362.

rience now know what that explosive word means for their permanent interests."[314]

Arana did, however, resettle a total of 6,719 Indians from Aranalandia to his other holdings (mainly near Pebas) between 1924 and 1930.[315] For his part, Father Pinnell moved the Huitotos in Güepí across the Putumayo River to Puerto Montclar on the Colombian side from whence they emigrated on their own volition to Salado Grande.[316]

Father Pinell made a second journey down the Putumayo to Manaos (from December 15, 1926, to May 31, 1927), accompanied by fellow Capuchin, Father Bartolomé Igualada. Things had changed immensely in Aranalandia since his 1918 visit. The two travelers passed through vast areas that were now entirely uninhabited. Diseases brought by whites had certainly had their morbid impact upon the indigenes, but vast numbers of them had either simply fled into the dense forests or abandoned the area altogether to escape the burden of rubber collection.[317] Others had been forcibly removed by the Casa Arana for relocation on its other holdings in anticipation of the pending Salomón-Lozano treaty that might transfer much of Aranalandia to Colombia.[318] Pinell was told by a veteran

---

314. Quoted in Bákula Patiño," Introducción," 91.
315. San Román, *Perfiles*, 192.
316. Restrepo López, *El Putumayo*, 21.
317. Camacho, *Holocausto*, 144–48. Pinell describes in vivid detail how, by his time, the whites located the increasingly furtive interfluvial tribes:

> Upon arriving at the mouth of a brook where it was suspected there might be Indians, four parties were organized; two of them were situated along the large river into which the brook discharged; one was sent downriver, a kilometer from the brook's mouth, and the other upriver the same distance. From there they were to enter the forest paralleling the brook's course. The other two parties were placed on its left and right banks, respectively, at a distance of 50 to 100 meters [from the brook]. The four parties would enter the forest in parallel fashion for at least a half day's trek. If in these penetrations no sign whatsover of some people was found, it was certain that no Indians were living there. All of these operations had the objective of overcoming the Indians' astuteness, as they take special care not to leave any sign in the mouths of the brooks where they live, if they do not wish to be found by anyone. But once they are within the forest, believing that no one will come looking for them there, they no longer take such precautions and leave sign everywhere, and even make paths to communicate among themselves, and trails for going hunting. (Pinell, *Excursión*, 156)

318. Ibid., 231. Then, too,

> The anti-patriotic and decolonizing program of exporting the Indians of the Putumayo to other regions was not an exclusively Peruvian undertaking; several Colombians who do not merit such a name also transported to Ecuador, Peru and Bolivia entire tribes, if not specifically from the Caraparaná and Igaraparaná, then from other rivers in the Putumayo region, attending only to their personal interests and ignoring the harm that such an inhumane practice did to their country. I personally saw in 1918, in several parts

cauchero that the region between Caucayá and Yubineto (on the Caquetá) was now depopulated. He mentioned a few Huitotos families from the Caraparaná and Güepí who, in order to shun contact with whites, were now living hidden in a remote region on the Ancusiyá.[319] In 1923, Alfredo White Uribe filed a report with the Colombian government that lamented the extreme depopulation from Puerto Asís down the Putumayo and along the neighboring Caquetá. Those who but twelve years earlier had witnessed a vibrant economy based upon rubber gathering and the activity of the river traders were simply shocked by the human wasteland.[320] In a 1924 publication, Capuchin Father Estanislao de Las Corts reported, "The Caquetá is like the Caguán . . . immense and empty . . . empty of people."[321] By then, it could be reported that the Indians were using vegetative contraception potions—preferring that their people simply die out rather than produce descendants to serve as future slaves.[322] It was impossible to calculate how many Indians remained in the Putumayo, but they certainly numbered no more than a few thousand.

Arana immediately instituted a new program to resettle "his" Indians within the territory remaining to him in the Putumayo. Many fled into the forests of the Caquetá and were pursued by Julio César's men. Others refused to move when ordered, even if the territory was now Colombian. There is evidence that one employee of the Casa Arana, Miguel Zumaeta, resettled two hundred Andoques closer to Iquitos.[323] There is also the report of the resettlement of sixty Muinanes Indian families from the Igaraparaná to the Cuemaní drainage, a tributary on the uncontested Colombian side of the Caquetá. The settlement was under the command of Julio Arana, brother of Julio César.[324] In any event, at the end of the day, 6,719 Indians (including 2,351 children) remained "in the claws of the Casa Arana."[325] Their continued forced labor, supplemented by debt peonage of non-Indians, allowed production of *balata* rubber from the Putumayo to remain relatively constant throughout the 1920s.[326]

---

of the Amazon, in Brazilian territory, some groups of Indians from the Caraparaná and Igaraparaná. (Ibid.)
319. Ibid., 79.
320. Cited in Gómez López, *Putumayo*, 31.
321. Fray Estanislao de Las Corts, "En busca de las tribus salvajes," *Revista de las Misiones* 20 (January), 1924, 384.
322. Stanfield, *Red Rubber*, 187–88.
323. Landaburu and Pineda, *Tradiciones*, 32.
324. Dominguez and Gómez, *Nación*, 186.
325. Lagos, *Arana*, 362–63.
326. Stanfield, *Red Rubber*, 198, 202.

However, there was the real problem that the era of South American rubber was essentially over, and the Casa Arana was simply not up to the challenge of effectively transforming its holdings into other forms of tropical production. Arana's real consolation seemed to be Article 9 of the treaty that guaranteed Colombia would respect the rights of any Peruvian within its territory. Notwithstanding that the measure was clearly designed for the Casa Arana, in 1930, Colombia simply expropriated Julio César's Putumayo holdings without any offer of compensation.[327]

On December 17, 1930, Leticia was handed over formally to the Colombians, and they immediately set about abrogating the terms of its status under the treaty. Astride the Amazonas River, rather than facilitating free navigation on the rivers of the Upper Amazon, the town became a Colombian chokepoint on Iquitos's river traffic. Its Peruvian residents were refused the right to leave and were forced to take out Colombian citizenship.[328]

That same year of 1930, Peruvian president Leguía was deposed by a military coup, imprisoned for a few months, and then died shortly thereafter in a military hospital.[329] During 1930–1931, Peru would have no fewer than six presidents, the last of whom was Luis Miguel Sánchez Cerro.

There would be yet one last spasm (indeed a dramatic one) of Loretan and Casa Arana resistance[330] to their mutual geopolitical fates. When it became evident that Peru's elected new leader, Sánchez Cerro, had no intention of repudiating the Salomón-Lozano Treaty, in Iquitos six persons formed a Patriotic Board. These included Julio César's son, Luis Arana

---

327. Lagos, *Arana*, 373.
328. Ibid., 372.
329. By this time Andrés A. Aramburú, the owner-editor of Lima's *La Opinión Nacional* newspaper, to whom Arana wrote his letter of gratitude for the editorial defense of the Casa Arana (1908), now owned the highly influential literary magazine *Mundial*. He editorialized that, given his advanced age and poor health, former president Leguía should be released from prison on humanitarian grounds. For his trouble, Aramburú was himself arrested and incarcerated. His own health broken, he was exiled to Chile, where he died in 1933.
330. Bákula Patiño hypothesizes that the Loreto's periodic flaunting of central authority stemmed from a history in which petitions for assistance from the remote Oriente were regularly ignored. Lima passed ill-conceived laws without consulting the locals and then sent out unprepared functionaries to implement them. The result was cynicism among Loretans regarding their relations with the Peruvian government. However, Bákula Patiño rejects the notion that this was ever true separatism; rather, the Loretans sought special treatment that accommodated their peculiar circumstances—in essence, federalism. Bákula Patiño, "Introducción," 107–108, 112–13.

Zumaeta,[331] and Pedro del Águila Hidalgo, husband of daughter Lily Arana Zumaeta. Julio César provided the Board with 44 carbines. On September 1, 1932, a force of 48 Loretans launched a surprise attack and defeated the 128 Colombians defending Leticia. The Peruvians occupied the town—demanding that the lands alienated under the Salomón-Lozano Treaty be returned to Peru. The Peruvian victory was then underscored by another successful attack on the town of Tarapacá in the Putumayo.

When 10,000 people in Iquitos gathered to celebrate, both Pedro de Águila Hidalgo and Luis Arana Zumaeta addressed the ecstatic throng. The Patriotic Board then informed Lima that it had reversed the treason of the Leguía years; Sánchez Cerro cabled back his congratulations to the Loretans for defending the national honor. This burst of patriotism spread throughout Peru.

Then, on September 17, 1932, Colombia tried to send its gunships anchored in the Putumayo to Leticia. The move was halted by the Peruvians, and, in reality, the two countries were once again on a war footing. Colombia needed to raise funds for the purchase of warplanes, but the world economy was now immersed in the Great Depression. There began a national campaign urging citizens to donate their jewelry. The president of the country and his spouse both gave their wedding bands. Four hundred kilos of gold were collected in this manner.

Colombia had the larger population and a relatively sounder economy, and its president was far more popular than was his Peruvian counterpart. Then, too, the Pan American Union refused to approve the Loretans' seizure of Leticia. Colombia cobbled together an air force but had to rely upon Germany to train the pilots. Still, Peru possessed certain logistical advantages. The war zone was far more accessible from Iquitos than from Bogotá, and Peru already had warplanes and gunboats stationed in the area.[332] Nevertheless, Colombia rejected any suggestion of arbitration.

Over the next several months, Peru retained Leticia while both countries continued the stare down and military build up. Colombia was already suffering casualties, since its recruits sent from the Andes to the Amazon were quite vulnerable to malaria and yellow fever. Colombia moved naval vessels to Manaos in anticipation of an eventual military assault in the Upper Amazon. U.S. Secretary of State Harry Stimson informed Lima of his country's disapproval of the seizure of Leticia and

---

331. He had been educated as a mining engineer at the Massachusetts Institute of Technology and had recently returned to Iquitos to enter business. He would have a successful career as a businessman, although on a much more modest scale than his genitor. Ibid., 364.
332. Ibid., 366–69.

suggested that it be turned over to Brazil pending the outcome of negotiations between the two belligerents scheduled to be held in that country. Peru demurred, while claiming that the invasion was prompted by Colombian abuses, and that its military preparations were simply in response to the disproportionate Colombian ones. Throughout the mobilization, Julio César resided in Lima, albeit maintaining constant telegraphic communication with Luis.

For the Aranas, war with Colombia was a godsend. Should Peru prevail, the Casa Arana's holdings in the Putumayo would be restored. Should there be a stalemate, at the very least, Arana stood a reasonable chance of being indemnified in any international arbitration or direct negotiations between the belligerents. However, for the indegenes, the conflict was devastating:

> The war brought more suffering to the Indians. Both sides impressed Indians as porters. . . . The Colombians attacked and destroyed the Peruvian military base at Güepi. German pilots flying Colombian planes killed many of the defenders. Sánchez Cerro ordered full mobilization of the country's forces. On April 30, while en route to a public revue in a Lima stadium of thirty thousand troops, the president was assassinated by a political opponent. It was not only the end of Sánchez Cerro but of the war with Colombia as well, not to mention Julio César Arana.[333]

The next president of Peru was (Basque-surnamed) General Oscar R. Benavides Larrea, of La Pedrera fame (and the former president of the country between 1914 and 1915). He assumed office in 1933 and made normalizing Peru's relations with its neighbor (and the international community) a priority. Peace was brokered by the League of Nations, and in May 1934, the Salomón-Lozano treaty was reinstated. The east bank of the Putumayo and the Leticia Trapezoid were returned to Colombia and remain its territory to the present day. In 1991, the Departamento de Amazonas was created and, with its capital in the city of Leticia, became one of Colombia's thrity-two internal divisions. This department encompasses all of the Colombian portion of what we have denominated as Aranalandia.[334]

---

333. Stanfield, *Red Rubber*, 204.
334. There is also now a Departamento de Caquetá, with its capital in Florencia, and a Departamento de Putumayo that includes all of the Alto Putumayo, and with its capital in Mocoa. The Colombians learned well from their former Spanish rulers how to confound and redraw both civil and ecclesiastical borders. In 1845, the Congress of Nueva Granada had renamed the territories known as Andaquí and Mocoa as the Territory of Caquetá. In 1887, the governor of Cauca created the Province of Caquetá. With the creation, in 1904, of the Department

There is the interesting report by (Basque-surnamed) Carlos Uribe Gaviria,[335] Colombia's minister of defense during the 1932 war with Peru. In 1933, he was dispatched by Bogotá to review the progress of the postwar initiative. The Colombians had fortified Tarapacá with a labor force of about eighty men, all of whom were Brazilians. One of their supply ships was named the *Río Putumayo*.[336] Several Huitotos had been transported to Bogotá for "care and education."[337]

His men were anxious to be sent home (they were reservists). Effort was being made throughout the region to drain swamps and treat standing water with oil to cut down on the mosquitos. The military mission had initiated considerable agriculture—particularly pasturage, bananas, and yucca—and was trying to recruit Indians for road construction and manual labor in general.[338] The Capuchins had already established a new mission at El Encanto, obviously intended as the headquarters for further proselytizing in the former Aranalandia. It was in 1934 that the Colombians established a naval base on the Putumayo, Puerto Leguízamo, the upstream limit of Huitocia. That same year the Navesur shipping company based three vessels in Puerto Leguízamo and began shipping merchandise on the Putumayo as far upriver as Puerto Asís and on the Caquetá between Mangalpa

---

of Nariño, the Putumayo Commissariat was made one of seven of the new department's provinces. A year later, the national government created two intendencies—the Alto Caquetá and the Putumayo. But the following year they were again incorporated into the Department of Nariño. Six years later, the Colombian government divided the Territory of Caquetá into the Commissariat of the Putumayo, with its capital in Mocoa, and the Commissariat of Caquetá, headquartered in Florencia. Meanwhile, in the ecclesiastical realm, in 1904, the Apostolic Prefecture of Caquetá was established that encompassed only a part of the Department of Nariño—the eastern Middle Putumayo was excluded. In 1916, there was a treaty between Colombia and Ecuador, after which part of the Apostolic Prefecture of Caquetá fell within the latter country. In 1924, the new Apostolic Prefecture of San Martín de Sucumbios incorporated the formerly "dismembered" part of the Apostolic Prefecture of Caquetá. The missions of the new prefecture were under the aegis of Monfortian missionaries—with permission from the Capuchin prefect at Sibundoy to celebrate the sacraments of matrimony and confirmation. In 1932, the prefect of the Apostolic Vicariate of Caquetá supported (in the interest of evangelizing efficiency) transfer of the eastern part of the Middle Putumayo/Caquetá (territory around the Apaporis and Vaupés rivers) to the San Martín de Sucumbios Apostolic Prefecture. Kuan Bahamón, "La Misión," 77–79. Whew!

335. Carlos Uribe Gaviria, *La verdad sobre la Guerra* (Bogotá: Editorial Cromos, 1936), vol. 2, 261–85 [*La verdad*]. He was a military reservist officer and served in the Colombian parliament and subsequently as Colombian consul in San Francisco, California. He was the son of General Rafael Uribe Uribe.
336. Ibid., 268.
337. Ibid., 274.
338. Ibid., 281.

and La Tagua.[339] So, the Colombians now controlled the Leticia Trapezoid and had a strong military presence at the upstream boundary of the Middle Putumayo/Caquetá.

During the 1930s, Julio César was forced to sell his assets in Iquitos to cover his living expenses in Lima. There is mention that he mounted, briefly, an unsuccessful placer gold-mining operation on an affluent of the Marañón River. As for the destiny of his remaining holdings in the Putumayo, there is some confusion. Victor Israel, an Iquitos Maltese Jew and Arana's former business associate, certainly ended up with them. In 1939, Julio César sold any remaining claims that he had in the Colombian Putumayo (3,553,600 hectares) to Israel for 800,000 soles, of which 300,000 (or about 40,000 dollars) were advanced at the time.[340] Israel, in turn, sold them to the Colombian government for a down payment, and then, in 1964, there was final settlement of 160,000 dollars. One version claims that Arana's descendants received some of that money. However, there is another that suggests that Arana was defrauded by Israel after he signed over a power of attorney, believing that his representative was negotiating a sale to North American interests. Ostensibly, Israel used it to sell Arana's claims to the Colombian government instead. It is said that a furious Luis Arana went to Israel's home to confront him.[341]

Eleonora and Julio César had five children, three girls—Lily, Angélica, and Alicia—and two boys—Julio César and Luis. They all had privileged childhoods with years of governesses and private tutors in France, Switzerland, and England. In Iquitos, they sat atop the social pyramid. Then came the particularly painful divisive impact upon the family of son Luis's marriage to a servant's daughter. The husband of sister Lily, Luis's fellow Loretan patriot Pedro de Águila Hidalgo, was from a socially prominent family. Lily enlisted her parents in the attempt to convince Luis not to marry beneath his station. It was to no avail and simply resulted in an irreparable rupture that would cut Julio César and Eleonora off from their son and eventually their only grandson.

Pedro de Águila Hidalgo won a seat in the Peruvian Senate in the 1950s, and the childless couple moved to Lima. Lily, however, was so appalled by her parents' poverty that she maintained her distance from them. Angélica, a bookish spinster, lived with her parents until she died in middle age.

Luis became a prominent businessman in Iquitos. He owned an import-export company called the Suramérica, as well as consider-

---

339. Restrepo López, *El Putumayo*, 28.
340. Lagos, *Arana*, 377–78.
341. Ibid., 378.

able real estate. However, his only son contracted poliomyelitis at age eight and was constrained to a wheel chair for life. His mother moved with him to Lima for medical reasons, and Luis soldiered on in Iquitos alone, when not visiting his wife and invalid child in the national capital. He served no fewer than nine terms as the city's mayor.

But then, in 1968, Luis became the object of a negative newspaper campaign for having given the unpopular order to fell several of the city's popular mango trees. He was denounced for his "aboricide." There was also an investigation into his handling of public funds that could even have led to a criminal prosecution. Whatever the combination of reasons, Luis committed suicide.[342] Initially, his family was able to maintain a good life style in Lima, but over time the Iquitos properties had to be sold off one by one.[343]

By the time he turned seventy-five, Julio César Arana and Eleonora were nearly penniless. They were constrained to move, along with their daughter Angélica, into a hovel in the miserable Lima suburb of Magdalena del Mar. His health, never good, failed him completely, and he endured constant physical pain. Eleonora, as always, cared for her man, surviving him by three years after his death at eighty-eight years of age on September 7, 1952. Julio César Arana was buried in a pauper's grave.[344]

A couple of modest obituaries appeared in Lima newspapers, recalling Arana's former glory as both the king of rubber and Peruvian parliamentarian. Nevertheless, Julio César's biographer Ovidio Lagos notes that, while Peru's other great rubber king, Carlos Fermín Fitzcarrald, lives on in the Peruvian popular imagination as a national hero, and is memorialized in many fashions (including the renaming of his province of birth for him), Julio César Arana

---

342. Ibid., 377, 387–88.
343. Eventually, mother and son were living as recluses in their own abject poverty. On September 27, 2002, Lima's Channel 5 television aired a documentary on their incredibly unsanitary home. She died shortly thereafter, and Julio César's biographer, Ovidio Lagos, interviewed the grandson Luis Arana Ramírez for his book. Arana Ramírez noted that his aunt, Alicia, was buried in Lima—but gave no other details regarding that nondescript descendant of Julio César. He then observed, before dismissing Lagos abruptly, "With me the Aranas are finished. I am the last of this species" (Ibid., 389).
344. It seems that the survivor of our tale with the longest longevity was Miguel Loayza. Rumrrill reports interviewing him about 1960, Loayza was ninety and living in the Punchana district of Iquitos. At first, he was very suspicious, but then opened up and recounted frankly his recollections of his time in the Putumayo. Rumrrill was astounded by Loayza's lack of remorse and attitude that the inhumane treatment of the Indians was to have been expected as simply "something natural" (Rumrrill, "El sueño," 3).

... has not even a street named after him let alone a province. It is as if he has been erased from the face of the earth and nobody, not in Iquitos, nor in Lima, nor in the rest of Peru admits to having any degree of kinship with him, no matter how remote, even when this is the case.[345]

---

345. Lagos, *Arana*, 59. *Au contraire*. There is a street named after Julio César in his birthplace, La Rioja, http://historiariojaperu.blogspot.com/2012/02/julio-c-arana-biografias-riojanos.html. Apparently, a principal street in the center of Iquitos called "Nauta" at present was once "Julio César Arana." David Crazy Rainbow, "The True Cost of Rubber," *Iquitos Times*, July 12, 2013, http://www.iquitostimes.com/julio.htm.

Part 5

Configuring New Indigenist Policy

It has been slightly more than a century since the London Hearings of the Select Committee exposed for posterity our scandalous tale—or, more accurately, its many versions. While we have considered subsequent developments in the Putumayo/Caquetá until the 1930s, as well as the personal destinies of many of our protagonists and the remaining years in the lives of Julio César and Eleonora until their deaths in the 1950s, I would be derelict if I failed to consider the fates of the main victims—the Indians. The past century has witnessed many developments in the circumstances of South American indigenes, including the protagonists of this narrative (or their surviving remnants). For our purposes, the most relevant indigenist policies are those within the national polities of Peru and Colombia.

We have noted that, by the latter part of the nineteenth century and early twentieth, the Putumayo/Caquetá had become for Colombians a destination for both refugees and exiles. It is clear that many of the original white and *mestizo* settlers of Aranalandia were displaced from elsewhere by Colombia's War of a Thousand Days and other civil conflicts. The Colombian government, particularly under Reyes, established prison camps along the Caquetá River to warehouse its political opponents. Such governmental incursions, however, were preceded by the phalanx of fortune-seeking *caucheros*, both small scale and large, as well as their sponsoring suppliers. Attempts to missionize the area had failed, so there was no countervailing religious brake upon secular greed. In short, Huitocia was beyond the pale, a no-man's land viewed from Lima and Bogotá, not to mention a nebulous borderland where the limits of their respective national hegemony had yet to be agreed and defined. Nevertheless, it was a region rich in rubber and, by Amazonian standards, densely populated by indigenes who might be enticed and/or coerced to gather it. For a calculating Julio César Arana, the Middle Putumayo/Caquetá presented an opportunity of the perfect-storm variety. To the extent that there was indigenist policy in his time, it was more in the form of criticism and denunciation by a few governmental officials and missionaries of the abusive treatment of the Indians of Huitocia; yet neither the two states nor the Catholic Church were able to impose effective oversight.

Since the mid-twentieth century, the evolution of indigenist policy in both countries has been nothing short of extraordinary—in addition to being remarkably similar. In both, the surviving Indians themselves have undertaken considerable initiative in organizing themselves and advocating for their rights, particularly with respect to territorial claims. These grass-roots associations have learned the advantage of strength in numbers and have therefore created regional, national and even international indigenous rights' coalitions. In this regard, they have been aided by pro-indigenist missionaries, governmental officials and academics (particularly, though not exclusively, my fellow anthropologists). Several international NGOs, not to mention the United Nations (1945–) and the Organization of American States (1948–), now presume to monitor the behavior of national governments regarding their indigenous peoples. The Catholic Church and individual countries in the developed world (notably Denmark, Canada, and the Netherlands) weigh in as well. There is also the world-wide environmentalist movement, whose concern with climate change focuses in significant part upon the preservation of the planet's remaining forests (mainly in the tropical areas that were/ are indigenous territories). In short, the indigenes in this account have moved from the margins of history toward its center stage. There still remains, however, the question of what happens next—both because of them and to them—as they stand blinking in the spotlight.

Today, both Peru and Colombia are democracies with liberal (progressive) constitutions that confer, in principle, equal rights on all citizens. Those of ethnic minorities and indigenous peoples are accorded specific mention. Both guarantee freedom of religion and political expression, near universal enfranchisement irrespective of gender, and the workers' rights to organize and strike. Both have evolved from constitutional arrangements in which little of the foregoing was guaranteed. Both have governments with three independent branches—the Executive, the Legislative, and the Judicial. Both current consitutions are quite recent (Colombia's 1991 and Peru's 1993) and were promulgated in the midst of violence on a national scale that bordered upon civil war among the state security forces, leftist rebels,[1] and ultra-rightist paramilitaries. Both constitutions prohibit members of their armed forces from voting or intervening in the political process—antidote to the legacy of the periodic military coups punctuating their common histories.

In their foreign affairs, with the occasional hiccup, both countries have been allied to the United States. Their constitutions echo the U.S. Constitution in most regards. Their social and economic legislation derives to a considerable degree from Roosevelt's New Deal and Johnson's Great Society. Both countries sided with the Allies in World War II and have supported the United Nations and Organization of American States from their beginnings. Both signed bilateral free trade agreements with the united states; neither has entered as a full-fledged member of the competing South American entity, MERCOSUR, though both are now associates of it.[2]

There are certain historical differences between the two countries of relevance to this account. Peru has been somewhat more secular than Colombia. In the latter, both the Catholic Church and its missionary orders have enjoyed greater influence than in the former throughout most of the twentieth century. It might also be noted that Peru has elected two presidents of partial Quechuan descent—Alejandro Toledo Manrique (2001–2006) and Ollanta Moíses Humala Tasso (2011–2016)—the former Latin America's first indigene ruler in all of its five-hundred-year history. However, as we shall see, this did not meet the indigenist expectations that it raised. Colombia has yet to elect an indigenous leader. Finally, since most of our study area ended up in Colombia, that country's indigenist policy is of greater relevance than is its Peruvian counterpart. I will therefore conclude with detailed consideration of it after discussing Peruvian developments.

---

1. The two most notorious groups were the militants of the Sendero Luminoso (Shining Path) movement in Peru and those of FARC, or Fuerzas Armadas Revolucionarias de Colombia (Revolutionary Armed Forces of Colombia). At their height both movements had activists numbering in the many thousands.

2. Established in 1988, when presidents Raúl Alfonsín of Argentina and José Sarney of Brazil accorded the Argentina-Brazil Integration and Economics Cooperation Program that was then formalized as a free trade zone and customs union that included Uruguay and Paraguay as well under the Treaty of Asunción (1991). MERCOSUR is a counterweight to U.S. policy in South America that favors bilateral agreements between the United States and individual countries. Argentina, Brazil, Paraguay, Uruguay, and Venezuela are the member countries; Chile, Peru, Bolivia, Ecuador, and Colombia are associates. The latter can conclude free trade agreements within MERCOSUR, but are not a part of the customs union.

Peru

In 1928, the leftist (Basque-surnamed) Peruvian journalist and social philosopher, José Carlos Mariátegui, published his collected essays on Peruvian social reality in which he described the circumstances of the indigenes as follows:

> While the Viceroyalty was a medieval and foreign regime, the Republic is formally a Peruvian and liberal one. Consequently, the Republic has responsibilities that the Viceroyalty did not have. Raising the condition of the Indian corresponded to the Republic. Nevertheless, contrary to this duty, the Republic has impoverished the Indian, aggravated his lot in life and exacerbated his misery. The Republic has supposed for the Indians the rise of a new dominant class that has systematically expropriated their territories. In a traditional race with agrarian souls, this displacement has constituted a cause for material and moral dissolution.[3]

Mariátegui was a founding figure of the Peruvian Communist Party and he perceived in indigene society an expression of primitive communism (as opposed to capitalistic materialism):

> The Indian, in spite of one hundred years of republican legislation, has not become an individualist. And this is not because he resists progress, as is claimed by his detractors. Rather, it is because individualism under a feudal system does not find the necessary conditions to gain strength and develop. On the other hand, communism has continued to be the Indian's only defense. Individualism cannot flourish or even exist effectively outside a system of free competition. And the Indian has never felt less free than when he has felt alone. Therefore, in Indian villages where families are grouped together that have lost the bonds of their ancestral heritage and community work, hardy and stubborn habits of cooperation and solidarity still survive that are the empirical expressions of a communist spirit. The "community" is the instrument of this spirit. When expropriation and redistribution seem about to liquidate the "community," indigenous socialism always finds a way to reject, resist, or evade the incursion.[4]

Peruvian President Benavides (1933–1939) resolved the border conflict with Colombia and embraced universal democracy for his citizenry, as

---

3. José Carlos Mariátegui, *Los siete ensayos de la interpretación de la realidad peruana* (Lima: Minerva, 1928), 42–43.
4. Ibid., 57–58.

well as respect for the rights of indigenous populations. Then, during the first of his two terms as president, Fernando Belaúnde Terry (1963–1968; 1980–1985) recognized the legal status of hundreds of Indian tribes. Nevertheless, Peruvian anthropologist Alberto Chirif underscores the checkered history of change and continuity in Peru's treatment of its indigenes. In the 1960s, for instance, the Peruvian Air Force bombed the Matsés tribe at the behest of the mayor of the nearby town of Requena who was frustrated when they opposed his initiative to exploit their timber. The bombing was justified on the grounds that the Indians were against progress and a threat to civilization.[5] Writing as recently as 2009, Chirif stated, "During the century that has transpired since the fall of rubber until now, in the imaginary of the dominant society, the indigenes have continued to be the same dangerous beings that the *caucheros* sought to civilize . . . "[6] Furthermore, the trafficking in Indians as virtual slaves still obtained through the system of *enganche*, or indenture, with the buying and selling of the contracts.[7]

In 1968, the anthropologist Stefano Varese published his work *La sal de los cerros* about a salt-mining operation's abuses of the Asháninkas. It imparted an historical and contemporary overview of the tribe, as well as of their mythology and world view. It became an instant classic, and Varese was recruited into the Velasco government (1968–1975) to work on Indian affairs. In 1969, some indigenes held the Congreso Amuesha, the first Peruvian one and only the second in all of South America designed to agitate for Indian rights. It was, in part, a reaction against the Peruvian government's policy of relocating the surplus population in the highlands (largely Quechua- and Aymara-speaking indigenes) in the "empty" rainforest.[8]

There was educational reform instituted in 1972 and designed to provide bilingual education to all indigenous Peruvians (nearly half the popu-

---

5. Alberto Chirif, "Cien años después del caucho: Cambios y permanencias en las relaciones con los pueblos indígenas," in Alberto Chirif (ed.), *Cien años después del caucho: Cambios y permanencias en las relaciones con los pueblos indígenas* (Lima: Tarea Asociación Gráfica Educativa, 2009), 212.
6. Ibid., 213.
7. Roger Rumrrill, a noted Indian writer, confirms this. He contends that, in the 1970s, Indians continued to be sold at public auction in Atalaya on the Tambo River. He also witnessed, in 2010, the semi-slavery of Indians in the Putumayo, Madre de Dios, Yurúa, and Purús regions! Rumrrill, "El sueño," 3.
8. María Elena García and José Antonio Lucero, "*Un País Sin Indígenas?* Rethinking Indigenous Politics in Peru," in Nancy Grey Postero and Leon Zamosc (eds.), *The Struggle for Indigenous Rights in Latin America* (Brighton and Portland: Sussex Academic Press, 2004), 166.

lace). At the same time, in 1975, Quechua was elevated to the status of Peru's co-official language. Peru thereby became the first Latin American country to give formal recognition to an indigenous language, although in practice it was seldom employed in official administrative matters. While the Velasco administration in a sense empowered indigenes, it also undermined their identity by lumping them with rural *mestizos* as a unified *campesino*, or rural, underclass. Ironically, the president launched his program by declaring June 24, 1969, to be the country's "Day of the Indian" while prohibiting future use of the term *indio* in national discourse.[9] Thus was born the notion of the indigenous *campesino*.

It was also about this time that pro-indigene elements within the Catholic Church weighed in. In 1972 Monsignor Gabino Peral de la Torre O.S.A., the Vicar Apostlic of Iquitos, founded CETA (Centro de Estudios Teológicos de la Amazonía or Theological Center for Amazonian Studies). Its purpose was ostensibly to refine missionary efforts in the Oriente, but it quickly became a focal point for indigenous studies and provided an overarching shield for secular NGOs interested in indigenous affairs. It sponsored lectures and courses on Indian subjects and compiled a magnificent library collection (La Biblioteca Amazónica) of twenty-four thousand volumes, easily the world's largest collection of material on Amazonia.[10] Then, in 1974, nine bishops in the Peruvian Oriente founded CAAAP (Centro Amazónico de Antropología y Aplicación Práctica or the Amazonian Center for Anthropology and Practical Application) to service marginalized indigenous populations while championing their causes, particularly economic development that does not endanger traditional ways of life. It publishes a scientific journal called *Amazonía Peruana* and has assembled its own library in Lima of five thousand items on the region. Both CETA and CAAAP have active publishing programs and have produced the majority of works on Peruvian Indians, often in cooperation with indigenous associations and international NGOs like IWGIA (International Work Group for Indigenous Affairs, funded by the Danish government).

Under Velasco there was legislation that recognized Amazonian indigene territorial claims and, in 1974, a law was passed that allowed for the legal recognition of such indigenous/*campesino* communities, while ac-

---

9. Ibid., 160, 162.
10. CETA continues to function to this day and in 2007 was awarded in a ceremony the prestigious Bartolomé de las Casas prize by Spain's Prince of Asturias (that country's current monarch) in recognition of the Center's outstanding record in Indian affairs. One might note, ironically, that CETA might be viewed as the Augustinians' revenge for their maltreatment a century earlier by Loretans, and Julio César Arana in particular.

cording them considerable self-governance. This quickly led to formation of local and regional indigenous associations. By 1978, many were interacting and, in 1980, they formally constituted their overarching organization, AIDESEP (Asociación Interétnica de Desarrollo de la Selva Peruana or Interethnic Association for the Development of the Peruvian Rainforest).[11]

In 1975, Velasco had passed a law giving Quechua equal status with Spanish as a Peruvian national language. Peru thereby became the first Latin American country to officialize and indigenous language. However, it was never put into affect and, indeed, the country's ingrained anti-indigene racial prejudice was one of the factors in overthrowing the Velasco government that year. Nevertheless, a new national constitution, promulgated in 1979, stipulated that Quechua and Aymara (with Spanish) would become co-official languages; but only where their speakers were in the majority. In the event, the enabling legislation was never passed.

In short, by this time there was change in the air (if not necessarily on the ground). Velasco had, in effect, both furthered and undermined indigenous rights simultaneously. Indigenous languages and territorial claims were to be respected, while the very category of *indio* was abolished in favor of *campesino*, thereby creating tension between ethnicity and class identity that continues within Peruvian domestic policy to the present day. It would also exacerbate the distinction between indigenous highlanders (Quechuas and Aymaras) and lowlanders.

In 1984, AIDESEP hosted in Lima the First Congress of Indigenous Organizations of the Amazon Basin, resulting in the founding of COICA (Coordinadora de Organizaciones Indígenas de la Cuenca Amazónica or Coordinator of Indigenous Organizations of the Amazon Basin) that included the key national indigenous rights organizations of Colombia, Ecuador, Bolivia, and Brazil.[12] Then, in 1987, there was a meeting of 350 indigenous delegates held in Lima at which they founded CONAP (*Confederación de Nacionalidades Amazónicas del Perú*—The Confederation of Amazonian Nationalities of Peru).[13] Its purpose was to coordinate revindication of indigenous rights with protection of the natural environment and inclusion of indigenous concerns and priorities in the Peruvian national debate concerning the country's economic development and social and political organization. The two leaders of CONAP had

---

11. Ibid., 214. By 2014, it represented 57 Peruvian Indian federations and territorial organizations from no fewer than 1,350 communities with 350,000 inhabitants and 16 different language families.
12. Ibid., 212.
13. Note the rhetorical employment of "nationalities" to underscore the claim to political autonomy, that is, control over one's own affairs.

attended classes on leadership in the Panama-based ICI (Instituto Cooperativo Interamericano or the Interamerican Institute for Cooperation).[14] There was immediate competition and bad blood between AIDESEP and CONAP, although they began eventually to coordinate their activities and agendas.[15]

In 1987, AIDESEP was instrumental in the creation of FECONACO (Federación de Comunidades Nativas del Corrientes or Federation of the Native Communities of the Corrientes), based near Pebas. The Field Museum of Natural History in Chicago teamed up with FECONACO and IBC (Instituto del Bien Común or Institute of Common Good) to create three interfluvial nature reserves in the Oriente, one of which included the Peruvian shore of the Middle Putumayo. Their purpose was to protect forest resources—guaranteeing their Indian inhabitants full control over them. Paralleling these developments, there was a movement among young Huitoto and Bora artists and intellectuals designed to salvage and express their cultural legacies while configuring their futures.[16]

Nevertheless, in 1987, famed novelist (and future presidential candidate) Mario Vargas Llosa published his book *El hablador* in which the narrative verbalizes the national misgivings regarding Peru's indigenist policy:

> Should sixteen million Peruvians renounce the natural resources of three-quarters of their national territory so that seventy or eighty thousand Indians could quietly go on shooting at each other with bows and arrows, shrinking heads and worshipping boa constrictors? Should we forgo the agricultural, cattle-raising, and commercial potential of the region so that the world's ethnologists could enjoy studying at first hand kinship ties, potlatches, the rites of puberty, marriage, and death that these human oddities have been practicing, virtually unchanged, for hundreds of years? ... Hadn't Marx said that progress would come dripping blood? Sad though it was, it had to be acceptecd. We had no alternative. If the price to be paid for development and industrialization for the sixteen million Peruvians meant that these few thousand naked Indians would have to cut their hair, wash off their tattoos, and become mestizos—or, to use the ethnologists' most detested word, become acculturated—well, there was no way round it.[17]

---

14. Founded by the Canadian Roman Catholic Scarboro Mission Society as a non-denominational training ground for future Latin American leaders.
15. San Román, *Perfiles*, 263.
16. Ibid., 211–12.
17. Mario Vargas Llosa, *El hablador* (Barcelona: Seix Barral, 1987). The quote is taken

Meanwhile, in 1989 the International Labour Organization issued a "convention" on the rights of the planet's indigenous peoples to retain their languages and cultures, known as ILO Convention 169. Between 1989 and 1993, the IWGIA managed to free hundreds of enslaved and indentured Peruvian Indians (particularly in the Upper Ucayali) and assisted them to become recognized communities with their own territories and land titles.[18] By this time, COICA was aligned with Oxfam, the planet's best-known overarching NGO committed to the eradication of human poverty. By 1991, Oxfam published a work in Peru that was practically a manual of how COICA should go about designing and realizing its many projects.[19] The United Nations became increasingly concerned about the cultural ethnocide and genocide of indigenous peoples. It declared 1993 to be the Year of Indigenous People, followed by the International Decade of the World's Indigenous People (1995–2005).

On December 30, 1992, a *Frente Cívico de Loreto* (Loretan Civic Front) was founded in Iquitos. A few days later two hundred thousand inhabitants throughout the Loreto demonstrated for restoration of their semi-autonomous government.[20] The platform of the Loretan Civic Front, issued April 25, 1993, was a call to all Peruvians to unite while respecting one another's regional differences and rights, in a campaign to decentralize political power in the nation. Articles 4 and 5 were of particular relevance to this study:

> 4. It is necessary to respect, within the interior of each of the regions, the indivisible unity of its distinct identities, above all the indigenes who form a substantial part of the national society and who guard its most genuine legacy from time immemorial.
>
> 5. The ecosystems of all of the regions should be respected, particularly in Amazonia, as it represents the greatest biological diversity of any world area, a patrimony that should be guaranteed

---

from the English translation—Mario Vargas Llosa, *The Storyteller* (New York: Farrar, Straus, Giroux, 1989), 21–22.
18. Ibid., 215.
19. Alberto Chirif Tirado, Pedro García Hierro, and Richard Chase Smith, *El indígena y su territorio son uno solo: Estrategias para la defensa de los pueblos y territorios indígenas en la Cuenca Amazónica* (Lima: Oxfam América, 1991). Nor were these many exercises without their internal frictions and schisms. See Alfredo García Altamirano, "FENAMAD 20 años después: apuntes sobre el movimiento indígena amazónica en Madre de Dios," in Beatriz Huertas Castillo and Alfredo García Altamirano (eds.), *Los pueblos indígenas de Madre de Dios: historia, etnografía y conyuntura* (Lima: Grupo Internacional de Trabajo sobre Asuntos Indígenas—IWGIA, 2003).
20. San Román, *Perfiles*, 265.

within any negotiation regarding sustainable development of the country and as a legacy for future generations.[21]

Nevertheless, in Peru there remained many obstacles to full indigenous rights. There was the impact of the *Sendero Luminoso* movement of the 1980s. Its revolutionaries took refuge in some of the remote territories of forest Indians—in some cases setting up narcotics operations by growing drugs on Indian *chagras*, as well as forcibly conscripting the local inhabitants into the guerrillas' ranks. This activity, of course, placed tribal areas in the crosshairs of the Peruvian military. Many indigenes fled their ancestral homelands. There is the sad case of the Asháninkas, who, upon their return after the Shining Path rebellion was put down, found that the Peruvian government refused to recognize their earlier land claims and instead initiated a program to resettle Andean *campesinos* in their tribal area.[22]

The nation's new constitution, promulgated in 1993 under the ostensibly populist (but de facto neo-liberal) President Fujimoro, promised to protect the cultural identity and communal lands of indigenous peoples. The Indians were accorded ownership of their agricultural lands, but not their subsoils or forests. Furthermore, the Constitution reserved considerable territory for economic development. If, in 1992, mining companies (largely foreign and in the main Canadian) had claims on approximately 4 million hectares, this would increase within a few years to 25 million hectares.

The organizer of the protests against the usurpation (and pollution) of this activity, Miguel Palacín, was harassed and falsely accused of criminal activity and went into hiding. Thanks to the Internet, Canada's First Nations learned of the matter and, having had their own complaints against the same Canadian mining interests, championed Palacín's case. Given such international clamor, Peru investigated the matter and Palacín was exonerated. Given his narrow escape from incarceration, in 1998, Palacín decided to organize CONAMACI (National Coordinator of Communities Adversely Affected by Mining). It was largely, albeit not exclusively, a highland Quechuan entity and funded entirely from its outset by Oxfam America.[23]

In 1997, there was a Cusco Conference attended by representatives of many NGOs and Peruvian indigenous groups. It brought the Amazonian AIDESEP and CONAP Indian associations together with the

---

21. Ibid., 267.
22. Chirif, "Cien años," 205–206.
23. García and Lucero, "Un País," 177–79.

pan-Andean CAN[24] and the CCP (Confederación Campesina de Perú or Confederation of Rural Workers of Peru). Together they launched what has come to be known as COPPID (The Permanent Coordinator of the Indigenous People of Peru). Its purpose was to provide "space" in which the NGOs and indigenes could discuss and organize initiatives regarding indigenous rights. In 1998 it held its first meeting. With the assistance of a Canadian NGO, the Institute of Democratic Rights, this was followed up with a series of workshops throughout Peru. COPPID then received the formal support of Oxfam and the Japan Development Agency. It publishes *Servindi* (*Servicio de Información Indígena*), which provides both Peruvian and international news regarding the world's indigenous peoples.[25] Clearly, the international concerns and activism of governments, NGOs, and the United Nations forced the Peruvian government to concentrate to a greater degree on its indigenous affairs than it would have without such external pressure.[26]

In February 2001, interim chief executive Paniagua received personally delegates from three Amazonian tribes and promised that their concerns regarding the status of land claims and the devastation of their area by the recent civil violence would be addressed by the nation. The subsequent accession to the presidency of Peru of a Quechuan (albeit Stanford-educated and a former employee of the World Bank), Alejandro Toledo Manrique, was welcomed by many indigenes. In July 2001, he expanded his inauguration to Machu Picchu, where he addressed the country in Spanish and Quechua while hosting the presidents of Bolivia, Ecuador, and Colombia. The four chief executives signed a Declaration of Machu Picchu committing themselves to the defence of indigenous rights. In November, President Toledo created CONAPA (Comisión Nacional de Pueblos Andinos, Amazónicos y Afroperuanos or National Commission of Andean, Amazonian and Afro-Peruvian Peoples), and appointed his Belgian, Quechua-speaking wife, Eliane Karp, to direct it. In 2002, he required that, in regional elections, 15 percent of the candidates be indigenes wherever there was a significant Indian presence. He also lamented the wealth disparity between indigenes and other Peruvians, and he pronounced that the Indians needed to be further incorporated into the mainstream economy and society for their own well-being and that of the

---

24. CAN refers to the *Comunidad Andina* (Andean Community), founded in 1969 by Bolivia, Colombia, Ecuador, and Peru to encourage regional industrial, agricultural, and trade cooperation.
25. Nancy Grey Postero and Leon Zanozc (eds.), *The Struggle for Indigenous Rights in Latin America* (Brighton and Portland: Sussex Academic Press, 2004), 180–81.
26. García and Lucero, "Un País," 171.

nation (shades of President Velasco). Some critics pointed out that this could become an assault on indigenous languages and cultures. Then, in 2003, Peru subscribed to the United Nations's Declaration of the Rights of Indigenous Peoples.

Despite all of these measures, indigenes quickly became disillusioned with the Toledo government. IWGIA noted that, by March 2003, CONAPA had met but three times, and then only over perfunctory matters.[27] The entity was directed by Karp in very autocratic fashion. In June, the coordinator of the Indigenous and Afro-American Peoples' Development Project was dismissed without justification (he had been elected by popular vote in a process supervised by indigenes).

The Indians' disappointment over CONAPA was now complete. The advice and concerns of indigenous associations were being ignored, and when CONAPA was reorganized in late 2003, the government simply appointed the indigenes' representatives on it (rather that having the tribes elect their own spokespeople).[28] At that autumn's national assembly of COPPIP, it was decided that CONAPA did not speak for indigenous interests. It was resolved to initiate a lawsuit against the Peruvian government to have all of CONAPA's decisions to date rescinded. A request would be submitted to the Inter-American Commission on Human Rights of the Organization of American States for precautionary measures to protect Peru's Indians. It also asked,

> To approve and support the Proposal for an Institutional System for Indigenous Peoples, approved by the Indigenous Consultation held from 12 to 14 April 2003, which proposes creating a Decentralised Public Body for Indigenous Peoples at ministerial level, a Development Fund as executing agency and a real space for dialogue and consultation at a high level between the State and indigenous peoples as equals.[29]

In 2004, CONAPA was abolished and replaced by INDEPA (Instituto Nacional de Desarrollo de los Pueblos Andinos, Amazónicos y Afroperuanos or National Institute for the Development of Andean, Amazonian and Afro-Peruvian Peoples). While its mission was identical to CONAPA's, IWGIA was cautiously optimistic regarding INDEPA's future, particularly since it was created by law rather than decree and would

---

27. Diana Vinding (ed.), *The Indigenous World 2002–2003* (Copenhagen: International Work Group for Indigenous Affairs, 2003), 133 [*Indigenous World 2002–2003*].
28. Vinding, *Indigenous World 2004*, 148.
29. Ibid., 150.

likely have better functionaries.³⁰ In December, the First Summit of Indigenous Peoples of Peru was held in Huancavelica and one of its agenda items was the indigenes' concern over the possible impact on their tribal lands of the country's evolving Free Trade Agreement with the United States.³¹ They were already actively opposing several logging, fossil fuel, and mining operations.³²

Then, beginning in 2005, under Toledo there began a direct assault upon Peruvian indigenous rights in the name of "development." Legislation known as the Lands' Law (*Ley de Tierras*) was passed that undermined the communal nature of Indian land titles by allowing individual tribal members to opt out, claiming a part of the communal patrimony that they were then at liberty to exploit, lease, or sell. Chirif underscores the irony in that a law designed to produce "development" (in capitalist terms, to be sure) actually led to the impoverishment of many indigenes. They were inadequately prepared for entrepreneurial activity and became the unsophisticated targets of Peruvian and foreign economic interests. Those who sold their land were likely to squander the proceeds quickly on poor investments and/or luxury goods—ending up landless, penniless, and without rights in any communal patrimony.³³

By late 2005, the stipulated nine Indian and Afro-Peruvian seats on INDEPA had been filled by elections in their respective communitiies. However, that year's IWGIA report concluded:

> Expectations of INDEPA are limited given that the government has thus far shown a lack of suitable politicians, professionals and technicians to promote indigenous issues. INDEPA could, in the future, benefit from the institutional powers it enjoys to revitalise indigenous issues if the new government, to be inaugurated in July of 2006, can appoint suitable people to run this body and if there exists a firm political will to support its initiatives. Nonetheless, the outlook is dismal, for any references to indigenous demands are few and far between in the proposals being made by the presidential candidates.³⁴

The report applauded intentions in the recent environmental legislation passed in 2005, but noted that, "it caused such a strong reaction

---

30. Diana Vinding and Sille Stidsen (eds.), *The Indigenous World 2005* (Copenhagen: International Work Group for Indigenous Affairs, 2005), 175–76.
31. Ibid., 177–78.
32. Ibid., 178–84.
33. Chirif, "Cien años," 216–17.
34. Stille Stidsen (ed.), *The Indigenous World 2006* (Copenhagen: International Work Group for Indigenous Affairs, 2006), 172 [*Indigenous World 2006*].

among the business sector that amendments were made that cancelled out all its practical effectiveness."[35]

Under Toledo Manrique's successor, the situation worsened. President Alan García Pérez was a neoliberal who had recently concluded the so-called Free Trade Agreement with the United States that opened his country's natural resources to foreign investors—petroleum, logging and mining interests in the main. There had been more than 100 legislative decrees undermining the territorial sovereignty of Peru's indigenes (one even conferred impunity on Peru's military and police forces from any civil or criminal consequences for the injuries or deaths inflicted on citizens, including protestors against official policies). In the words of one jurist, the Indians now faced

> ... a whole process that is cumulatively increasing the degradation of their rights until they are wiped out of this country's legal framework. This package gives no justification as to why the mining or oil companies should enter indigenous territories as they please.[36]

In response, AIDESEP organized a major protest on August 8, 2008, or the International Day of Indigenous Peoples. Thousands of Amazonian Indians participated in a peaceful protest designed to paralyze the region's transportation system on the rivers, roads and in its towns. It also shut down two hydroelectric plants. The government's initial response was to denounce the manifestations as the work of political agitators of the foreign NGOs, but when the Peruvian press demonstrated that they were populists fighting for their peoples' interests, Peru's Congress blinked. Over the objections of President García and his prime minister, Jorge del Castillo, it repealed Decrees 1015 and 1073—the most onerous.[37] It promised to review the others.

While encouraging, this addressed but some of the issues. Clearly, illegal loggers were pushing into the territories of "uncontacted peoples" on the tributaries of the Ucayali and Madre de Dios rivers. With the collusion of corrupt Peruvian officials, there was considerable export of illegal mahogany (some by U.S. and Asian interests). A Canadian petroleum company was allegedly polluting a river, and there were several worrisome hydroelectric projects on the drawing boards.[38] Then, too, a clandestine

---

35. Ibid., 174.
36. Efraín Jaramillo Jaramillo, "Peru," in Kathrin Wessendorf (ed.), *The Indigenous World 2009* (Copenhagen: International Work Group for Indigenous Affairs, 2009), 162.
37. Ibid., 164. Chirif, "Cien años," 221.
38. Jaramillo Jaramillo, "Peru," *Indigenous World 2009*, 168–71.

tape recording surfaced of an important member of the APRA Party, Rómulo León, negotiating bribes from foreign oil companies in return for concessions to drill in the Oriente—ultimately landing León in prison.

In 2009, the Free Trade Agreement with the United States went into effect. The García government issued new decrees to accommodate its commitment to open up the Amazon region to foreign investment. Beginning in April, there were sixty-five straight days of protests by indigenes. They occupied many facilities. In June, the government sent in the Army and there were two bloody clashes.[39] On June 5, at least thirty-two people were killed near Bagua (nine policemen and twenty-three Indians) when the armed forces attempted to dismantle a road block erected by five thousand protesters. An indigenous leader, Alberto Pizango, director of AIDESEP, accused the police of massacre of unarmed indgenes by firing on them from helicopters. The police were said to have burned Indian bodies to hide the death toll. García denounced the criminal indigenes.

The next day, nine more police officers were killed while attempting to rescue thirty-eight of their comrades held hostage at a Petroperú facility. The president's reaction was to label the Indians "terrorists" who were being incited by foreigners. García refused to meet with AIDESEP and Pisango was granted political asylum in Lima's Nicaraguan embassy. The Indians vowed to continue their protests until the offending decrees were repealed. On June 18, a nervous Congress abolished the two worst ones and the protests were over. However, the García government then proposed dissolution of AIDESEP, which by now, with sixty-five member tribes, was easily Peru's largest indigenous association.

Activist liberation theology priests in Loreto were also formidable critics of the government and were targeted by the García government. In 2010, there was an attempt to expel a British missionary for organizing Indian protests in and around Iquitos. An Italian priest was accused of instigating Indian rebellion after the Bagua massacre. Daniel Hurley, an Augustinian priest from Chicago, and serving as bishop of Chulucanas, was targeted, as were Francisco Muguiro, a Jesuit and vicar general of Jaén, Cajamarca, and the Basque José Luís Astigarraga, the bishop of Yurimaguas. Liberation theology remained alive and well in Peru, despite opposition to it by the highest ecclesiastical authorities in Lima. García himself had labeled them "false Christs" standing in the way of the country's economic"progress."[40] On April 16,

---

39. One might assume that the military were on high alert since earlier that April *Sendero Luminoso* had ambushed a patrol and killed fourteen soldiers.
40. Peru Support Group, "Editorial: Will No-One Rid Me of These Turbulent Priests?" Update 140, June/July, 2010.

2011, in a ceremony in Bilbao, Bizkaia, Augustinian friar Monsignor Miguel Olartua was consecrated (in his own homeland) as the new bishop of Iquitos. The city's retiring activist bishop, Julián García Centeno, was present, and it was presumed that Bishop Olartua would follow his lead in speaking out against injustice.[41]

In 2011, Ollanta Moíses Humala Tasso became president of Peru. His ancestry was half Quechuan and half Italian. Like his immediate predecessors, this president remained committed to economic development at the expense of environmental concerns and indigenous rights. In 2012, there were 24 civilian deaths and 649 injuries in 227 conflicts throughout the country, of which 148 were socio-environmental (most involving indigenes). Despite "its initial progressive image," President Humala Tasso's government "has, like its predecessors, responded to social protest by brutally suppressing it and criminalizing the social leaders."[42]

In the Amazon region, many tribes had formed the Platform for Amazonian Indigenous Peoples United in Defence of Their Territories (PUINAMUNT), calling for cancellation of oil and mining concessions in their territories that were awarded by the Peruvian government without prior consultation with the indigenes. The indigenous Púrus Federation of Native Communities was opposing construction of what it labeled the "highway of death" that was projected to slice through the Alto Púrus National Park with its several uncontacted tribes. At the same time, the image of AIDESEP was tarnished by the news that its leader had accepted a payment of 73,000 dollars from the Brazilian oil company Petrobras for agreeing to a clause in its concession whereby the indigenes forewent any redress of disputes in the local, national and international court systems.[43] The following year, it could be reported that the Peruvian government was instituting sentences of twenty-five years to life for agitators (leaders) of social protests, and was employing its armed forces in such confrontations with greater frequency. The recent civilian toll was twenty-nine fatalities.[44]

Then, on September 1, 2014, four Indian leaders were killed while trying to expel illegal loggers from their communal forests.[45] We are told:

---

41. http://augustinians.net/index.php?mact=News,cntnt01,detail,0&cntnt01articleid=2&cntnt01returnid=39.
42. Jorge Agurto, "Peru," in Cœcilie Mikkelsen (ed.), *The Indigenous World 2013*, (Copenhagen: International Work Group for Indigenous Affairs, 2013), 140.
43. Ibid., 147.
44. Jorge Agurto, "Peru," in Coecilie Mikkelsen (ed.), *The Indigenous World 2014* (Copenhagen: International Work Group for Indigenous Affairs, 2014), 159–66.
45. Ibid., 163–64.

In conclusion, a concern to prioritise extractive activities over and above defending thelives of peoples in isolation explains the State's lack of interest in implementing the framework of protection contained in Law No. 28736. The five existing territorial reserves do not have Protection Plans but President Ollanta and some of his ministers have been happy to project a social welfare image by personally delivering plastic cradles and nappies to the Nanti of Camisea, where chronic malnutrition affects 67.3% of children under five, five times the national average; acute malnutrition is double the national average.[46]

According to Chirif, the assault upon the Indians goes well beyond land rights—and here the missionaries are key. Indians of the Peruvian Oriente have been subjected to renewed "modern" missionary activity by Dominicans and Franciscans who have constructed schools along with churches—thereby continuing the tradition of making over indigenes into Spanish-speaking vassals of Peruvian culture.[47]

Chirif frames his discussion of the way that the term "modernize" has replaced "civilize" in white discourse regarding indigenes by first quoting Pablo Zumaeta's (or Rey de Castro's?) justification of the policies of the Peruvian Amazon Company in the Putumayo:

> This much is obvious: the Indians who find themselves in a transitory state from savage to civilized life, cannot and should not be considered capacitated, because they have not the slightest notion of the law and what is considered to be good, and, consequently, of that which is punishable or not, anymore than what is licit and illicit, besides the fact that their total ignorance and abasement before the white and civilized man puts them in a situation in which they scarcely understand their own personalities, since, for that to happen, it is first necessary to instruct them so that they can have self consciousness and come to esteem civilization's advantages.[48]

---

46. Ibid., 170.
47. Chirif, "Cien años," 205, 217. Then, too, there are the Protestant fundamentalists who, through the activity of the Summer Institute of Linguistics, translated the bible into many languages in order to proselytize more effectively. In 1945, the Ministry of Public Education signed an agreement with the Summer Institute of Linguistics that conceded it juridiction over missionary activity in the country, but under the aegis of the ministry (one can only imagine what the country's Catholic hierarchy thought about that deal). Since then, the SOL has established schools in most of Peru's Indian settlements (San Román, *Perfiles*, 224). Novelist Mario Vargas Llosa penned a particularly probing account of this activity and its pernicious impact on the Indians. See Vargas Llosa, *The Storyteller*, 70–72.
48. Zumaeta, *Las cuestiones*, 12–13.

Chirif sums up the situation from his early twenty-first century perspective as follows:

> The manner in which the State and part of civil society justify this process of aggressions has amazing similarities with those that have been used to defend the outrages against the indigenous peoples throughout history. In effect, during the Colony, the rubber epoch and now, the measures taken, according to those behind them, were motivated by the negative characteristics of indigenous societies (their savage and barbarous state and atheism, in the past, and their poverty, at present), from which the former colonial power and today's republicans assumed their feigned inspired historical role as saviors of such damned beings. Racism is an element present throughout the relations of the State and civil society with indigenous peoples, a characteristic that is profoundly rooted in Peruvian society and which expresses itself in diverse ways. One of them is to assume the weight of development, of progress. If their territories are invaded and degraded, their organisms affected by the pollution of heavy metals, their women raped, their lives impoverished economically and morally, and in the end their social relations destroyed to the point that they face internal strife and are obliged to pursue disorientated agendas, it is because such is progress. "It is the price of progress," is the phrase that is heard frequently, without those who pronounce it ever questioning why it is always the indigenes who should pay this bill. It is as if destiny had established that that is how things are and against fatality nothing can be done but to accept it as the cost of a greater good: progress.[49]

Sardonically, Chirif asks if the same policies and politics would obtain regarding its extraction in the name of development if oil were found beneath the streets of Lima.[50] He then concludes:

> If the State is truly interested in the poor, its first priority should be to concentrate its efforts where the majority of poor in the country are to be found, the cities populated by the halt and lame, by street people, the unemployed and underemployed (automobile vigils, street musicians, the "will-work-for-anything" pleaders), criminals without connections (those with them tend to be well-protected and are scarcely impoverished) and, in sum, a long list of persons who have had to opt for the most diverse work strategies

---

49. Chirif, "Cien años," 219.
50. Ibid., 220.

to survive in a hostile society, a condition that does not obtain in indigenous societies.[51]

Shades of Mariátegui's primitive communism and Paredes's Samarem!

Colombia

Enrique Olaya Herrera assumed the presidency of Colombia in August 1930. He had fought on the side of the liberals during the War of a Thousand Days and was therefore an enemy of Rafael Reyes. As a journalist for the newspaper *El Autonomista*, owned by Rafael Uribe Uribe, he was an acerbic critic of Reyes and his presumed sellout to the United States over the loss of Panama. He was also highly critical of the rise of conservative, authoritarian and clerical power after promulgation of the Constitution of 1886. At one point, he founded a newspaper called *El Soldado Cubano* (The Cuban Soldier) out of admiration for the anti-colonialist revolutionary José Martí.

By then, Friar Monclar had resigned and returned to Europe (1928) due to poor health, and was succeeded by Fray Gaspar Monconill de Pinell. Then, in 1930, the Vatican elevated the Apostolic Prefecture to the higher status of Vicariate with Pinell now the vicar apostolic. By this time, the policies of the Capuchins had totally transformed the Sibundoy. If, in 1906, there were 32,600 Indians in the prefecture, compared to 2,200 white settlers, by 1933 the indigenes had declined to 13,997 and the settlers numbered 21,587.[52]

Fray Pinell was replaced as head of the Sibundoy mission by Fray Bartolomé de Igualada. While the Capuchins sometimes championed indigene rights against settler abuses, the missionaries themselves were engaged in enhancing their own holdings at the expense of their charges. Also, when stocks (the *cepo*) were outlawed throughout Colombia in 1931, the Capuchins refused to give them up, while arguing that the lack of prison facilities in their territory would thereby make it impossible to punish malefactors.[53]

Liberal politician Enrique Olaya Herrera ultimately sought to improve relations with the United States as foreign minister during the 1920s, and gained a reputation as a compromiser working toward a more amiable working partnership with conservatives in the Colombian national interest. In 1930, he was elected president as a liberal, but with the support of moderate conservatives. He immediately appointed people

---

51. Ibid., 224.
52. Bonilla, *Servants*, 186.
53. Ibid., 193–94.

from both parties to his cabinet. His victory alarmed the Capuchins, fearing that the Church's (and their) priveleges and subsidies might be terminated. Nothing of the sort happened. Nevertheless, in 1932, a conservative and Catholic colonel in the Peruvian Army sent a confidential letter to the Minister of Industry denouncing "the appalling persecution of the Mission which calls itself Christian."[54]

That same year, Sibundoy Indian leader Tomás Jacanamejoy wrote directly to the president of Colombia. He denounced the fact that the Capuchins were exacting eight days annually from the Indians of uncompensated manual labor for their projects. The missionaries accepted no excuses for non-compliance and were still meting out considerable physical punishment:

> The punishments are terrible: they do not represent a sanction that ordinary justice imposes upon the greatest criminals, rather it approaches limits that are inconceivable in a civilized and Christian people. Flagellation, the forceful blow, the barbarous torment of the stocks and a thousand more martyrdoms that the missionaries employ to castigate my brethren, to the point of making them bleed, along with the most atrocious sufferings and the most painful complaints. Aside from the fact that the system of obligatory work is now abolished in Colombia; aside from the fact that it deprives my companions and their families of their bread during the time of their service; aside from the fact that the [missionaries] do not impart truly useful instruction as the teaching of manual arts and scientifically based agriculture would be, they continue to punish so cruelly as to be labeled criminal. We are, Your Excellency, under the aforementioned conditions, truly unfortunate beings, worthy of living in Cochin China or in the Congo, but not in a country, such as ours, that is ruled by wise laws.[55]

Olaya Herrera would preside over Colombia's war with Peru and was in his final year in office when the peace was signed in Rio upholding the Salomón-Lozano Treaty that finalized their border. President Olaya Herrera was succeeded in office, in 1934, by Alfonso López Pumarejo (1934–1938), again of the Liberal Party. He was practically unopposed and continued the so-called *Revolución en Marcha* (Revolution on the March) that legalized labor unions and, in 1936, passed Law 200 that called for expropriation of private properties in the interest of social

---

54. Cited in Bonilla, "Servants," 192.
55. Cited in Dominguez and Gómez, *Nación,* 44.

well-being. This gained him the support of the Communist Party, but also prompted opposition from more moderate liberals who thought his policies too extreme. He initiated several modification of the 1886 Constitution and land reforms that entailed expropriating large estates. He was a great admirer of the New Deal and the government-interventionist arguments of John Maynard Keynes. Indeed, he had a personal friendship with Franklin D. Roosevelt and became a Latin American spokesman for the American president's "Good Neighbor" policy. President López Pumarejo, came from the landed elite, so, as happened to the populist Roosevelt in the United States, he was considered by the Colombian oligarchy to be "a traitor to his class."[56]

Throughout the decade of the 1930s, the Capuchins moved to consolidate their control over the Sibundoy Valley, in part through land leases and purchases from the unsophisticated indigenes and also by usurping lands vacated by disillusioned ones who abandoned the area, died of disease or committed sucide.[57] This was particularly true under the leadership of Fray Marceliano de Vilafranca, appointed head of the Capuchin parish in Sibundoy in 1938. Dubbed *Fray Manga*, or "Brother Strongarm," he had fought with the Franco rebel forces after his natal city of Barcelona was overtaken by a "red tidal wave." While in the Sibundoy, his sermons were often about Spanish politics and exaltation of their savior, Francisco Franco.[58] The continued advance of Capuchin interests in the Sibundoy Valley prompted some of its white settlers to decry the usurpation of the country by "foreigners," to the exclusion of both its indigenes and *colonos*. One of their leaders, Benjamín Caipe, incited the Indians to file lawsuits over contractual inequities. Surprisingly, they prevailed, thereby incurring the enmity of the Capuchins. As for the *colono* instigator,

> Caipe himself saw some of his own land fall into the hands of his neighbours at the Mission. Still now, in 1970, he recalls how on 28 September 1940 Fray Marcelino took advantage of his absence to effect a decree of expulsion against him. At the head of a small band of *peones* and local officials, the priest destroyed his house, took his wife (though 'suffering from lung trouble') to prison, and left his invalid son lying in a field; after which he left fifty calves loose among his plantations, which he later sprinkled with salt.[59]

---

56. H. W. Brands, *Traitor to His Class: The Privileged Life and Radical Presidency of Franklin Delano Roosevelt* (New York: Doubleday, 2008).
57. Bonilla, *Servants*, 196–99.
58. Ibid., 207.
59. Ibid., 210.

Bonilla tells us that not all of the Capuchins were interested in the order's material gains. Fray Marcelino de Castellví, one of Fray Manga's subordinates, took seriously the mandate to evangelize the entire Vicariate, but recognized the impossibility of doing so given the lack of Old-World missionaries. He therefore established a seminary in Sibundoy to produce home-grown priests—indigenous and white. It proved to be a failure as only one creole became a priest over the next twenty years and one Sibundoy Indian became a lay brother after leaving the seminary. Twenty years after its founding, an indigene from Sibundoy obtained his doctorate in anthropology from a secular university in anthropology (over considerable Capuchin opposition).[60]

There was also the interesting case of the Kamsá Indian, Pedro Juajibioy, that American ethnobotanist Timothy Plowman met in 1974 in Sibundoy while searching for hallucinogenic plants[61] in the Valley and its hinterlands. Of Pedro, Plowman reports:

> He's Kamsá. Schultes met his family in 1941. By then the Capuchins owned the valley. Pedro went to their schools and became a devout Catholic. They made him an altar boy. He went to church twice a day and might have become a priest if Indians had been welcome. Then Schultes came along. Pedro's father was worried about malaria, but he let Schultes take his son into the lowlands as far as Mocoa. That's how Pedro got turned on to plants. Now he's a botanist. He's worked with everyone, all of Schultes's students, anyone who comes to Colombia. His feet are in both worlds. His dream is to build a herbarium right in Sibundoy. He already has a garden of medicinal plants all labeled with scientific names. Among the Kamsá he is known as a household healer, an expert at treating the ailments of the ordinary world. He has even trained among shamans of the lowlands.[62]

Meanwhile, there was considerable population growth and economic expansion in the Loreto. The 1940 census listed 180,000 *"colonos"* residing in the Department of Loreto of whom 38 percent were indigenes.[63] Interestingly, the former non-indigene modest *caucheros* and many lower class newcomers established small farms along the river banks, carving a hand-

---

60. Ibid., 214–15.
61. Wade Davis notes that the Sibundoy Valley has the greatest variety in all of South America of hallucinogenic plant species (Davis, *One River*, 139–40).
62. Ibid., 164.
63. San Román, *Perfiles*, 184.

ful of hectares out of the forest for a few crops and feed for some livestock. They adopted the lifestyle of subsistence agriculturalists characteristic of the "mestizo Indian."[64]

Their urban focal point was Puerto Leguízamo. It was quickly becoming a regional servicing center of considerable importance, projecting transportational infrastructure into its hinterland. A 25-kilometer road was contructed between it and La Tagua on the Caquetá, thereby linking the commerce of the two rivers. In 1943, it was extended to Mocoa, and, ten years later, to Puerto Asís. In 1953, both Puerto Asís and Puerto Leguízamo received air service.[65] Between 1952 and 1956 the number of settlements (*fincas*) in the area increased from 5,388 to 17,911.

It was in 1946 that Fray Pinnell died suddenly. Colombia's change of government from liberal to a conservative one about this time proved fortuitous, given that the Capuchins were coming under increasing criticism. It was even rumored that they had transferred tens of millions of pesos to Spain to assist Franco's reconstruction of that country. However, now the Capuchins' title to their *haciendas* were recognized. While they no longer used the stocks, the friars continued to be denounced for employing corporeal punishment (whippings) of the Indians.[66]

Meanwhile, given the massive immigration of creoles from elsewhere into the Putumayo/Caquetá, the Capuchins' capacity to conduct their work within their Apostolic Prefecture was stretched. In 1951, the Prefecture was divided in two with the Caquetá region becoming is own Vicariate and the rest of the territory, to the shores of the Amazon, being designated its own Vicariate of that name. Capuchin Fray Marceliano de Vilafranca was appointed its prelate, based out of Leticia.

In the Sibundoy Valley, there was a new influx of creoles seeking farms. A commission was appointed that recommended that the land of about four hundred Indian families held in common in a reserve be privatized into about two-hectare parcels, ostensibly in the interest of alleviating their poverty by allowing them to lease or sell them. This quickly resulted in their loss and the greater impoverishment of these marginal indigenes. Bonilla viewed the measure as part of a new national campaign to abolish all of the existing *resguardos* in the country.[67]

---

64. Ibid., 186.
65. Restrepo López, *El Putumayo*, 31.
66. Ibid., 220–21.
67. Ibid., 226–27. By the end of the 1950s, there were 32,500 settlers in the valley. The indigene population was a quarter of the size that it had been when the Capuchins first entered the area (Ibid., 229).

In 1960, the Colombian government launched an agency, the Department of Native Affairs Division (DIA) of the Ministry of the Interior, ostensibly to protect its "surviving" indigenes. By this time, the estimated total of Indians, counting both those on *resguardos* and the ones still living in the forests beyond the government's control, was 350,000. Bonilla believes that far from recognizing any intrinsic value in indigene cultures and languages, it was meant to enculturate the remaining Indian into the national culture, but "condemned to take the lowest place in a society which could only see him as an inferior kind of person." He summed up the situation as follows:

> That is why, contrary to the claims of some South American governments, one cannot look to a speeding up of the inevitable process of trans-culturalization for any improvements in the life of the Indians, any profit to the nation's economy, any modernization or any progress towards a hypothetical 'national unity'. This 'programmed integration' of Indian minorities would seem in Colombia, as elsewhere, to be no more than at best a Utopia—with the hope of opening new consumer markets to industry—and at worst merely an excuse for taking the last scraps of their land way from them, or then, again, it might look like the result of the childish attempt of the Indo-American majorities to gloss over the fact—which they so hate to recognize—that most of us have Indian ancestors.[68]

In 1961, Congress passed Law 135 that created INCORA, the Instituto Colombiano de la Reforma Agraria (the Colombian Institute of Agrarian Reform). It was basically charged with acquiring land through purchases and allocation of "empty areas" for redistribution to impoverished peasants. It made specific mention that INCORA would determine and improve the circumstances in indigenous and Afro-Colombian communities with regard to territorial issues. It was to be an independent agency whose board was presided over by the minister of agriculture with membership from the heads of various agencies charged with aspects of rural economic development. Peasant organizations were to hold two seat and the various indigenous associations had one. The livestockmen and farmers also had representation. The general manager would be an appointee of the president and would attend INCORA meetings, but without a vote. In sum, the indigenes now had a voice at the table, albeit a tiny one, and their interests were being recognized.

Section 18 of Article 12 of the empowering legislation charged INCORA with "Studying the land needs of the indigenous communities and

---

68. Ibid., 232.

constituting, amplifying, improving and restructuring their reserves . . ." Albeit, there remained grounds for worry in the possible interpretation of terms like "restructuring" and "empty lands." Not to mention that the main purpose of the initiative was to adjudicate and ameliorate the intrinsic conflict of interests between the landed oligarchy and the rural agriculturalists (given that, ostensibly, the objective was to settle the latter on lands currently in possession of the former).[69]

INCORA entered into a series of disputes with the Capuchins regarding their stewardship of Putumayo/Caquetá. In 1962 there was a Supreme Court ruling to the effect that all lands within the *resguardos* were ultimately under Indian ownership, declaring any historical alienations of them to have been illegal. INCORA also appointed an investigative commission that, in its 1964 report, declared that all of the alleged abuses of the indigenes by the Capuchins were true. For their part, the Capuchins denounced every INCORA initiative within its jurisdiction as preparing the establishment of Castroist Communism in Colombia. While such developments were of great concern to the Capuchins, in point of fact they did not result in concrete policies and measures. The Capuchins' control of their mission territory, and of the educational system within it, remained intact.[70] Furthermore, by 1970, church and state in colombia were allied over mutual concern with the growing popularity of liberation theology among the younger generation of Catholic priests, particularly their propensity to assume the causes of the oppressed.[71]

Nevertheless, by this time the president of Colombia was the Liberal Party's Carlos Lleras Restrepo (1966–1970). Among the several new agencies created during his administration were the environmental Instituto de Recursos Naturales no Renovables (The Institute to Protect Non-Renewable Resources) and the Instituto Colombiano de Cultura "Colcultura" (the Colombian Institute of Culture "Colcultura") to protect the country's cultural patrimony and diversity. Certainly, one of the influences leading to them was publication of the book by Víctor Daniel Bonilla in 1968 that we have been citing under its translated title *Servants of God or Masters of Men?* (1985).[72] Originally published in 1968,

---

69. This eventuated in the creation of right-wing paramilitaries, the most notorious being the AUC (Autodefensas Unidas de Colombia or United Self-Defense Forces of Colombia) — aligned with the oligarchy and, at times, with factions within Colombia's Armed Services. The paramilitaries produced thousands of fatalities in the rural underclasses without discriminating among the *mestizo* peasantry, indigenes and Afro-Colombians.
70. Ibid., 235–43.
71. Ibid., 252–53.
72. What follows is based upon an interview with him conducted by Alejandro Cuevas Ramirez entitled "¿Quién es realmente Víctor Daniel Bonilla, el autor de Siervos de Dios

Siervos de Dios y amos de indios: El Estado y la misión capuchina en el Putumayo[73] was an exploding bombshell within Colombian intellectual life. He was an indigene that, given the subject of his book, many assumed to be a Sibundoy. In fact, he was born in Cali and raised on a farm near Popayán. Educated in sociology and economic development at the National University in Bogotá and the University of Paris, he pursued a career in Colombia as an investigative journalist. That brought him to Leticia to do a story on a fugitive Nazi engineer who was running the physical plant of that city's Capuchin mission. It was along that particular way that he learned of the Capuchin history in the Upper Putumayo as well—hence the book.

In addition to the first edition's two thousand copy press run (sold out in four months), there were twenty satin-covered luxury copies. Bonilla sent dedicated ones to Pope Paul VI, Colombian president Carlos Lleras Restrepo and the chief administrator of INCORA. Rights to publish the second edition, somewhat to Bonilla's surprise, were purchased by the La Salle Christian Brothers. They printed three thousand copies. There would ultimately be a third edition of more than ten thousand copies issued by the University of Cauca.

The work, which tellingly appeared the same year as Varese's exposé of maltreatment in Peru of the Asháninkas, was translated immediately into several languages. The Vatican dispatched a special delegate to look into the situation and recommend revisions in the Colombian Concordat and its Convention regarding the missions. Bonilla now had an international reputation and was invited to lecture in several European venues, including, in 1969, to the convention of the World Council of Churches.

For their part, however, the Capuchins launched a campaign against him in the Colombian press for his "calumny." They called upon the mission headquarters in Catalonia to send out its best historian, Ramón Vidal, to prepare a rebuttal. In 1970, the Catholic Church in Pasto published that brief work and the outraged congregration demanded that Bonilla be sued. But then an eminent Jesuit jurist volunteered to defend Bonilla and he was exonerated. Since the book and its hoopla, Bonilla has dedicated his life to Colombian indigene causes, helping many tribes to make their case and petitions for assistance—particularly in the Cauca, where, in 1971, the indigenes formed CRIC (Regional Indigenous Council of the Cauca Department). Accordingly,

---

y amos de indios?" http://pagina10.com/index.php/culturas/item/1772-¿quién-es-realmente...I-bonilla-el-autor-de- siervos-de-dios-y-indios#.WGUz_TvvFBx, np.

73. Víctor Daniel Bonilla, *Siervos de Dios y amos de indios: El Estado y la misión capuchina en el Putumayo* (Bogotá: Ediciones Tercer Mundo, 1968).

The main indigenous protagonist of that period, CRIC, emerged fighting alongside peasant organizations for a change in the distribution of rural land and for democratization of society. CRIC advocated not only indigenous territories, but also for indigenous rights, the maintenance of indigenous customs, and the importance of cultural diversity in the country. It became the model for regional indigenous organizations. From its beginning in 1971, Cauca's indigenous movement attracted attention because of the strength of its mobilizations even in times of repression. . . . Cauca's indigenous movement opted for establishing a dialogue with state institutions. After the mid-1980s—when the armed conflict became more acute and the Cold War crushed many social movements and trade unions—the CRIC tried to convince the government, the Catholic Church, and even the military to implement the civil option to solve the conflict. At the national level it achieved the creation of the Department of Indigenous Affairs at the Attorney General's Office for the investigation of abuses committed by the military and the police in indigenous territories.[74]

Between 1971 and 2000, two national and thirty-six regional indigenous associations were founded in Colombia. In 2000, an indigenous candidate, Floro Tunubalá, was elected governor of Cauca Department.[75] At the same time, during the period approximately six hundred indigenous leaders have been assassinated and most of the perpetrators remain unpunished.[76]

The 1970s also witnessed considerable unrest among the country's impoverished *colonos* regarding the lack of government support. This included several urban manifestations by thousands of demonstrators—particularly in Florencia.[77] By the mid-decade, FARC was active on the fringes of the Putumayo/Caquetá, even taking over some the the towns. From 1978 to 1982, there were many clashes between the leftist guerrillas and the government forces in what was labeled the Guerra del Caquetá (the Caquetá War). In March 1984, M19 (another leftist militant group) captured Florencia and FARC was active in the Caguán River drain-

---

74. Theodore Rathgerber, "Indigenous Struggles in Colombia. Historical Changes and Perspectives," in Nancy Grey Postero and Leon Zamosc (eds.), *The Struggle for Indigenous Rights in Latin America* (Brighton and Portland: Sussex Academic Press, 2004), 112, 113.
75. Ibid., 106, 109.
76. Ibid., 112.
77. Oscar Arcila Niño, Gloria González León, Franz Gutiérrez Rey, Adriana Rodríguez Salazar, and Carlos Ariel Salazar, *Caquetá, construcción de un territorio amazónico en el siglo XX* (Bogotá: Instituto Amazónico de Linvestigaciones Científicas Sinchi, 2002), 63–64.

age. The government eventually exerted control over the main towns but the guerrillas controlled much of the countryside—the ongoing conflict prompting an exodus of impoverished colonos and indigenes from the countryside to the urban areas.[78]

It was also in 1975, thanks in no small measure to Bonilla, that the Concordat between the Vatican and Colombia regarding the extraordinary powers of the Catholic Church in the country was abolished. Thenceforth, Colombia was to be a secular state. This broke the Capuchins stranglehold on the Putumayo, of course, as well as that of the Church on Colombia's educational curriculum. In 1976, the Capuchins turned over to the government their extensive church, school and hospital holdings, and then leased some of them back.[79]

During the 1980s, several Colombian tribes began organizing associations to protect their interests. In 1982 the vast of these created an umbrella organization, ONIC (Colombian National Indigenous Organizations) to represent their collective interests. By mid-decade, an anthropologist and former student of Gerardo Reichel-Dolmatoff, Martín von Hildebrand (1943-), emerged as both a defender of indigenous rights and of the rainforest environment. In 1972, he had founded the Amazonian branch of the Colombian Anthropological Institute in La Pedrera. It brought together a multidisciplinary team to enable indigenous peoples to assert their territorial rights, while devising their own path for economic development consistent with their cultural values. In 1979, von Hildebrand completed his doctoral studies in ethnology at the Sorbonne. In 1981, the presidential campaign of candidate Alfonso López Michelsen asked him to prepare a position paper on the environment and indigenous rights. He did so in consultation with Roque Roldán, an attorney specialized in Indian affairs.[80]

From 1986 to 1990, von Hildebrand served as Colombia's Head of Indigenous Affairs. He was designated Colombia's delegate to a 1989

---

78. Ibid., 66–67.
79. Cuevas Ramirez, "¿Quién?" np.
80. See http://gaiaamazonas.org/historia/?lang=es. In 1999, Roldán co-authored a compendium and pro-indigene commentary on Peruvian legislative history regarding its indigenes. See Roque Roldán and Ana María Tamayo, *Legislación y derechos indígenas en el Perú* (Lima: CAAAP, 1999). The Colombian NGO COAMA was a co-publisher. The following year, he authored a similar work on Colombia. See Roque Roldán Ortega, *Pueblos indígenas y leyes en Colombia. Aproximación crítica al estudio de su pasado y su presente* (Bogotá: Tercer Mundo Editores, 2000). The latter work was published under the imprimaturs of the *Fundación Gaia* and COAMA and was subsidized by a grant from the International Labor Organization.

gathering of Amazon Pact nations,[81] where he advocated successfully for creation of a special commission for indigenous affairs and another on the environment. Under his aegis, Colombia recognized 200,000 square kilometers of Amazonian rainforest to be *resguardos*. Colombia was one of the earliest signatories of the International Labour Organization's 1989 declaration regarding respect for the rights of the planet's indigenous peoples.[82] In 1990, he hosted Vincent Brackelaire, consultant on "socio-environmental" affairs to the European Community Commision in Brussels, for an extensive visit to the Colombian Amazon. His report ultimately galvanized the involvement of several European governments and NGOs (notably from Austria, Denmark, the Netherlands, and Sweden) to provide support to the projects of Amazonian Indians.[83]

In 1989, as well, Martín co-founded COAMA (Consolidación Amazónica or the Amazonian Consolidation) that brought together some NGOs and indigenous associations into a network advocating for indigenous rights.[84] He also founded the Fundación GAIA Amazonas (Amazonas GAIA Foundation) designed to advise (willingly complicit) Indian leaders on how to manage relations between indigenous associations and government agencies, as well as potential international sponsors of indigenous initiatives. GAIA would channel considerable foreign support into indigene projects and has sponsored indigenous peoples interested in environmental issues.

In 1991, Colombia held its Constitutional Convention, including members of the former rebel guerrilla group M19 (but not FARC) and adopted an extraordinarily progressive new constitution. The human rights of all citizens were to be safeguarded most emphatically—especially those of women. The environment would be protected. Same-sex mar-

---

81. The Amazon Pact was signed by in 1978 by Bolivia, Brazil, Colombia, Ecuador, Guyana, Peru, Suriname, and Venezuela. Its purpose was to set up national oversight commissions of economic development of the rainforest. The initiative was designed to prevent foreign governmental and entrepreneurial interests from dictating policy in the vast region. Unfortunately, the Pact led to neither an overarching structure nor enforcement mechanisms. Consequently, by 1989, most world financial institutions were refusing to lend funds to Amazonian projects for failing to meet international standards for environmental protection. It was the purpose of the 1989 encounter between the member nations to put such measures in place.
82. Ibid., 21.
83. The two men would collaborate on a book. See Martín von Hildebrand and Vincent Brackelaire, *Guardianes de la selva: Gobernabilidad y autonomía en la Amazonia colombiana* (Bogotá: Fundación Gaia Amazonas, 2012).
84. Since its beginning, 60,000 indigenes from 52 peoples have gained their autonomy and control of their own governance (See COAMA website).

riage would be approved by Congress; as would abortion when a mother's health was at risk. All political parties and movements would be permitted and protected. Workers had a right to collective bargaining and to strike. Freedom of the press was ensured. The judiciary would be protected from political interference. Presidents would serve a single four-year term and were not eligible for reelection.

The promulgation of the new charter for Colombia was applauded by its indigenous groups. Article 7 of the document's Fundamental Principles expounds that "The State recognizes and protects the ethnic and cultural diversity of the Colombian nation." There is a sense, then, in which the plight of the indigenes was linked to that of the Afro-Colombians (the descendants of African slaves), seemingly the two most proscribed ethnic groups in the country. Henceforth, their destinies would be intertwined within Colombia's domestic affairs' political deliberations. It is also clear that by this time Indian rights had their advocates across a broad spectrum ranging from local indigenous organizations, their regional associations, their Colombian intellectual advocates (governmental, NGOs, academics, and missionaries), overarching regional organizations that embraced more than one Amazonian nation, and both private and public international supporters that included such multinational entities as the International Labor Organization, the Organization for American States, and the United Nations. In 1995, OPIAC (Organization of the Indigenous Peoples of the Amazonian Basin) was established and I soon became a part of the international group COICA.

Nevertheless, in its 2002–2003 report, the International Work Group for Indigenous Affairs observed that the recent breakdown (on February 20, 2002) of the negotiations between the Colombian government and FARC had created an intolerable situation in rural Colombia. FARC was driven into more marginated areas of the country, where it resumed its active armed resistance with a vengeance. It produced a series of massacres, assassinations, disappearances of leaders and threats (directed particularly at indigenous associations that might provide a counterbalance to FARC authority). It could be stated that, "it is a war in which we, along with Afro-Colombian and peasant populations, are the main victims."[85] By way of examples, the government forces had failed

> ... to act in a timely and effective manner in the case of such publicized massacres asthat of Alto Rio Naya, in which around hundred indigenous Paece, Afro-Colombians and settlers were murdered by paramilitaries; that of Bojayá, in which 127 Afro-

---

85. Vinding, *Indigenous World 2002–2003*, 114.

Colombians seeking refuge in a church, were killed by a cylinder bomb thrown by FARC guerrillas; or the massacres perpetrated by paramilitary groups in the Sierra Nevada de Santa Marta where, during 2002, more that one hundred indigenous Kankuamo and Cogí were killed.[86]

By now, indigenous leaders were being assassinated regularly by FARC, other leftist militants, and right-wing paramilitaries like AUC. FARC also continued to press young indigenes into service in their ranks—sometimes with persuasion and at others by force.

It was in 2002 that (Basque-surnamed) Álvaro Uribe Vélez won the presidency. The indigenes viewed Uribe's election as a dangerous shift to the right, "the authoritarian alternative." It also seemed to portend approval of the pending United States' Plan Colombia[87] and a proposed free trade agreement that threatened to open up indigenous lands to foreign investment and development. The growing U.S. military presence in Colombia, "to eradicate drugs trafficking and terrorism hark[s] back to the early days of the Vietnam War"; it all "encourages a violation of human rights of indigenous people and other rural communities leading, among other things, to further displacements of communities from their territories when mega-projects or the exploitation of natural resources is planned."[88]

The relative incapacity of the indigenes to resist such powerful forces with force of their own led them to conclude:

> Out territorial roots, our community cohesion around traditional authorities, our organisational strength and tradition of struggle as a social movement, along with the fact that we have ended up the victims of all armed players, are the reasons why we indigenous have opted for a strategy of peaceful resistance, within our territories, to all players in the war: State, rebels and paramilitaries.[89]

---

86. Ibid., 116.
87. This was a U.S. initiative, launched in the late 1990s, that called for cessation of talks with leftist guerrillas and the complete eradication of coca production (with enhanced aerial fumigation if necessary). This was problematic from several standpoints: by then hundreds of thousands of *colonos* and indigenes were dependent upon coca growing or the ancillary employment that it provided to feed their families. The disparity between coca prices and those of any alternative crop was enormous and the Plan Colombia had no provisions for bridging that gap. Furthermore, it essentially aborted the substantial progress that had been made by several Colombian government agencies and NGOs in seeking a negotiated settlement to the conflict. For an extensive analysis of the situation viewed from the trenches of the Caquetá Department see Arcila Niño, et al., *Caquetá*, 149–217.
88. Vinding, *Indigenous World 2002–2003*, 117.
89. Ibid., 120.

The following year's IWGIA report was even more pessimistic. It declared that Colombia's poor were now legend and that their circumstances were worsening as Uribe pushed forward his neoliberal agenda. This was no longer simply an error, it was now a crime.[90] This was the year in which President Uribe presented his complex referendum to the Colombian electorate. He wanted overwhelming approval as ratification of his autocratic program of "Democratic Security" and his neoliberal economic moves. It was a gross miscalculation. Indigenous leaders, along with those of other social sectors, urged their voters to abstain. On fifteen of the sixteen proposals the referendum failed to garner the minimum 25 percent of the electorate required to make the exercise official. This was deemed a serious political defeat for President Uribe.

That same IWGIA report proclaimed that the vast expenditure of resources in the fight against the leftist militants had failed to defeat FARC. It had merely driven it underground to await the exhaustion of the State before resuming its militancy. The indigenes also criticized the Uribe administration's program of absolving AUC and other paramilitary activists of their crimes should they agree to abandon the struggle. The Indians had been the main victims of the paramilitaries and indigenous rights were being ignored. In 2003, 139 indigenes had been murdered and one thousand displaced from their homes.[91]

The National Indigenous Organisation of Colombia (ONIC) issued its own report maintaining

> That the internal armed conflict has had devastating effects on indigenous areas. These are very vulnerable peoples who, although not involved in the country's armed conflict, are its main victims. Peaceful by nature, these peoples are often forced by the armed players to participate in a war they do not understand and is not theirs. This violates their rights to life, autonomy, to their own culture and territory, free from violence. This can be observed in massacres, selective crimes, threats, intimidation, "informing," collective forced displacements or confinements, forced recruitment, restrictions on the entry of food and medicines to their communities and attacks on the communities' cultural and spiritual objects.[92]

As for the neoliberal measures,

> In implementing megaprojects on indigenous territories, undertaken with national and transnational capital, the state takes no

---

90. Vinding, *Indigenous World 2004*, 121–22.
91. Ibid., 127–29.
92. Ibid., 127.

account of the ILO's requirement to duly consult indigenous peoples and their organisations before handing over the exploration and exploitation of hydrocarbon and other natural resources on indigenous territories to voracious companies, without considering the environmental and social consequences these activities cause, with the aggravating circumstances that multinational companies are involved in human rights violations.[93]

Then, in 2004, Uribe proposed legislation that would accommodate "squatters' rights," the acquisition of legal title to lands that had been acquired through displacement. In short, this meant that title to lands seized by right-wing paramilitaries might be normalized. At the same time, the state would expropriate land on which drugs were being grown (hundreds of thousands of hectares) and auction them off. This would be another legal means of displacing peasants and indigenes. It could be contended that the leftist militias had lost considerable support among their base and would soon be negotiating the terms of their demise; whereas the paramilitaries were gaining political, social, and economic influence in addition to their military impact.

Rodolfo Stavenhagen, the first UN Special Rapporteur on the human rights and fundamental freedoms of indigenous peoples, visited Colombia in March of 2004. His report underscored "the enormous gap between Colombian indigenous legislation (among the most progressive in the world) and the state's failure to guarantee these rights in practice."[94] The massacres, murders, forced disappearances, torture, threats, displacement of communities, recruitment of children and youths of both sexes, and the rape of women were being carried out against indigenes by the paramilitries and guerrilla groups alike:

> The report particularly advises the state that it must cancel the informer network, "peasant soldier" and "soldier for a day" programmes for children and youths, all features of the Democratic Security policy of Álvaro Uribe Vélez's government.[95] Between September 12 and 16, 224, 60,000 indigenes, including delegations from Nariño and the Putumayo, descended on Cali shouting, "This country is ours and now we are reclaiming it". "We want neither guerrillas, paramilitaries nor soldiers on our territories."

The 2005 IWGIA report on Colombia concludes:

---

93. Ibid., 128.
94. Ibid., 149.
95. Ibid., 151.

> What the country and our territories are experiencing is serious, there is no time to be lost, we must act rapidly. The critical situation that we excluded people are experiencing is due to a lack of respect for our rights, in order to smooth the path to neoliberalism and globalisation, in which our economies—supportive and respectful of nature—are an obstacle. Now it is no longer just our rights that are in danger. Life itself is at risk. For this reason, now more than ever we must call for unity, solidarity and dignity to defend what is ours.[96]

But then, in 2004, and bolstered by the president's popularity with the electorate, Uribe's supporters in Congress sought to have the non-consecutive-presidential-terms provision of the 1991 Constitution changed. While Uribe had pledged to respect the provision, he now did an about face. With the reelection of U.S. president Bush, Uribe's prospects for changing the Colombian law to allow him a second term soared. There was, however, a sense of urgency on the part of both presidents to sign the final Free Trade Agreement. The U.S. congressional mandate to do so was scheduled to expire in late 2005.[97] Early that year, the U.S. National Intelligence Council issued a report identifying the next fifteen years' threats to peace in Latin America. It stated that populist presidents, like Evo Morales, Rafael Correa, and Hugo Chávez, were using anti–North Americanism as a rallying call for socialist agendas. It specifically cited the danger of "militant indigenism" and touted President Uribe as the representative of North American interests in the Andean region.

Meanwhile, the peasants, indigenes, and Afro-Colombians were now the Uribe-Vélez regime's main opponents. There were two hundred thousand refugees from these sectors in Ecuador alone. Particularly, in the Cauca, there were indigenous protests that threatened to become more widespread; its indigenes held a popular consultation on the Free Trade Agreement and 50,000 of the 51,330 who participated voted "No."[98] So, it could be concluded:

> Through these protests, the indigenous—now joined by black, peasant and other popular sectors—have highlighted the exclusive policies in the Colombian agricultural sector. In one decade these policies have concentrated landholding and displaced two million peasant, black and indigenous men and women, giving

---

96. Ibid., 153.
97. In the event, it would be extended and Colombia and the United States signed their FTA the following year on November 22, 2006.
98. Stidsen, *Indigenous World 2006*, 138–43.

rise to an agrarian counter-reform that has placed 4 million hectares of land in idle hands, expanding livestock ranches that create neither employment nor development. This dispossession is now in the process of being legalized by Uribe's Justice and Peace Law. And, given this vision and mandate, the indigenous are beginning to occupy ranches in a number of zones in Cauca. After several weeks of clashes, including violent evictions and injuries along with further occupations, on 13 September 2005 Uribe was finally forced to address land demands in Cauca and publicly recognize the country's serious land distribution problems.[99]

The change in Colombia's constitution was approved by Congress and then the Constitutional Court (in October of 2005). Uribe won re-election in May 2006 and rewarded one of his supporters, Juan Manuel Santos Calderón, by appointing him as the minister of national defense (2006–2009). This minister carried out successful military campaigns against both the leftist rebels and the rightist paramilitaries.

In 2002, the Colombian Ministry of Education had recognized formally that the AATIs (*Asociaciones de Autoridades Indígenas* or Associations of Indigenous Authorities)[100] could take charge henceforth of education and health matters in accord with their traditional values and practices. So, in 2007, GAIA worked with twelve AATIs to initiate autonomous indigenous educational programs throughout the Department of Amazonas.[101]

Nevertheless, the 2009 IWGIA report had disheartening observations about Colombia. It noted that, under the six years of President Álvaro Uribe, the rights of Indians and Blacks had experienced a major setback in the interest or opening up the national territory, and particularly its natural resources, to foreign investors interested in petroleum exploration, logging, and agri-business.[102]

In October 2008, thousands of Indians from around the country marched in protest (arriving in Bogotá in November) demanding protec-

---

99. Ibid., 144.
100. The 1991 Constitution empowered ETIs (Entidades Territoriales Indígenas or Indigenous Territorial Entities) considerable autonomy in administering their own affairs. A 1993 decree called for creation of AATIs and an expansion in size and number of *resguardos* under their jurisdiction. A 1995 decree (Number 2,164) further refined the law, ascribing to ATTIs and/or *cabildos* (councils) constituted by indigenes themselves the right to administer *resguardos* according to their desires and traditional practices.
101. See http://www.iaf.gov/our-work/results/ex-post-assessments/indigenous-education-rights-move-forward.
102. Jaramillo Jaramillo, Efraín, "Colombia," in Kathrin Wessendorf (ed.), *The Indigenous World 2009* (Copenhagen: International Work Group for Indigenous Affairs, 2009), 124–34.

tion of their resources in the interest of "liberating Mother Earth." Uribe committed the Colombian military to intimidate and control the demonstrators in the interest of democratic security and investor confidence. He also launched a public relations campaign that denounced the Indians, who represented 3 percent of the national population and still controlled 27 percent of the national territory, as unreasonable. The indigenes' leaders viewed this as a blatant attempt to turn the Colombian lower class against them. They countered that their demand was not just from Indians of the Cauca (Uribe's home region) and that

> Another important aspect of our *minga* [exposé] is to put a stop to what we have called "Legislation for dispossession." We are also demanding that the Law on Rural Development, the Mining Code, the water plans and all laws that should have been submitted to prior consultation in line with ILO Convention 169 be revoked.[103]

They also demanded that President Uribe sign the 2007 UN Declaration of the Rights of Indigenous Peoples (which he had yet to do) and incorporate its provisions into Colombian national law. The response of Uribe's minister of agriculture was that, on his watch, not another square meter of national territory would be turned over to indigenes or Afro-Colombians—without addressing the real issue of the calculated erosion of their existing rights.[104]

There would be a new initiative to change Colombia's Constitution to allow Uribe a third consecutive presidential term. The president let his supporters lead the charge and they were able to get the measure through the Congress. However, on February 26, 2010, the lead justice of the Constitutional Court announced its decision that the measure was unconstitutional. It also contended that presidents could serve only two terms, period. That precluded the possibility of Uribe Velez running again in 2014. He announced that he would respect the Constitutional Court's decision.

It was on May 23, 2009, that Juan Manuel Santos Calderón, Colombia's minister of national defense (2006–2009) resigned. He announced that he might run for president, but only if the initiative to allow Uribe a third term failed. After the ruling of the Constitutional Court, Santos became a candidate. He won in the second round, victorious in thirty-two of the thirty-three electoral districts and polling a record 69 percent of the popular vote. He assumed office on August 7, 2010.

---

103. Ibid., 129.
104. Ibid., 130.

Juan Manuel Santos Calderón came from an illustrious family that owned the prestigious newspaper *El Tiempo*. He was an economist and worked at times as a journalist, having studied at the University of Kansas, the London School of Economics, and Harvard. In 1981, he was a Fulbright scholar at Tufts University. In 1991, Santos was appointed by President César Gaviria as Colombia's first minister of foreign trade and served until 1994. In 1992, he was named president of the VIII UN Conference on Trade and Development. In 1994, he founded the Good government Foundation that was designed to increase the effectiveness and credibility of Colombia's government. It advocated peace talks with FARC. From 2000 until 2002, he was President Andrés Pastrana's minister of finance and public credit. It was Santos who founded the U Party, in 2005, to support President Uribe and spearhead the fight for the amendment that would allow the president a consecutive term.

After Santos won the presidency, Uribe joined the U Party and initially supported his protegé. But the two men soon fell out when Santos announced that, since September 2012, he had been engaging in secret peace negotiations in Havana with FARC. The encounters were sponsored by both the Cuban and Norwegian governments and had the endorsements of Venezuela and Chile as well.

Since Uribe believed that FARC was on the run due to the assiduous security measures of his administration, he feared that such (premature) negotiations would likely give the militants both legitimacy and breathing space. So, Uribe Velez became Santos's most outspoken critic. In 2013, the former president left the U Party to form his own Centro Democrático (Democratic Center) political party. He then ran as its candidate for a Senate seat in the Congress in 2014. He won and entered the legislative body along with eighteen other Democratic Center senators and twelve representatives. Uribe was now the main voice of the opposition.

There was initial leeriness on the part of the indigenes, given that FARC had a history of favoring reform to establish Colombia's peasantry on the land without promising to respect indigenous territorial claims. This was potentially a huge issue. Now, FARC seemed to be committed to respecting indigenous rights, although indigenes had no place at the negotiating table. Of more immediate concern was the fact that some indigene leaders were negotiating possible concessions of mineral resources on their tribal lands to interested outsiders as a means of raising funds for social programs and economic development. There was fear that this would play into the obvious governmental campaign to undermine its commitment to prior consultation with the indigenes before awarding such concessions. By empowering weak Indian political hierarchies in the targeted areas, the extractors and government were trying to make it easier to "buy" Indian concurrence.[105]

There was, however, push back. A coalition of Colombian tribes declared that they were the proper custodians of the country's Amazonian lands and that "irreversible harm to the natural environment can only be described as a crime against humanity."[106] Then, too, when the Colombian government, and those who had newly-acquired estates, the squatters, tried to overturn the country's Rural Development Law and Forestry Law (which required prior consultation with the indigenes before awarding concessions on their lands) the Constitutional Court declared the new legislation to be unconstitutional. Re-emboldened, the indigenes held a National Indigenous Communication Forum in Cauca, attended by more than seven hundred persons, including delegations from Mexico, Peru, and Ecuador. A film festival with the slogan "For life, Images of Resistance" was held in Bogotá, attended by artists from fifty countries. In all, there were sixty-five films, and the successful festival was then repeated in Medellín.[107]

The following year, there were two weeks of Indian protests held in Cauca, Valle, Antioquía, Risaralda, Tolima, and Huila (most of the places of origin of the original white and *mestizo* Colombian settlers in the Putumayo/Caquetá) involving no fewer than forty thousand indigenes. They were demonstrating against governmental erosion of the prior-consultation laws and their exclusion from the peace negotiations with FARC (there was lingering concern that provisioning of land to landless Colombian peasants would be at the expense of the country's indigenist and Afro-Colombian reserves). The government blinked and signed a series of accords.

There was then the IXth COICA Congress that was attended by the Colombian minister of the interior, (Basque-surnamed) Dr. Aurelio Iragorri Valencia, from one of the "highest ranking families" in the Cauca. He praised the indigenes for conserving the Amazon and stated: "It is we who are the devastators, the whites, as opposed to the age-old preservation maintained by the indigenous peoples." While speaking, he wore a feathered headdress and an indigenous necklace. He promised "to establish non-aggression pacts for the ecosystem and to promote preservation and protection policies to care for the Amazon." The comment in the IGWIA report was as follows: "We do not know what opinion these words elicited from his indigenous countrymen and women from Cauca (quite possibly

---

105. Jaramillo Jaramillo, *Indigenous World 2013*, 112–13.
106. Ibid., 113.
107. Ibid., 113–14.

surprise)." At the COICA Congress, the Peruvian Huitoto leader Edwin Vásquez was re-elected coordinator.[108]

President Santos prevaricated regarding running for a second term in 2014, given that his popularity in the polls plummeted under the acerbic Uribe criticism as the talks in Havana drug on. However, just before the six-month deadline for filing, Santos announced that he would be running again, citing the importance of the "unfinished business" (the Havana negotiations). The race was close and the Democratic Center's (Basque-surnamed) Óscar Iván Zuluaga (highest vote getter in the first round with 29.25 percent to Santos's 25.69 percent) seemed headed for victory, when, a week before the runoff, there was a scandal. A secret tape was released of Zuluaga discussing how to use to his political advantage information on the Havana talks gleaned from illegal wiretaps of them. Zuluaga at first denounced the tape as doctored, but it was not. When a public media newscaster then revealed that he had been offered the tape by Zuluaga's campaign manager, (Basque-surnamed) Luis Alfonso Hoyos Aristazábal, the latter resigned. Santos won re-election in the second round with 50.95 percent of the vote.

The year 2014 was dominated by the peace talks in Havana between the president's representative and FARC leaders. Indigenes were both encouraged and discouraged by developments. On October 7, 2014, President Santos signed Decree 1953 which authorized temporary activation of indigenous self-rule in their territories until the Congress could finalize the stalled Organic Law on Territorial Regulation.[109] Furthermore, during the ongoing Havana negotiations, FARC was committed to working democratically within Colombia's existing political framework, as defined most recently by the 1991 Constitution that proclaimed the nation to be multiethnic and pluricultural. Presumably, FARC would now respect indigenous rights. Indeed, of greater concern was former President Uribe and his allies, as well as the remaining paramilitaries. It could be said that

> The many years of so-called narco-paramilitaristic presence and its impact on local andregional politics (although also reaching into national politics) have penetrated the country's institutions, making them malleable to the ebbs and flows of money and giving a "Sicilian" flavour to the country. Moreover, it is noticeable that the economic interests that benefited from the armed conflict are able the move easily within this scenario. These interests do not approve of and will certainly oppose any agreements to return the

---

108. Jaramillo Jaramillo, *Indigenous World 2014*, 120–21.
109. Jaramillo Jaramillo, *Indigenous World 2015*, 123–24.

lands, assets and resources of the peasant farmers, Afro-Colombians and indigenous peoples, acquired illegally and at very little cost.[110]

Then, in 2015, there were renewed Indian protests in Colombia, centered upon developments in the Cauca Department. Ten thousand Indians occupied four estates, sixty were wounded by the government's *Fuerza Pública* security force, defended by Uribe's ultra-right parliamentarian ally, Senator Paloma Valencia, herself a Cauca latifundist. The leader of the indigenes, Feliciano Valencia of the Regional Indigenous Council of Cauca (CRIC), demanded that the government Decree 982 (1999) that mandated distribution to the Indians of thousands of hectares of land be implemented rather than ignored by the Colombian government.[111]

Meanwhile, the negotiations continued in Havana, as well as in Oslo. FARC, clearly bruised[112] and beleaguered, was disposed to negotiate honorable terms for cessation of its hostilities. An obvious key issue was the fate of its militants. Events took a bizarre turn when internationally famed Bengali Indian musician and Hindustani classical music composer Ravi Shankar met in Havana with FARC leaders and convinced them to forsake violence à *la* Gandhi. On July 8, 2015, FARC announced a unilateral ceasefire, along with a pledge to work within the political system. On June 23, 2016, there was an announcement in Havana of a preliminary accord, and then, on September 26, FARC Commander Rodrigo ("Timochenko") Londoño and President Santos Calderón signed a final agreement in Cartagena. Several world leaders were in attendance, including UN Secretary General Ban Ki-moon, U.S. Secretary of State John Kerry, and Cuban President Raúl Castro. Secretary Kerry pledged 390 million dollars in U.S. aid to implement the agreement. It had been praised but a few days earlier by President Obama in his final address to the United Nations.

Under the terms of the accord, FARC militants were to turn their arms over to a UN delegation within 180 days and assemble in safe zones where their individual cases would be heard by a Special Peace Jurisdiction tribunal consisting of a wide array of Colombian judges and officials, as well as some foreigners. In general, FARC militants would be exonerated from "sedition." Those who admitted to committing war crimes and being repentant were to be given light sentences (about five to eight years) that were, in effect, probationary since they would not entail jail time.

---

110. Ibid., 125.
111. J. Ignacio Lacasta Zabalza, *La memoria histórica* (Iruña/Pamplona: Pamiela, 2015), 30–31.
112. Its former supreme commander, Alfonso Cano, was killed by government security forces on November 4, 2011.

Rather, they would be served in designated restricted zones. FARC would become a normal political party, but with an initial guaranteed five seats in the House of Representatives and five seats in the Senate. Rehabilitated FARC militants were eligible to run for office, including Congress. The victims of both sides would eventually be compensated, partly from FARC funds. The spirit and purpose of the agreement were encapsulated in the preamble: "Respect for human rights in every corner of the national territory is a state purpose that should be promoted; economic development with social justice and in harmony with the environment is a guarantee of peace and prosperity." Key to everything was acknowledgment that the country needed land reform and more democratic institutions—the FARC's stated objectives when launching its resistance fifty-two years earlier. The agreement also invited the participation of others; neither ELN (Ejército de Liberación Nacional or National Liberation Army) nor the remaining paramilitaries had been parties to the negotiations.

There were, of course, likely be many devils in the details. At the very least, for the agreement to become the basis of the new political reality would require considerable legislation by the Congress. Also, President Santos had vowed to submit the final agreement to the Colombian electorate for approval. Since the likely positive vote was leading by double digits in the polls, this seemed to be a shrewd means of consolidating the achievement. However, it was opposed by Uribe and his political allies, as well as by most of the large-scale landowners. Uribe launched a vitriolic campaign that criticized the agreement's specifics while characterizing it as tantamount to issuing amnesty to criminals. When the plebiscite was held on October 2, the "no" vote prevailed by a slim 50.2 percent. This triggered a spate of violence across the country in which several political and social leaders were killed.

Acknowledging his defeat, President Santos announced that, with FARC's concurrence, the Havana negotiations would resume. A new accord was signed on November 24. The agreement was tightened—notably, among other things, it now limits considerably the movements of persons (FARC) serving sentences of "restriction of liberty," the accord would not be incorporated (as FARC had wished) into the Colombian Constitution, nor would there be foreigners on the Special Peace Tribunal hearing cases and its existence was reduced to ten years (with the possibility of a single five-year renewal) and FARC would provide an immediate accounting of all of its assets (which would then be dedicated to retribution of victims).

Interestingly, there were no changes whatsoever in the Chapter 6.1.12.1, which regards the rights of ethnic peoples. It is introduced by the statement

> That the National government and FARC-EP recognize that the ethnic peoples have contributed to the construction of a sustainable and lasting peace, to progress, to the economic and social development of the country, and that they have suffered unjust historical conditions, product of colonialism, of enslavement, of the exclusion that has dispossessed them of their lands, territories and resources, and that besides they have been gravely affected by the internal armed conflict, and they should be accorded the maximum guarantees for the full exercise of their human and communal rights within the context of their own aspirations, interests and world views.

It is also stated, "In no case will the implementation of the agreements proceed to the detriment of the rights of the ethnic peoples." The accord specified particular territories that FARC agreed to return to their indigenes the territorial homelands that had been usurped by it. Both parties to the agreement agreed to work for a program that would resettle the refugees dispossessed by the conflict to their homes. It was also envisioned that a high-level authority, with representation of the government, FARC, and the relevant indigenous and Afro-Colombian associations, would oversee implementation of the accord insofar as it affected ethnic peoples.

President Santos (fearing another political debacle) then took the position that the modified document need not be subjected to national referendum, but rather could be approved instead by Congress. In late November 2016, there was unanimous approval in both chambers, with Uribe's Democratic Center Party and allies walking out before the vote.

As of this writing, there is still considerable indeterminacy about the outcome. It is unclear what the Uribe Vélez forces plan to do. The strategy of the Santos administration to "fast track" the agreement through Congress, along with the plethora of legislation that will be required to implement its details, is currently under review by the Colombian Constitutional Court. It could all be declared invalid—thereby subjecting the accords to months or even years of delay. That is a disturbing prospect, since there is little likelihood that FARC militants would endure their present legal limbo indefinitely. Nor is it clear that President Santos could sustain the country's political will through the process—particularly if Uribe launches a new assault on it.

Over the fifty-two years of its existence, the conflict between FARC, the paramilitaries and the Colombian government had killed at least 260,000 people (by some estimates as many as half a million) and dispossessed six to seven millions from their homes. For his "transiton from hawk to dove" and his tireless efforts on behalf of a permanent resolution of the conflict, on December 10, 2016, President Santos Calderón accepted that year's Nobel Peace Prize in Stockholm.

Aranalandia Revisited

It is difficult to determine how many Huitocia indigenes remain in our study area (and elsewhere) at present. Clearly, their numbers have been reduced dramatically from the time that Crisóstomo Hernández first settled in the future Aranalandia. Virtually all of the estimates of that era postulated the numbers of Huitotos alone in the tens of thousands, and there were a number of other tribes. Calculations of the magnitude of the subsequent decline of this population range as high as "80 or 90 percent" and tend to be stated in such vague terms. As we have seen, some Indians were murdered or died from other forms of abuse, and illness took a significant (if largely undocumented) toll. There are also reports regarding psychological malaise that produced infanticide and an unwillingness to bring children into such a troubled world. Then, too, Indians were sold as slaves or indentured beings to enterprises located elsewhere in Peru, Colombia, and Brazil. Once it became likely that Aranalandia would transfer from Peruvian to Colombian control, the Casa Arana itself forced several thousand to resettle on its other holdings (notably the ones near Pebas). Many indigenes before and after that process fled the area on their own to resettle beyond the perceived oppression in Huitocia.

We have the account by Carlos Uribe Gaviria in his report to the Colombian minister of war sent on May 9, 1933:

> I transported in the gunboat to this place one hundred and two Huitoto indigenes picked up at Inonias and Florida, where they were kept naked, hungry and subject to forced labor by the inhumane employees of the Casa Arana . . . It is exactly the situation that I described with regard to the Algodón River, as I reported opportunely to General Rojas: hundreds of Indians vilely exploited, treated like animals, worse than slaves; reduced to a lamentable state of physiological penury by too much work, scarcity of food, of clothing, drugs. . . . The physical and moral suffering of these poor people was to such a degree, such was the intensity of their tuberculosis, their malaria, their tropical anemia and venereal diseases, that it was necessary to isolate them completely to shield our troops against contagion. . . . In the year 1933 everything was the same as at the beginning of the century.[113]

---

113. Uribe Gaviria, *La verdad*, 208.

There were subtler processes at work as well. It seems clear that some indigenes embraced white culture and became self-motivated players within it. They learned skills that qualified them as river mariners and urban contruction laborers, not to mention agricultural hands. They sought employment on their own terms, and often relocated idiosyncratically to find it. Such persons could end up just about anywhere. And then there was another way for "Huitotos" to disappear. Overshadowed by the accounts of sexual abuse in the many testimonies that we have considered were the consensual sexual unions. Hernández was given two Indian women by *caciques*, and it was reported by some that Indian parents were honored if a white chose to marry their daughter. Such unions, of course, produced *mestizo* children who might or might not nurture their indigene identity and heritage, particularly if the family relocated to a place like Puerto Leguízamo, Leticia, or Iquitos. Consequently, ethnocide, as distinct from genocide, was a factor as well.

We have the missionary accounts of the vast depopulation of Aranalandia by the late teens and 1920s. By 2004, the Colombian Constitutional Court had listed the Huitotos as an endangered people.[114] The national Censo DANE (Departamento Administrativo Nacional de Estadística) of 2005 placed the total population of Huitotos [Uitotos] in Colombia at 6,444, of whom 3,725 (or more than half) lived in the Department of Amazonas. Urban Huitotos (including Bogotá) numbered 1,652.[115] The 2006 report on the Resguardo Indígena Predio Putumayo noted that the entire population of the vast La Chorrera region was but 2,343 inhabitants, of whom only 47 percent were indigenes. The urban nucleus of La Chorrera itself had 324 people.[116]

By mid-twentieth century, the remnants of Huitocia's former tribes were scattered among the region's population centers and dispersed among whites and *mestizos* throughout the rural districts. There were examples of indigene neighborhoods and agricultural communities with some degree of socio-economic and political control over their own affairs. For the most part, these were "inventions" in that they tended to be multi-

---

114. It did so in its report Auto 004/09, issued on January 26, 2009, and entitled *Corte Constitucional — Protección de derechos fundamentales de personas e indígenas desplazados por el conflicto armado en el marco de superación del estado de cosas inconstitucional declarado en la sentencia T-025/04 de 2004, después de la sesión pública de información técnica realizada el 21 de septiembre de 2007 ante la Sala Segunda de Revisión* that can be consulted on the Internet. The document includes a detailed listing of the violence inflicted on indigenes by the Colombian conflict.

115. As reported by the Ministerio de Cultura, República de Colombia, *Características de los pueblos indígenas en riesgo* (Bogotá: Ministerio de Cultura, 2010), 249–50.

116. Ministerio de Ambiente, Vivienda y Desarrollo Territorial, *Resolución*, 13.

ethnic, composed out of components from several of the former tribes. Their challenge, then, was to forge a new collective cultural and political order out of the shards of their individual traditions, all on a Colombian national political playing field upon which they were neither experienced nor skilled players. They were impoverished, uneducated, and in ill health. Publishing as recently as 1985, Father Restrepo López said of the Huitotos around Puerto Leguízamo that there were

> Two or three hundred in all. Their beautiful language was now disappearing, and they are abandoning their traditional dress. . . . The enormous "malocas" (multi-family dwellings) are now rare.[117]

The indigenes were not entirely on their own, however. By the middle of the twentieth century, a combination of certain Colombian governmental agencies and missionaries was preparing a new generation of indigenes (largely males) to take their place in the regional and national economies, although the Huitotos remained largely near-subsistence agriculturalists. Then, in the 1980s, a number of NGOs became active in the Putumayo/Caquetá, training locals to be activist political leaders in protecting the environment, while retaining their cultural and linguistic legacies. They also served as role models and windows on the wider world, including life in Bogotá. There began a degree of migration of young male Huitotos to that city.[118]

We have the interesting Luisa Fernanda Sánchez account regarding Huitotos in the national capital. Beginning in the 1950s, young Huitotos were exposed to outsiders from some governmental agencies, academics (particularly anthropologists), and representatives of NGOs from the Corporación Araracuara, Tropenbos and the Fundación Gaia. At the same time, these outsiders served as role models and bridges to Bogotá. So, until the 1980s, there was a trickle of Huitoto emigration to the national capital.[119] These were in the main young males in search of higher education, facilitated in some cases by missionaries and possibly under the patronage of Bogotá's Fundación Katuyumar (established to help Indian students to find lodgings in the city).[120]

---

117. Restrepo López, *El Putumayo*, 54. The situation with other groups was worse. In 2014, it could be said regarding the Okainas, "It's just one old man and his five sons. . . . When he dies the language will go with him." http://fucaicolombia.org/blog/colombian-woman-hopes-to-solve-century-old-mystery.
118. Luisa Fernanda Sánchez, "Paisanos en Bogotá," in Margarita Chaves and Carlos del Cairo (eds.), *Perspectivas antropológicas sobre la Amazonia contemporánea* (Bogotá: Instituto Colombiano de Antropología e Historia, 2010), 129–52.
119. Ibid., 133.
120. Ibid., 134–35.

Meanwhile, in the 1980s, an airstrip was constructed at La Chorrera, facilitating frequent air service with Bogotá. In 1986, a national college scholarship program was initiated for bright indigene students of both sexes, and several Huitotos obtained this financial aid to study in Leticia or Bogotá. They were thereby exposed to many cosmopolitan influences that broadened their personal tastes well-beyond indigene, and even Colombian national, ways of life. The women pursuing higher education were exposed to feminist ideas and aspirations along the way.[121]

If Peruvian President Fujimori pretty much ignored the ILO Convention 169, promulgated in 1990 (or the same year he first took office), Colombia was far more impacted by that international declaration. This provided considerable stimulus to Indians and their fellow travelers to embellish their formal organizations in order to protect indigenous rights within the Colombian political process—ranging across the entire political spectrum from the municipal through the departmental to the national levels.

In order to become active on this new political playing field, the indigenes had to constitute *cabildos*, or councils, which might encompass a whole village or simply the neighborhood of a town. The process involved a certain amount of invention and "re-ethnicization," as it were, since many of the traditions, customs and crafts had been fragmented or lost altogether. Many of the new *cabildos* adopted names that invoked their indigenous past, and new rituals were constituted as echoes of former ones dimly remembered. The imaginative "magic show" was as much a performance for insiders recreating their own identity as a means of projecting it to the outside world.[122] Furthermore, the Indians remained a distinct minority[123] and, in order to be noticed politically, they needed to constitute larger associations with a multiplicity of *cabildos*. In 1987, the Huitotos formed the Organización Regional Uitoto del Caquetá, Amazonas, y Putumayo (the Huitoto Regional Organization of the Caquetá, Amazonas and the Putumayo) headquartered in Orocapu, near Florencia. However, it met with strong resistance from the non-Indians of the Department when it sought its own *resguardo*).[124]

---

121. Ibid., 136–42.
122. Margarita Chaves Chamorro, "Cabildos multiétnicos e identidades depuradas," in Clara Inés García (ed.), *Fronteras, Territorios y Metáforas* (Medellín: Hombre Nuevo Editores, 2003), 121–26; 130.
123. By the early twentieth century, they constituted but 8.2 percent of the total population of the combined departments of Caquetá, Putumayo and Guaviare. Carlos del Cairo, "Las encrucijadas del liderazgo político indígena en la Amazonia Colombiana," in Margarita Chaves and Carlos del Cairo (eds.), *Perspectivas antropológicas sobre la Amazonia contemporánea* (Bogotá: Instituto Colombiano de Antropología e Historia, 2010), 192.
124. Ibid., 191–200.

Within the nation's liberal intellectual circles, it now became fashionable to be an indigenist. There emerged a notion that the indigenes were the authentic precursors of the nation and descendants of ancestors who possessed privileged knowledge regarding the natural environment. For indigene leaders and environmentalists alike, Indians were potentially the wisest (as in most experienced) stewards of "Mother Earth."[125]

Such idyllic attitudes and their philosophical impacts upon political processes notwithstanding, in the trenches there were the standard human shortcomings. Since indigenes were in the distinct minority in most political venues, it was often the case that lip service alone was paid to their rights that were in fact ignored or given low priority in the real decision-making process. Unsophisticated aboriginal leaders might easily be coopted or corrupted. In the Middle Putumayo, several were indicted by the government for taking bribes and selling favors.

In 1973–1974, there was an intense military campaign to rid the country of its leftist rebels. It inflicted considerable casualties but without finishing off FARC. It did, however, impact our study area by driving the rebels into more remote areas of the country. In some cases, FARC appropriated political control, even of Indian communities, assassinating their leaders. The rural parts of the Middle Putumayo/Caquetá fell under the control of FARC, with several consequences for the remaining indigenes. FARC forcibly conscripted many Indians into its ranks (shades of Peru's Sendero Luminoso). It exacted an onerous revolutionary tax. In principle, FARC was committed to clean government and so it assassinated what it perceived to be corrupt local leaders, including some indigene ones. FARC also began to cultivate coca, at times in forest reserves and at others on the smallholdings of indigenes. The very presence of FARC drew in paramilitaries and government forces that were at times their own scourge of the countryside.

Then, too, there was a considerable influx of *cocaleros*, primarily from the Nariño region, whose coca fields there had been devastated by the joint Colombian and United States program of spraying their crop with toxins from the air. These newcomers, as well, usurped indigene lands and communities. This transpired particularly in the Upper Putumayo/Caquetá, where the eleven thousand indigenes were increasingly beleaguered in their homeland.[126]

---

125. Del Cairo, "Las encrucijadas," in Ibid., 190; 198–99.
126. Restrepo López, *El Putumayo*, 33; 56. It also crossed the border into Peru where there was much the same dynamic with the Shining Path movement. By the early 1990s, there were 200,000 hectares of *coca* under cultivation in Peru and it was, after rice, the country's most-extensively seeded crop. San Román, *Perfiles*, 256.

It was also about this time that the Putumayo/Caquetá was the epicenter of a petroleum boom. Beginning in 1965, Texas Petroleum began developing what would prove to be rich fields in the Alto Putumayo at Orito, La Hormiga, Acaé, La Dorada, San Miguel, and El Tigre. Soon, there were a hundred producing wells and migrants from elsewhere (particularly the Nariño) flocked to the new job opportunities. The population of Puerto Asís, the primary servicing center for the petroleum economy, tripled in three years' time. It acquired all of the characteristics of a boomtown, including countless taverns and ubiquitous prostitution. New roads were opened and agricultural colonists from elsewhere in Colombia settled along them. The district's (the Comisaría's) budget went from 230,000 pesos annually before the petroleum to 175,000,000 pesos in 1975. The municipal budget of Puerto Asís had been 100,000, yet, in 1975, or just a year before ownership of the oil passed to State-owned Ecopetrol, Texas Petroleum paid the town a 25-million peso royalty.[127] By the mid-1980s, the more than 150 wells in the Orito oilfield would be producing 30 percent of the nation's total petroleum output and the pipeline from Orito to Tumaco was the country's longest.[128]

All of these factors accelerated the flight of many Indians, particularly Huitotos, from their rural settlements into nearby and distant urban centers (including Bogotá). During the 1990s, then, both out of fear and in search of economic opportunity, there were many Huitoto job seekers of both sexes who left the Putumayo. Some effected family reunions with former migrants to the city. Most found employment in largely menial jobs. There was a demographic bias favoring male job-seekers in the migration.

There were, however, notable countervailing forces in play. Huitotos tended to associate with fellow ethnics in the city. They had an interest in being "professional Indians," as a way of commanding respect within the new constitutional reality (after 1991), and even as a means of networking for educational and employment opportunities (or what has become known as "affirmative action" in the United States regarding its own minorities).

Yet there remained the "eternal nostalgia," the intention of ultimately returning to the ethnic homeland in the Putumayo/Caquetá. Preferenital endogamy was one means of doing so. If a man married a Huitoto woman, she was likely a speaker of his indigenous language and a bearer of his indigenous culture. Such a marriage clearly facilitated the intended return "home."[129]

---

127. Ibid.
128. Ibid., 61.
129. Ibid., 147–50.

The Leticia Trapezoid became the focal point for human settlement in the Lower Putumayo/Caquetá region, centered upon the twin cities of Leticia on the Colombian bank of the Amazon River and Tabatinga on the Brazilian shore. By the early twenty-first century, the estimated population of the area was no fewer than 130,000 to 150,000 persons.[130] While the physical boundaries among the three countries bordering the Trapezoid were determined in the 1930s, the social and cultural ones remain a work in progress—albeit configured in the main by much of the history that we have considered in this text. Ironically, the area has become a vibrant liminal and ephemeral zone hosting a complex interplay of several social and ethnic identities.

Anthropologist Jean-Pierre Goulard provides rich ethnographic detail in this regard. There are the "whites," some born locally and of mixed racial origin and others (the *real* whites, as in persons of European descent) present as migrants from distant parts of Colombia and Brazil. Some of these are fugitives from nebulous pasts, while others are the agents of external governmental agencies and NGOs. Then, too, there are those in search of opportunity within the "frontier" economy. While few in absolute numbers, the whites regard themselves to be the vested social and political elite. They tend to maintain external Colombian regional loyalties, socializing primarily with persons of like regional background. In any event, the whites tend to interact but little with other social and ethnic groups.[131]

Then there are the *mestizos*. This is a broad category that includes locally born persons—descendants of *caucheros*, manual laborers from other parts of Brazil, Colombia, and Peru in search of employment in the vibrant economy (including its logging industry and drug trade), *caboclos*, and migrant *indígenas* from elsewhere who have ceased to claim an Indian identity. Among all of these groups, the "outsiders" retain an awareness that their ancestral homeland lies elsewhere and their Trapezoid roots are shallow.[132]

The *indígenas* encompass more than half the population and are persons of diverse tribal origins. The Ticuna are the region´s endemic and most numerous group (many thousands). They are mainly agriculturalists settled along the river banks. However, both Colombia and Brazil have encouraged other indigenes to settle within the Leticia Trapezoid, includ-

---

130. Jean-Pierre Goulard, "Un horizonte identitario amazónico," in Margarita Chaves and Carlos del Cairo (eds.), *Perspectivas antropológicas sobre la Amazonia contemporánea* (Bogotá: Instituto Colombiano de Antropología e Historia, 2010), 293.
131. Ibid., 291–92.
132. Ibid., 292–93.

ing a few thousand Huitotos. Some of the latter are urban workers and students, while others have settled in rural districts—even erecting *malocas* under the authority of traditional *capitanes*. There is a certain awareness among the Huitotos that they are outsiders whose homeland is elsewhere. Some came from Huitocia and others first fled it to the Pebas area near Iquitos before resettling in the Trapezoid.[133]

Another element in the mix is provided by the "Israelites," the followers of the prophet Ezequiel Atacausi. Founded in the Peruvian highlands, this reclusive, indigenist, cult-like messianic movement is a syncretism of Judaism, Christianity, and "Incaism." Its proponents converse in Spanish and Quechua. Many resettled in successive waves on the left bank of the Yavarí River. Today, they constitute almost a tenth of that region's population and control the mayorship of several of its towns. They do not, however, interact much with other ethnic groups and do not claim a particular ethnic identity.[134]

Superposed upon all of the foregoing is a certain sense of national origin. Those with roots in Peru invoke their "Peruvianity" and are perceived in such terms by others. The same is true of persons with Colombian and Brazilian roots. As if all of this complicated interplay of identity were insufficient, Goulard notes that a new one of "Amazonian Man," based in the shared Leticia Trapezoidal experience, is emergent.[135] Goulard sums up the situation:

> This reading of the Colombian Amazonian Trapezoid reflects the social fragmentation of its inhabitants, who, despite everything, share in a common legal system known to everyone, that is managed by each of the classes according to its particularities. To a national legal system are added regional law and local law as they affect each class, which in the latter case allows each to negotiate the terms of its relations with all others. Consequently, in the Amazonian Trapezoid one lives a constant reconstruction of identities, in accord with specific criteria and the requirements of each class.[136]

---

133. Ibid., 293, 94.
134. Ibid., 294–95.
135. Ibid., 289. I am more than a little fascinated by all of the foregoing. In the 1970s, I co-authored two essays with sociologist Stanford M. Lyman in which we argued that ethnic identity is not a fixed and exclusive feature of an individual's persona, rather, most people have access to multiple ones that they manipulate in the course of their everyday social intercourse depending upon the situation and their interlocuters. See Stanford M. Lyman and William A. Douglass, "Ethnicity: Strategies of Collective and Individual Impression Management," *Social Research* 40, no. 2 (1973), 197–220. I can scarcely imagine a better context than the contemporary Leticia Trapezoid in which to test our hypotheses.
136. Jean-Pierre Goulard, "Cruce de identidades El Trapecio Amazónico colombiano,"

There is also considerable transnational interaction in the Leticia Trapezoid. Given that it is a frontier borderland for three countries whose indigenous (and non-indigenous) ties transcend the formal boundaries, it has been said of the Ticunas, for example, that

> Actually, in the frontier region of Brazil, Colombia and Peru, the daily life of the Ticuna Indians manifests constant transfrontier interaction, in terms of the sociocultural practices characteristic of this group: frequent visits to parents and friends, participation in the adolecence menarche ritual..., attendance at religious events, commercial exchange, establishment of marital ties, traditional medical practices, sporting matches and participation of indigenous authorities and leaders in the events organized by the different supralocal Ticuna political organizations.[137]

Claudia Leonor López Garcés notes that the Ticunas' sense of national identity in Peru and Brazil (countries with little Catholic influence in governmental affairs) derives mainly from military service,[138] whereas in Colombia it is fomented primarily by the missionaries working in the indigenous communities.[139] It might also be noted that the three states now facilitate interaction among all of their citizens in adjacent borderlands.[140]

The Indians themselves see sub-ethnic differences among themselves derived from their differing national experiences. The Brazilian Ticunas are said by Colombian and Peruvian ones to be "little civilized," aggressive trouble makers and given to witchcraft. Nevertheless, their curers are the best and are sought out by their transborder neighbors. The tension between Colombian and Peruvian Ticunas is even greater, and derives both from the conflict between the two countries in the 1930s and general Colombian pejorative views of Peruvians as a whole. So the Peruvian Ticunas are said to be "little thieves, hypocrites, and traitorous. They even smell bad!"[141]

---

in Clara Inés García (ed.), *Fronteras: Territorios y Metáforas* (Medellín: Hombre Nuevo Editores, 2003), 100–101.
137. López Garcés, "Etnicidad," 151.
138. San Román, *Perfiles*, 194–95. He tells us that military service also provides Peruvian youth with the cultural experience in the wider world that also leads them to migrate to the cities.
139. Claudia Leonor López Garcés, "Etnicidad y nacionalidad en la frontera entre Brasil, Colombia y Perú. Los Ticuna frente a los procesos de nacionalidad," in Clara Inés García (ed.), *Fronteras, Territorio y Metáforas* (Medellín: Hombre Nuevo Editores, 2003), 147–50.
140. Ricardo Zuluaga Gil, "Régimen jurídico de las entidades territoriales de frontera," in Clara Inés García (ed.), *Fronteras, Territorio y Metáforas* (Medellín: Hombre Nuevo Editores, 2003), 351–61.
141. López Gárces, "Etnicidad," 154–55.

In any event, despite the lengthy litany of pain and penury experienced by the Indians of Aranalandia, as well as the considerable depopulation of their area, the flame has not gone out entirely. The Huitotos, in particular, demonstrated considerable staying power—reconstituting communities out of the remnants of the shattered ones within their ancestral homeland and as dispersed migrants in greater Huitocia and beyond.

There were also interesting developments in the heart of Aranalandia itself. In 1939, the Colombian government had authorized its Banco Agrícola Hipotecario (Agricultural Mortgage Bank) to acquire the Casa Arana holdings in the Putumayo/Caquetá, with a downpayment of $40,000, prior to final implementation of the Salomón Lozano agreement. Presumably, this was the Victor Israel transaction referred to earlier; whereby he acquired the rights from the Aranas.

Then, in 1954, the government authorized its Caja de Crédito Agrario, Industrial y Minero (Agrarian, Industrial and Mining Credit Bank), to liquidate the failed Agricultural Mortgage Bank. In 1959, the government declared its Amazonia region to be a forest reserve, and, by the 1970s, it was creating indigenous reserves around the country. It was in 1975, or under President Michelsen's administration, that the Colombian state approved creation of several such reserves within the Putumayo region. We have noted that some Huitotos fleeing Aranalandia had resettled up the Putumayo and Caquetá rivers near the future Puerto Leguízamo. In 1975, 40 families (193 individuals) were given a 5,000 hectare reserve at Jirijiri (near La Tagua) and the Huitotos of Samaritana, or about one hundred persons, received 2,000 hectares as their reserve.[142]

By the late 1970s, the state's failed Agricultural Mortgage Bank was undergoing liquidation and, in 1980, the Agrarian, Industrial and Mining Credit Bank acquired its assets in the Putumayo. In 1983, and again in 1985, the head of the Caja visited the area for meetings with indigenous leaders to explain to them his intentions. In 1986, the Caja announced its ambitious scheme, called the La Chorrera Plan, to protect the region's ecology and inhabitants in collaboration with the latter. It also proposed that within the former Aranalandia 5,879,000 hecatares be declared an Indian Reserve. There were 10,335 persons inhabiting the area at the time, a significant minority of whom were indigenes. In 1987–1988, the Caja negotiated a sale of the entire territory to INCORA, holding out 802.5 hectares at the La Chorrera site, home of a number of indigenes, for continuation thereon of its research projects.[143] The indigenes opposed the

---

142. Restrepo López, *El Putumayo*, 59.
143. It seems that the plan was to transfer this land ultimately to a private investment company that planned to grow rubber and other tropical products on it (it was later dis-

hold out (to no avail) on the grounds that it might be another dispossession of them in the making. In 1988, *the Resguardo Indígena Predio Putumayo* was formalized—intended to be both an Indian and environmental reserve.[144] During the ceremony creating the new *resguardo*, its president, Virgilio Barco Vargas, said:

> Faced with the need to conserve the Amazonian rainforest, we consider the indigene approach to be a valid proposition for coexistence with the ecosystems. The industrialized countries . . . have an ecological debt with humanity. In less than two centuries, the model of exploitation of natural resources in the industrialized nations not only wiped out the fauna and the forests of Europe and North America, they practically brought to the brink of extinction their indigenous peoples and autoctonous races . . . these countries destroyed the bulk of their renewable resources, without giving any consideration whatsoever to the consequences and costs for humanity of their lack of ecological consciousness. The destruction of the ozone cap, the contamination of the seas, acid rain, climatic change, the nuclear tests and disasters are just the most evident effects of unfettered development in the industrialized nations. They are indebted to us, to all of humanity. We should join forces in a new equitable and global formulation regarding treatment of threatened environments that could work against the [well-being] of humanity. One of the means by which the industrialized countries could pay their ecological debt with the Third World and humanity would be to take concrete measures to ensure that vested interests do not continue violating the planetary ecosystems while providing indispensable resources so that the alternative models of sustainable management of Amazonia can be a reality.[145]

In 1993, the twenty communities of the Igaraparaná River drainage formed AZICATCH (La Asociación Zonal Indígena de Cabildos y Autoridades Tradicionales de La Chorrera or The Chorrera Zone Indigenous Association of Cabildos and Traditional Authorities), declaring its Gen-

---

closed that the director of the *Caja* had an interest in the latter). Colombian Ministerio de Ambiente, Vivienda y Desarrollo Territorial, *Resolución Número* (1947), September 29, 2006, 12–14.

144. [Colombian] Ministerio de Ambiente, Vivienda y Desarrollo Territorial, *Resolución*, 12–14.

145. Asociación de Cabildos y Autoridades Tradicionales de la Chorrera—AZICATCH, *Plan de Vida de los hijos del tabaco, la coca y la yuca dulce y Plan de Abundancia Zona Chorrera 2004–2008* (La Chorrera: AZICATCH, 2006), 47.

eral Assembly to be the highest political authority within the area (while respecting the right of each component to manage its affairs).[146] In 1994, and with the intervention (solicited by AZICATCH) of Colombia's first lady, Ana Milena Muñoz de Gaviria, the ruined buildings of the Casa Arana at La Chorrera were leased for ten years by the Caja to the Department of Amazonas for creation of a secondary boarding school for indigenes. However, when the lease ran out in 2004, the property was put up for sale. The chagrined Indians protested successfully to President Uribe and the minister of the interior. The sale to the private sector was blocked and public funds were allocated to purchase the facility and turn it over the Indian authorities of the *resguardo*.[147]

In 2006, AZICATCH published an extraordinary "Plan of Life" (and "Plan for Abundance") of the "Children of Tobacco, Coca, and Sweet Yucca" as the inhabitants, constituted as they were by four separate tribes, now self-identified both as individuals and collectivities. They were all protected and safeguarded by Moo Uai, the father creator who, from the beginning of time, mandated that their mission was caring for and respecting the environment and all of the creatures living within it. Every territory had its protector spirit and each being a protector mother. Employing the principle of environmental respect, the chiefs of each clan, in partnership with the spirits, administered their territory.[148] The Plan of Life came from the ages and is defined by the laws of tobacco and yucca, which together constitute the Law of Life.[149]

The document was signed by more than one hundred persons, including the leaderships of the *cabildos* and the esteemed elders of the Traditional Authorities of the four tribes. It begins with an extraordinary litany of concerns that I reproduce as appendix 7. It links the past, present and future with loyalty to indigenous traditions, but without eschewing the benefits of modernity. In the words of Vincent Martinez, representing the Traditional Authority of the Huitotos, the most numerous people in the La Chorrera Zone:

> Our grandparents, owners of these resources, in the past cared for them and protected the creatures of each clan and of each people, under the word of tobacco, coca and sweet yucca from which derives the spiritual acquiescence for the proper use of space and time [which is] the foundation of social equilibrium. . . . Today

---

146. Ibid., 19.
147. Ibid., 13.
148. Ibid., 10.
149. Ibid., 2.

we are the grandparents of the present generation. We safeguard and protect, but in order to continue safeguarding and protecting we need to find new resources, like money, in order to guarantee economically this public work that is carried out from the *malocas* and the *mambeaderos* [scattered farms]. In order to realize our dreams [the Plan of Abundance], it is urgent that we have the real support of the state with its insitutions, NGOs, and international cooperation as well.[150]

The 2006 text also provides some interesting insight into various aspects of life in the La Chorrera zone. It claims that, by then, there was 100 percent literacy among the Indians, as well as 100 percent attendance at primary schooling and 70 percent at the secondary school level. There were fifteen bilingual teachers providing instruction in all of the Chorrera's Indian languages. Each student studied his/her own indigenous tongue for three years. Many graduates had gone on to institutions of higher learning—particularly the Universidad Pontífica Bolivariana of Medellín and the Universidad del Bosque. Students from La Chorrera received financial aid from both universities.[151]

While the *manguaré* was still employed for communication, and should be sustained as a critical component of traditional culture, some communities had a radio-telephone and there was a telephone link with public booths at La Chorrera. That community also had the Chorrera FM radio station with a range of 300 kilometers, broadcasting local news and information about indigenous culture. There was passenger air service once weekly and a cargo plane landed at La Chorrera biweekly.[152]

Finally, the document declares a philosophical position:

> For indigenous Amazonian cultures no man is backward or advanced, it is not necessary to "advance" towards a superior state of well-being at the expense of others or the environment. This philosophical position is in opposition to the capitalist thinking that nurtures a vision of economic growth and development, and in which there cannot exist societies that do not embrace an interest in the economic growth that ultimately determines the futherance of their level of development.[153]

---

150. Ibid., 47–48.
151. Ibid., 22; 26; 37.
152. Ibid., 30.
153. Ibid., 28.

We might punctuate the uncertainty surrounding the future of indigenous affairs by considering two works in progress among the Huitotos and their fellow indigene travelers within our study area. The two tales illustrate how most of the dynamics and tensions among the protagonists can play out in the real world. Were I to characterize them in Dantesque terms, the first would be "Inferno," or at least "Purgatorio," and the second "Paradiso." Once again, *avanti*.

Our first narrative regards intra-ethnic indigenous relations in the Leticia area, as well as with the dealings of these indigenes with wider worlds, played out against the backdrop of certain assumptions of international and national NGOs, Colombian government officials, and well-intended indigenist missionaries and academics (particularly anthropologists). To wit, the goal is to preserve the cultures and languages of Indian peoples—the individual surviving tribes. Policies and programs are designed to underpin "traditional" life styles, which means propping up the remains of former social arrangements, practices (dress, dance, handicrafts, etc.), beliefs (i.e., world view), and leaders. Each tribe therefore constitutes its own ethnic group. The thinking underpinning the creation of *cabildos* was that they should be ethnically monolithic. Furthermore, they should be rural rather than urban. Nevertheless, *cabildos* often encompass more than one Indian group and possibly non-indigenes as well. Some are suburban. This "mixing" tends to be regarded as less than ideal and one more measure of the degradation of authentic indigenous culture—creating the distinction between "authentic" and "false" Indians. The desire to preserve an indigenous group in amber, as it were, is shared by some tribal members and many of the protagonists of outside agencies as well. One obviously disturbing consequence is the tendency of indigenous youth to migrate, with many simply disappearing after being absorbed into the wider world. That obviously threatens the group's prospects for continuity.

There is, however, the competing anomaly that all involved are preoccupied by the relative poverty and political marginalization of most of the country's Indians. So there are also in play the twin goals of economic development and political involvement in order to protect the indigenes' interests. How then to walk the divide between preservation of traditional ways of life while modernizing them. It is a razor's edge that has the capacity of cutting right through shoe leather.

In 1994, Leticia had 22,866 inhabitants, 17,758 of whom lived in the urban nucleus. Of them, 1010 were indigenes from nine different tribes, largely concentrated together in their own barrio called El Piñal. Despite the fact that this was originally Ticuna territory, and that group is by far the largest in the immediate hinterland, Ticunas (250) are not in the majority in Leticia itself. Rather, the largest group is Huitoto (334), and there are Boras (72) ,and Andokes (19) as well.[154]

None, of course, is residing on a *resguardo* or reserve. This is by preference, since there now exists one called the Resguardo Tijuna-Uitoto Km. 6–11. Reference is to its distance from the city's limit along the Leticia-Tarapacá highway. Founded in 1978, the *resguardo* has several small nucleated Indian *cabildos* (as well as disseminated households) within its 7,560.52 hectares. Indigenist missionaries provided the impulse to create it, given that the Indians had little say in Leticia's municipal politics, dominated as they were by the community's "whites." Whites also ran commerce; the indigenes were the low-paid, unskilled workers.

The main Indians in the foundation of the *resguardo* were several local Tikuna clans and Huitotos and Boras who were taken from Aranalandia to Iquitos and Pebas between 1910 and 1929. From 1930 to 1937, many aspired to return to their original homeland on the Igaraparaná and Caraparaná rivers, now a part of Colombia. So they migrated to Leticia as a way of becoming established in that country as part of the homeward odyssey. Some completed it, but a significant number found employment in the city and settled there, entering into intense interethnic relations with the Tikuna and sharing with them a common indigene fate in their dealings with non-Indians. By about 1990, the total indigene population of the *resguardo* was 442 persons. The largest settlement is San José-Nuevo Milenio, where Tikuna predominate. There are several other smaller ones.

It was during the 1980s that coca-growing was displaced from other parts of the country to remote regions like the Putumayo/Caquetá. So, Indians from various backgrounds (including new Huitotos and Boras), as well as farmers and *peones*, arrived from throughout Colombia and from Brazil and Peru to take advantage of the employment opportunities. A few invaded the *resguardo* to establish farms on unoccupied land and some of its established Indian agriculturalists switched to the new crop.[155]

In 1991, about forty disaffected Indians of various tribes, but mainly Huitotos, still residing in the city decided that they wanted to move to

---

154. Marco Alejandro Tobón, "Asentamientos multiétnicos: conflictos y distinciones en Leticia," in Margarita Chaves and Carlos del Cairo (eds.), *Perspectivas antropológicas sobre la Amazonia contemporánea* (Bogotá: Instituto Colombiano de Antropología e Historia, 2010), 216.
155. Ibid., 218–19, 222. With their presence, the population of the *resguardo* increased to around 1,200 inhabitants. Givanna Micarelli, "Pensar como un enjambre: redes interétnicas en las márgenes de la modernidad," in Margarita Chaves and Carlos del Cairo (eds.), *Perspectivas antropológicas sobre la Amazonia contemporánea* (Bogotá: Instituto Colombiano de Antropología e Historia, 2010), 491.

the countryside. They formed an association called called Renacer Huitoto (Huitoto Rebirth) and began lobbying for land. It was then that the Huitoto *cabildo* known as Nimaira Naimeki at km. 11 in the *resguardo* offered to accept them. There was tension almost immediately between the newcomers and the old-timers and, in 1993, the former left to found their own satellite settlement at km. 9.8. that they called (no less of a mouthful than) Manaida Nairi Isuru Jussy Monifue Monilla Amena.

There, too, immediate tension arose between those who wanted a rural residence while commuting to their employment in the city and those who wanted to farm—a division perceived to be between the "rurals" and the "urbanites."The latter aspired to political leadership on the grounds that they were more sophisticated and had better contacts in the non-indigeous world that would facilitate protection and furtherance of the *cabildo*'s interests. So, there was a new schism, supported by the leadership of Cabildo Nimaira Naimeki, that produced another settlement but 400 meters from the original *cabildo*. The latest move of residents, who were Huitotos, Boras, Mirañas, Cocamas, and Macunas, was led by a sophisticated Bora. He understood Colombian national law well and tied his discourse to the nation's recent commitment (constitutionally) to being multicultural and pluri-ethnic. It was he that proposed the name of the new settlement—Multiétnico—that was adopted, in 1994, by its residents. It quickly attained its own *cabildo* status and its lands were distributed among the residents.[156]

But then, *plus ça change, plus c'est la même chose*. Some of the residents wanted to maintain a residence in Leticia and work there; others were more disposed to residing and earning their livelihood in Multiétnico. The latter began to promote display of traditional culture as a way of authenticating the *cabildo's* ethnic claims to be included within "the" Amazonian indigene identity. They constructed houses out of forest materials, employed traditional cures, performed traditional dances, expounded indigene environmental lore, cultivated *chagras*, and so forth. Nevertheless, they deployed their "authentic Indian" status to qualify for all kinds of national and international support for development projects, most of which were failures. Micarelli describes the *cabildo* as follows:

> ... the indigenous community Nimaira Naimeki Ibiri Muina-Murui, known as Kilometer 11, surrounds sporting fields strung along a gully of fetid water whose flow was interrupted years earlier by an ill-conceived bridge. The community's terrain is covered with the ruins of a series of failed development projects: the clinic,

---

156. Ibid., 221.

the livestock enclosure, fish-farm tanks, the poultry farm, the flour plant, the experimental garden, the latrines, the aqueduct, and others.[157]

Meanwhile, non-indigenous officials in Leticia erroneously thought that the dispute in Multiétnico stemmed from its tribal mix, that is, interethnic tensions. They were under orders from above to apply Colombia's new multiculturalist policy, but without realizing its inherent anomaly. That is, the law favored constituting individual tribal entities in rural areas, not mixed urbanized or semi-urbanized ones, that could then lay claim to privileges and respect within multicultural national life. The process ignored the "New Indian" reality of progressive indigenes like some some of those in Multiétnico, not to mention that it also dismissed the historical ties that existed in the past among various Indian tribes.[158]

It was then that the traditionalists in Multiétnico established their own settlement that they called Kaziya Narai. The leaderships of the two entities remained at sword point. The tension increased as "eco-ethno-tourism" became the prime source of income for the *resguardo*, prompting the traditionalists to intensify their self-indigenization. They received federal funding from the Ministry of Culture and a subsidy from the NGO Fundación Gaia to construct a *maloca*. For the residents of Multiétnico, this usurpation of the consummate symbol was the supreme insult. It underscored their weak political position with regards to the outside world and called into question their very claim to be indigenes.[159]

So, on a December day of 1999, a contingent from Multiétnico burned down the *maloca*. Shortly thereafter, they drove its main proponent from the *resguardo* with the threat of force. There followed a period of mutual insults, challenges to duels and exchanges of shotgun fire between the inhabitants of Kaziya Narai and Multiétnico. With the matter at an impasse and the violence escalating, and particularly after the local media began publicizing the events, the antagonists had recourse to peacemakers. They were not traditional *capitanes* or shamans, nor missionaries, but rather secular state authorities. A commission was constituted consisting of functionaries of the Office of Indigenous Affairs of the Ministry of the Interior, the regional government's Human Rights' Defense and the Attorney General of the Amazonas Department. In order to dissimulate their hegemonic role, the external commissioners convened a meet-

---

157. Micarelli, "Pensar," 495.
158. Tobón, "Asentamientos," 223.
159. Ibid., 225–26.

ing with the leaders of the several *malocas* within the *resguardo* and the Council of Elders of Nimaira Naimeki. The Commission and the Council handed down a joint verdict that Multiétnico would be disenfranchised and dissolved as an entity, but that its residents could continue to live in the *resguardo*. The people of Multiétnico rejected the idea that they could continue to have access to their lands, but would not be allowed to petition collectively for future benefits and subsidies.[160]

Meanwhile, the institutionalization of traditional culture within the *resguardo* proceeds apace. Visiting its several *malocas* is a tourist attraction promoted by several travel agencies in Leticia.[161]

The second narrative is set at ground zero of our study area. Between October 6 and October 12, 2012, culminating (ironically) on the *Día de las Américas*, or Columbus Day, a series of conferences and events were held in the rehabilitated original house and new buildings that now constitute the college at La Chorrera,[162] created after the tribes clawed back ownership of the 802.5 hectares originally held out for the Caja Agraria when *the Resguardo Predio Putumayo was* established in 1988. The complex was known as the "House of Knowledge" (Casa de Conocimiento). According to Chirif and Cornejo Chaparro, "... the place of death and suffering has now been converted into one of study and hope." Furthermore,

> In the same sense, wherein were kept the jail cells of the enterprise, the place of the ancestors' suffering, today are housed the nutriments that give life to the students and professors of the centers of learning. The house hosts a primary and secondary school, whose teachers are all indigenes formed in the country's universities, as well as a program of university training that functions in accord with the Indigenist Zonal Association of Municipal Councils and Traditional Authorities of La Chorrera (Azicatch), in conjunction with the Pedagogical University of Colombia. There, some twenty young people of both sexes are pursuing a career in biology.[163]

---

160. Ibid., 228–29.
161. Giovanna Micarelli, "Fortalecimiento del Consejo de Autoridades Tradicionales Indígenas del Resguardo Tikuna Uitoto Km. 6 y 11 para la construcción participativa de estrategias de gestión territorial y para su difusión (Leticia, Amazonas, Colombia)," (2009).
162. Reported jointly by two participants, the pro-indigenist Peruvian anthropologists. See Alberto Chirif and Manuel Cornejo Chaparro, "Reencuentro de familias separadas por la barbarie del caucho: Abriendo el canasta de la abundancia," *Servindi* (Servicios de comunicación Intercultural), October 20, 2012. Downloaded from http://www.nacionmulticultural.unam.mx/mezinal/docs2926.pdf taken from http://www.servindi.org/actualidad/75193.
163. Ibid., 4.

Two Huitoto artists from Peru had just completed a 10-by-4 meter mural on the site called "The Shout of the Sons of Coca, Tobacco and Sweet Yucca." It depicts how the indigenes went from a peaceful state to one of terror, violence, and death at the hands of the *caucheros*, followed by a future in which they will regain freedom and abundance. Its theme was debated and approved by the traditional tribal elders before they granted permission for the painting. During the October event, the artists were thanked publicly by the organizers, given that the mural would ". . . remain here as a memory, as a legacy, as a root and as a hope. This is your womb, this is your territory, you are a part of us because you have our own blood in you. Thank you for having visualized this shout of pain and hope."[164]

Several discussions were held among the many indigene associations in attendance that considered the historical impact of the rubber boom on their ways of life. Each was asked to present its views and agenda for the future. They jointly called for agreements with all of the relevant outside agencies, ones encompassing indigene plans for the organization of their own communal lives. The encounters should be held "right here in the middle of the jungle." They also called for reconciliation between Peru, Colombia, and Brazil over any residual frontier differences, as well as a tripartite agreement and commitment by them to facilitate future maintenance of indigenous societies, "while reconstructing the social fabric torn asunder by the quest for rubber."[165]

The main day of the festivities, with delegates from various governmental and religious agencies in attendance, several Indian dance groups performed and then a "basket of sadness" was covered over to symbolize the end of the tragic memories of the rubber epoch and a "basket of abundance" (containing fruit) was uncovered to symbolize hope in the future.

Attendees from the Colombian government were a member of the House of Representatives, a functionary from the Victims' Agency of the Colombian presidency, a presidential minister and director of the Presidential Program for Indian Peoples, a representative of the Colombian Center for Historical Memory and, finally, someone from the Ministry of Culture. Speakers included Mr. Tood Howland, representing the UN High Commission of Human Rights, Msgr. Aldo Cavalli, apostolic nuncio to Colombia, and John Deew, British ambassador to Colombia. Representatives of the indigenous associations in attendance read a letter from Pope Benedict XVI "to the Colombian indigenes," in which he made reference to this being the one hundredth anniversary of the encyclical *Lacrimabili statu indorum* of Pope Pius X, motivated as it had been precisely

---

164. Ibid., 5.
165. Ibid., 7.

by the Putumayo scandal.¹⁶⁶

There was also a letter read aloud from President Santos of Colombia in which he lamented not being able to attend personally because of a health issue.¹⁶⁷ It went on to denounce in the strongest of terms the ethnocide and genocide committed by the *caucheros*. It stated:

> In the name of the Colombian state, to the communities of the Huitotos, Bora, Okaina, Muinane, Andoque, Nonuya, Miraña, Yukuna and Matapí peoples, to all of you I ask pardon for the deaths, for the orphans, for the victims in the name of a business, of a government, of pretended 'progress' that failed to understand the importance of saving every indigenous person and culture as an essential part of the society that today we recognize with pride as being multiethnic and pluricultural.¹⁶⁸

Also asked to speak were Manuel Cornejo, of Peru's Amazonian Center of Anthropology and Practical Application, and his fellow Peruvian anthropological consultant Alberto Chirif. They denounced the history of the *caucheros* in the area. There were letters from the president of the Colombian Congress and the Peruvian minister of culture. The former noted that she had spoken in Congress about her profound shame ". . . over the thousands of deaths of Indian brothers and sisters of the Putumayo that we were unable to protect in their moment of need." She also pledged to Colombian authorities that she would create rapport with the Peruvians in order that the two states could construct a new reality for their indigenous peoples. For his part, the Peruvian minister of culture wanted to "honor all of those indigenous brothers and sisters who died due to the rubber fever." He underscored that "As Peruvians we must ask

---

166. Ibid., 7–8.
167. On October 3, the Huitotos *cacique*, Marcelo Buinaje, had traveled to Bogotá to try and meet with President Santos to give him several items: a basket symbolizing abundance, a clay jar as symbol of food, ground coca leaf for understanding and wisdom and chewing tobacco for relaxation. Buinaje wanted to implore Santos to send a high-placed functionary to attend the ceremony. The president was recuperating from recent prostrate surgery and did not receive the Huitoto leader. Buinaje related to the Bogotá press that it was the centenary of the *caucherías*' holocaust that, beginning in 1912 [*sic*] claimed eighty thousand Indian lives. The chains that had once bound the Indians had been thrown into the Igaraparaná some twenty-six years earlier by the Casa Agraria when it was remodeling the buildings at La Chorrera. Some nights, especially when it stormed, in the middle of the forest one could still hear strange cries and the weeping of men and children (the victims of those earlier times). See http://www.lapatria.com/nacional/la-casa-arana-una-historia-para-no-repitir-16823 and http://eltiempo.com/archivo/documentos/CMS-12288557.
168. Ibid., 8.

heartfelt pardon for the suffering that as a society, as a country, we were unable to prevent."[169]

So, precisely one hundred years after the London Hearings, literally the panoply of protagonists in this account—governmental officials of both Peru and Colombia (including the latter's president), the Catholic Church (including the pope himself), the United Nations, NGOs, indigenist academics, and the leaders of relevant indigene associations—came together at La Chorrera to excoriate the past and celebrate the promising future of the Putumayo's peoples—and particularly its Indians. The rhetoric was truly magnificent; the reality, unfortunately, less so.

---

169. Ibid., 9.

Part 6
Analysis

Sources

How to make sense of all of the foregoing? Clearly, our tale is rife with problems—a veritable bundle of controversies and contradictions. For Arana's critics, the indigenes were noble savages, while Julio César and his fellow rubber barons were savage nobles. For Arana's defenders, the indigenes were barbarians and the barons were their benign civilizers. It is certain that no two protagonists of this complicated historical narrative would agree entirely upon a single version. They all recount a Manichaean clash of epic proportions between heroes and villains.

The "primary" sources regarding the alleged atrocities in the Putumayo are provided by Saldaña Rocca, Hardenburg, Whiffen, Paredes, and Casement. Yet it might be noted at the outset that none of them ever *witnessed* the atrocities that they were reporting! Furthermore, the later reports were informed by the earlier ones. Hardenburg leaned heavily upon Saldaña Rocca's newspaper articles, depositions, and the statements of former Casa Arana white and *mestizo* employees, as did Whiffen (while recycling Saldana Rocca's material as well). Casement took a manuscript copy of the Hardenburg account into the jungle with him. Judge Paredes read all three of his predecessors before embarking upon his own investigation in the Putumayo. Genocchi was privy to the accounts of the other five. Judge Valcárcel digested his judicial colleague Paredes's report before issuing his arrest warrants. All were polemical, seeking to advance their case by alternating between threnodic and outraged denunciations of the plight of the indigenes. In short, there was enormous cross-fertilization (not to mention redundancy) among the several accounts of the atrocities and the responsibility of the villains—the Aranas, Zumaetas, Rey de Castro, Julio Egoaguirre, certain section heads in the Putumayo like Armando Normand and Victor Macedo, and, to be sure, the directors of the Peruvian Amazon Company.

To the extent that Casement's two Putumayo reports informed and underpinned[1] the proceedings of the House Select Committee and the wider international scandal, they were based upon their author's zeal and established reputation as a leading humanitarian. Casement's investigation of Leopold's Congo had been the prime catalyst leading to the king's divestment of his interests there, not to mention the subsequent reforms by the Belgian government in its new African possession. Yet, there is a curious parallel in the British consul-general's descriptions and conclusions regarding the two areas. In both, the ruthless overseers were drawn from the criminal dregs of white society. In both, there was a middle management class of recruited black outsiders (from Zanzibar in the Congo and Barbados in the Putumayo) administering a force of somewhat westernized and privileged natives (think *muchachos de confianza*) who captured indigenes of tribes other than their own to enslave as rubber gatherers. It was they who pursued absconders with a vengeance, making examples of them through torture and execution. In both venues, the very thongs used (literally) to whip the native work force into shape were not of the pedestrian cowhide variety. Rather, they were made out of hippopotamus skin in the Congo and that of the tapir in the Putumayo—both ostensibly capable of inflicting far deeper lacerations than ordinary bovine rawhide. I do not suggest that Casement was acting in bad faith, but the parallels in his reports on the two continents are striking. At the very least, he must have had an enormous sense of *déjà vu* when visiting the Putumayo—nevertheless, there is a paucity of explicit comparison with the Congo in his description of the latter.

In point of fact, every one of the protagonists was an enigmatic and polemical figure in his own right, praised uncritically by his admirers and subjected to *ad hominem* vilification by his detractors. For some, Saldaña Rocca was a courageous crusader willing to take on the en-

---

1. Recall that his Blue Book was reproduced in both the proceedings of the House of Commons' Select Committee's Hearings and the work *Slavery in Peru* published by the U.S. Department of State in response to a resolution by the House of Representatives. Casement not only had Grey's ear in London, but visited Washington, D.C., where he met with the highest government officials on his way home from his second (and last) visit to the Putumayo. It is also clear that Casement's sterling reputation as the consular official who had produced the report that brought an end to the atrocities in the Congo preceded him and gave him unusual credibility with the British and world press. That was a key factor in Grey's decision to appoint Casement as Great Britain's official representative to accompany the Commission constituted by the Peruvian Amazon Company to investigate the Putumayo atrocities.

tire Arana machine, and nearly paying for it with his life. For his detractors, he was a vindictive, anti-capitalist muckraker who turned against Arana after being denied financial support. For many, Hardenburg was a larger-than-life, saintly, and altruistic humanitarian willing to take extraordinary risks to expose gross injustice. For his critics, he was a sore loser who, when deprived of his golden opportunity to become Serrano's partner in a rubber enterprise, became a blackmailer and forger to boot.

For some, the judicial heroes of the story were the two judges Paredes and Valcárcel—the former providing the evidence of the atrocities and the latter the courage to act upon it. By taking on Julio César and the Casa Arana, both became instantly unpopular in Iquitos. According to their detractors, Paredes, as leader of the political faction in Iquitos opposed to President Leguía and the Casa Arana, produced a false report designed to embarrass the establishment, while Judge Valcárcel was unduly taken in by Paredes and was of dubious character himself—witness the charges that he abused his judicial powers and then abandoned his post.

Casement, too, could easily be regarded either as the great humanitarian who brought both King Leopold and Julio César Arana to their knees, or a self-promoting publicity seeker who served up a recycled Congo report in his investigation of the Putumayo in order to secure more fame and even a knighthood.

Enter Taussig

Easily the most influential (as in widely read) scholarly treatment of the Putumayo atrocities is *Shamanism, Colonialism, and the Wild Man: A Study in Terror and Healing*,[2] by medical anthropologist Michael Taussig. Arguably, Taussig's book is a milestone in Anglo-American social anthropology's recent journey into postmodernist deconstruction. His investigation and analysis borders upon an exercise in "magico-realism."[3] Indeed, regarding the tales that the whites and Indians told about one another, Taussig comments:

> Far from being trivial daydreams indulged in after work was over, these stories and the imagination they sustained were a potent po-

---

2. Taussig, *Shamanism*.
3. Taussig is a skilled and compelling writer, and he was publishing about Colombia at the very height of the popularity of Colombian novelist Gabriel García Márquez, whose book *Cien años de soledad* (1967) had by then been translated into dozens of languages and sold millions of copies, becoming the iconic work of the foundational Latin-American branch of the magico-realism literary movement.

litical force without which the work of conquest and of supervising rubber gathering could not have been accomplished. What is crucial to understand is the way these stories functioned to create through magical realism a culture of terror that dominated both whites and Indians. The importance of this colonial work of fabulation extends beyond the nightmarish quality of its contents. Its truly crucial feature lies in the way it creates an uncertain reality out of fiction, giving shape and voice to the formless form of the reality in which an unstable interplay of truth and illusion becomes a phantasmic social force. All societies live by fictions taken as real. What distinguishes cultures of terror is that the epistemological, ontological, and otherwise philosophical problem of representation—reality and illusion, certainty and doubt—becomes infinitely more than a "merely" philosophical problem of epistemology, hermeneutics, and deconstruction. It becomes a high-powered medium of domination, and during the Putumayo rubber boom this medium of epistemic and ontological murk was most keenly figured and thrust into consciousness as the space of death.[4]

Inspired by Walter Benjamin, Taussig eschews "facticity" for the exploration of understanding of the respective perceptions of whites and Indians that informed both their collaboration and mutual fear of one another—their confluence constituting "epistemic murk." The stage on which this action transpires was the rubber station. Taussig states:

There were several such way stations where Indian and white society folded into the assumed otherness of each other, where what was taken to be an Indian practice met with what was taken to be a white one, where assumed meanings met with assumed meanings to form strange codependencies and culture itself—the culture of colonization. There were the rubber traders living with Indian "wives" (who bore strangely few children, according to Rocha); missionaries "baptizing" Huitotos with Christian names while Huitotos carried out lavish rites "baptizing" whites with Huitoto names (and let's not forget titles, a doubly circulated process in which Indians applied to whites' [sic] titles that whites applied to them, with changes where appropriate, of course—as with the informal leader of the Colombian rubber traders, Gregorio Calderón, ("Captain General of the Rationals"); whites going to Indian medicine men; whites (so the tale goes) like Crisóstomo Hernández out-orating the Huitoto orators and thereby bending

---

4. Taussig, *Shamanism*, 121.

them to the whites' will; the great Indian festivals where the "advances" of trade goods were exchanged for rubber; and, of course, the host of interwoven Indian and colonist assumptions about the rights and duties built into the debt-peonage relationship itself. These were vitally important practical affairs. They were also ritual events. As such they were in effect new rituals, rites of conquest and colony formation, mystiques of race and power, little dramas of civilization tailoring savagery which did not mix or homogenize ingredients from the two sides of the colonial divide but instead bound Indian understandings of white understandings of Indians to white understandings of Indian understandings of whites.[5]

For Taussig, then, the Putumayo atrocities derive from a colonial legacy that regarded indigenes as inferiors (even subhuman) and the mutual fear and hostilities of the Indians and whites, reflected in the *correrías* designed to capture adult male rubber gatherers and their subsequent ill treatment (whippings, torture, death) by overseers who in turn feared the periodic violence of their charges, not to mention their black magic and cannibalism. In short, the system rested upon coercion with terror by those also terrorized; creating self-perpetuating, negative epistemic murk between the protagonists/antagonists.[6]

According to Taussig, the gatekeepers of this exchange were the *muchachos de confianza*:

> The debt-peonage system established by the Putumayo rubber boom was more than a trade in white commodities for india rubber. It was also a trade in fictitious realities, pivoted on the *muchachos* whose story-telling bartered betrayal of Indian realities for the confirmation of colonial fantasies.[7]

Furthermore, any narrative is set in its own historical moment. He tells us:

> The connection between history and memory here invoked would seem to have little in common with the historicist view of events unfolding progressively over time. On the contrary, we are startled by an image from the past, a magically empowered image flashing forth in a moment of danger. . . .[8]

---

5. Ibid., 109.
6. Taussig cites the Paredes report but briefly, when, in point of fact, the Iquitos judge himself developed even more fully than did the anthropologist the "culture of terror" concept. See the extensive quotes from Paredes reprinted in the present text (pages 93-96).
7. Taussig, *Shamanism*, 123.
8. Ibid., 367.

It might be asserted that the most significant contribution of postmodernism (at least in my view) was its call for the deconstruction of every construction—that is, that "facts" are always human constructs and are therefore neither self-evident nor eternal truths. Furthermore, such constructs are always the result of perceptions, biased to be sure, each of which is a form of witnessing. At times, the individual perceptions and their subsequent accounts of facts ("history" when set in motion) are sometimes, albeit not necessarily, irreconcilable. How can we possibly reconcile the disparity between the Colombian and Peruvian accounts of the history of white settlement of Aranalandia, let alone either with that of Aquileo Tobar?

Then, there is the figure of the arch-villain, Armando Normand. He seemingly epitomized malevolence in the Putumayo. We have considered Roger Casement's extraordinary *ad hominem* characterization of the man, even before their first meeting. Sir Roger had already accepted as truth the allegations of Normand's crimes in the Saldaña Rocca and Hardenburg testimonials. Yet we have Normand's autobiography, as related to American journalist Peter MacQueen (appendix 4), that differs in many respects from Casement's account of their meeting. Taussig seems to have been unaware of this source. What would he have made of the disparity between Normand's and Casement's description of their encounter?[9]

We have many other founts of epistemic murk undermining our attempts to understand what is, in so many ways, incomprehensible:

There was the mapping war. Our Colombian protagonist, Rafael Reyes, mapped the Putumayo and destroyed the unacceptable (to him) boundary marker erected by Peru and Brazil after their agreement on their common borders in the Javarí River drainage. Julio César Arana spent considerable money and effort in mapping Aranalandia. He subsidized the French explorer Robuchon after specifying the precise boundaries of both his cartography and ethnography. He contributed mightily to the construction of a Peruvian military infrastructure designed to define and defend the boundaries of *his* empire. The rubber boom, of which the Putumayo was a part, exacerbated an already centuries-old panoply of border disputes between the Lusitanian and Hispanic spheres of influence within Northwest Amazonia, on the one hand, as well as among all four of the Hispanic players (Bolivia, Ecuador, Peru and Colombia), on the other. To be sure, this generated a trove of mutually irreconcilable maps. Then, too, there was the endemic internal tension between Iquitos and Lima that

---

9. For a discussion of Normand's multiple images, see Carlos Guillermo Páramo Bonilla, "Un mónstruo absoluto: Armando Normand y la sublimidad, del mal," *Maguaré* 22 (2008), 80.

threatened, at times, to redefine the configuration and sovereignty of the Loreto. In short, our story amply illustrates that the making of maps is as much (or more) a political exercise than a geographical one—the purpose often being to produce epistemic murk disguised as precision.[10]

There was the artifactual conflict. King Leopold used his vast wealth strategically to create museum displays, international exhibits, and botanical gardens in support of his case that he had shouldered one considerable white man's burden in civilizing the Congo while educating the civilized world. These were all ways of creating and pandering to the public's curiosity about exotica, while projecting an image of benign and objective scientific interest in our planet's diversity. The British certainly engaged in the same exercise. Casement was given to near poetic descriptions of Amazonian nature in his journal, not to mention his many sketches of its aspects. He collected its butterflies and sent back to Europe specimens of its exotic fauna. He clearly understood the impact of visual evidence when he produced, during the London Hearings, the shoddy trade goods employed by the Peruvian Amazon Company to "conquer" and indenture the Indians. Then, too, there was the "show-and-tell" inherent in the decision to bring Omarino and Arédomi to Europe for display.

All of the contending sides sought to harness the new technology known as photography to make their case. There were the famed photographs of Eugene Robuchon that might have led to his murder. Rey de Castro used certain Robuchon photos (while omitting others) in his editing of the explorer's text in order to demonstrate that the Indians were in good health and spirits. When Rey de Castro and Arana joined the American and British consuls on their trip to the Putumayo, they brought along Brazilian photographer Silvino Santos and staged Indian dances and festivities at every turn to be able to capture photographic evidence of unscarred, well-nourished, and blissful indigenes. Arana would subsequently pay for Santos's cinematic studies in Paris, and then underwrite a documentary of Putumayo Indians favorable to the Casa Arana's interests. Casement took his own photos and drew sketches of abused victims. Then, too, there was the particularly damaging image for Arana of the corpse of an emaciated woman lying dead and abandoned in her hammock that circulated widely in the international press, even though it is not entirely clear that the photo was even taken in the Putumayo.[11]

The literary representations and advocacy of literary titans regarding both the scandals in the Congo and the Putumayo have been nothing

---

10. See Jerry Brotton, *A History of the World in 12 Maps* (New York: Viking [Penguin Group], 2002).
11. Chaumeil, "Guerra," 48–49.

short of extraordinary. The campaign against Leopold galvanized such leading lights as Arthur Conan Doyle, William Butler Yeats, Joseph Conrad, George Bernard Shaw, and Mark Twain. The relationship between Conrad and Roger Casement, and the accounts of the latter shared with the former, may have inspired some of the writing in *Heart of Darkness*, a mainstay of the western literary canon and arguably the most damning critique of white imperialism in Africa. We have discussed its Latin American counterpart, the novel *La vorágine*, by Colombian author José Eustacio Rivera, that depicts in the strongest of terms the abuses of the Casa Arana in the Putumayo. Published in 1924, it has become a canonical work that is still taught regularly throughout the world in university courses on South American literature. That same year, (Basque-surnamed) César Uribe Piedrahita traveled to Aranalandia and, in 1933, published his semi-autobiographical, widely acclaimed novel *Toá: Narraciones de caucherías* (or *Toá: Stories of Rubber Properties*).[12] It provided a detailed history of Casa Arana activities in the Putumayo—a most nefarious and pro-Colombia narrative to be sure.[13] Sir Roger Casement certainly continues to fascinate to the present day. In 2010, or the same year in which he won the Nobel Prize for Literature, Peruvian novelist Mario Vargas Llosa wrote *El sueño del celta* ("The Dream of the Celt"), an only slightly fictionalized account of Casement's life and tragic death.

Then, too, there are the journalistic and academic narratives—and it is here that we encounter considerable disparity between Colombian and Peruvian sources. The Putumayo scandal is a topic of perennial and continuing concern in the former, including bursts of publication during the periodic times of crisis between the two countries. This literature is cumulative and now spans more than a century. Beginning with the work of Cornelio Hispano, *De París al Amazonas: Las fieras del Putumayo* [From Paris to the Amazon: The Wild Beasts of the Putumayo] (1913), including Ricardo Gómez A.'s *La guarida de los asesinos: El relato histórico de los crímenes del Putumayo* [The Murderers' Hide-out: An Historical Account of the Crimes of the Putumayo] (1924, reissued in 1933), and ending with that of such scholars as Roberto Pineda Camacho's *Holocausto en el Amazonas: Una historia social de la Casa Arana* [Holocaust in the Amazon: A Social History of the Casa Arana] (2000) and the massive two-part work

---

12. César Uribe Piedrahita, *Toá: Narraciones de caucherías* (Manizales: Arturo Zapata, 1933).

13. The book has never been translated into English, however, it appeared in Russian translation with an astounding press run of twenty thousand copies. See John Wilson Odorio, Prólogo, *Toá: Narraciones de caucherías* (Medellín: Colección Bicentenario de Antioquia, Universidad CES, 2013), np.

Putumayo: La vorágine de las caucherías: memoria y testimonio [Putumayo: The Whirlpool of the Rubber Enterprises: Memory and Testimony] (2014–2016) compiled by Augusto Javier Gómez López. One need only contemplate the titles of these works to understand the Colombian point of view. At some juncture during her education, a Colombian student is likely to be exposed to this history, one that ascribes to the Casa Arana a particularly grim image.

Conversely, in Peru, Collier's book was not published in translation until 1981.[14] Guido Pennano, historian of Peru's rubber industry, penned its introduction that he terminates with the words:

> As regards the novelistic-history that concerns us herein, we should alert the reader that its author clearly takes a pro-British and pro-Colombian stance, and for greater detail I invite you to read it.[15]

During the first decade of the twentieth century there were several outstanding publications compiled and edited by pro-indigenist Peruvian anthropologists. Alberto Chirif was involved in two significant ones regarding the Putumayo/Caquetá. In 2005, he compiled and edited the work *La defensa de los caucheros*[16] that reprinted works published nearly a century earlier by both sides in the Putumayo scandal and, in 2009, he co-edited (with fellow anthropologist Manuel Cornejo Chaparro) the work *Imaginario e imágines de la época del caucho: Los sucesos del Putumayo* (or Imaginary and Images of the Rubber Epoch: The Putumayo Happenings), which makes no apology whatsoever for the Casa Arana's nefarious history there.[17]

Andrew Gray[18] provides an interesting analysis of the literary qualities of the texts by Richard Collier and Michael Taussig:

> When we contrast the style of Collier with that of Taussig we can see that they complement one another. Both process the horror

---

14. Richard Collier, *Jaque al barón: La historia del caucho en la Amazonía* (Lima: Centro Amazónico de Antropología y Aplicación Práctica, 1981).
15. Ibid., vi.
16. Rey de Castro, Larrabure y Correa, Zumaeta, and Arana, *La defensa*.
17. They also included a graphic article by Colombian anthropologist Juan Álvaro Echeverri, "Siete fotografías: una Mirada obtusa sobre la Casa Arana," which shows the evolution of the Casa Arana headquarters at La Chorrera from its original purpose to the current tourist accommodation and cultural center dedicated to freedom (42–57).
18. Andrew Gray was a graduate student at Oxford University who presented a paper in 1990 entitled "The Putumayo Atrocities Revisited" in a seminar at his university. It was never published in English but was translated into Spanish and included by Alberto Chirif as the "Introducción" to his compiled collection of essays on the Putumayo, see Rey de Castro, Larrabure y Correa, Zumaeta, and Arana, *La defensa*, 15–50.

of the Putumayo through romantic means. Collier's narration is novelistic and seeks its solution in ethics. The narration of Taussig is somewhat more poetical and searches for resolution in aesthetics—torture and terror are forms of ritualized art. Collier's book is disturbing because the atrocities are described as real, nevertheless it is comforting because good wins out in the end. Taussig's book is comforting because it displays a studied skepticism that the atrocities actually happened. Nevertheless, it disturbs because his aesthetic approach leaves one with the fear that he could be precisely seeking to evade the reality of one of the most horrible mass crimes of the century.[19]

Not everyone is prepared to eschew historicity for Taussig's and Collier's flights of literary fancy. Fernando Santos Granero and Frederica Barclay provide an extensive[20] deconstruction of Taussig's "economy of terror,"[21] predicated as it is upon the nearly inexplicable and seemingly counterproductive abuses of the very indigenes who constituted the only real source of wealth in the Putumayo. For Santos Granero and Barclay, there *is* historical facticity that better explains the situation once the full geographical and temporal detail is taken into account. According to Taussig's explanation that the system was simply persistence of the Iberian colonial mindset, the earlier Colombian settlers in the Putumayo should have instituted the same punitive practices as did the Casa Arana and the system would have prevailed more or less evenly throughout the entire Loreto and Upper Amazon.[22] Such was not the case.

Furthermore, Santos Granero and Barclay reject the notion that the *mestizos* and indigenes pertained to two separate economic systems—the former being participants in debt peonage and the latter the enslaved victims of the *correrías*. Rather, historically, the *correrías* were targeted more at women and children, who could then be bought and sold, rather than at male rubber gatherers. Nor were the *correrías* a white invention. Prior to the intrusion of Europeans into the Amazon Basin, many tribes organized slave raids against their neighboring enemies in which the object was to kill the adult males and enslave the women and children, either for assimilation into their conquerors' tribe or for use as trade objects within the intertribal barter system. The *correrías* of the whites and *mestizos* continued this tradition. It was only in the late nineteenth and early twentieth

---

19. Ibid., 30.
20. Santos Granero and Barclay, *La frontera*, 61–91; Santos-Granero and Frederica Barclay, *Tamed Frontiers*, 34–55.
21. Actually, Taussig describes a "culture of terror," rather than an "economy" of one.
22. Ibid., 35.

centuries, when the rubber boom was full-blown and labor most scarce, that some raids were organized for the purpose of enslaving potential adult male rubber gatherers.[23]

Mestizos and indigenes who pushed back were both subject to physical abuse—whippings and punishment in stockades. If a *mestizo* fled his contract, it was to another part of the "civilized" world, where, under the formal legal system he would likely be captured and punished, or forced to return to his *owner*.[24] That the punishment of Indians was at times more severe reflected the fact that they were prone to default on their debts and had the option of retreating into the forests to flee their creditors. Hence, the torture, and even the murder, of indigene absconders after punitive *correrías* to recapture them. It was all meant to serve as an example to others.[25]

The potential flight of the Indian debtors also explained, in part, the decision to provide them with the cheapest of trade goods, given that the *cauchero* might lose all of his extended credit. There was also a practical security reason for providing potentially hostile indigenes with the cheapest of firearms—shotguns rather than rifles, the so-called "trade guns" that were useless after about fifty shots.[26] In short, while the goods sold to Indians might have been inferior to those provided to *peones*, and the enforcement of repayment more severe, both were participants in *enganche*.

There are also the open questions of the extent and efficacy of the terror in creating the economy. If the system rested exclusively upon the whites' capacity to simply terrorize the indigenes, how do we explain such practices as the Casa Arana's continued dependence upon Indian *caciques* as intermediaries with, and organizers of, the rubber gatherers?[27] Why bother with the demanding bookkeeping and rhetoric that the Indians were incurring legitimate debts for goods that had to be paid off in rubber? Stated differently, if the indigenes were not part of the system of *enganche*, why maintain the fiction of it?[28] In effect, "slavery" versus "*enganche*" was a distinction without a difference.

That the worst abuses and atrocities occurred in the Putumayo was more conjunctural than coincidental or aleatory. They derived from Julio

---

23. Ibid., 40–41.
24. Ibid., 39.
25. Santos-Granero and Barclay, *La Frontera*, 62–63.
26. Santos-Granero and Barclay, *Tamed Frontiers*, 40.
27. It is also clear that errant *capitanes* were likely to be treated more harshly than anyone else, subjected to far more lashes in a public whipping before being killed. One suspects that the message was intended more for other *capitanes* than their subjects.
28. Santos-Granero and Barclay, *Tamed Frontiers*, 88.

César's decision to take the Casa Arana public in London. Consequently, the overseers on the ground were given participation incentives that both reflected their greater personal risks with respect to an increasingly restive Indian population and the need for greater output. It is therefore scarcely surprising that the level of coercion and abuse of the Indians increased.[29] Then, too, that very restiveness derived from the shift from the collection of *caucho* from Castilloa trees to that of the weak *Hevea* product as the former disappeared. This required a radical change in work habits that was unacceptable to the idigenes, one that could only be effected through harsh discipline by the overseers. In short, it was not perverse sadism that underscored the system (and changes within it); rather, there was always an underlying economic rationale.[30]

In my view there remain further unexamined aspects. My own take begins with criticism of the tendency of critics on all sides of the issues to deal in one-dimensional stereotypes of *the* indigenes, whites, *peones*, *caucheros*, section chiefs, *muchachos de confianza*, and missionaries. All agree that these groups were components of an unadulterated capitalist system, albeit one at times couched in the guise of both a religious and secular Christianizing/civilizing undertaking that *ipso facto* justified the subordination and destruction of Indian societies and cultures "for their own good." Both missionaries and *caucheros* imposed upon indigenes notions of individual responsibility and accountability—in the guise of a work ethic and "morality." Both felt endangered and/or offended by the indigenes reputed sloth, cannibalism, nakedness, communal living (the *maloca*), threatening communications (the *maguaré*) and ritual. There is a sense in which all of these protagonists are presented as caricatures, that is, without much (indeed usually no) appreciation of their internal human heterogeneity.

One clear manifestation of this is the scarcity of sensitivity to the allocation of merit and gain, as well as blame and punishment, within the capitalist system itself. If we look at the literatures of patently egregious inequalities, such as that between master and slave in the North American Black experience, galley slaves and their overseers, chain gangs in the American South and the inmates of concentration camps and their guards,[31] we find that all had their own systems (both formal

---

29. Ibid., 79–83.
30. Santos-Granero and Barclay, *Tamed Frontiers*, 69.
31. One might cite, by way of example, there was system of leasing Black chain gangs to coal mines in Alabama where, "Prisoners valued being treated like human beings . . . [and] skilled prisoners could capitalize on their proficiency at coal mining to gain status, rest, extra food, and even dignity." Mary Ellen Curtin, *Black Prisoners and Their World,*

and informal) of reward and punishment. Consequently, not all of the subordinated rebeled, some tolerated their overlords while others actually collaborated with them. The latter, in turn, meted out certain benefits to encourage compliance and punishments, not to mention to dissuade dissent. We can only imagine that there was a similar dynamic operative in the Putumayo.

So it was that some indigenes fled into the forest, while others resisted violently. There were the *correrías* to bring back the former and armed military expeditions against the latter. Nevertheless, we also get glimpses of indigenes voluntarily interfacing with the whites. If some who failed to produce their quota of rubber might be beaten, placed in stocks, or worse, one can only imagine that those who excelled as gatherers were rewarded in ways other than simply greater accounting-book credit toward the purchase of merchandise. Being the best gatherer in a section must have been a point of pride for the individual and extolled as an example for others by the overseers, particularly the latter who were themselves working on commission. We also get glimpses of differences among section chiefs in their treatment of the Indians. Not all of those working for Arana made the lists of criminals compiled by Casement and Paredes, indeed only a minority did.

Of particular interest are the uses of the concepts of "contact" and "cannibalism." I find them to be related weapons and, like most swords, they are double-edged. The credence of whites that many of the Indians beyond their control (i.e., uncontacted) were cannibals provided justification of the civilizing/evangelizing secular and ecclesiastical missions that conveniently entailed subordination and domestication of them—with all of the potential for abuse that we have contemplated in this text. Conversely, many indigenes fled the consequences of the white juggernaut, crossing into the liminal territory beyond the pall, as it were, becoming over time "uncontacted." Their reputation for being cannibals served not only as an excuse for, but as a shield from, white conquest. It was, after all, at least somewhat dangerous to enter the territory of bestial, warlike bar-

---

Alabama, 1865–1900 (Charlottesville and London: University Press of Virginia, 2000), 104. Fyodor Dostoevsky provides an essentially autobiographical exegesis of life in a Siberian prison in nineteenth-century tsarist Russia (Fyodor Dostoevsky, *The House of the Dead* [New York: Dell Publishing Company, 1959]) and Aleksander Solzhenitsyn engaged in a similar exercise in Soviet prisons—cf Aleksandr Solzhenitsyn, *One Day in the Life of Ivan Denisovich* (New York: Farrar, Straus and Giroux, 1991) and Aleksandr Solzhenitsyn, *The Gulag Archipelago, 1918–1956* (New York, Evanston, San Francisco, London: Harper & Row, 1974), 3 volumes [*Gulag*]. Both of these famed novelists provide detailed descriptions of prison life that encompasses the "human" aspects, including reciprocity between inmates and guards, that were so essential to the formers' survival.

barians who would not only kill you but might then mutilate and eat your corpse. Along these lines, Wade has a fascinating description of the Waorani of Ecuador. They were a highly isolated, warlike group who viewed all outsiders to be enemies and therefore as candidates for extermination (they believed that all of the indigenes surrounding their territory were cannibals and were therefore sub-human). We are told:

> If the Waorani were uncertain about the world beyond their borders, the people on the outside viewed the Waorani as savages, symbols of the demon heart of the world. Although the tribe before contact never numbered more than five hundred, Ecuadorian officials regularly estimated a population of several thousand, partly because of the vastness of the Waorani lands and partly due to sheer hysteria. Waorani raiding parties regularly covered as much as forty miles a day through the forest. Within a week the same Indians might be responsible for incidents occurring two hundred miles apart, killings that the government attributed to different groups of raiders. What's more, each Waorani carried several spears and thrust more than one into a victim. No corpse was left with fewer than eight. In 1972 the body of a cook for an oil company was discovered stuck with eight spears; the authorities suggested that twenty Waorani had been involved but, in fact, the killing was the work of only three.[32]

Then there is the casual, yet unrelenting, use of such loaded terms as "holocaust," "hecatomb," "atrocities," "slavery," "assassination," "rape," "torture," "devil," and so forth. They inevitably lend a hyperbolic flair to the accounts while masking such human complexity as agency, volition and formulation of life strategy. Yes, indigene women were raped, enslaved, discarded, and/or murdered, yet we have had glimpses of Indian parents offering their daughters to whites, not to mention that of white males marrying and caring for their indigene spouse and their children. Yes, we have examples of Peruvian soldiers killing Colombian settlers, yet we also have glimpses of some who refused to do so and paid with their careers when not their lives. Yes, we have examples of Arana employees, including the Barbadians, following orders to torture and kill indigenes, yet we have glimpses of some who refused to do so and suffered severe consequences. Yes, we have examples of missionaries and lay priests and bishops colluding with the *caucheros*, yet some of our clearest denunciations of the abuses were provided in mission reports and the Genocchi investigation.

---

32. Davis, *One River*, 272.

Yes, we have examples of Loretans entirely beholden to Arana, yet some businesses in Iquitos purchased advertising in *La Felpa* and *La Sanción*. Yes, we have evidence of the dependency of judges and politicians on the Casa Arana, yet we had Paredes and Valcárcel, as well as some prefects, who were acerbic critics of the system.

Then, too, we have glimpses of indigenes adapting to the new realities, learning trades, and pursuing them well beyond their ancestral homeland. Many, by choice, eschewed their indigenous cultural background, came to self-identify as Peruvians or Colombians, married non-Indians, and raised their children as *mestizos* and *peones*. Furthermore, we have had glimpses of indigenous "stayers," "leavers," and "returnees" vis-à-vis ancestral homelands (in short, in some senses there are as many indigene narratives as there are indigenes).

There remains the question of why it was Arana's fateful destiny to be singled out as such an arch-villain by history? There is no simple answer, rather we are dealing with conjuncture of multiple factors—all necessary yet none sufficient in itself as an explanation.

There was the element of simple bad luck in that the Casa Arana's prime critics (with the exception of Whiffen) were all crusading idealists. Saldaña Rocca, Rómulo Paredes, Roger Casement, Walter Hardenburg, and Giovanni Genocchi were all, each in his own way, humanistic socialists. The only one identified as such by the Arana forces was Saldaña Rocca. We have, of course, considered the ode to socialism, likely written by Saldaña Rocca himself, on the first page of the inaugural issue of *La Sanción*. Paredes would write pro-socialist novels, plays, and poems during his career—the prime example being his scandalous poem to Samarem. Casement denounced in the strongest terms the Anglo-American commitment to Mammon; the socialist Fabian Society agitated to have his death sentence commuted.[33] Hardenburg contributed several articles to the Canadian socialist journal *The New Review* and authored the booklet entitled *What is Socialism?*[34] Genocchi was a champion of the oppressed and acerbic critic of his fellow clergymen who collaborated with exploitative capitalism or regarded it with a blind eye. In effect, Arana was a quintessential capitalist at what was arguably a robust and optimistic period for proponents of Marxism.

---

33. Angus Mitchell, *Roger Casement* (Dublin: The O'Brien Press, 2013), 295.
34. Curiously, none of these "leftists" denounced, specifically, the circumstances of Upper Amazonia's indentured *peones* ("workers of the world"), an economic arrangement likened to slavery by some observers. All of these Arana critics remained highly focused upon the plight of the indigenes—framed within a discourse on the exploitation by ostensibly Christian, superior, white civilizers of non-white, sub-human, pagan savages.

Then there was that decision to incorporate the Peruvian Amazon Company in London, thereby unwittingly exposing Julio César to the assuaging of their guilt by the Occidental imperial powers regarding their own treatment of the planet's indigenes. Transatlantic slavery had been abolished but a few decades earlier. The Congo scandal was winding down as the Putumayo one geared up—the crusading figure of the international celebrity Roger Casement being central to both. The United States was living the consequences of its imperialistic victories in the Spanish American War and its machinations in Panama, particularly vis-à-vis a leery Colombia. That, it turn, made Peru suspicious of possible U.S. appeasement of its neighbor, including by now respecting Colombian territorial claims in the Oriente after having ignored them in Panama. Then, too, Anglo-American private sector activity as investors and developers in Upper Amazonia's rubber boom made all of the involved Latin American nations nervous. And what was there to think regarding the contemporary implications of a Monroe Doctrine that persisted after being proclaimed nearly a century earlier? Did it protect the New World from Old World designs or simply establish North American hegemony in Latin American affairs? All of these factors were in play as the world unwittingly careened towards its first global conflict. In short, despite its physical isolation and territorial liminality, the Putumayo/Caquetá was a kerfuffle of domestic and international political intrigue.

Things were no less conjunctural on the religious front. New World governments struggled to transcend their colonial Catholic legacy, a church with vast economic and political interests directed from the Old World by autocratic church fathers and in the Americas by a sclerotic ecclesiastical hierarchy. Then, too, the interests of mission orders dependent directly upon Rome could easily clash with those of a particular country's secular clergy. Nevertheless, by the end of the nineteenth century, change was in the air. American bishops were ordered by the Vatican to reform their mission policies (or neglect). Meanwhile, several Protestant churches were contesting Catholic hegemony throughout the region. Our discussion of the Putumayo's missions, the many inititives and failures, contain the seeds of the contemporary pentacostalism among both Catholics and Protestants as they both contest hegemony of the traditional Catholic Church throughout Latin America, not to mention that other challenge within it from liberation theologists (a movement actually started in Peru). Indeed, it would be one of the hallmarks of the latter to accuse the continent's hierarchies of being aligned, indeed often through family and kinship ties, with the oligarchs who exploited a particular country's lower classes and indigenes.

Finally, I would situate my anthropological discipline within this epistemic murk. I would note the intrinsic tension between most anthropologists and most missionaries. We target the same populations, but with diametrically opposed agendas. Aside from the turf issue, anthropologists purport to describe the lives of indigenous peoples in their own terms; missionaries seek to change those lives in fundamental fashion. The distinction has sometimes been characterized as the incompatibility between cultural relativism and the natural law and moral absolutes that presumably underpin universal values. That particular debate is beyond the ken of this book.

There is, however, a relevant distinction between Michael Taussig's and my social anthropology with that of our Latin American colleagues and counterparts. Taussig's book was both radical and iconic in the postmodern deconstruction of Anglo-American anthropology, albeit somewhat conservative and marginal within the thinking of Latin America's own practicitioners. The Anglo-American approach emphasized intellectual distancing from one's subjects in the interest of maintaining at least some degree of objectivity when it came to analyzing and reporting their social and cultural realities. This did not preclude outsider empathy, and even sympathy, with informants that often translated into support of the indigenes' right of cultural survival—should *they* so desire.

In sharp contrast, commitment to, and advocacy of the causes of the indigenes that one studied were the driving force of the new generation of Peruvian and Colombian anthropologists. If those who were British and North American-trained[35] studied the "Other" by going abroad, usually to parts of the less-developed world, the new generation of home-grown Latin American anthropologists were likely to select a field site within their own national boundaries. They were studying their "Other Self," as it were, and advocacy for their informants was an essential personal and professional responsibility. The majority of these academics were leftist in their personal politics, and therefore viewed the plight of the Indians in terms of a disempowered element within the national class struggle. While such "Action Anthropology" was not entirely absent by this time within Anglo-American anthropology as well, it had relatively few practitioners.

The genealogy of "Colombian" anthropology is extraordinary in its own right. It had two "foreign" founding figures, as it were. In the 1940s, the Capuchin Fray Marcelino de Castellví embarked upon his own remarkable, if bizarre, career as a self-taught anthropologist. He founded

---

35. One could add French anthropologists as well—think of Claude Lévi-Strauss's field work in Brazil.

in Sibundoy an organization called CILEAC (Center for Linguistic and Ethnographical Research in Colombian Amazonia). He concentrated upon the peoples within the Vicariate of Caquetá, defining their study himself under such rubrics as "Colombian hierology," "Indo-Colombian glottology," "Amazonian museology," "Ethnomorphic aesthetics," "Ethno-pharmacology," "Ethno-therepeutics [sic]," and so on. He gleaned his information without ever leaving the Sibundoy, interviewing indigenes, missionaries, travelers, explorers, military officers, and so forth, while collecting and collating all kinds of secondary sources. He represented Colombia at many anthropological conferences and was a member of nearly three-dozen scientific societies around the world.[36] In the 1940s, he published five issues of a journal called *Amazonia colombiana americanista*. It mingled scientific information regarding indigenous cultures with apologies of the Capuchin missionary activity in the region. The first issue concluded with the words, "Blessed be God, and blessed be Franco."[37] Needless to say, he is a controversial figure, when not ignored altogether.

Then there is the more mainstream protaganism of Austrian-born anthropologist and archeologist Gerardo Reichel-Dolmatoff (1912–1994) and his fellow-anthropologist spouse, Alicia Dussan. He left Europe during World War II and emigrated to Colombia. He and his wife conducted many of the pioneering anthropological and archeological studies among several indigene peoples. By the mid-1940s, the Dolmatoffs were living in Santa Marta where he headed the Instituto Etnologico del Magdalena. They created an ethnographic and archeological museum in the city. In the late 1950s, the Reichel-Dolmatoffs moved to Cartagena where he taught classes in medical anthropology. In 1960, they were in Bogotá where he held the nation's first chair of anthropology at the University of the Andes. Both he and his wife taught there, and, in 1964, created Colombia's first department of anthropology at that university.

---

36. Bonilla, *Servants*, 215.
37. Ibid., 216. Fray Castellví was extraordinarily reclusive and possessive regarding the holdings of CILEAC's library and archive, denying access of other investigators to them. When, in 1951, his request to the Colombian government for a subsidy of CILEAC was denied he sent part of his holdings to the Capuchin headquarters in Catalunya with stringent restrictions on their use and kept the rest to the attic of the personal residence that he came to occupy in Bogotá. In the 1970s, the latter remained inaccessible to anyone. Ibid., 218. As if the international anthropology influencing this story were not bizarre enough, It is also possible that controversial Carlos Castaneda of *The Teachings of Don Juan* fame was related to Julio César Arana. Castaneda was a Peruvian national from Cajamarca whose birth names were César Arana. Robert Marshall has written one novel about Castaneda (*A Separate Reality*, 2006) and is researching another. He suspects that the two Aranas were likely related (personal communication).

Gerardo Reichel-Dolmatoff's forty books and four hundred articles, his influence on the formation of a generation of Colombian anthropologists, and his substantial involvement in the world's key anthropological associations made him a renowned authority with a stellar international reputation. He was a visiting professor at Cambridge University and UCLA. In 1975, he was awarded its prestigious Thomas H. Huxley medal by the Royal Anthropological Institute of Great Britain. Throughout his decades in Colombia he was an ardent defender of indigenous rights and acerbic opponent of their treatment as second-class citizens. His many studies underscored the alternative, yet equally valid, sophisticated cosmologies of indigenous peoples (as opposed to the ostensibly "civilized" western world view).

Nevertheless, Reichel-Dolmatoff's legacy is now overshadowed by controversy. In 2012, at the 54th International Congress of Americanists held in Vienna, Augusto Oyuela-Caycedo (University of Florida) and German anthropologist Manuela Fischer (and others) presented a paper in which they claimed to have evidence of Reichel-Dolmatoff's involvement with the Nazis. Oyuela-Caycedo has since claimed to have documentation that at age fourteen, or in 1926, Reichel-Dolmatoff joined Hitler's Youth Group, was then a member of the NSDAP and served in the SS between 1932 and 1935, becoming an SS sergeant at Dachau (1934–1935). He was expelled from the SS due to mental disorders and left Germany for Paris in 1937. He studied at the Sorbonne in 1937–1938, and then emigrated to Colombia in 1939, becoming a citizen there in 1942.[38]

There were problems with the activist approach of all outsider indigenists. They created the intrinsic dilemma that their support of the traditionalists, and particularly their leaders, implied the survival (i.e., conservation) of past cultural traditions in the present and preservation of them into the future—a conservative position and a possible formula for perpetuating the social and economic irrelevance of the people in question. Yet coopting the voice and initiative in the name of progress and social justice was to, in effect, hamstring the traditional leadership, thereby hastening the group's social disintegration. Furthermore, given that the messages of the well-intentioned missionaries, government and NGO agents and academics were more likely to appeal to the restive young than to their elders, at times outsider intervention was capable of producing generational conflict within indigenous societies.[39]

Such "outsiders" easily gravitated to the strategic point when it came to formulating and implementing new policy. After all, it was

---

38. See https://en.wikipedia.org/wiki/Gerardo_Reichel-Dolmatoff.
39. Del Cairo, "Las encrucijadas," 190, 194, 200, 206.

they alone who possessed the sophistication and contacts within the wider national, and even international, contexts essential for converting hypothetical planning into concrete programs and achievements. Nevertheless, there was the evident dilemma that they were thereby undermining not only traditional authority, but that of the new leadership generation as well within indigenous societies. In some cases, then, the outsider activists incurred the resentment of all of a particular tribe's leaders—traditionalists and progressives alike.[40]

There was the added problem that some Indian leaders, like many Colombian non-indigene politicians, were themselves guilty of clientalism, nepotism and corruption.[41] Indeed, a few were assassinated by FARC for these very reasons.

We now turn to that greatest conundrum of all in our account—Arana himself.

Who Was Julio César Arana?

Consider the newspaper article published on September 12, 1908, in Lima's prestigious newspaper *La Opinión Nacional*:

> It seems inexplicable that in the middle of the jungle, there, where there is not felt any governmental influence, thousands of Indians have been wrenched out of savagery and nationalized, to the degree of instilling in them a love of Peru and its flag, in whose defense they have spilled their blood, drawing a line before the invader that attempted to rip away with force this rich piece of national territory. How did this solution come about? That which the government could not accomplish was done by a single man; and we have the satisfaction of proclaiming the name of this good Peruvian, who is none other than the King of rubber in Peru, Mr. Don Julio C. Arana.[42]

Hardenburg had a slightly different opinion:

> And under the magic wand of wealth, Julio Arana, the quondam peddler, the erstwhile bare-footed vendor of Panamanian hats, quickly became a "gentleman." Under its polishing and refining influence, he soon accustomed himself to boots and

---

40. For a discussion of all of the foregoing see Jean Jackson, "La política de la práctica etnográfica en el Vaupés colombiano," in Margarita Chaves and Carlos del Cairo (eds.), *Perspectivas antropológicas sobre la Amazonia contemporánea* (Bogotá: Instituto Colombiano de Antropología e Historia, 2010), 235–37, 238–265, and elsewhere.
41. Del Cairo, "Las encrucijadas," 208.
42. Quoted in Lagos, *Arana*, 150.

the other conventionalities of contemporary society. Grinding under his calloused heel the helpless Indians of the Putumayo forests, upon their bleeding backs he mounted the swaying pyramid of capitalism, there to take his seat along with the other vampires who feed upon the blood and sweat and tears of the world's workers.[43]

Given Julio César's prominence and that of the Casa Arana/Peruvian Amazon Company at the vortex of the scandal, it is astounding how little we actually know about him. He left practically no documentation about himself other than the brief work (69 pages) *Las cuestiónes del Putumayo: Declaraciones prestadas ante el comité de investigación de la Cámara de los Comunes y debidamente anotadas* (1913), supposedly authored by him, but almost certainly ghost-written by Rey de Castro. Most of it is simply a self-serving, *ex-post facto* polemical reply to the questions put to Julio César during the London Hearings. It does, however, begin with a brief autobiographical statement detailing his early years as a river trader, the move to Iquitos, and then a rather sketchy account of his dealings with the Colombians in the Putumayo.[44]

We do know that Julio César was descended from a prominent highland Peruvian clan and was related to some influential contemporaries, yet we have almost no glimpse of his dealings (if any) with his more distant Peruvian relatives. Given Arana's direct and deep involvement in Loretan politics, one might have expected some sort of tie with his uncle, Benito Arana, former governor of the Loreto. Yet any evidence of such remains undiscovered in some public or personal archive. Nevertheless, on the ground in Iquitos, Julio César's web of close family ties (including affinal) could scarcely have been denser—Aranas or Zumaetas (and their spouses) occupying most of the critical posts in the Casa Arana's inner circle.

There is no question that Julio César preferred striking a low personal profile. In a city where the wealthy sought to display their success in flamboyant lifestyle and fancy bricks and mortar, Arana was a modest and private man by comparison. While everyone agrees that he wielded immense economic and political power, Julio César seems to have been more the consummate backroom kingmaker than seeker of public plaudits (unless, of course, they suited his purpose). Until finally becoming a senator at a relatively late age and effectively moving his residence to Lima, he spoke and acted through his publicist Carlos Rey de Castro, his political minion Julio Egoaguirre and his

---

43. Hardenburg, "Story of the Putumayo Atrocities 2," 688–89.
44. Arana, *Las cuestiones*, 7–11.

brother-in-law, business partner and sometimes mayor of Iquitos—Pablo Zumaeta.

In light of the public positions that Arana held in Iquitos and, subsequently, in Lima, there is an astounding dearth of photographs of Julio César. There is one formal posed portrait of him that has been reproduced so frequently that it seems to have been unique. There are only a couple of group photos in the literature that include Arana. Regarding Eleonora, the wall of privacy is even higher; there are no known photographs of her. The same may be said for the children. To my knowledge, no investigator has come up with a family portrait.

Ovidio Lagos, in his *Arana, rey del caucho* (2007), provides our most thorough biography of Julio César, and we also have the excellent work *The Devil and Mr. Casement* (2010) by Jordan Goodman. Both rely heavily upon Richard Collier's *The River That God Forgot* (1967), a point acknowledged by Lagos, while labeling the American author a "panegyrist."[45]

Collier begins his work with a sketch of Julio César's character, set in Manaos:

> Briefly, conscious always of the dignity that set him apart from other men, Arana braced himself, leonine head held high, white well-manicured hands gripping the silver-topped cane that supported him. Then, with one thrust of his powerful shoulders, he breasted through the nightclub's swinging doors. An awed silence fell upon the room. The sad burden of the pianola's "O Sole Mio" ground to a standstill. In this Brazilian cosmopolis, there was scarcely a man or woman who did not fear the advent of Julio César Arana, the Peruvian rubber baron, a secretive, iron-willed man who was known to live only for his work and his family. And Arana, his eyes taking in the tawdry throng, thought, as he had so often thought before: This city is doomed. It's time to get out. . . . But it was no desire to convert Manaos' free spenders that had brought Arana thither; his concern, as always, was both personal and private. One man bolder than the rest shouted, "Hey, Don Julio, how about a drink?" but one glance of Arana's cold hooded eyes silenced him. Moving delicately, despite his bulk, avoiding the dive's ornate porcelain spittoons, the Peruvian made a beeline for a table in the room's far corner. The sight he had seen numberless times over the past three years now met his eyes again: his young brother, Lizardo, disheveled and ashen-faced, passed out cold on the marble top. With arms entwined about the young man's neck

---

45. Lagos, *Arana*, 337.

was a woman of the town, solicitous yet watchful of her chances. Grimly Arana fought to keep his anger in check. To him his sleek, pomaded younger brother, in his natty belted plaid jacket and jaunty bow tie, was a symbol of all that ailed this city. Three years earlier, at a salary of £2,500 per year, Julio had appointed him manager of his Manaos branch, yet despite the pressures of business Lizardo time and again arrived late at the office, fuddled with champagne, his white linen collar soiled with rouge. So long as the Peruvian remained remote and aloof, no man could penetrate his armor. Not deigning to speak, he glanced at the street woman. One peremptory jerk of his head was enough; she melted discreetly into the crowd. His dark piercing eyes bored into Lizardo; his voice was pregnant with unspoken menace; "Get out of here *now* and clean yourself up. I want to see you in the office in half an hour—better, make it twenty minutes." Still dazed with liquor, Lizardo groped from a haze of hangover to mutter an unabashed apology. Fumbling to fix his collar and tie he sent a waiter speeding for ice and black Brazilian coffee. Without vouchsafing him a further glance, Arana shouldered back through the crowd, out into the city's gathering heat. Behind him the voices and laughter rose again, but uncertainly now; the pianola once more jangled into life, but with an air of seeming bravado. In the twenty-three years since Julio César Arana had arrived on the Amazon, few men had opposed him and lived to talk about it.[46]

Collier subsequently describes the exchange between Julio César and Eleonora on his visit to her in Switzerland just before giving his testimony at the House Select Committee's London Hearings:

> And though Julio and Eleonora conferred only behind closed doors, out of the earshot of the children, he wondered how often they heard her cry out: "Now it's too late, Julio—I warned you back in Yurimaguas and you would never listen, and now it's too late. The rubber and this terrible Casement have poisoned our lives—why didn't you get out before it was too late?"[47]

These are certainly gripping scenes, replete with extraordinary detail and color. The problem is that they are pure invention, an exercise in what has come to be called "faction"—the fictionalizing of historical data. Published in 1968, Collier's book is a contemporary of the two foundational classics of the genre—Truman Capote's *In Cold Blood* (1965) and Alex

---

46. Collier, *The River*, 18–20.
47. Ibid., 256.

Haley's *Roots* (1976). It is no exaggeration to say the Collier *created* his Arana protagonist/antagonist, albeit one based upon an actual person.

From an historian's viewpoint, *The River That God Forgot* is a maddening text, devoid as it is of footnotes. Furthermore, there is its blatant advocacy. Collier dedicates his book:

To the memory of
WALTER ERNEST HARDENBURG
Son of Liberty
1886–1942

According to Collier, Arana became the sixth largest taxpayer in Manaos,[48] which would suggest that he was one of the city's richest men. Like most treatments of him, the superlative transcends reality. While Julio César no doubt had important connections in Manaos, it is doubtful that he was one of the city's prime movers. We have the excellent study by António José Souto Loureiro of the city's rubber elite and *aviadores*, including many foreigners, and Arana is conspicuously absent.[49] He is also well down the list of rubber exporters, having exported none from Manaos in either 1908 or 1911, after having shipped 15,423 kilos in 1909 and 68,557 kilos in 1910 to Liverpool.[50]

Similarly, if Julio César would definitely be an important player in the Loreto for decades to come, there is some reason to believe that his hegemony may have been overstated by his supporters and critics alike. While the Casa Arana was certainly prominent in Iquitos's economic (and political) life, it was scarcely preeminent. Prefect Hildebrando Fuentes[51] provided statistics regarding the city's registered vessels[52] and major importers[53] for 1906. Arana y Hermanos is one of many on both lists, but

---

48. Ibid., 58.
49. António José Souto Loureiro, *A Grande Crise (1908–1916)* (Manaos: T. Loureiro & CIA., 1986), 187–92.
50. Ibid., 210. We should remember, however, Arana's well-crafted strategy to circumvent both Peruvian and Brazilian tariffs and taxes.
51. Hildebrando Fuentes Núñez del Prado (1860–1917) was an accomplished author (several books), journalist, newspaper editor, professor, civil servant, and politician. A political conservative, he held positions within the Peruvian military and, at various times, was elected to the Peruvian Congress. He served in Lima in a number of administrations and was appointed prefect of the Loreto by interim president Serapio Calderón (who assumed office on May 7, 1904) and then was relieved of that post by President José Pardo in the spring of 1906.
52. Fuentes, *Loreto* 2, unpaginated tables "Tercio Naval de Iquitos" and "Chatas ó Alvarengas de la Matrícula de Iquitos" after page 10.
53. Ibid., 29–30.

without heading either. Prefect Alayza Paz y Soldán commented that in Iquitos, local government, the Beneficent Society and the Chamber of Commerce were all dominated by the merchant class, "the majority of whom were foreigners."[54] So for Julio César, if the Putumayo was *his* river, Iquitos was not *his* city in the same sense. In point of fact, many foreigners were influential in the development of Peru's Oriente. The investments of foreign interests in the rubber industry and river transport—particularly British, French, German, and American—dwarfed those of Peruvian nationals.[55]

If it is generally difficult to penetrate the veil of privacy and secrecy that Julio César erected around himself, this is particularly true of his finances. We have seen that he was active in many businesses in several venues—including Brazil, Bolivia, and parts of Peru outside the contested Putumayo. Our understanding of the nature and extent of such activity is fragmentary at best. There is also an unexplained conundrum. Recall that Jerome informed Casement that the source of Arana's political influence and protection stemmed from the fact that many prominent Peruvians and government officials, including possibly the president himself and his inner circle, had investments with Arana. Really? By all accounts, the attempt to sell the shares of the Peruvian Amazon Company was a failure. All but a few thousand remained in the hands of Arana's underwriter and bankers. Arana and his closest associates (brother Lizardo and in-laws Pablo Zumaeta and Abel Alarco) held about 80 percent of the stock, and Julio César's approximately 70 percent ownership was pledged as collateral to his European bankers. So, just exactly what was available for prominent Peruvians, let alone in quantities that would influence them to protect the Casa Arana? There remains, of course, the probability that the officialdom was simply bribed instead.

Arana's penchant for privacy means that we have the most disparate renditions of Julio César the man. If all agree that he had scant formal education, according to some he came to own the best library in the Ama-

---

54. Alayza Paz y Soldán, *Mi País*, 126–27.
55. José A. Flores Marín, *La explotación del caucho en el Perú* (Lima: Concejo Nacional de Ciencia y Tecnología—CONCYTEC, 1987), 77–79. [*La explotación*]. According to this author,
> As might be supposed, rubber with its enormous possibilities for industrial use, aroused a voracious appetite among the industrialized nations, and England in particular. The [Peruvian] jungle fell within the orbit of English colonial dominion. Of course later other foreign interests were introduced as well such as; French, German, Portuguese, Chinese, North American, etc. The jungle was converted in this fashion into an extractive enclave of the great foreign monopolies. (Ibid., 73–74)

zon. Nevertheless, Judge Valcárcel scoffed at the notion that Arana was even literate. According to the judge, most, if not all, of Arana's letters and publications were written for him by Rey de Castro, given that Julio César was, "incapable of writing four words in mediocre Spanish."[56]

Nevertheless, Arana clearly employed his language skills strategically, frequently underscoring his ignorance of any other than Spanish, insisting upon using it exclusively in critical dealings whether in business or the London Hearings. Others maintain, to the contrary, that Julio César was an accomplished polyglot. The latter seems quite probable, given his extensive personal history in Manaos (Portuguese), Biarritz and Switzerland (French), and London (English). One suspects that his use of a translator during his testimony in the London Hearings was as much a ploy as a necessity. In reading his responses to his interrogators, Arana clearly used the translation process itself to control the tempo (and even mood) of the proceedings. It gave him the opportunity to formulate carefully measured replies to questions that were often intended to knock him off balance. Yet anyone who reads his testimony, particularly during the first two days in which he supposedly[57] surprised his interrogators and the London media alike, cannot fail to appreciate the man's skillful defense.

Then, too, if his critics depict Arana as evil incarnate, there is another side. Julio César regularly loaned money (not necessarily altruistically) to a huge array of public officials and private citizens in an Iquitos that sometimes lacked a functional banking system, and where the funds from Lima to meet the public payroll were frequently delayed for months at a time. There was the civic leader who advocated public schools and hospitals for both his town and country. There was the aviation and telecommunications visionary who understood their importance to Peru's national destiny. There was the patriot who expended private resources in the defense of Peruvian sovereignty (again, even if not altogether altruistically). There was the man who insisted upon defending his honor by appearing voluntarily at the London Hearings.

Was Julio César duplicitous and mendacious? The very question reeks of Taussig's epistemic murk. Its answer requires a certain faith in historic-

---

56. Valcárcel, *El proceso*, 269.
57. It is Collier (*The River*, 265) who tells us that the London papers had anticipated a bumbling performance from a South American bumpkin. There is, of course, a greater dramatic effect in portraying Arana as a shrewd witness who then becomes flummoxed and unglued when informed that his accuser, Hardenburg, was present in the hearing room and would be called as a witness. There is nothing in the transcripts of the Hearings that support this contention.

ity, that is, absolute standards of what constitutes facts and truths. There is the obvious problem of presentism. It would seem, however, that duplicity and mendacity were standard operating procedure among Loreto's elite. Our main Peruvian critics, the jurists Paredes and Valcárcel, contended as much. Certainly, Arana manipulated his own PAC Board of Directors. He withheld critical information from them when it didn't suit his purposes (such as the Saldaña Rocca accusations against the Company) and often seemed willing to say whatever it might take to convince his interlocuters. His very audacity in insisting upon testifying before the Select Committee was evidence of his confidence in his own powers of suasion and manipulation. Recall that his first act during his appearance was to try to introduce several documents into the proceedings. Whether genuine, contrived or both, they were clearly designed to discredit his accusers.

A key part of Arana's (and Rey de Castro's) defense throughout had been to impute ulterior motives to each of Julio César's critics. Whenever possible such *ad hominem* attacks were produced by supposedly well-meaning third-party observers, such as legislators in the Peruvian Congress, governmental and judicial functionaries (particularly in the Loreto) and concerned citizens writing letters to the editors of key newspapers. While some of this might have been sincere personal opinion, clearly a significant portion was orchestrated. Another tactic favored by Arana was to cast himself as both good and bad cop in his dealings with critical others—critical in both senses of being important to the success of Julio César's agenda and as outspoken opponents of it. The message was that there were benefits for those who cooperated with the Casa Arana; serious adverse consequences for those who did not.

No doubt Julio César was chagrined when Chairman Roberts dismissed his documents out of hand and proceeded to the interrogation. Arana then testified to have known nothing about the Putumayo atrocities (based upon his infrequent trips there and the self-serving silence of his section chiefs so as to avoid self-incrimination). That was simply preposterous. It requires us to accept that all of the Zumaetas and Aranas, both in the Putumayo and Iquitos, were in on a conspiracy to keep their kinsman/boss in the dark. There is not a scintilla of evidence (other than Arana's claim while under pressure on the witness stand) that Julio César was that gullible and/or out of the loop regarding the administration of his various business affairs. Again, the claim that, if indeed crimes had been committed in the Putumayo, they were perpetrated by Colombians and before Arana's time, is belied by Julio César's multiple business associations with partners such as the Larrañagas and Juan B. Vega. It requires us to believe that he was totally ignorant of how rubber was being extracted in the Putumayo by his Colombian customers (before he supplanted them there). It is equally certain that the practices continued under Arana's ownership.

Possibly the most succinct assessment of Julio César, the rubber baron, is provided by a French anthropologist, André-Marcel d'Ans:

> In effect, the particular genius of Arana consisted in knowing how to slip his ambitions into the interstices among nationalities, using to best advantage his capacity as a Peruvian to exploit, in Colombian territory, and with English capital, a product that he exported through Brazil, without first passing it through [the controls of] his own national territory! The success [comes] from Arana's transfer of the headquarters of his firm from Iquitos to Manaos, later from there to London, in 1908, when, having the money in hand he knew how to find in that city respectable *gentlemen* disposed to be strawmen. Thanks to them the *Peruvian Amazon Co.* became a legal English firm. It being well understood that the only thing demanded of the English administrators was to close their eyes, in the same fashion that eyes were always closed in Iquitos and Lima regarding everything that concerned the conditions under which Amazonian rubber had been extracted.[58]

There remains the question whether Julio César was evil. Since the time of the ancient Greeks western philosophers have debated whether *absolute* good (and hence, by implication, evil) are human possibilities. Socrates was reported to have said, "no one would choose evil knowing it to be such."[59] This statement has bothered western philosophers and historians of philosophy alike. Anthony Gottlieb muses:

> Not only did Socrates conveniently ignore impulsiveness and irrationality, he apparently declared that willful immorality was simply impossible. He seems never to have met a fallen man, let alone a fallen angel.[60]

Furthermore, Gottlieb asks rhetorically,

> Was he then just naïve? Nietzsche wrote of the 'divine naïveté and sureness of the Socratic way of life', but what he seems to have had in mind is the clear-eyed focus of Socrates' vision, not any

---

58. André-Marcel d'Ans, *L'Amazonie Péruvienne indigène: Anthropologie écologique: Ethnohistoire: Perspectives contemporaines* (Paris: Payfatesot, 1982), 181.
59. Jonathan Barnes (ed.), *The Complete Works of Aristotle* (Princeton: Princeton University Press, 1984), vol. 2, 1900.
60. Anthony Gottlieb, *The Dream of Reason: A History of Philosophy from the Greeks to the Renaissance* (New York and London: W. W. Norton & Company, 2000), 154.

merely foolish innocence. Nietzsche thought long and hard about Socrates' habit of expressing himself in apparently naïve propositions, and concluded that it was in fact 'wisdom full of pranks'.[61]

Well, hold on folks. I believe that Socrates meant what he said irrespective of the attempts by the squeamish to sanitize his words. I also believe that the modern debate over the concept of absolute evil has been influenced enormously (both disproportionately and inevitably) by Hannah Arendt's canonical work on Hitler's genocide of the Jews.[62] Indeed, mention of the German *führer* is quite relevant, since Arana has been equated to him and, as we have seen, such words as "genocide" and "holocaust" are regularly invoked when describing the Putumayo atrocities—even well before the existence of Auschwitz's gas chambers.

Were the Nazi (and Putumayo) exterminators of sound mind or simply crazy? Framed thusly, they are either fully cognizant and responsible for their actions or not. In the case of the former, it is tempting to condemn the perpetrators of "the ultimate crime against humanity"—genocide—to be absolutely evil. This would apply to such modern poster boys in the debate as the Marquis de Sade, Joseph Stalin, Mao Zedong, Adolf Hitler—and Julio César Arana.[63] Without seeking to exonerate any of them of personal responsibility, I would contend that every human being, no matter how seemingly malicious to both his contemporaries and future biographers, believes that there was an (ideo)logical justification for his or her acts. Aleksander Solzhenitsyn tells us in his epic *Gulag Archipelago*,

> Macbeth's self-justifications were feeble——and his conscience devoured him. Yes, even Iago was a little lamb, too. The imagination and spiritual strength of Shakespeare's evildoers stopped short at a dozen corpses. Because they had no *ideology*. Ideology—that is what gives evildoing its long-sought justification and gives the evildoer the necessary steadfastness and determination. That is the social theory that helps to make his acts seem good instead of bad in his own and others' eyes, so that he won't hear reproaches and curses but will receive praise and honors. That was how the agents of the Inquisition fortified their wills: by invoking Christianity; the conquerors of foreign lands, by extolling the grandeur of their Motherland; the colonizers, by civilization; The Nazis, by race; and the Jacobins (early and late), by equality, brotherhood,

---

61. Ibid.
62. Hannah Arendt, *The Origins of Totalitarianism* (New York: Schocken Books, 1951).
63. One could add many names to the list, such as the former Serbian and African leaders brought recently before the International Criminal Tribunal.

and the happiness of future generations. Thanks to *ideology* the twentieth century was fated to experience evildoing on a scale calculated in the millions. This cannot be denied, nor passed over, nor suppressed.[64]

Obviously, a presentist reading of our narrative would condemn Julio César Arana as a white supremacist and racist. However, neither his attitudes nor his behavior were extraordinary for his times. In his day, as we have seen, the majority of Latin Americans viewed the indigenes as inferior savages in need of Christian morals and civilizing—as in being imbued with a proper work ethic and materialist mentality. There was not a single country in which the indolent and semi-nomadic Indians were not regarded as impediments to economic development and nation building. Indeed, the same can be said for the actions and policies around the globe of Arana's critical European imperialists (not to mention customers and investors) in *their* capacity as both overlords and settlers. Furthermore, it is possible to demonstrate that Arana was not all that unique[65] among the South American rubber barons in the mistreatment of Indians, nor was he necessarily the worst. Even the Select Committee allowed as much:

> ... in the course of the Inquiry your Committee have been impressed with the fact that the ill-treatment of the Indians is not confined to the Putumayo. It appears, rather, that the Putumayo case is but a shockingly bad instance of conditions of treatment that are liable to be found over a wide area in South America. No doubt there are special features peculiar to the Putumayo problem, such as the dispute over the territorial sovereignty, which would not occur elsewhere. But the spirit of the "conquistador" appears to be at work on other rivers.[66]

One example among the multitude that could be cited regards the denunciation to the British Foreign Office of the Bolivian Luis Suárez for his alleged crimes—murder, rape (including of his own daughter), floggings, arson at the properties of others that he intended to usurp, and enslavement of the Indians. The letter began:

> The outrageous crimes committed in the rubber regions of the Putumayo district in Peru which were published in most of the English newspapers, are mere trifles compared to the ones I here

---

64. Solzhenitsyn, *Gulag*, Vol. 1, 173–74.
65. See sppendix 8 for a comparison of his life and career with that of the Bolivian Nicolás Suárez and the Colombian Rafael Reyes.
66. British Parliament, *Report*, xix.

relate. Those of the Putumayo were committed by various individuals, but here only one criminal has satisfied his bloodthirsty lust.[67]

Then, too, in 1914, or *after* the London Hearings of the Select Committee and despite the oft-repeated view that Colombians were more humane than Peruvians in their treatment of the indigenes, it was stated, in a denunciation sent to Bogotá authorities by a Colombian citizen, Sr. Riascos:

> The quotidian punishment of the Indians is the whip and it is applied for the slightest fault that they commit, or unjustly at the whim of the owners or their employees. The Angarica firm has a "Peruvian" named Lóver Guzmán who is like the official lasher since he is employed expressly for this duty or purpose. He is the salaried executioner. In the diferent sections of the aforenamed firm, as well as in the remaining firms of others, this is the common procedure. . . . When it regards the treatment of the Indian women that they have in the main house, Sr Borrero says that before they are whipped they are trussed up naked before all of the employees to shame them more. "This is what is done by those who presume to call themselves advanced guardians of civilization in these places." One of the ways that they punish vengefully those Indians, who, hounded by mistreatments try to rebel against the authority of their bosses and their representatives, consists in taking away their wives or their daughters, who are then raped first by the bosses and then by their employees. In this regard is practiced immorality in all of its most grotesque manifestations.[68]

There is evidence that Arana was truly befuddled by much that befell him. He certainly never foresaw the danger of an international scandal arising out of his activities in the Putumayo. Nor do I think that he was an intrinsically malevolent person who counted on the obscurity and physical isolation of the Middle Putumayo to insulate his evil agenda from scrutiny. Indeed, I would argue that Arana believed that his actions were justified. He was a sincere and loyal Loretan regionalist and Peruvian patriot. While both postures at times served his political and economic interests, there were simply too many examples in which he incurred grave risks while furthering those agendas. He personally financed and fueled Loretan rebellion against Lima on occasion, while defending Peruvian

---

67. Bigoberto Canab to Charles Gosling, English ambassador in La Paz. May 6, 1914. Bodleian Library MSS Brit. Emp S-22 G332.
68. Quoted in Dominguez and Gómez, *La economía*, 193.

hegemony in the Putumayo against Colombia on others. As a senator in the Peruvian parliament, Arana stood nearly alone in opposition to his own President Leguía's Treaty of Salomón Lozano that conceded most of the Putumayo to Colombia. Nevertheless, once it was approved he counseled calm, if grudging, acceptance of it by his fellow Loretans.

Then, too, Julio César opted of his own free will to go to London to testify before the House Select Committee. Again, while (with the benefit of hindsight) a miscalculation, surely a piece of Arana must have believed that it would clear his name once his side of the story was told. And again, his letter to a Roger Casement sitting on death row, would suggest that Julio César believed *himself* to be an innocent victim in the Putumayo scandal. Otherwise, would he have truly expected the contrition he asked of Casement at Sir Roger's moment of truth? If Arana went to his grave a broken and forgotten old man, he left Iquitos for Lima as the Loreto's popular hero and its electorate's chosen representative in national affairs. Nor is there evidence that Arana was ever repentant, or even regretful, of his behavior in the Putumayo. If the man had trouble sleeping at night, it was because of his physical ailments from his injuries and illnesses from years as an *aviador* on tropical rivers—not from a guilty conscience.

Peruvian neglect of the figure and legacy of Julio César Arana, transformed into silence and denial, is palpable in the tardy (1981) publication of the translation of Collier's book, as well as the review by Luisa Elvira Belaunde of the book *Imaginario and imágines de la época del caucho: Los sucesos del Putumayo* (2009) by Peruvian anthropologist Alberto Chirif and his Colombian colleague Manuel Cornejo Chaparro. No less than a Peruvian anthropologist herself, she states in her review:

> Certainly, it was through a reading of the book by Taussig, while I was studying in England, that I myself learned of the Putumayo happenings, because while in college in Lima none of it had ever been mentioned to me, nor at home either, despite my belonging to a family that had been directly implicated in the debates regarding the Putumayo and the negotiation of the Colombo-Peruvian border after the loss of the territories in question between the Putumayo and Caquetá rivers.[69]

Novelist Marie Arana, daughter of a Peruvian father and American mother, was Julio César's first cousin twice removed. Her great grandfather, Pedro Pablo, and Julio César's father Martín, were siblings. She places the founding ancestor of her Peruvian Arana clan in

---

69. Laura Elvira Belaunde, "Reseña," *Anthropológica*, 2010, XXVIII (28), Suplemento 1, 339.

Cajamarca, "where Pizarro and the Incas first came face to face" (she seems to imply, but does not contend outright, that her ancestor was a part of that confrontation). According to Marie, one Arana stayed in Cajamarca and remained in precious metals. Martín—Julio César's father—"settled in Rioja and made his fortune in the Panama-hat boom." A third, Benito, went to the Loreto and entered politics. Her direct ancestor, Gregorio, went south to Ayacucho and Huancavelica in the highlands where he engaged in mercury and silver mining.[70] She was three and living in Lima when Julio César died, yet he never saw her. For that matter, it seems unlikely that he ever met her father, Jorge, either.

Marie's great grandfather was a "Peruvian hero" who had "led the last known populist uprising against the military in 1895," heading up a force of three hundred rebel horsemen who swept down from the highlands during the successful insurrection of Huancayo.[71] He served in the Peruvian Senate and, in 1907, on the eve of the Putumayo scandal, Pedro Pablo was appointed prefect, or governor, of Cusco by civilian Peruvian president Manuel Pardo. But then the *New York Times* published an article on the Putumayo atrocities. She writes:

> I imagine my great grandfather, Pedro Pablo, reeling, stunned, back and forth from Lima to Cusco to his estate in Huancavelica, trying to get a grip on his life. He had been a patriot, a warrior, a hero, a public servant, no more than a cousin to the rubber baron; he had not been prepared for this blot on his family name. He had not anticipated the jungle splatter. . . . Pedro Pablo began trying to salvage what good name he had. He cut off all contact with his extended family in Iquitos. He stepped down from the Cusco governorship and retreated to Huancavelica. He refused to take questions about the "Devil of the Putumayo," When asked, he responded simply: I have no siblings or ancestors. Not one.[72]

His twenty-five year-old son, Victor Manuel (Marie's grandfather), had been educated as an engineer at Notre Dame University in Indiana and was then residing in Maine. He was ordered home by his father as a part of the family's disengagement. Once back in Peru, Victor Manuel threw himself into his work—imbued with his father's reformist zeal and determined to raise his country's standing in the world. For two decades,

---

70. Marie Arana, *American Chica*, 40–42.
71. Ibid., 41.
72. Ibid., 50. Marie Arana does not provide specific sources, noting only that she did extensive reading on her family's history in the Stanford University Library while on a summer fellowship, in 1996, at the Hoover Institute (Ibid., 39).

he maintained a hectic and active career and social pace, both as a practicing engineer and academic at the University of Lima. But then,

> ... somewhere along the way his star began to dim. He withdrew from his work. Few knew why, and among those who did, no one wanted to say. Come the '30s and a worldwide Depression, he stepped into his study, switched on the lights, and sat there for forty years.[73]

There was also a sense in which the Arana curse was upon him. Marie recounts that when her grandfather approached the parents of Rosa Cisneros y Cisneros to ask for her hand in marriage, there was the question,

> Was he one of those Aranas? No, certainly not! Following Pedro Pablo's directive, Abuelito pushed his relations away with such force that one day it propelled him in the opposite direction—out of society, out of career, up to a second-floor limbo.[74]

For Marie Arana, her relative Julio César was a phantasmagorical figure, yet an oddly pervasive presence. When, as a child, she first heard of the (in)famous *cauchero*, and asked the adults in her family if she was related, the answer was, "Oh, there are so many Aranas, Marisi. He has nothing to do with you."[75] And yet,

> All my life, strangers had asked me about the rubber baron Julio César Arana, and I'd always given the rote response: no relation, no connection, not me. So a shadowy figure had been responsible for a human hecatomb in the jungle? Well, that story had played out at the turn of another century, at the hearth of another family; it had little relevance to me. But Julio César crept into my life anyway.[76]

Marie grew up in a household in which her Arana grandfather, Victor Manuel Arana Sobrevilla, was a recluse. She says of him, "He was consumed by the idea of honor pricked by some unnamed remorse."[77] And again,

> My grandfather's demeanor was lordly. He walked with his chin in the air. But it was a backward trajectory, a voyage inward, a solemn recessional, as if something cankered his heart.[78]

---

73. Ibid., 22.
74. Ibid., 54.
75. Ibid., 41.
76. Ibid., 39.
77. Ibid., 20.
78. Ibid., 23.

For Marie, then, the "Mark of Arana" is not the scar upon a Huitoto's body, but rather the stain on her family,

> It is this bridge—steeped in yesterday, wrapped in guilt, shut in stone—that brings to mind my father's deep history. Like Mother, he had been molded by the past, but his was a past he had not made and was unaware of, a legacy inherited before he'd seen the light of day. It was the Mark of Arana: as real as a shriveled leg, a maimed hand, a welt from shoulder to shoulder. It had reverberated from jungle to mountain, from one side of the Aranas to another. It had spun into every branch of the family, stung his grandfather, stifled his mother, chased his father up the stair. Nobody spoke of it, no one acknowledged it, nor did anyone really care to track the circuitry, but for the Aranas the past had been toxic, and shame spilled through generations like sap through a vine.[79]

A Basque Palimpsest

I began this book with the words "I write of the Basque Arana." Really? Well, maybe. Throughout the telling of this story, I have indicated whenever one of its protagonists possessed a Basque surname. It is this perspective and criterion that I have sought to bring to bear upon an analysis of Julio César Arana, his kinsmen and close associates—as well as several of the further removed players in this story. Therefore, throughout the narrative, whenever we encounter a Basque-surnamed individual I have highlighted that fact.[80]

As I stated in the introduction, historians of Latin America are far too prone to treat *peninsulares* or "Peninsulars," the Spanish-born, monolithically and as constituting their own ethnic group. This overlooks the significance of internal ethnic differences within Spain, and their capacity

---

79. Ibid., 39.
80. For historians engaged in the detective work of teasing Basque elements out of larger Hispanic narratives, surnames provide an initial and often determinative clue. Given the uniqueness of the Basque language, there is a corresponding universe of Basque surnames that are quite distinctive compared to other Spanish and French ones. Whenever they occur throughout the historical record, for the Basque scholar they are flashing beacons marking a path worthy of exploration. However, the exercise is far from perfect. Individual bearers of Basque surnames may, in fact, be unaware of, or indifferent to, their Basque cultural legacy. Conversely, not every culturally Basque (and even Basque-speaking) person has a Basque surname. In short, the surname evidence can distinguish candidates for further consideration, but, ideally, there should be additional evidence of Basque cultural awareness informing some aspect(s) of the individual's behavior before consigning him/her to the "Basque" category.

to inform behavior both in Iberia and throughout the Spanish empire. For many historians, colonial figures like Catalina de Erauso (the transvestite "lieutenant nun"), the explorer Juan Bautista de Anza or the prelate Juan de Zumárraga are usually treated as "Spaniards"; and the Basque-descended liberator, Simón de Bolívar, is a New-World-born Creole of Spanish ancestry. While this certainly captures an essential part of their personae, by exploring their Basque cultural and political heritages, personal loyalties and social networking a more complex and complete picture emerges.[81]

Indeed, like many of my colleagues in Basque diaspora studies, I could have told the present story as a Basque narrative. Julio César Arana was born in a town part of whose founders were of Basque descent. He was of Basque descent himself through both his paternal and maternal lineages. He married the Basque-surnamed Eleanora Zumaeta. His closest business associates were his own Arana siblings and his Zumaeta siblings-in-law. His attorney and political protégé, Julio Egoaguirre, was Basque-surnamed, as was Carlos Zubiaur, the captain of the *Liberal*. His key business partnership during his incursion into the Putumayo was with the Basque-surnamed Colombian, Benjamín Larrañaga. Several of the vilified white overseers in the Putumayo mentioned specifically in the recounting of the atrocities, were Basque-surnamed (Martín Arana, Bartolomé Zumaeta, Elías Martinengui, and Luis Alcorta). Arana's political fortunes prospered most during the two administrations (1908–1912; 1919–1930) of Basque-surnamed president of Peru, Augusto Leguía. Finally, there was the extended residence of Julio César's family in Biarritz in the French Basque Country.

Nevertheless, along with all of the other enigmas surrounding Arana's persona, his possible "Basqueness" is equally problematic. It is not underscored by any of his interpreters—Ovidio Lagos, José Eustasio Rivera, Mario Vargas Llosa, Marie Arana, Richard Collier, Jordan Goodman. The first four (an Argentine, a Colombian, and two Peruvians) were all from countries with significant Basque immigrant traditions; yet all treat Julio César as a Peruvian of "Spanish" ancestry. In Lagos's case, even Arana's decision to settle his family in Biarritz is devoid of Basque significance—dismissed as a courtesy to Eleonora who could then cross the border into

---

81. See Eva Mendieta, *In Search of Catalina de Erauso: The National and Sexual Identity of the Lieutenant Nun* (Reno: Center for Basque Studies, University of Nevada, 2009); Donald T. Garate, *Juan Bautista de Anza: Basque Explorer in the New World, 1693–1740* (Reno and Las Vegas: University of Nevada Press, 2003); and José Mallea-Olaetxe, "The Private Basque World of Juan Zumárraga, First Bishop of Mexico," *Revista de Historia de América* 114 (1992), 41–60.

the Spanish Basque area and employ her Spanish while purchasing familiar Hispanic foodstuffs. In short, there is indeed no evidence that either Julio César or his family took any particular shared ethnic interest in their French Basque cultural and linguistic setting. Rather, it seems that they formed part of the discernible South American expatriot community in Biarritz. We should recall that Arana also settled his family for extended periods of time in Paris, Switzerland, and London (all venues preferred by expatriot South Americans).

Consequently, at one level, this book should serve as a cautionary tale for my fellow Basque diaspora scholars. It would seem that the Aranas, the Zumaetas, and the Larrañagas were far too distanced generationally from their Old-World roots to retain any meaningful sort of Basque identity—let alone one that informed their daily lives and behaviors. Clearly, their shared Basque surnames did not spare Benjamín Larrañaga from Julio César's wrath; anymore than it spared the life of Manuel Erazo, assassinated by Bartolomé Zumaeta. If any of these men were aware of their Basque ancestry, they remained notably silent regarding it.

And yet I am not willing to dismiss altogether a far more subtle, indeed certainly subconscious, Basque influence as operative within Julio César's character—my own epistemic murk if you please. Stereotypically, throughout much of Hispanic Latin America, Basques are regarded as hardworking, entrepreneurial, loyal, and faithful in their personal lives and business affairs. That description certainly fits Arana. It is also true that every one of his interpreters, including his enemies, regarded Julio César as "different" from the typical Latin American tycoon. In short, Julio César Arana may actually have been an unconscious hyphenated Basque-Peruvian, while undoubtedly more the latter than the former.

Finally, I can't resist extending my Basque palimpsest to a comparison between Roger Casement and a different Arana—Sabino Arana Goiri. Both ill-fated men might be viewed as youthful martyrs to seemingly lost causes. Both began life in families within the establishments that they ultimately came to contest (Ulster Protestantism in Roger's case and Carlism in Sabino's). Both came to believe that they belonged to an oppressed race (for Roger, Irish Celts, and for Sabino, Basques). Both became prime movers within their respective ethnic nationalist movements—Casement emerging as a leader of Irish independentists and Arana as the founding figure of contemporary Basque nationalism. Both were imprisoned for their activities. Roger was executed as a traitor at age fifty-four by Great Britain, and Sabino died at age thirty-eight shortly after serving the prison term that shattered his already fragile health. While neither man lived to see it, the causes of both were somewhat vindicated (part of Eire is now the Irish Free State while four of the five presidents of the Basque Autonomous Region of Spain since Franco's death and promulgation of the country's present constitution belonged to the Basque Nationalist Party founded by Arana). Both men are vilified to this day by their enemies (British and Spanish nationalists) and idolized by their successors (Irish and Basque nationalists). Each in his own way was a Samarem, but with the distinction that instead of metaphorically descending from the mountains down river to reform and then rule over the overarching political oppressor (Great Britain and Spain), each sought to remove his mountainous reign from the larger whole.

So there were the two Aranas—Julio César and Sabino. While contemporaries and separated by but a few miles during the former's Biarritz period, to my knowledge they never met. Both sought at the very least autonomy, if not downright independence, for their respective constituents (when Julio César was wearing his Loretan rather than Peruvian hat). Both struggled with issues surrounding genocide, ethnocide and ethnic cleansing. Julio César privileged Peruvians over the Colombians that he sought to ethnically cleanse from Aranalandia, not to mention civilized whites in general over barbarous indigenes. Sabino privileged Basques over the Spaniards that he sought to ethnically cleanse fom his Basque homeland. Both were racists. Sabino believed in Basque racial superiority over Spaniards;[82] Julio César regarded whites to be superior to indigenes. In any event, had the two men met, I would have wished to have been a fly on their wall.

Conclusion

My concluding remarks are more self-confessional and admonitive than either theoretical or substantive. It is hard for me to imagine a better case study than the Putumayo story for postmodernist treatment—given the fascination of its central figure, Michel Foucault, with discourse analysis of texts regarding marginated subjects (the victims) as a means for understanding power relations.[83] Yet I do not regard myself to be a consummate postmodernist—despite sharing many of their concerns and even premises. To wit: I am skeptical (as are they) of what might be called "mindless positivism," a "scientific" search for truth, predicated upon the assumption

---

82. See William A. Douglass, "Sabino's Sin: Racism and the Founding of Basque Nationalism," in Daniele Conversi (ed.), *Ethnonationalism in the Contemporary World: Walker Connor and the Study of Nationalism* (London and New York: Routledge, 2002), 95–112.

83. Michel Foucault, *Power*, edited by James D. Faubion, *Essential Works of Foucault 1954–1984*, Vol. 3 (New York: New York Press, 1994).

that it is out there to be discovered and that investigators are capable of a truly objective search for it. I am equally skeptical of the quest for universal "laws" informing and determining human behavior.[84]

In most respects, I remain the product of Clifford Geertz—one of my mentors while a graduate student at the University of Chicago. Like Geertz, I began my intellectual career in literature (his English and mine Spanish) before moving on to anthropology. We have both dabbled in writing fiction. Consequently, like him,[85] I have blurred disciplinary boundaries throughout my career[86] and have been sympathetic to the discourse-analysis of literary critics such as Jacques Derrida. I am sympathetic (to a degree) with the subject positions of Vico and Nietzsche when they contend that there are no facts but just interpretations, and that we can only understand the world that we ourselves have created. Both underscored the importance of literary persuasion (Vico's *fantasia* and Nietzsche's Homerian poetics) in the formulation of historical narrative.[87] In two of my previous

---

84. In this regard, I would side with Stephen Jay Gould in his debate with Edmund O. Wilson concerning the nature and capability of scientific methodology and epistemology, that is, that there is a qualitative difference between the natural sciences and the humanities (including the social sciences) in that they are two distinct, yet complementary, ways of attaining knowledge. See Stephen Jay Gould, *The Hedgehog, the Fox and the Magister's Pox: Mending the Gap between Science and the Humanities* (Cambridge, MA: The Belknap Press of Harvard University Press, 2011), 189–260. Gould also maintained that science and religion are two different ways of understanding reality. See Stephen Jay Gould, *Rock of Ages: Science and Religion in the Fullness of Life* (New York: Ballantine Books, 2002), which led him to coin the acronym of NOMA (Non-overlapping magisteria) to describe their differences. Wilson conceived the term "scientific humanism" and invented the field of sociobiology—arguing that all human behavior is genetically determined; therefore it is biologically based and subject to discoverable natural laws (one can only imagine what his "explanation" of the protagonists and events in the Putumayo might have been). It is certainly ironic that this debate on the nature of human affairs devolved between two natural scientists specialized in the study of land snails (Gould) and ants (Wilson)!
85. Clifford Geertz, "Blurred Genres: The Refiguration of Social Thought," *The American Scholar* 29, no. 2 (Spring 1980).
86. I began my first book, *Death in Murelaga* with the appropriate epigraph from Thomas Mann: "It is a fact that a man's dying is more the survivors' affair than his own," and quoted from Hermann Hesse's last novel, *The Glass Bead Game*, as the epigraph for my work *Emigration in a South Italian Town: An Anthropological History*. To wit:

> To study history one must know in advance that one is attempting something fundamentally impossible, yet necessary and highly important. To study history means submitting to chaos and nevertheless retaining faith in order and meaning....the writing of history—however dryly it is done and however sincere the desire for objectivity—remains literature. History's third dimension is always fiction.

87. David W. Price, *History Made, History Imagined: Contemporary Literature, Poiesis, and the Past* (Urbana and Chicago: University of Illinois Press, 1999), 38–41.

forays into history,[88] I have commenced the journey with the admonition of Hayden White (like Hesse and consistent with both Nietzsche and Vico) that histories are authorial inventions:

> But in general there has been a reluctance to consider historical narratives as what they most manifestly are: verbal fictions, the contents of which are as much invented as found and the forms of which have more in common with their counterparts in literature than they have with those in science.[89]

He goes on to say that:

> ... a historical discourse should not be regarded as a mirror image of the set of events that it claims simply to describe. On the contrary, the historical discourse should be viewed as a sign system which points in two directions simultaneously: first, toward the set of events it purports to describe and second, toward the generic story form to which it tacitly likens the set in order to disclose its formal coherence considered as either a structure or a *process*. Thus, for example, a given set of events, arranged more or less chronologically but encoded so as to appear as phases of a *process* with a discernible beginning, middle, and end, may be emplotted as a Romance, Comedy, Tragedy, or what have you, depending upon the valences assigned to different events in the series as elements of recognizable archetypal story-forms.[90]

Neither Geertz nor I confuse our literary influences,[91] let alone our own forays into fictional writing, with our anthropological re-

---

88. See Joseba Zulaika and William A. Douglass, *Terror and Taboo: The Follies, Fables and Faces of Terrorism* (New York and London: Routledge, 1998), 16; 65; 156; and William A. Douglass, "In Search of Juan de Oñate: Confessions of a Cryptoessentialist," *Journal of Anthropological Research* 56, no. 2 (2000), 143.

89. Hayden White, *Tropics of Discourse: Essays in Cultural Criticism* (Baltimore and London: The Johns Hopkins University Press, 1978), 82.

90. Ibid., 106. White had attempted to actualize said process by considering in a synoptic table the four tropes, modes, emplotments, types of argumentation, and ideology underpinning the historical narrative of key works by four noted historians—Michelet, Tocqueville, Ranke, and Burckhardt, as well as four philosophers—Nietzsche, Marx, Hegel, and Croce. While an enormously influential work, unfortunately, the exercise drowned in its own "epistemic murk" when White had to admit that none fell into the neat compartments of his table since each contains elements of more than one of his tropes and emplotments. See Hayden White, *Metahistory: The Historical Imagination in Nineteenth Century Europe* (Baltimore: The Johns Hopkins University Press, 1973).

91. My own literary luminaries are Hermann Hesse, Jorge Luis Borges, Gabriel García Márquez, Umberto Eco, V. S. Naipaul, and, above all, Italo Calvino. All were consummate novelists deeply immersed in fanciful historical fiction.

search. I endorse and practice his call for "thick description" in the search for central themes and symbols that encapsulate and project the uniqueness of a particular culture—its worldview as it were. A prime Geertzian example would be his analysis of the cockfight as public performance of Balinese cultural norms[92] and mine that of the funerary rituals of a Basque village.[93] In this regard, we are both what might be called "symbolic anthropologists," committed to explicating the "dignity" of the cultural differences among different peoples. We are both the products of "Kantianism without a transcendental subject, Hegelianism without an absolute spirit" and thereby committed to "the scientific study of cultural diversity by profession and of postmodern bourgeois liberalism by general persuasion."[94] At the same time, we are both students of the grand designs of Max Weber and Talcott Parsons—each of whom sought explanatory theories of human behavior.

I am a "relativist" in a methodological sense as well. While I believe that absolute truth is unattainable, relative objectivity is the objective. I believe that each history is unique (relatively) from every other, and that each is a product of a combination of human intent (both individual and institutional) as reflected in decisions and their implementation, outcomes that become, at least in part, cumulative (i.e., the past configures the present and the present prefigures the future). In short, cultures are human creations and their histories are, at least in part, the cumulative outcomes of individual and collective agency.

Therefore, while I believe myself in this book to be relating a "story," I do not regard it to be entirely fictional (unlike Foucault who famously characterized *his* texts to be novels). If I have "created" this metanarrative of the Putumayo events and their protagonists, it has referents to something(s) that happened; to my mind there *was* a there there. In presenting a narrative, arranged in the main chronologically and constructed out of the stories (texts) of others, I have made critical decisions regarding the reliability of such accounts. In order to capture the flavor of the dispute, I have included certain mutually irreconcilable versions of the events and descriptions of their protagonists. I have also edited (or "erased" a la Michel de Certeau)[95] through exclusion of both extraneous text and

---

92. Clifford Geertz, "Deep Play: Notes on the Balinese Cockfight," *Daedalus*, 101, no. 1 (Winter 1972), 1–37.
93. William A. Douglass, *Death in Murelaga: Funerary Ritual in a Spanish Basque Village* (Seattle and London: University of Washington Press, 1969).
94. Clifford Geertz, *Available Light: Anthropological Reflections on Philosophical Topics* (Princeton: Princeton University Press, 2000), 73, 74.
95. Price, *History Made*, 111.

whole accounts that were too implausible (in my judgment, of course) to be of any use. I would also underscore that these stories are hugely biased in favor of the Occidental point of view, given that we have an extreme paucity of indigene discourses. Furthermore, the latter's viewpoints are oral historical in the main, rather than textual, and remain further limiting by the fact that they are reconstructed (and even consciously invented) rather than contemporary discourse now converted into a legacy of their past.

I differ from Hayden White (as reflected in the quote above) in one fundamental way. My Putumayo narrative has been a composite of many others, and *none with a beginning, middle, and end*. Rather, we have visited and then departed several ongoing streams of "history." Broad-stroke are the white colonization and subsequent evangelization of South America. But, as we have seen, these can both be disaggregated into a myriad of sub-streams, such as: relations among the various countries contesting political hegemony of the Putumayo, their internal conflicts, the dealings of each with external European powers and the United States, the religious struggles between traditional Catholic secular hierarchies and the Church's mission orders, the confrontations of both constellations with their own liberation theologians and fundamentalists, the competition between Catholicism and Protestantism for religious hegemony of the continent, white capitalistic exploitation of the environment versus indigene collectivist custodianship of it, the impact of the contemporary forces of globalism and environmentalism upon much of the foregoing, and so forth.

Finally, I believe that there is a sense in which no two investigators of the ostensibly same phenomenon share the same intellectual formation and/or concerns, any more than do any two lectors of this interpretation; consequently, both the writing and reading of this story are equally solitary exercises. Nevertheless, that solitude is itself relative in that we all have enough in common for me to write this book and you to read it.

In sum, South America's (and the world's) indigenous peoples are no longer passive. We have considered only a portion of the acronymic alphabet soup of the many indigenous associations and those of their fellow travelers from the perspective of just two countries—Peru and Colombia. Nevertheless, the new willingness to take on the world underscores just how formidable are the interests arrayed against the various tribes. And what a world it is. Foreign and domestic governments, not to mention the array of multinational and domestic corporations bent on developing tribal resources for their own gain, are near invincible in any ultimate sense. Today's victory is often simply the prelude to tomorrow's battle.

Furthermore, it should always be kept in mind that the many of the forces configuring the battlefield are beyond the control of the contending protagonists. The domestic and international political climates are never static givens; nor is the play in the fields of the Lord.[96] Over the past century, the political pendulum in both Peru and Colombia, not to mention the rest of Latin America, has swung wildly back and forth between the radical left and ultra-right extremes of the political continuum. Entrenched oligarchical interests are challenged both by the *nouveau riche* aspiring to enter the ranks of the landed elite and the new activism of indigenous peoples, ethnic minorities, and impoverished peasantries who themselves can share at times common interests or at others end up at cross purposes. Powerful military dictatorships have emerged all along the political spectrum.

Since the London Hearings, the world has contemplated the dramas surrounding such larger-than-life Latin American and Caribbean figures as Emiliano Zapata, Pancho Villa, Rafael Trujillo, Jacobo Árbenz, Papa Doc Duvalier, Fidel Castro, Juan Perón, Anastasio Somoza, Augusto Pinochet, Daniel Ortega, Hugo Chávez, and Evo Morales—to list but a very few. There have been two world wars, Hiroshima, decolonization of much of Asia and Africa (often triggering bloody civil wars), the Korean and Vietnam Wars, a Cold War with its nuclear balance of terror, Islamic jihads, a post-9-11 unending War on Terror (with its War on Drugs subplot), most of the Middle East in flames, an aggressive Russia in the West and assertive China in the East, a wobbling European Union in the wake of BREXIT, tens of millions of displaced political and economic migrants and refugees (and counting), global warming, and a plethora of other environmental challenges. There is now an impulsive president of the United States, arguably the most powerful man on the planet, who has zero experience in domestic politics and international affairs. As fellow denizens of this world, we all share a common fate with the indigenes of this narrative. In this regard, our planet is now truly globalized in most respects, including the gruesome ones.

How any of this will turn out, no one knows. Indeed, I rather suspect that we are all living out our personal and collective epistemic murk from

---

96. See Peter Matthiessen's novel *At Play in the Fields of the Lord* (New York: Random House, 1965), a story of the interaction between missionaries and Indians (who are bombed by the Brazilian Air Force) in the Brazilian Amazon. The novel was made into a motion picture in 1991. Our discussion of the Putumayo's missions, the many initiatives and failures, contain the seeds of the contemporary Pentecostalism among both Catholics and Protestants as they both contest hegemony of the traditional Catholic Church throughout Latin America.

which nothing will really "turn out." The muddle in itself seems interminable. In the words of the novelist Kurt Vonnegut, Jr.[97], referring to the attitude of Allied prisoners of war, unsure of their fate as they cringed in their basement bomb shelter during the Allies' firebombing of Dresden,

"So it goes."

---

97. Kurt Vonnegut, Jr., *Slaughterhouse Five or the Children's Crusade* (New York: Delacorte Press, 1969), passim.

Appendix 1

Memorandum 25

(This document is translated from the Spanish. It is taken from a file of the papers belonging to the Company. It would appear to be a document used for the promotion of the Company. Some of its phrases can be traced to the Draft Prospectus of the Company.)

Sketch of the Putumayo
J. C. Arana & Hermanos

The river Putumayo is an affluent of the Amazon, navigable at all seasons of the year by vessels drawing five feet and during the rainy weather by large steamers.

The rivers Igaraparana and Caraparana are nearly parallel, and flow into the Putumayo on the left bank.

The business of Julio C. Arana & Hermanos comprises practically the possession of the territories formed by the extreme points "Junin," "Delicias," and all the river Caraparana and its affluents on the north; "Arica," the mouth of the river Pupuna and the river Cahuinari and its affluents on the south; the river Japura or Caqueta on the east, and the rivers Campuya, Algodon and Tamboraco on the west.

In all that region are distributed more than forty commercial agencies [stations] which depend, some on "Chorrera" or on the "Encanto," and others directly on the house of Iquitos.

The immense quadrilateral whose boundaries we have delineated, is inhabited by sixty to seventy thousand uncivilized Indians, of whom more than half are reduced to obedience, thanks to the efforts put out by the house of Arana.

We will give a short sketch of the history of the river Putumayo.

The Peruvian rubber dealers, since the year 1880, the period at which rubber was for the first time exploited in the west of Peru, have explored the river Putumayo on various occasions, the expeditions of Deofanto Reategui and Francisco Capa in the launch "Tahuayo," of Pablo Zumaeta in the "Galvez," and many others, having failed.

Source: Sketch of the Putumayo. Memorandum No. 25. Copy in the Bodleian Library MSS Brit. Emp. S-22 G338.

The failure came from the limited resources with which they began that arduous enterprise, in which besides they had to subdue great numbers of cannibals, whilst in the neighbourhood of Iquitos and in the rivers Purus, Javary, Jurua, Ucayali, etc. the rubber was found in abundance and without the drawbacks of the Putumayo. In the year 1896, Jose Maria Mori Ramirez, one of the clients of J. C. Arana, of Iquitos, formed a partnership with Benjamin Larraniaga, freighted a launch, loaded it with goods and started for the river Putumayo with sufficient men, and, following the affluent Igaraparana, they established the commercial house called "La Chorrera" at the foot of a beautiful cascade.

The forests were immense, with abundant rubber, and everything promised the most successful working if sufficient civilised men could be brought to the place. But this was very expensive, and the house of Larraniaga, Ramirez & Co. had a very heavy outlay, and they became debtors to Mr. Arana for a very large amount.

The difficulties increased with the obstacles that the Brazilian authorities put in the way of the free navigation of the Peruvian flag through the Brazilian Amazonas on the way to the Putumayo river. These circumstances discouraged Morey, who transferred his share to Mr. Arana, and the house of Larraniaga, Arana & Co. began work.

With the capital brought in by Mr. Arana, and with the credit which his name gave to the house, the working of the rubber trees began in a serious manner, and the Indians who were before so dangerous were used for the work. Strict orders were given to treat them well and to try and obtain their friendship, and this policy helped to obtain the submission of the Indians. They were made to understand that in exchange for the rubber they would get goods which would be useful to them, and advances were made to them in clothes, implements and so forth with the condition that they would pay in rubber or in work.

In one of his voyages from Iquitos to Chorrera, freighted with merchandise and carrying on board a large number of men, the launch "Putumayo" was seized by the Brazilian authorities at the frontier and taken to Manaos, denying to the Peruvian flag the right of free transit by the Brazilian Amazonas on the way to the Putumayo. Then followed a long and costly litigation in which were invoked the existing treaties, celebrated in the year 1854 between Brazil and Peru with respect to the river Putumayo, and after many debates in Manaos and in Rio de Janeiro the Peruvian government obtained that Brazil would put into execution the treaties to which we have referred; since then the Peruvian flag encounters no obstacles in the Brazilian Amazonas on its way to the Putumayo.

In 1903 Mr. Larraniaga died, and his heir was his son Rafael, who continued in the business, associating himself with Mr. Juan B. Vega, changing the name of the firm and commencing the present firm of Arana, Vega & Co. Since then Julio C. Arana & Hermanos have bought from Mr. Rafael Larraniaga his share.

The Indians work for preference in the Putumayo a kind of rubber known in the European and American markets as "Rubber Tails," and they give the preference to working for this rubber on account of the easy manner in which it is obtained, consisting as it does only of making incisions in the trees and to squeeze the gum with the hands, in a rough way, to form the tails.

The Indian is generally naturally submissive. Eight or ten civilized men are enough to control 300 or 400 of them.

One thing worthy of note is that there exist some 30 odd distinct tribes, with distinctive languages, and without any connection between them, who, having lived always in constant warfare, now for some reason seem disposed to wage war again. That diversity of tribes and that antagonism between them always impedes a rising *en masse*, which it would be difficult to suppress on account of the number of Indians.

The firm has in all this region more than 1,500 civilised employees who maintain order and regulate the work of the Indians; the occupation of the employees reduces itself to that of armed vigilants, their general arms being Winchesters which are being changed for modern automatic arms.

The house of Arana has monopolized all the business of the region which we have already delineated; either for purchases made from other industrial concerns of their possessions or marked out according to the forest laws of Peru (Ley de Montanas) other points considered commercially strategic, and Mr. Arana is now obtaining a special concession by which the firm asks for a final title to certain lands and various privileges in the way of taxes, etc., in exchange for roads and other facilities that they would put at the disposition of the government. . . .

For the use of this business the firm has eight steamships, two of them being 120 tons and the other five [sic] smaller; they possess besides a great number of rowing vessels, and others that are towed by steamers.

In all the Amazon region there is not a better or healthier climate that than of the Alto-Putumayo, the best proof of which is the great number of Indians found living there. Whilst in the other rivers there are none, or if there are any, their numbers are but few.

All of the enterprises formed for the exploitation of rubber in the Amazonas have to face the grave difficulty of the scarcity of labour. In the Putumayo this problem does not exist, as the Indian is a splendid worker, sober, of very few needs, accustomed to the woods, and extremely cheap.

Another difficult problem for those works is the food of the workers, food that almost always is of canned goods imported at some cost. This, however, does not apply in the Putumayo, where the Indian cultivates his vegetable garden.

Suppose for a moment that there should be a great fall in the price of rubber, that the price should fall so low that it would be impossible to work it with any result. What would be the result? There would undoubtedly be a great crisis in the Amazonas which has the fictitious life given to it by rubber; but the region of the Putumayo, having navigable rivers which put it into communication with all parts, with a splendid climate, with thousands of square miles of land that Humboldt said would be the emporium of the world, with thousands of people to apply themselves to agriculture, woodcutting, etc., it would be, in all time, a stupendous store of wealth.

Appendix 2

Translation of Testimony of Felipe Cabrera M.

My name is Felipe Cabrera M. and I left Florencia, Department of Tolima, in the month of September of 1899 with several companions and as employees of the commercial firm of Pizarro and Gutiérrez to set myself up in a place called Puerto de los Monos, situated on the right bank of the Caquetá River, more or less four days distant from the Araracuara Falls. In that place the house of Pizarro and Gutiérrez had an important outpost engaged in the extraction of rubber. While I was in that place Sr. Emilio Gutiérrez, along with thirty companions and a large quantity of merchandise, passed through. He was an important Colombian trader who was involved in the extraction of rubber and who also had important outposts established throughout the region. In our conversation he informed me that he was journeying to Brazil to find a commercial firm that would advance him abundant goods to stimulate his business, given that he [Gutiérrez] had relations with a large number of Indians of the Andoques, Muinanes and Muitufichanes tribes. He told me that he hoped to find this operation with Brazil to be advantageous as opposed to continuing to bring in his merchandise from where it was very dear due to the high transportation costs. He also asserted that with this operation he wanted to free himself from the constant requests of the Julio C. Arana commercial firm, Peruvian merchants established in the Putumayo, who wanted him to enter into business with them in order to receive from them the goods that he needed. Sr. Gutiérrez carried out his projected trip and, at a spot where the Caquetá flowed into the Amazon, he came across the establishment of a Brazilian trader who extended him the credit that he was seeking. On his return voyage to his own agencies he was accompanied by Sr. Ferreira Da Silva, who is known as "Chicon," who came with Sr. Gutiérrez as far as the point called El Palmar, on the Caguinarí [sic—Cahinarí] tributary of the right bank of the Caquetá River. It was there that Sr. Gutiérrez encountered part of his work crew and from there that Sr. Ferreira Da Silva returned to Brazil. He was to come back during the epoch agreed upon with Sr. Gutiérrez to receive the product [rubber] in

Source: Taken from a document in a Colombian archive quoted in Dominguez and Gómez, *La economía*, 187-89.

exchange for the consigned merchandise. When Sr. Gutiérrez arrived at his holdings I was on a round up (*correría*) looking for rubber trees and Indian workers. After completing that mission I ascended to Tres Esquinas, at the confluence of the Orteguaza River and the Caquetá. From there within a short time I returned to the place that I had first come [Puerto de los Monos] and it was then that I learned from the mouth of one of his workers who had accompanied Don Emilio on his trip to Brazil that a few days after having arrived at one of his residences Sr. Gutiérrez was assaulted by a gang of workers of the Peruvian firm of Julio C. Arana y Hermanos. They murdered him, his wife and almost all of his employees, comprised of more than sixty white workers. I also learned that the rubber that Sr. Gutiérrez had in his agencies and the goods that he had brought from Brazil were stolen by the aforesaid workers and taken to La Chorrera, the principal agency of the mentioned Peruvian firm. Upon my return to Tres Esquinas the employee who had replaced me in the firm where I worked, Sr. Jesús Torres, showed me an order of the La Chorrera agency in which it intimated that we abandon our houses and the Indian territories that we occupied or else said agency (La Chorrera) would come to remove us with force. Neither Sr. Torres nor I obeyed said order because we did not have the authorizations of our employers, Srs. Pizarro and Gutiérrez. Business matters obliged me to undertake a new journey to Tres Esquinas; upon my return I saw that the intimidation of the Casa Arana had been fulfilled; all of our buildings and fields had been sacked and burned. I found Sr. Torres and several of his companions to be fugitives in the bush. I encountered them during the trajectory from the burned house and Puerto de los Monos and they recounted to me the assault carried out by the employees of La Chorrera. I returned with Sr. Torres and the survivors of the attack to Tres Esquinas in order to inform Srs. Pizarro and Gutiérrez of this sacking and arson in the Huitocia. For almost five years I lived in that part of the Caquetá working for myself, transporting rubber to Tolima and returning from there with merchandise that I needed for my businesses. In the first days of the month of September in 1906, I left Florencia, accompanied by the Srs Feliciano Muñoz, Pascual Rubiano, José de Paz Gutiérrez, Bonifacio Cabrera, Jorge Carvajal, Carlos María Silva, Heliodoro Anturi, Crisanto Victoria, Roso España and two women named Elena Polo and Eudosia Silva, supplied by the commercial firm of Sr. Urbano Gutiérrez and that which we had purchased with cash, in more or less ten craft—counting row boats, rafts and small canoes. Since I had some consignments of rubber in the Upper Caquetá I also invested the proceeds from their sale in goods that I bought from the firm of Calderón Cuéllar, Manuel Berrío and Braulio Cuéllar,

Colombian busninessmen established on the Caquetá and who also extended me credit for some merchandise. Provisioned now with all these elements we continued our journey and we set ourselves up close to the Araracuara rapids on the left bank of the Caquetá among the Andoques tribe of Indians. We began constructing our houses and clearing our fields when, on the first day of January of 1907, we were surprised while engaged in these tasks by an assault of more than thirty Peruvians armed with rifles. On the third day of the same month Sr. Normand, leader of the Peruvians, arrived accompanied by thirty more men. The Peruvians who came on January first took as prisoners Pascual Rubiano, Feliciano Muñoz, Roso España, Eudosia Silva, thirty Indians and me; Paz Gutiérrez and the rest of our companions were on the other side of the river. The day that Normand arrived, and after killing in my presence two Indians with a machete, he ordered me to tell Paz Gutiérrez and the rest of our companions to hand over their arms, threatening to kill us if they failed to do so. Given the situation that we were in, I acceded to his desires and Normand gave the order to our companion Roso España who was serving as intermediary. Normand embarked with more or less half his Peruvians in our craft. When they reached Paz and his companions, and seeing white men and without me among them, they grabbed their carbines in anticipation of an attack. Roso España shouted to them not to shoot and that he bore an order for them to hand over their arms; they did so and when the Peruvians were in possession of them they surrounded the house and field and opened fire on the Indians who were working there. I calculate that somewhat more than fifty persons were killed by Normand's companions, including every Indian working on the roof ridge of the house and most of those working in the fields. They threw many Indian women into the river and then used them for target practice. They even killed some Indian children, shoving them head first into the holes that had been dug for the house's posts. After finishing their assault they took me to where we were finishing another house and that night we slept tied each to a post and with a watch guard. They took all of our goods, arms and belongings that we had brought from Tolima and the fourth day in the morning we were taken as prisoners by the Peruvians to the place called Matanzas, Normand's residence. In this place and in our presence they killed three Indians that they had captured along with us, including the *Cacique* Jaidui, Captain of the Andoques. The death of these unfortunates was carried out with a gunshot after having given them each more than 100 lashes. In this Matanzas site we were fellow prisoners with more than 200 Indians in all counting men, women and children, who had been placed in stocks, mouth down, all of their bodies atrociously mutilated by their daily whippings. We stayed in Matanzas for a day and a night, and in such a short time I personally saw them remove the bodies of three Indian children from the stocks, dead from the lashings and from hunger. From Matanzas on the eighth of January at one in the afternoon they took us to the Sabana agency that was under the Peruvian Aristides Rodríguez. We stayed there for one day from whence they took us to the Oriente agency where the head was the Peruvian named Fidel Velarde. We remained in Oriente for close to ten days. They had taken us there with our hands bound and now they exchanged our bindings for the stock and the chain. During our stay in Oriente we were the victims of savage abuses: when the three section heads of the Casa Arana came together they spent their time in constant orgies and in slapping and kicking us. They wanted to screw two of the white women, one the wife of Paz Gutiérrez and the other of Pascual Rubiano, and for not giving them their pleasure the women too were put in stocks. The majority of our companions stayed in Oriente; they took Paz Gutiérrez and me, chained together, to the Port of Palitos, and in a row boat they transported us to Santa Julia on the River Igaraparaná. From Santa Julia we were taken to the agency of Abisinia where we were held prisoner, chained up, for nineteen months. The first six of which we were in stocks and chained; the rest of the time we were in chains only. The agency of Abisinia at that time was under the Peruvians Abelardo Agüero and Augusto Jiménez Sensiriario. When we arrived Jiménez was occupied in what are called in the Putumayo the round ups (*correrías*) of Indians. More or less a month after we arrived in Abisinia, the said Jiménez returned bringing somewhat more than one hundred Indians. With the excuse of his arrival and possibly to celebrate the success of his expedition, he, Agüero and the remaining white employees gave themselves up to drink. That afternoon and in our presence they killed thirty Indians, including four women in their number. Here, as in Matanzas, and during the nineteen months of my imprisonment I saw Indians being whipped daily and corpses being extracted from the stocks frequently. I also heard, from the mouths of the employees of the same firm, as well as from the aforementioned Agüero, of the death of all the Colombians who worked in that region. News of such events were the motive in Abisinia for great rejoicing. All of the employees got drunk and broke into shouts of vivas to Peru and death to Colombia. The names of the murdered Colombians which I learned in that time included, among others, those of the Srs. David Serrano, Ildefonso González, Gustavo Prieto, Justino Hernández, Juan de Dios Escobar, all crimes carried out by the employees of the firm of Julio C. Arana, captained by the Section Heads. July 13, 1908, they took off my chains and sent me to the Morelia agency under the command of Bartolomé Zumaeta; my companion Paz Gutiérrez and his wife remained prisoners in Abisinia. I am not certain how long he was held prisoner and what happened to him there, although in fact someone told me that he later saw Gutiérrez on the streets of Manaos. And of all the companions that remained in Oriente I only saw again Roso España, it would be two years ago in Puerto Córdoba, on the Caquetá River. Fourteen days I was in Morelia obliged to work under the orders of the Section Head, Sr. Zumaeta; July 28 of that same year a festival was celebrated in Abisina to which Zumaeta was invited. It is to this circumstance that I owe my freedom. I was able to win over his replacement, a Brazilian named Juan Bautista Brag, and I escaped from Morelia. By way of the Caguinarí River I reached the Caquetá where I have worked for more than two years in the firm of Srs. Jaramillo, Mejía y Cía. Whenever I ask Sr. Agüero the reason for my imprisonment he replies that he was simply following the orders of his Chief, Sr. Julio C. Arana.

Appendix 3

The Flower of the Selva
A Tale of the Upper Amazon.
(Based upon Facts.)
W. E. Hardenburg

I

The clear crystal waters of the River Urubamba splashed noisily against the numerous boulders that obstructed our progress, as we wearily pushed our way up-stream. The burning rays of the tropical sun poured down upon us, unrelieved by the slightest breeze, for the steep high banks of the narrow stream shut out every breath of air. Still the Indian boatmen continued their patient poling, while I—somewhat exhausted by the tedium of the journey—dozed off in a sweet, dreamy slumber, from which I was presently awakened by the voice of Acate, the *popero*.

"Wake up, *Señor*. We have reached the Mashino rapids and must unload the canoe."

I arose lazily from the box of condensed milk, which had been my seat, and, stepping ashore, fastened the canoe, while the five Indians began to unload the cargo and carry it up the steep bank. Acate, the headman had just shouldered his load and was preparing to follow the others, when a yell from the last man caused us to look up. A huge boulder had detached itself from the bed and was leaping down the precipitous bank straight for Acate. I shouted to him, but, encumbered by the heavy load strapped to his back, the unfortunate man was unable to turn aside in time, and the boulder, with a final, malignant leap, struck him full in the chest and knocked him into the seething whirlpool at the foot of the rapid.

Aroused from my languor by this catastrophe, I hastily grasped a coil of rope, one end of which was fast to the canoe, and plunged in after him. After severe exertions, I succeeded in getting him out and, finding that he was still alive, within a few minutes restored him to consciousness. A few draughts of my only remaining bottle of Scotch soon fixed him up, except for the immense bruise where he had been struck by the boulder.

Source: Unpublished manuscript. Bodleian Library MSS Brit. Emp. S-22 G346.

That night, when our frugal supper of *farina*, *danta*-steak and coffee was over and I was sitting by the fire, smoking my last pipe before turning in, Acate approached my *rancho*. I could see that he wanted to say something, so I spoke.

"Well Acate, how are you feeling now?'

"Quite well, thank you, *Señor*. You have saved my life and . . . I must tell you something."

"Well, out with it. I'm going to bed directly."

"*Señor*, I'm going to leave you to-night."

"The devil you are. You forget that you are engaged for the whole trip and that I have advanced you half the money."

"Listen, *Señor*, I had no need to tell you, for I was going to desert to-night, but since you saved my life, I can't leave you that way, without a word of thanks, so I beg you to let me go. You'll have four men left and they'll be quite enough, now that the dry season has begun. Here, I return to you the money you advanced me."

I paused to reflect. If he was bent to go, it would be best to let him have his way, for, if I had to keep him by force, he would be more trouble than help. Still I rather liked Acate and did not relish the idea of losing him, for he was worth any two of the other men. At last, I growled:

"What's your reason? Why didn't you tell me of this before?"

"*Señor*, I can't give you my reason . . . I didn't tell you before because I knew you wouldn't take me, if I told you I couldn't go all the way, and I had to come."

"Look here, if you can't tell me the reason, you can't go. I suppose you want to leave me thinking that you'll get more money somewhere else. Is that it?"

The Indian was silent a moment. His tall, slim figure—covered only by his long, cotton *cushma*—was silhouetted in bold relief against the moon-lit sky, as he stood before me, apparently deep in thought. Then turning to me with a look of sudden resolution, he said:

"*Bien, Señor*. You're an *Inglés*, so I'll tell you, and you shall be my judge."

Then, still standing before me, Acate told me his tale. Sometimes during his narration, his voice would break and he would be compelled to stop to master his emotion: again, he would raise his head, which, covered with its long, black hair, resembled a lion's in its savage majesty, while his dark eyes gleamed with a diabolical light.

II

"Ten years ago, *Señor*," he began, "on a night like this, I was the happiest man in the Campa tribe. Why? *Señor, Señor*, have you never been in love?—but no, white men cannot love; their hearts are too cold."

"For years, I had watched the Princess Cuma grow up—from a little, laughing *guegua* to a woman. She was the daughter of old Guema, our chief—the flower of the *selva*, we called her, for never, in all our tribe, was there a girl like her. Oh, I see her now before me, as she was then—those bright, black eyes, that smiling mouth, that laugh, like the murmur of the river, her hair, long, black, luxuriant, her figure, graceful as a fawn's . . . Oh God, why was she born?"

"But I must be calm, *patrón*. It's a long story, but I shall cut it short. I had just passed the test as a warrior and that very afternoon we had finished my little hut. The chief had consented to my suit and on that fatal evening, we sat together, Cuma and I, planning for the morrow, when she was to become mine. Everything had been made ready; great jars of *chicha* and *masata*; meat and *fariña* there was in abundance; a place had been cleared for the dances; everything was there, all ready for the three-days celebration that would signalize the marriage of the chieftain's daughter with young Acate."

"Until a late hour, we sat there, apart from the rest, talking as only lovers can. Then old Guema's voice was heard calling her and so, with a last, fond kiss, we parted, and I made my way to the new hut, which stood in the forest, a little out of the common clearing. Here I lay for hours, thinking of my great happiness, until, at last, I fell asleep."

"But what was this? Thunder, lightning, fire, yells, shrieks and cruel laughs! I dashed out of my hut to the main camp. A horrible scene met my eyes. There was a band of the dreaded white men, all dressed in clothes, whom we had heard of but never seen. The air was thick with smoke, but lighted continually by the terrible lightning of the whites and the blaze of our burning camp. As I rushed forward, spear in hand, I saw numerous corpses lying on the ground, while there, in the distance, struggled a group of women and children in chains. Among them I saw Cuma—my Cuma, together with old Guema and several others. I leaped forward to the rescue, but once more the awful thunder of the white men was heard and I knew no more."

You understand, *señor*, at that time, we knew nothing of firearms. We thought the whites were fiends, sent by the Spirit of Evil himself, who had supplied them with thunder and lightning for our destruction. Thus, such was our surprise and terror that, paralyzed by fear and unable to escape, over two thirds of our tribe of seventy-odd families were destroyed that night—the men killed and the women and children kidnapped to serve as slaves for the extraction of rubber—that accursed rubber, which has been the ruin of us all.

"I was awakened, *patrón*, by the vultures plucking at my eyes. It was mid-day, and the tropical sunlight poured in a dazzling stream over the ghastly scene that met my gaze. There lay the corpses strewn about among the ashes of the camp, picked at and torn about by an army of *gallinazos*. *Ay, Dios mío, señor*, never shall I forget that spectacle. There lay old Guema a body, completely disembowelled by the carrion birds, intent upon their awful feast...."

"I arose weakly and, finding that the wound in my breast had started again, thru the pecking of the vultures, bound it up with a few *piri-piri* leaves and then, looking once more upon the sickening sight around me, I dropped upon my knees and registered a vow of vengeance against the fiends responsible for this night's work."

"*Señor*, I left my people, my blood, my soul; I left everything. I wandered five days, five centuries, without arms, without food, without anything, alone, in the midst of the *selva*. On the afternoon of the fifth day, I stumbled and fell, and, overcome by my weakness, I fainted. How long I lay there, I know not; at last, I awoke, and there, bending over me, chafing my hands, stood a white man, but apparently, of a different tribe, for he wore a long, brown robe, something like our *cushma*s, and on his feet, instead of shoes, were sandals. I tried to crawl away, but in vain, for, lifting me in his arms, he carried me a few paces and we entered his *rancho*."

"That holy man was a Capuchin father—Padre Estanislao—and for five long weeks he stopped there, nursing me and curing my wounds. During this period, he taught me a little Christianity, some Spanish, the use of fire-arms and the ways of the white men. Then came the time when I was well and, at last, we parted. It cheers me to think that years afterward, I had the pleasure of repaying him, but that's another story."

As I say, I left the *Padre*, *Señor*, and for the long years, I wandered thru Eastern Peru, searching for my Cuma. Sometimes, I thought I had found a trace of her, but no, it resulted in nothing. During this time, I stopped at nearly all the rubber camps of this region, and, oh God, the sights I have seen in some of them—tortures, flagellations, chains, slavery and lust—all for the fatal black gold of the Amazon."

"At last, wearied and discouraged, I arrived at Iquitos. This, *patrón*, was just before I met you. I found her there."

III

"It was on a hot, burning afternoon, when I was walking down the Calle de Prospero, that I met her. A *chola*, carrying on her neck an immense roll of laundry, was being pulled about by a group of boys. Chasing them away, I turned to the slave and saw—Cuma, but, good God, how changed. Her hair, once abundant, long and glistening, was now thin and mangy; her eyes, before, like two deep pools of the black water of the forest lakes, were now dull and bleared her figure—but, oh *señor*, I can't go on . . . you understand."

"I turned to her quite calmly and said: 'Cuma, I have found you at last. Do you remember me?'"

"She started; then, with a glad cry, she rushed towards me. But stopped half-way and hesitated."

"'Oh, Acate. Acate. Is it really you? . . . I thought you were murdered with the rest.'"

"'I escaped, Cuma, and have been looking for you ever since, and now . . . I've found you. But we can't talk here—where are you going?'"

"She lead the way, still carrying her laundry, to a little stream, not far off, that emptied into the Amazon, and there, while she—my Princess Cuma—washed the dirty clothes of her white owner, I listened to her story."

"I will not shock you with her narrative, *patrón*, for it would give me too much pain. It will suffice for you to know that on the very night of their arrival at the white man's house, my Cuma, in spite of her appeals, was ruthlessly violated by the chief, who afterwards kept her as the favorite of his harem for nearly three years, when he tired of her and sold her to a friend for forty pounds. During this time, she bore him one child, whom he kept when he sold its mother."

"Her new owner suffered from a most repugnant and infectious disease, very common in the Amazon, but, nevertheless, he, too, made her his victim. This lasted about four years, when, becoming short of money, while on a trip to Iquitos, he sold her as a common *chola* to her present master, a human brute, who flogged her into a state of insensibility, when he learned of the dread malady she had contracted. Continual cruelty had been her portion ever since."

"I ascertained the name of the wretch who had made the raid and who was thus responsible for all this, and turning to the repugnant wreck of her, who, long ago, had been the pride of the Campas, I said:

"'Cuma, our lives are wrecked, wrecked by that monster. I have found you, but, alas I have found you ruined. In this life, we can never come together; perhaps, in the next, we can. What do you say, Cuma? Shall I end all this for you? I'll follow you as soon as I've settled with the wretch that caused it all.'"

"She lay silent in my arms. Then presently, she murmured: 'Give me one last kiss, Acate, and I'm ready.'"

"So for the last time, I kissed those lips, once so pure and sweet, but now scarred and disfigured by the foul disease of which she was the victim."

"'Follow me soon, Acate,' she whispered."

"'I will, Cuma,' I answered, and five minutes later, her spirit had entered the land of Rest."

"For hours I sat there, holding in my arms the blood-stained corpse of her whom I had loved so well. I was thinking—thinking of the happy days of long ago, of my Cuma, as she was then—pure and spotless as the snowy peaks of our Andean *páramos*—before the coming of the 'civilizers', for as you know, *señor*, these fiends say they are 'civilizing' us. I sometimes doubt that there is a God, *patrón*, for what sins have we poor people of the *selvas* ever committed, that He should deliver us to such punishments? *Señor*, I know; we are a doomed race. A few years more, and no longer will the foot-steps of the red man be seen within these *selvas*; no longer will the forest ring with the laughter of our children; no longer will our dug-outs navigate these broad rivers. No, *señor*. These vast *selvas*, where we have lived so long, will form the sepulcher of the remnants of my race. The falling leaves of the forest trees will soon cover the rotting bones of the last of the Indians."

"But enough, *señor*. As I say, I sat there, on the banks of the little stream, until the last, red rays of the setting sun hid themselves behind the dark forest across the river. Then, arousing myself with an effort, amidst the fast-falling dusk, I carried the stiff, cold copse of the *selva*'s daughter to the mouth of the quiet stream and then, pressing a last kiss upon those cold lips, I threw her into the Amazonas."

"Shortly after that, I came with you."

IV

"Well, *señor*, my tale is finished." He paused a moment, then: "What do you say, *patrón*?"

"Go, Acate, and good luck to you," and I gave him my hand as I asked, "Who's the man?'

"Jose I. Fonseca," he replied calmly.

"What? Fonseca, the big *cauchero* at Retiro on the Maná?"

He nodded. I knew the man—a wretch, who had accumulated a fortune by the sweat, the sufferings, yes, the very life-blood of his fellow-creatures; a wretch proverbial, even in this vast and tragic theater of lawlessness and crime, for his rapacious cruelty toward the unhappy human beings who had the misfortune to become his victims.

So I arose and, unpacking the boxes, gave him a supply of *farina*, some dried meat, a bottle of *aguardiente* and a few other necessities for his journey. Then, clasping once more his slim, bronze hand, I said:

"Good-bye and good luck, Acate."

"Good-bye, *patrón*. God will not mark this down against you; He will understand."

And his dark figure disappeared in the gloom of the forest. That was the last I saw of Acate.

Six months later, I returned to Iquitos and, happening to glance over some local nespapers, which had accumulated during my absence, I saw the following interesting piece of news:

"It is with the utmost regret that we inform our readers of the horrible death of the popular and enterprising *cauchero* of the River Maná, Don Jose Inocente Fonseca, one of the foremost explorers of the remote regions of this department."

"It appears that one day, while out inspecting his *cholos*, at work extracting rubber, he was assaulted in a lonely spot by one of these savages, and choked into a state of unconsciousness. Then the assassin dragged his helpless victim several kilometers into the heart of the forest and, altho the latter was still alive, proceeded to crucify him by nailing him to a tree."

"Five days later, the horribly mutilated body of the unfortunate *caballero* was found by one of his employees. The eyes had been plucked out by the vultures, which had also disemboweled him. Not far off was a *rancho*, where, judging by the ashes, the Indian had stopped several days to gloat over the prolonged agony of his victim. The criminal was nowhere to be found, having probably made good his escape."

"*Señor* Fonseca was one of the first to explore and open up the Maná district, and bring the aborigines into the ways and customs of civilization and Christianity. It is sad to think that one of the people he had spent so many years in civilizing, can have repaid him in this manner."

"This is only one proof more of the innate savagery and barbarism of the Indians, and their total unfitness for civilization. Either the Indian must yield to the forward march of progress, or submit to extermination, and it has long been our opinion that the latter is the only true solution to the problem."

"As the lamented sr. Fonseca has no other heirs, his large fortune—

made in the rubber business during the last eight or ten years—will go to his brother, sr. Fortunato Fonseca."

So Acate balanced the account at last.

Do you blame him, reader?

Appendix 4

Letter of Julio César Arana

Sir:

I have maintained my silence, for more than six years, regarding the incessant defamatory campaign sustained against the rubber businesses that, through great efforts and no few sacrifices, I succeeded in implanting in the zones washed by the rivers Putumayo, Caraparaná, Igaparaná [sic], etc.

Despite the continuous pleas of friends and associates to raise my voice and engage that campaign, showcasing its authors and denouncing their maneuvers, I understood that I should let time and the representatives of the judicial system do their work and shine the light necessary for the truth to triumph.

I was scarcely able to maintain my silence when, in 1910 "El Comercio" of Lima published and critiqued in its editorial columns a letter from Don Enrique Deschamps, and the fiscal of the Most Excellent Supreme Court of Peru, Doctor Salvador Cavero, formulating denunciations of the happenings in the Putumayo. But it was sufficient for one of the directors of The Peruvian Amazon Company to tell me—not out of loyalty, as I would later learn—that he considered it neither opportune nor decorous to engage in polemics in the media, and he convinced me to abstain from effecting any kind of publicity.

Having recourse to the media to reveal the miseries and to highlight all that is petty and wretched in the conduct of those who, not satisfied with denigrating me, have ended up dragging their own country to universal pillory and presenting it—in a gathering of *maniacs* and foreign speculators [the London Hearings]—as the womb of the worst criminals that the human species has ever aborted, was at odds with my sincere desire not to call upon myself the attention of others while offending my sentiments as a [sober] working man, adverse to passions and quarrels.

The conjuration of multiple related interests, responding to the same narrow and unspeakable yearnings and acting in various duplicitous ways, [out of] spite or vengeance, have served as an irremediable obstruction to the clarifying of the truth; to such an extreme that arrest warrants have

Source: Translation of the missive sent to *El Comercio* newspaper of Lima.

been issued against my partner and brother-in-law, Pablo Zumaeta, and against me, without any recourse to the laws nor the minimal consideration of just decorum.

These arrest warrants, although rescinded after deliberate examination, issued by the most illustrious superior court of Iquitos, were enough, as shall be seen, to force me to renounce my silence, in the defense of my honour and of the good name of my children, (I) will provide whatever light that is necessary so that the Putumayo process, of world-wide repercussion, will be known in its most intimate aspects and not continue to be the ignominious infamy for Peru nor a threat to those who have met their obligations, always attentive to the mandates of patriotism.

Other events, after those arrest warrants and which transpired in London, oblige me with even more imperious force, to speak out without residual blame or complacencies: various members of the House of Commons, constituted as a Select Committee, have opined regarding the same events that I am broaching—after a series of sessions that I will analyze in due time—in a manner opposed to truth, logic and the supposed rectitiude of English justice, such that my silence today would be more than inexcusable, it would rather be criminal.

As a consequence, I am editing an extensive exposition of all the events, incidences and circumstances related to the Putumayo questions; and I propose to clarify not only my activities in those zones, but rather demonstrate, with testimonies of startling probatory force, that until now, the only thing that has been clarified perfectly is the series of *chantages* [blackmailings] with which to obtain [from me] abundant personal benefits, even when to carry them out meant scoffing at the most sacred obligations and trampling under foot the clearest rights.

In order that public judgement be best informed regarding the points that I esteem to be significant, I have the pleasure of remitting, attached, the following pamphlets:

—*Perú y Colombia en el Putumayo*, by señor Carlos Larrabure y Correa;
—*Las cuestiones del Putumayo*, memorial of señor Pablo Zumaeta, and
—*Las cuestiones del Putumayo*, second memorial of señor Pablo Zumaeta.

The author of the first of these pamphlets is, as is well known, chief of Peru's information offices in Europe; and with the independence of character, stainless patriotism and vast knowledge that distinguish him, he has proven, in incontrovertible fashion, Peru's rights to the zones in which I founded my rubber businesses.

No one, even an obfuscater or lunatic, would dare to insinuate, after knowing the work of señor Larrabure y Correa, that the Aranas or their partners have wounded, in any fashion, the territorial rights of Colombia or damaged the Colombians who, with cunning maneuvers, tried to take over the Putumayo. The existence of that work, destined to come out in "The Times," of London, poses the challenge to the spokesmen of the chancellery of Bogotá that they are unable to destroy the strong line of reasoning of the illustrious advocate of the Peruvian cause.

The two remaining pamphlets contain the first and second memorials that señor Pablo Zumaeta has deemed appropriate to dedicate to the Putumayo questions, which will serve, in my view, to begin to remove not only the veil but the thick and dark cloak that has thus far interdicted their examination by impartial people.

Señor Zumaeta is a man who has spent his life, since the most tender infancy, dedicated to work; it being possible to consider him one of the most active and honorable pioneers of our eastern mountains. He has merited, through his energy, his probity and loyal sentiments, the appreciation and support of all the social classes of Iquitos, of all well-intended persons in the capital of the Loreto, which is more than enough to have accorded him the opportunity to have exercised, and to exercise at present, despite the iniquitous propaganda of certain persons, positions of highest authority and responsibility in the management of public affairs of a local nature, that is to say, all of those posts that require—in addition to intelligence and culture—abnegation and love of the country in which one has been born, given that none of these offices offers the incentives of the most insignificant monetary compensation, worthy of note being the direction of the Beneficent Society, the presidency of the Chamber of Commerce and of the Commerical Club, the mayorship of the municipality, etc., etc.

In the exposition that I am editing and which, despite its synthetic nature, will require a certain amount of time to bring together and put into circulation due to the many topics that it covers, the following will be clearly demonstrated, with dazzling clarity:

> 1. That the majority of the accusers of my businesses and The Peruvian Amazon Company Ltd., if not all of them, have handed down opinions that are entirely alien to their proclaimed sentiments [of] altruism and philanthropy.
>
> 2. That the denunciations of Saldaña Roca, Julio Murriedas, E. W. Hardenburg, captain Whiffen and doctor Rómulo Fernández[1]— around which so much noise has been made—lack legal and moral authority, since they are based neither upon duly proven facts

---

1. Likely Paredes.

nor do the character and backgrounds of the accusers make them worthy of respect.

3. That the posture of the English chancellery, intransigent and even illegal with respect to Peruvian sovereignty, has corresponded to egoistic goals, one of almost personal convenience.

4. That the visits of Mr. Cassement [sic] to the Putumayo and Iquitos, perverted when put into practice, and in violation of the most elementary principles of international respect, reflected a plan worked out in advance with the tripartite proposition of salvaging the reputation of an official of the English army, linked to the magnates of his country; protecting the interests of the journal *Truth*, and leaving intact the amour-propre of Mr. Grey.

5. That the process implemented by Mr. Cassement rests entirely upon declarations that are nullified from any point of view, [since] they were given by some negroes from Barbados, without any foundation whatsoever and under circumstances that not only make them refutable, but lend them criminal overtones.

6. That as much the English chancellery as Mr. Cassement—particularly the latter—have allowed themselves to be influenced, beyond a doubt, by the official agents and the meddlesome and Colombia, all of whom have exploited English vanity and pride and the sick fanaticism of certain persons of the United Kingdom, of whom the suffragists serve as a perfect example.

7. That some of the English directors of the Peruvian Amazon Co. Ltd, have assisted in the disparagement of the company, anxious to annul my action and obtain complete control of the business in order to favor Colombia's plans in exchange for customs' fees, concessions, etc.

8. That until now there has been no process inaugurated nor has any trustworthy proof come forward, valid and legal, regarding the commission of the crimes with which there has been an attempt to slander Peru's name and harm the honour of some of its sons.

9. That the reports of Mr. Michell and Mr. Fuller—in the event that the latter is attributable—after the visit they effected in the Putumayo in the months of August and September of 1912 in their capacities as consuls of S. M. B. and the United States, respectively, depart completely from the truth, that being easy to demonstrate, to such extremes that would bring down upon them great ridicule, with the simple reproduction of the numerous pho-

tographs that were taken during the visit by order of the Peruvian representative [Rey de Castro].

10. That the conduct of the government of the United States over the past several months, contrary to the posture it maintained at the beginning of England's intervention in the affairs of the Putumayo, pose alarming incentives for the Peruvian public spirit and should cause our government leaders pause for thought.

In order to give my exposition the probatory value demanded by the circumstances I will be forced to publicise documents, letters and diverse testimonies, ones capable of hurting more than one person and of diminishing the reputation of many others; I will have to speak out loud and clear, dispensing with euphemisms and reticences; but I am confident that I will not be accused, for this reason, of either indiscretion or emotion.

As I said before, I have remained silent and serene, for more than six years, in the face of a campaign of disparagement that only grew in step with my silence and serenity, as if they could be interpreted as a demonstration of my weakness or proof of my delinquency.

I have put up with everything, from the aggression against my person to the diminishment, perhaps irreparable, of my standing; and if they can blame me for anything, I believe it is for an excess of tolerance, never for precipitous anger.

The least that should be permitted to a man whose honour has been questioned, one who has been wounded gravely in his economic situation and who has been dragged before the criminal bench, is that he not remain silent, impassively removed from everything that gives meaning to his existence and his prerogatives of being a conscious human being.

I take this opportunity to extend to you this protestation of my special appreciation and consideration.

Your attentive Servant
Julio C. Arana

Appendix 5

A Criminal's Life Story:
The Career of Armando Normand
Peter MacQueen

Some months ago the civilized world was astonished and horrified by the revelations of murders and other atrocities perpetrated by the agents of the Latin-American rubber houses, who manage the great parties of seringeros who are now obliged to traverse the most remote wildernesses of the Amazon and its tributaries. Exposed to all kinds of perils, diseases and hardships, and the constant attacks of savages and the emissaries of competing concerns, the life is at best a very hard and precarious existence, in which the loss of a life is too often an ordinary incident of the calling. As it was on the Congo when human greed begot inhuman cruelty, so, it is alleged, prison, fire, torture and violence did their full work in the mad competition for the rubber treasures of the Amazonian wilderness. Following the investigations of Sir Roger Casement in 1910, one Armando Normand was arrested as the arch fiend of the jungle inferno, and confined in the city prison.

While in Peru Mr. Peter MacQueen went to the Guadelupe gaol, in company with Mr. C. N. Griffis, also an American journalist, to interview Normand. This man has been accused of murder, treachery, cruelty and bribery in connection with South American rubber interests, and the two visitors looked forward with tense interest to meeting so desperate a criminal. Having been admitted to the dingy prison, their first surprise was in the appearance of the prisoner. "Normand was brought forward," writes Mr. Griffis, "a slight, active figure. His features were clean cut and determined, his eyes bright and intelligent, and his manner fearless and at the same time courteous." Normand then briefly outlined the story of his life, giving an explanation of the events that led up to his arrest and imprisonment, which follows:

By Armando Normand

I was born in Cochabamba, Bolivia, thirty-two years ago and spent the first twenty years of my life in that city and vicinity. My father is a Peru-

Source: Peter MacQueen, "A Criminal's Life Story: The Career of Armando Normand," *The National Magazine: An American Illustrated Monthly*, 1913 (38), 942–46.

vian and my mother a Bolivian. I consider myself a citizen of the latter republic. Our family was one of the first in the Province of Cochabamba, and I was afforded excellent opportunities for securing an education. After graduating from the Seminario in my native city, I spent two years studying law, but finally abandoned the course and went to the Argentine. I attended the National School of Commerce in Buenos Ayres and graduated from that institution as a public accountant. Altogether I remained about two and one-half years in Buenos Ayres.

In 1903 I went to London and studied for a few months at the Pitman School in Russell Square in order to improve my knowledge of bookkeeping and modern business methods. While in London I was several times a guest at the home of the Bolivian minister, Mr. Avelino Aramayo, and had the pleasure of meeting many prominent members of the Bolivian and Peruvian colonies there.

In 1904, with a letter of introduction from Mr. Armayo to Mr. Carlos Larrañaga of the Para branch of the well-known Bolivian rubber firm of Suarez Hermanos, I went to the Brazilian port in search of employment. Mr. Larrañaga had nothing open just at that time, but kindly gave me letters of introduction to Mr. Julio Arana and others and advised me to go up the river to Manaos.

I left Para in January, 1904, and went to Manaos and there found employment with the Peruvian consul, Mr. Manuel Pablo Villanueva, staying at his residence during the ten months I spent in the latter port. My work there was chiefly in connection with the publication of La Union, the newspaper owned by Mr. Villanueva at Manaos, and I also gave lessons in English and bookkeeping and did translation work for the merchants of the port.

In October of the same year I accepted a commission from the Arana firm to accompany Mr. Abel Alarco as interpreter to Barbados where he was going to secure Barbadians to work in the rubber regions. Mr. Alarco was one of the managers of the firm. We contracted for thirty-six men to return with us and arrived back in Manaos some time in November. It is worthy of note that on the entire voyage the Barbadians constantly complained about the quality and the quantity of every accommodation given them. They were always quarrelling among themselves and could not be trusted. We remained at Manaos a week, where a young Argentine, named Glieman, who had worked for some time in the rubber regions, joined the party. At this point Alarco placed Glieman and myself in charge of the Barbadians and directed us to take them further up river to Iquitos. We left Manaos on the steamer Javari and went as far as the mouth of the Putumayo, where we met one of the Arana Company's steamers and received orders to transfer our men on board. Mr. Juan Vega, a member of the Arana firm, and Mr. Ramon Sanchez, both Colombians, were on this steamer and took charge of the expedition. We proceeded up the Putumayo River to its branch, the Igara-Parand [sic], and thence to La Chorrera, the principal establishment of Arana-Vega Compania, as the Peruvian Amazon Company was known at that time. It took us five or six days to reach this point. The manager of the station was Mr. Victor Macedo and the submanager Miguel Loayza. The Barbadians were here equipped with guns and we started off in a northwesterly direction into the wilderness to erect a new station. Sanchez and a man named Lizcano, both Colombians, were first and second in command, Alcorta, a Peruvian, who died in Lima about four years ago, was third in command. Glieman and myself went along as interpreters. On the way we passed two company stations Atenas and Entre Rios. We received constant warnings regarding the activity of the Indians the savage Andokes. In May of the same year Sanchez had led an expedition of twenty-eight white men into this same region and twenty were killed by the savages, Sanchez escaping back to Iquitos with only eight men. Such incidents were common in the Amazon country, and yet they ask in London why we carried guns.

At the end of about seven days on the trail we came to a deserted Indian house. The Amazon Indians of that region do not live together in villages, but apart in lonely huts in the forest. At this point we decided to build our station, naming it Matanzas. Altogether we had about forty-five men. Several attacks were made on us at that time by the savages, who were armed with guns taken from the various expeditions that had fallen into their hands.

The trouble with the Barbadians about food and work continued. They were lazy and constant grumblers. Because I was an interpreter and the intermediary between Sanchez and the men, they always blamed me if anything went wrong or they failed to secure what they wanted. It finally grew so bad that Sanchez was obliged to whip some of the negroes to get any work out of them at all. The whippings, as far as my knowledge goes, were light.

During this time, because of the enmity of the Indians, we had been unable to secure any rubber. In February, 1905, one of our men, Crisóstomo, a Peruvian, was killed by the Indians. I heard at camp that the savages made the attack because Sanchez had whipped two of their number. I never had any proof of this. In June, 1905, because of the Company's dissatisfaction with his work, Sanchez was discharged. His brother was made chief of the station and remained there one or two months, when he was recalled to Iquitos. I understand that the first Sanchez was killed in a quarrel with a fellow Colombian on the Jivari River a couple of years later. He was not in the employ of the Company at the time of his death.

During all of this time we were constantly having small encounters with the Indians. We kept sending our Indian interpreters, those who were able to speak Spanish, into the "bush" in attempts to induce the natives to bring rubber to the station and receive goods in exchange. At the time the last Sanchez brother left Matanzas, I received a note from Vega at La Chorrera informing me that I, together with another man at Matanzas, whose name I cannot recall just at this moment, would be placed in charge of the station. We were to receive, besides our small monthly salary, three soles on every fifteen kilos of rubber collected. This partnership lasted until about February or March, 1906, when the other man was dismissed because of his attitude towards the Indians. He attempted to coerce them into bringing rubber to the station, and would go out into the forest threatening them with various punishments if they failed to secure the amount he demanded. I knew the Indians, and the only result of such methods would be to drive them away from the station and we would get no rubber at all. I went to Vega at La Chorrera and stated the conditions to him and was then placed in full charge at Matanzas.

In November, 1906, we heard that the Indians were laying plans to attack the station in force. I wish you would emphasize the point that the grudge the natives held against any white men, or blancos, the very old one was that they were being dispossessed of land that was rightfully theirs, and so far as my experience teaches, the question of their treatment is only a minor issue. I often asked the Indians why they were conspiring against us and the answer invariably was that we were taking their lands away from them.

As soon as I heard the rumors of a general rising, I sent my interpreters into the forest to find out what the trouble was. To three of the interpreters I had given the names Roosevelt, Antonio and Pablo. The interpreters surprised a party of Indians and brought thirty-five of them, whom they found armed with guns, back to Matanzas. They reported that the savages had given a big war dance at which the main feature was three long poles about ten feet high, placed in the ground, and the middle pole bore on its top the mummified head of a white man who had been killed some time previously by the natives. We placed the rebellious natives who had been brought to camp in the stocks and gave two or three lashes to the more recalcitrant ones. Considering the fact that they had planned to wipe out our entire station, I consider this to be very light punishment. During the two or three days we held the Indians prisoners, the other Indians in the forest sent them food which we unsuspectingly allowed the prisoners to eat. It turned out, however, that the food had been poisoned by the Indians because they believed those held at Matanzas had deserted them. This is probably the origin of the stories told that I have poisoned many Indians. Eventually, the rising came to nothing and conditions became normal again.

The real cause of the uprising planned I believe was due to the agitation of a party of Colombians who, a short time previously, had come down to the Colombian border and across the river Caqueta, and were, we understood, laying plans to drive us out of La Chorrera rubber district and take possession of it themselves. I had received a warning note from headquarters to move very carefully in the matter. We knew that a number of Indians were armed by these Colombians for the purpose of attacking us. It will not be necessary to go into the details of this affair. I finally managed to meet several of the Colombians and arrange an entente with them, although a slight skirmish occurred during the negotiations and two of my interpreters were killed. That is, I crossed the Caqueta river with two of the Colombians and my men to meet the main party of Colombians. My men were ahead and as they approached the Colombian camp were fired on before they had an opportunity to explain their mission. At the camp I narrowly escaped death when a Colombian drew his revolver and was about to fire when one of my men knocked his gun into the air. But, as I said before, the matter was finally arranged satisfactorily.

The most important event during 1907 was a big fever which swept over the district, about fifty men dying in my section. I did everything possible to alleviate their suffering, sending to La Chorrera for quinine and other medicines. I had the bodies of a number of those who died cremated in order to prevent the spread of the fever. I am probably charged with the murder, by fire or poison, of all these people. What is there to say?

It was in 1908 that Captain Whiffen came to my station and stayed for five or six days. He told me that my station was the best of all of those he had visited and that my men were in the best condition. I gave him the best quarters in the settlement and treated him as well as I could. He told me that he was exploring in the Amazon for pleasure, and he also told me of his high connections in London and of his great importance in his own country. Quien sabe?

In the year 1909 there was a small rising in which I lost a number of men. A short time previously I had placed a Colombian, named Buchelli, in charge of a sub-station several miles from Matanzas so that when the rubber was brought in the natives would not have to carry it so far. There were four Peruvians with him. The Indians fell upon this house and looted it, slaughtering the men there.

I got wind of this affair just in time to prevent the same fate befalling Matanzas. After the looting of the small house there was a quarrel between the Indians as to sharing the spoils and in this quarrel an interpreter, named Segundo, was killed.

Some time in the latter part of 1910, Sir Roger Casement made his visit to Matanzas. He was there four or five days. I do not know where he gained his title Sir—I know an English gentleman when I see one, but Casement most certainly was not one of them. Mr. Tizon of the Arana Company was with him, and there were several other white men in the party. The only evidence I saw of judicial procedure was Sir Roger Casement taking two of my Barbadians into a room, one at a time, and bullying them into making untruthful charges against me for while they were in the room I often heard his voice raised in threatening tones, and once I heard him say "jail." I suppose he threatened them with imprisonment if they refused to give him the testimony he had come for. I questioned the Barbadians concerning the actions of Sir Roger Casement and they replied that he was attempting to force them to say that I had killed Indians—many Indians. That they could not truthfully say. He took the Barbadians away with him. Why did he not take their testimony before me at Matanzas, or before some other witness in the station? I was condemned by the testimony of the Barbadian negroes taken in a closed room. Such justice!

Then Roger Casement wanted me to lend him forty Indians to carry his baggage back to La Chorrera. I told him that my Indians were tired out and that within a day or two must carry a big consignment of rubber down to La Chorrera. I then had about 220 men. He grew angry and I also grew angry. I told Mr. Tizon that the Company might discharge me, but I saw no reason why I should burden my Indians with the Englishman's baggage, with its large consignment of whisky. I wonder now how many murders I was charged with because of my plain speech. But Sir Roger did not get my Indians for his baggage. He was forced to send to another station for the men, and left my station early one morning at six o'clock, without saying good-bye or by your leave. That day my Indians were going down to La Chorrera with rubber. You know his story of that march—all lies. The Indian women who went along carried only the food for themselves and their husbands, they did not carry rubber. Half way down I met one of the other Englishmen in the party. He offered me whisky, but I did not take it. Several Indians were taken ill on the march. Do you know of any tropical country where fever does not worry the line of march? I placed those who were ill in deserted houses along the trail and gave them medicine. But enough—my refusal to favor Sir Roger Casement had determined my fate as far as it lay within his hands.

I will not deny that I was often forced to place my Indians in the stocks—*cepo*—for short periods. What would you do with men who would not work, with men who steal and lie, with men who would murder you in cold blood if you gave them half a chance? I worked, often all day long, in the pouring rains; I suffered from chills and fevers and much sickness. I never was cruel to the natives. Considering the hard conditions under which we lived I treated them as kindly as I possibly could.

Early in 1910, before Sir Roger Casement came and before anyone had heard of his coming, I had asked to be relieved. I was often ill and had symptoms of the dread beri-beri. Victor Macedo, who was then manager of La Chorrera, said, however, that the Company had no one to replace me and asked me to stay several months longer. I did this under protest for I had spent five years in the great jungle without seeing anything of the outer world or having any of its comforts. The fact that the Company retained me nearly a year after I asked to be relieved indicated the standing I had with them at that time. Later my father received a letter from Julio Arana in which the latter had only praiseworthy things to say of me.

I left Matanzas in the latter part of 1910, a month or two after Sir Roger Casement left. From there I went to Manaos, Buenos Ayres, Valparaiso and then to Antofogasta, where for two years I made a business of selling Panama hats. During this time I made two trips to Ecuador and was in Lima a number of days. I did not know of the charges pending against me and I have always traveled under my own name. It was not until December of last year when I went to my home, at Cochabamba, that I learned of the lies circulating against my name. In Cochabamba I had started the business of buying and selling Chilean horses. When Sir Roger Casement's report was brought to my notice I immediately wrote a long letter to the Minister of Justice here in Lima, denying one and all of the charges against me, and offering to present myself at any time before the Peruvian courts for trial. The answer was an order for my arrest and extradition. Today you see me here in prison, but I am not afraid to face my accusers. You ask me to allow a picture to be published, but I do not desire that because I am now branded as a criminal and the world will not believe my word against that of the men who falsely accused me. After my trial in Iquitos I shall be glad to send you my picture, for then I shall have vindicated myself in court and cleared an honourable name.

Appendix 6

Poem by Rómulo Paredes (1918)

To Samarem
Distinguished compatriot,
sovereign of an impenetrable jungle,
sovereign of a powerful tribe,
sovereign of a hundred turbulent rivers
that stretch, twist and coil
that stretch, twist and coil through the forest
—constricting and epileptic—like boas . . .
Distinguished compatriot,
wandering pilgrim of untrodden wilderness,
hunter of horrible beasts, that you dominate
with your strength, your skill and with your fearlessness
and your disdain for a beneficent death,
that have made you so fearsome
and have given you such fame and such glory . . .
Samarem, beloved chief:
Samarem, master of the forest: distinguished compatriot:
I have long wanted to speak with you, I have long desired
To transmit to you my ideas and my anguish.
You have left the mountain, and walking along the edge of the precipice,
With a loving look, you have placed around my shoulder,
your potent and virile arm that parts the waters with an oar,
and as a colossal titan, cuts a path through the forest.
I wish to speak with you this day as a friend and as a brother,
beautiful day, grand date[1] on which the people commemorate
fifty years of existence, that have gone by
like black clouds galloping above your frond.[2]
The moment has arrived [to divulge] my saddest secrets,

---

1. Reference might be to the 1868 Arica tidal wave and earthquake (at an estimated 9.0 one of the strongest in history) that killed an estimated twenty-five thousand people. The following year marked the beginning of the War of the Pacific (1869–1872) that resulted in the loss of a significant part of Peru (including Arica) to Chile. Paredes may have been invoking one or both of these twinned natural and political disasters.
2. Probably reference to a palm leaf being used to shade Samarem from the sun.

of my complaints and my many painful confessions;
of which I speak to you from my soul, standing before the river that you are skimming and before the forest that you inhabit;
where only truths—the truths—predominate,
that, albeit bitter, on the other hand, are lifesaving.
And it is for this reason that by your side I speak gravely,
Samarem, beloved chief,
Distinguished compatriot:
Samarem, master of the forest:
Samarem: listen and weep . . .
Egoistical and ambitious, we have lost half a century
dominating audaciously the remotest wilderness,
and only one ideal has moved our hard hearts,
one ideal: that of the rubber;
and in pursuit of it we have struggled with tenacity and bravery,
writing in each place a grand and heroic account;
but the crime, always the crime spoiled these odysseys,
and the milk,[3] white and pure, we have turned into red blood.
Many Indians—your countrymen—have fallen
to satisfy ambitions in the disastrous struggle,
defenseless human flesh, that shames and makes one blush,
always hungry and naked, without affection and without justice,
more than slaves, more than pariahs, they are spectrums, they are helots . . .
Samarem, don't be embarrassed,
Samarem, listen and weep . . .
Samarem, the river deceives. The current of its turbulent and treacherous waters,
is not the directive force. Dig in, dig in, do not descend: downriver is hell,
downriver men are destroying themselves in Europe.[4]
downriver are your masters—wise monkeys in frock coats—,
vain and infatuated with laughable decorations,
with embodied decorations,
while the blood of my race, once strong, is now destroyed;
they are vain and infatuated with their positions,
poor positions that are flaccid and dishonorable, that they garner by fawning,
with vileness and through connections,

---

3. Latex sap.
4. Reference to World War I.

forgetting that history
gravely tells us, that a horse had been Consul:
Caligula's horse when he was the lord of Rome.
Samarem, master of the forest;
Samarem, sigh and weep . . .
Never come, never descend this treacherous river: let these lands
    be yours forever, for you, forgotten lands,
which calumny, the vilest mercantilism,
abuses and idleness whip mightily.
What would you find should you come?, your would find another
    tribe,
one inferior to that which you lead: a hypocritical tribe of idiots,
without morals and without conscience, only imitating in the
superior peoples, their wickedness and their clothing. . . .
there is not a single man who leads us today in our journey:
no one, nobody faces up to our most sacred problems;
wretched politics sap our energies completely,
they seduce us, they push us, they drag us, they immolate us.
Dig in, dig in. Never descend, Samarem, beloved chief:
It is better that you never know this sewer.
Downriver are the vices, downriver is the injustice,
downriver are the evils and the dregs . . .
Samarem, don't be embarrassed;
Samarem, listen and weep . . .
Samarem, master of the forest,
distinguished compatriot:
When you come, come haughtily, to demand of us the justice
that we deny to the Indians that are exploited today,
—yesterday strong, yesterday grand and happy,
and today slaves and helots.
If you conquer we aspire, come, conquer us tomorrow;
come and pull up our history, whose pages are filled
with embarassments and defeats.
The government of arrows that dominate vileness;
the government of your strong arm, let it be imposed;
and that way there would be a good country, a great country;
that way there would be peace, work, progress and reforms;
only in that way will the awaited and new beautiful aurora
be attained atop the great pile of ruins and disasters and plunders . . .
Samarem, master of the forest, distinguished compatriot;
Samarem, do not be bashful.

Spanish Original

A Samarem
Distinguido compatriota
soberano de una selva impenetrable,
soberano de una tribu poderosa,
soberano de cien ríos turbulentos
que se estiran, se retuercen y se enroscan
que se estiran y retuercen en el bosque
—constrictores y epilépticos—como boas . . .
Distinguido compatriota,
errabundo peregrino de los montes inhollados,
cazador de horribles fieras, que las domas
con tu fuerza y tu destreza, con tu arrojo
y el desprecio por la muerte bienhechora,
que te han hecho tan temible
y te han dado tanta fama y tanta gloria . . .
Samarem, querido jefe:
Samarem, amo del bosque: distinguido compatriota:
Hace tiempo que quiero hablarte, hace tiempo que deseo
transmitirte mis ideas y contarte mis congojas.
Has dejado la montaña, y de pie sobre el barranco
has tenido sobre mi hombro, con mirada cariñosa,
tu potente y viril brazo que abre el agua con el remo,
y como un titán coloso, taja el bosque con la trocha.
Quiero hablarte en este día como amigo y como hermano,
día hermoso, fecha magna en que el pueblo conmemora
cincuenta años de existencia, que han pasado como pasan
nubes negras cabalgando por encima de tu fronda.
Ha llegado ya el momento de mis tristes confidencias,
de mis quejas y mis muchas confesiones dolorosas;
de que te hable con el alma, frente al río en que te meces
frente al bosque donde moras;
donde sólo predominan las verdades –las verdades,
que aunque amargas, son, en cambio, salvadoras.
Y es por eso que a tu lado te platico gravemente,
Samarem, querido jefe,
distinguido compatriota:
Samarem, amo del bosque:
Samarem: escucha y llora . . .
Egoístas y ambiciosos, medio siglo hemos perdido
dominando con audacia la montaña más remota,

y un ideal sólo ha movido nuestros duros corazones,
un ideal: el de la goma;
y en pos de ella hemos bregado con tesón y con bravura,
escribiendo en cada sitio una acción grande y heroica;
pero el crimen, siempre el crimen malogró esas odiseas,
y la leche, blanca y pura, la volvimos sangre roja.
Muchos indios —tus paisanos— han caído
por saciar ambiciones en la lucha desastroza,
carne humana desvalida, que avergüenza y que sonroja,
siempre hambrientos y desnudos, sin cariño y sin justicia,
más que esclavos, más que parias, son espectros, son ilotas . . .
Samarem, no te avergüences,
Samarem, escucha y llora . . .
Samarem, el río engaña. No es la fuerza directiva
la corriente de sus aguas turbulentas y traidoras.
Surca, surca; no te bajes: río abajo está el infierno,
río abajo están los hombres destrozándose en Europa;
río abajo están tus amos —monos sabios con levita—,
vanidosos y enfatuados con insignias irrisorias,
con insignias encarnadas,
cual la sangre de mi raza, antes fuerte, y ya en derrota;
vanidosos y enfatuados con sus puestos,
pobres puestos de molicie y de deshonra,
que arrancan adulando, con bajezas e influencias,
olvidando que la historia
gravemente nos afirma, que un caballo ha sido Cónsul:
el caballo de Calígula cuando fue señor de Roma . . .
Samarem, amo del bosque;
Samarem, suspira y llora . . .
Nunca vengas, nunca bajes ese río traicionero:
siempre sean estas tierras, para ti, tierras ignotas,
a las cuales, la calumnia, el más vil mercantilismo,
las injurias y la holganza fuertemente las azota.
Qué hallarías si vinieras, hallarías otra tribu,
inferior a la que guías: tribu hipócrita y de idiotas,
sin moral y sin conciencia, imitando solamente
de los pueblos superiores, las maldades y las ropas. . . .
Ni un solo hombre que nos guíe hasta hoy en la jornada:
los problemas más sagrados nadie, nadie los afronta;
la política mezquina nos absorbe por completo,
nos seduce, nos empuja, nos arrastra, nos inmola . . .
Surca, surca. Nunca bajes, Samarem, querido jefe:

es mejor que a este pueblo de albañales no conozcas.
Río abajo están los vicios, río abajo la injusticia,
río abajo, las maldades y la escoria . . .
Samarem, no te avergüences;
Samarem, escucha y llora . . .
Samarem, amo del bosque,
distinguido compatriota:
Cuando vengas, ven altivo, a exigirnos la justicia
que negamos a los indios que hoy se explotan,
—ayer fuertes, ayer grandes y felices,
y hoy esclavos e ilotas.
Si conquistas aspiramos, ven, conquístanos mañana;
ven y arranca nuestra historia,
cuyas hojas están llenas
de vergüenzas y derrotas.
El gobierno de flechas que domine las bajezas;
el gobierno poderoso de tu brazo, que se imponga;
y así habría patria buena, patria grande;
así habría paz, trabajo y progresos y reformas;
solo así se alcanzará hermosa sobre el gran montón de ruinas
y desastres y despojos, la esperada y nueva aurora . . .
Samarem, amo del bosque,
distinguido compatriota;
Samarem, no te avergüences.

Appendix 7

WORDS OF THE ELDERS OF TOBACCO, COCA, AND SWEET YUCCA

IT IS OUR RESPONSIBILTY TO REMAIN ON GUARD

The worried elders searched for:
the beginning of illness,
the beginning of chaos,
the beginning of confusion,
the beginning of contempt,
the beginning of weakness,
the beginning of shame,
the beginning of falsity,
the beginning of the broken word,
the beginning of imbalance,
the beginning of envy of others,
the beginning of inequality.
In the midst of chaos,
the grandparents, guided by the
Tobacco, Coca and Sweet Yucca plants,
searched for the beginning of chaos,
the beginning of crisis,
the beginning of sadness,
the beginning of death.
From there was where they watched
Out of the spirit of illusion
The white man appeared,
there appeared ambition,
there appeared slavery and exploitation,
there appeared many things,
there appeared television,
radio . . . the computer.
There appeared projects,
the institutions, the researchers . . .

Source: Asociación de Cabildos Autoridades Tradicionales de la Chorrera–AZICATCH, *Plan de Vida*, 3–4.

There appeared many nice things
and the heart slept,
thought was damaged
and the soul became confused.
For this we must remain on guard
While the spirit of the wise slept,
The spirit of this woman said
I have for you a modern plan:
if you live with me
I am going to show you true progress,
the true food,
the true clothing,
true wisdom,
the true history,
true custom,
the true world,
true education,
true culture,
You will enjoy progress,
You will speak a different language
life will be easy,
there will be no suffering there.
Hearing this the spirit of our grandparents slept,
the thinking of the wise became confused,
the Word of Tobacco was forgotten,
the Word of Coca was forgotten,
the word of Yucca was forgotten,
the Law of Life was violated.
For this we must remain on guard.
The sacred plants showed us
the spirit of the foreign things,
there they discovered the cause of illnesses,
there was the root of wanting to be like others,
the march of things,
the whites' way, the way of confusion,
the way of sadness,
the path to ethnocide, of unhappiness.
the way of treason, of egotism.
All of this was known by our leaders
but their heart slept.
Searching for the roots of illnesses,
the children of Tobacco, Coca and Sweet Yucca

discovered the true path
the way of the Law of Life,
the way of the Plan of Life,
the plan that has no beginning or end,
that has existed always,
that does not change, even if we change,
that remains even if we are no longer here,
the the plan of the Law of Life
that is the wisdom of Tobacco and Coca.
The same plan that the creators had
when they made the day and the night,
the son, fire, the plants and animals,
that plan that gave birth to the basket and the *maloca*.
The Plan of Life is the way to be ourselves
the plan for recognizing ourselves as Children of Tobacco,
Coca and Sweet Yucca disposed to respect
its laws and precepts.
The plan for recovering the way
the plan for finding order and for knitting from the heart
our thinking,
the plan for continuing to be peoples
culturally differentiated,
with clearly constituted families,
the plan for having our own government,
organized, honorable and capable of representing
with dignity its people,
the plan to love and defend our homeland
as our mother, from within whom
we were brought forth into this world.
The plan to love and defend our natural resources,
and continue living in harmony and equilibrium
with the remaining beings of nature,
the plan for maintaining our alimentary sovereignty,
the plan to have an education
managed by ourselves
the plan to have our own justice system,
the plan for living healthily
with traditional medicine
and adequate western medicine as well.
To walk this path
we must sift and turn over [matters],
just like the intestine makes a selection

and chooses that which the organism needs
while expelling that which is useless in a natural manner and without violence
the things that come from the outside we should filter.
This filter should be the people's heart,
There should be selected that which is useful and what is not.
But it is not enough to filter,
In addition we can turn around these spirits
as with the spirit of anger that we carry within
we can turn them around and in a positive manner
in the guise of cultivation,
in the guise of art or of sport,
in the same fashion
we can turn to advantage the spirit of things:
A community radio station
to spread our customs
a computer to reinforce our languages,
communal projects to attain
and maintain our life plan,
Agreements with institutions
that respect our autonomy.
Some things put into our basket,
some things filtered and linked
that help us to reestablish
equilibrium and harmony.
Then the thinking of our leaders
Clarified itself and order returned.
But all of this will not be accomplished without sacrifice,
all of this will not come about without effort.
this will not be achieved without [our own] diets, without self-rule,
all of this will not be possible without allies,
all of this will not be done without risks,
it involves opposing the difficult with the easy,
creation with destruction,
order with disorder,
light with darkness.
For this we must remain on guard.

Appendix 8

Configuring and Comparing Rubber Barons

We have considered in considerable detail in the main text the biographies of the Colombian Rafael Reyes and the Peruvian Julio César Arana. We might now add another rubber baron, the Bolivian Nicolás Suárez, before comparing the striking similarities in the three men's lives. Bolivia, like Peru and Colombia, was a latecomer to the Amazonian rubber boom. In 1880, North American explorer Edwin Heath ascertained for the first time that the Beni River was a tributary of the Mamoré River and that the the Beni's lower stretches were inhabited by relatively benign indigenes. This opened up new opportunities for the expansion of Bolivian rubber stations into virgin areas, as well as the eventual exportation of their harvest downriver to the Amazon and beyond.

It was a young adventurer and youngest son of a ranching family, Nicolás Suárez, living in the town of Reyes in the Upper Beni region, who was quick to act. He paddled himself down the lower Beni and recognized the strategic importance of a place where there was an obligatory portage on one bank around an impassable waterfall. This would become the location of his future rubber station and the eventual settlement of Cachuela Esperanza. By 1881, the thirty-year-old Nicolás, had constructed his first buildings on the site that became the headquarters of the established Suárez Hermanos firm.[1] There was almost instantaneous expansion of the established Bolivian rubber gathering in the Upper Beni and within months there were one to two thousand aspiring *caucheros* in the Lower Beni. Suárez concentrated initially on supplying these clients in return for rubber.[2]

Francisco Suárez, the eldest sibling in his family, went to London in 1871, or well before the Bolivian rubber boom. That year he was appointed Bolivian Consul-General in London. By 1878, he had established his own London trading house, F. Suárez & Co. Brother Rómulo was charged with setting up Brazilian Suárez agencies just below the Madeira-Mamoré rapids, in Manaos and in Pará. He then returned to Beni to the family ranches near Trinidad to raise livestock and grow crops to supply the family's distribution network based in Cachuela Esperanza. Gre-

---

1. Valerie Fifer, "The Empire Builders: A History of the Bolivian Rubber Boom and the Rise of the House of Suarez," *Journal of Latin American Studies* 2 (2) (Nov. 1970): 121–125.
2. Ibid., 125–26.

gorio Suárez was assigned to the agency just below the Madeira-Mamoré rapids to supervise the transshipping and portaging activities (nearly five hundred indigene employees) that moved supplies upriver to Cachuela Esperanza and exported its products from Bolivia.[3]

In 1897, Francisco died suddenly in London, but in 1900 was succeeded as Bolivian Consul-General by Gregorio's son, Pedro Suárez. So the firm London foothold of Suarez Hermanos & Co. remained intact.[4]

In about 1904, brother Gregorio Suárez and his men were transporting supplies up the Beni in three boats to Cachuela Esperanza and were enticed ashore by familiar Caripúnas Indians. They all engaged in a friendly competition regarding marksmanship, but when Gregorio had used up the rounds in his revolver, an Indian turned and shot him in the chest with an arrow. The indigenes fell upon the rest of the party and killed them, all but two men who managed to escape by hiding under their boats. Nicolás, informed by the survivors, led an armed force that found the Indians' camped deep in the forest. They were drunk on cognac and were dividing up their loot; Gregorio's severed head displayed on a spear. Nicolás attacked and annihilated every one of the Indians.[5]

Nicolás Suárez managed to buy the estate of his prime rival (and customer), Vaca Diez, based out of Riberalta at the confluence of the Madre de Dios River and the Upper Beni. Both were ruthless in this regard as they stitched together personal empires. When Bolivia and Brazil faced off over control of the Acre region, Nicolás placed his private army of mercenaries in the service of his country's military.[6]

Then, in 1908, brother Rómulo was murdered on his ranch by a disgruntled employee.[7] By then, Cachuela Esperanza was a thriving town of two to three thousand residents. Nicolás had hired many European managers, particularly Swiss, and the community had its own library and hospital. Suárez Hermanos & Co. had extended its purview fifty or sixty miles into what would ultimately become Peruvian territory after Bolivia and Peru settled their boundary differences (1909–1912). The portage around Esperanza waterfall was now conducted by rail.[8]

Between 1909 and 1925, Nicolás and his spouse spent about half of their time in England. They educated their offspring and nieces and nephews at English finishing schools. When the Putumayo scandal (1912–1913)

---

3. Ibid., 128–29.
4. Ibid., 139.
5. Ibid., 135–36.
6. Ibid., 134.
7. Ibid., 136.
8. Ibid., 136–37.

erupted in London, Pedro Suárez was pressured to provide information on the organization and policies of the Suárez Hermanos & Co. Ltd rubber operations. He was considerably more successful in his defense than was Arana and his supporters. Pedro invited full scrutiny by British authorities; albeit when the "rubber met the road," and Minister Gosling showed up the Pando in 1913, he was hosted in Riberalta and Cachuela Esperanza, but pretty much denied access to the rest of the Suárez empire.[9]

Upon establishing his own presence in London, in 1909, Nicolás consolidated three Suárez enterprises (including the ranching one) into a single firm, Suarez Hermanos y Co., and issued shares in it to all of his family members and in-laws.[10] When he decided to leave London some fifteen years later, Nicolás bought back the interests of many successors in order to reconsolidate the company under his ownership. He was said to have refused an offer for it of 10 million pounds in 1924 in London. By then, Suaréz Hermanos accounted for no less than 60 percent of Bolivian rubber production, most of which was now destined for the North American market.[11]

It was in 1925 that Nicolás and his wife returned permanently to his beloved Cachuela Esperanza. From there Nicolás continued to preside, through the 1930s, over an empire that included, in addition to rubber, production of Brazil nuts and the raising of half a million head of cattle on his Beni ranches. He became a major shareholder in the national airline *Lloyd Aereo Boliviano*. He outlived his many critics, dying at eighty-nine years of age on January 11, 1940. He died a rich man. Nevertheless, the heirs initiated costly litigation against one another that dissipated completely the wealth of Suárez & Co. Ltd. When it was liquidated in London on February 25, 1961, it had but 1,500 pounds in assets. If Julio César Arana died as a forgotten pauper, Fifer concludes her biography of Nicolás as follows,

> The Bolivian Oriente has neither produced nor attracted any other comparable driving force on the scale required to come to terms with such isolation and physical difficulty as are found in Pando and Beni. Nicolás Suárez was a supreme opportunist, produced by, and, in retrospect, to some extent limited by his own intensely personal and ruthless struggle with the environment. He earned his place in the bitter history of the South American rubber boom as one of the few giants of its organization, and unquestionably as the most powerful Bolivian of the period. Significantly, he remained

---

9. Ibid., 139–40.
10. Ibid., 141–42.
11. Ibid., 142–43.

throughout his life a man of the Oriente and, as such, Suárez achieved the unique distinction of becoming the only entrepreneur of international stature that lowland Bolivia has ever produced.[12]

The commonalities in the lives of Reyes, Arana, and Suárez are nothing short of startling—a pattern that one suspects characterized, a little or a lot, the lives of many other Amazonian rubber barons. All three were explorers on the outer edge of the known civilized world in their respective areas. All three carved out personal empires in the "no man's land" of territory contested by several countries. All three used their economic clout to smash and absorb their competitors in the process of furthering their imperial designs. Arana and Suárez both employed paramilitary forces to protect and expand their interests; Reyes became a general and then president of his nation, roles that permitted him to further Colombia's agenda and his own. All three created family corporations, distributed their shares among close family members and relied upon these siblings and in-laws as their prime *hombres de confianza* in the conduct of their businesses. All three sought an Amazonian outlet for their forest product as an alternative to the cost-prohibitive option of exporting it over the Andes to a Pacific port and then around the southern horn of South America to North American and European markets. Consequently, all established a presence in Manaos and/or Pará. All three were Anglo-oriented and marketed their product mostly in North America and Europe—particularly the British Isles. Arana and Suárez both resided for a time in London and both were forced to defend their activities in the public hearings, courts, and media underpinning British and world public opinion. All three became larger-than-life figures in their respective nations. Reyes presided over his, Arana ran the Loreto and served in Peru´s parliament and Suárez, while personally shunning public office, exerted his considerable political influence through his office-holding close relatives. All three were advocates of their nation's infra-structural development; and particularly imporved communication between the national capital and their respective Amazonian Orientes. While Reyes predeceased dhe development of the planet's commercial aviation, both Arana and Suárez were pioneers and chamions of it in Peru and Bolivia.

In sum, the three lives underscore the considerable degree to which the private, home-grown, Latin American rubber barons (as opposed to the European and North American interlopers) were stamped out of a single mold.

---

12. Ibid., 146.

Acknowledgments

In writing a book like this, any scholar learns from people who are long dead in order to inform those who have yet to be born. However, along the way anthropologists like myself normally have some degree of firsthand contact with their venues, subjects, and collaborators. Such has not been my case. I have never been to the Putumayo. While I have passed through Lima, Iquitos and Bogotá, it was never as the intending writer of this work. I have not elicited oral histories from any of the descendants of the whites or Indians that inhabit these pages. Indeed, this text is of the post-modern, miraculous variety whereby an author can "research" from home with the aid of interlibrary loan services and the book deliveries of orders placed over the Internet. Hence my dedication of this book to Mark Lucas, the long-suffering interlibrary loan administrator at my university who humored a pre-modern scholar's incessant requests (as well as a plethora of anonymous drivers delivering my dozens of book purchases). Consequently, my personal wisdom regarding the Putumayo scandal rides in part on the wheels of UPS and Federal Express—not to mention the wings of Wikipedia.

While I have never met a single one of my fellow interpreters of this tale, and my attempts to correspond with several of them along the way went largely unanswered, I want to underscore my intellectual indebtedness to some in particular. It goes without saying that without the crusading zeal of Benjamín Saldaña Rocca, Walter Hardenburg, Roger Casement, Rómulo Paredes, Carlos A. Valcárcel, and Father Giovanni Genocchi, Arana's story would have likely been as little noticed as those of most of his fellow Amazonian rubber barons. While I have criticized Richard Collier's book *The River That God Forgot*, there is no doubt that it brought several threads together in a fashion that informs at some level every subsequent account and analysis of these events. Again, I would single out the Argentine writer Ovidio Lagos for his marvelous *Arana, Rey del Caucho*. It has the particular merit of focusing upon Julio César as the key protagonist, rather than his nemesis Roger Casement, the prime subject of the majority of treatments. My copy of the Lagos book is particularly dog-eared by now.

I have further boundless respect for four outstanding investigators whose works are central to this one. I refer to Michael Edward Stanfield whose *Red Rubber, Bleeding Trees: Violence, Slavery, and Empire in Northwest Amazonia, 1850–1933* is simply the canonical English language text on its subject. While I may have quoted excessively from this work, I also feel the frustration of having oversimplified his magnificent treatise. Then there are the many books and articles of Catalán historian Pilar García Jordan. Together these two scholars have relieved authors like me of years of archival research into the complexities of the Loreto's political and religious history. My third hero is English historian Jordan Goodman for his excellent book *The Devil and Mr. Casement*. While we may disagree that Julio César was satanical, this text is the best and most extensive English-language treatment to date of Arana. One could also mention Angus Mitchell's stellar research regarding Casement, work that reproduces much of Sir Roger's own writings while also examining the man as a tragic historical figure in his own right.

Unsurprisingly, several contemporary Peruvian and Colombian scholars (particularly my fellow anthropologists, given their professional interest in indigenous peoples) have analyzed this story. I would single out two as having contributed most significantly to my text. On the Colombian side there is Roberto Pineda Camacho, particularly with his book *El holocausto en el Amazonas: Una historia social de la Casa Arana*. Then there are the many articles and book chapters of Peruvian anthropologist Alberto Chirif Tirado. There is also his editing of two fundamental collections of documents and essays *La defensa de los caucheros* and *Imaginario y imágines de la época del caucho: Los sucesos del Putumayo* (with Manuel Cornejo Chaparro).

In sum, this particular dwarf wishes to acknowledge the giants upon whose shoulders he has stood while contemplating the distant Putumayo from Reno, Nevada.

Bibliography

Agurto, Jorge. "Peru." In *The Indigenous World 2013*, edited by Cœcilie Mikkelsen. Copenhagen: International Work Group for Indigenous Affairs, 2013.

———. "Peru." In The Indigenous World 2014, edited by Coecilie Mikkelsen Copenhagen: International Work Group for Indigenous Affairs, 2014.

Allison-Booth, W. E. *Hell's Outpost: The True Story of Devil's Island by a Man Who Exiled Himself There*. New York: Minton, Balch & Company, 1931.

Arana, Benito. *Lima al Amazonas*. Lima: Imprenta y Librería de San Pedro, 1896.

Arana, Julio C. *Las cuestiones del Putumayo*. Barcelona: Imprenta Viuda de Luis Tasso, 1913.

———. *Las cuestiones del Putumayo: Folleto No. 3*. Barcelona: Imprenta Viuda de Luis Tasso, 1913.

———. *El protocol Salomón-Lozano, o el pacto de límites con Colombia: al Congreso Nacional*. Lima: La Tradición, 1927.

Arana, Marie. *American Chica: Two Worlds, One Childhood*. New York: The Dial Press, 2001.

———. *Bolívar: American Liberator*. New York: Simon & Schuster, 2013.

Arcila Niño, Oscar, Gloria González León, Franz Gutiérrez Rey, Adriana Rodríguez Salazar, and Carlos Ariel Salazar, *Caquetá, construcción de un territorio amazónico en el siglo XX*. Bogotá: Instituto Amazónico de Linvestigaciones Científicas Sinchi, 2002.

Arendt, Hannah. *The Origins of Totalitarianism*. New York: Schocken Books, 1951.

Arens William. *The Man-eating Myth: Anthroplogy and Anthropophagy*. Oxford and New York: Oxford Univesity Press, 1979.

Asociación de Cabildos y Autoridades Tradicionales de la Chorrera—AZICATCH, *Plan de Vida de los hijos del tabaco, la coca y la yuca dulce y Plan de Abundancia Zona Chorrera 2004–2008*. La Chorrera: AZICATCH, 2006.

Bákula Patiño, Juan Miguel. "Introducción." In *Colección de leyes, decretos, resoluciones y otros documentos referentes al Departamento de Loreto*, vol. 1, edited by Carlos Larrabure y Correa. Iquitos: Monumenta Amazónica, CETA, 2006.

Barnes, Jonathan, ed. *The Complete Works of Aristotle. 2 vols.*. Princeton: Princeton University Press, 1984.

Belaunde, Laura Elvira. "Reseña." *Antropológica* 28, supplement 1 (2010): 339.

Boehner, Heinrich. *The Jesuits: An Historical Study*. Philadelphia: The Castle Press, 1928.
Bonilla, Victor Daniel. *Servants of God or Masters of Men? The Story of the Capuchin Mission in Amazonia*. Middlesex: Penguin Books, 1972.
Bonilla, Víctor Daniel. Siervos de Dios y amos de indios: El Estado y la misión capuchina en el Putumayo. Bogotá: Ediciones Tercer Mundo, 1968.
Brands, H. W. *Traitor to His Class: The Privileged Life and Radical Presidency of Franklin Delano Roosevelt*. New York: Doubleday, 2008.
British Parliament, House of Commons. *Report and* Special *Report from the Select Committee on the Putumayo together with the Proceedings of the Committee, Minutes of Evidence and Appendices*. London: Wyman and Sons, Limited, 1913.
Brotton, Jerry. *A History of the World in 12 Maps*. New York: Viking, 2002.
Bryant, William H. *Roger Casement: A Biography*. Lincoln, NE: iUniverse, 2007.
Cairo, Carlos del. "Las encrucijadas del liderazgo político indígena en la Amazonia Colombiana." In *Perspectivas antropológicas sobre la Amazonia contemporánea*, edited by Margarita Chaves and Carlos del Cairo. Bogotá: Instituto Colombiano de Antropología e Historia, 2010.
Calderón, Abel. *Viajes de Caquetá y Putumayo*. Bogotá: Imprenta Hernando Santos, 1904.
Calderón, José Gregorio. "Importantes documentos históricos: Contestación del señor José Gregorio Calderón," *Huila histórico* 11 (September 1933): 272.
Cameron, Catherine M., Paul Kelton, and Alan C. Swedlund, eds. *Beyond Germs: Native Depopulation in North America*. Tucson: The University of Arizona Press, 2015.
Casement, Roger, Sir. "The Keeper of the Seas." In *The Crime against Europe: A Possible Outcome of the War of 1914* by Sir Roger Casement. Philadelphia: The Celtic Press, 1915.
Castro Caycedo, Germán. *Mi alma se la dejo al diablo*. Bogotá: Plaza & Janes, 1982.
Chaumeil, Jean-Pierre. "Guerra de Imágenes en el Putumayo." In *Cien Años después del caucho: Cambio y permanencias en las relaciones con los pueblos indígenas*, edited by Alberto Chirif. Lima: Tarea Asociación Gráfica Educativa, 2009.
Chaves Chamorro, Margarita. "Cabildos multiétnicos e identidades depuradas."Iin *Fronteras, Territorios y Metáforas*, edited by Clara Inés García. Medellín: Hombre Nuevo Editores, 2003.
Chirif, Alberto. "Cien años después del caucho: Cambios y permanencias en las relaciones con los pueblos indígenas." In *Cien años después del caucho: Cambios y permanencias en las relaciones con los pueblos indígenas*, edited by Alberto Chirif. Lima: Tarea Asociación Gráfica Educativa, 2009.

———. "Imaginario sobre el indígena en la época del caucho." In *Cien años después del caucho: Cambios y permanencias en las relaciones con los pueblos indígenas*, edited by Alberto Chirif. Lima: Tarea Asociación Gráfica Educativa, 2009.

———, ed. and annotator. "Los informes del Juez Paredes." In *Cien años después del caucho: Cambios y permanencias en las relaciones con los pueblos indígenas*, edited by Alberto Chirif. Lima: Tarea Asociación Gráfica Educativa, 2009.

———. "Ocupación territorial de la Amazonia y marginación de la población nativa." *América Indígena* 35, no. 2 (1975): 265–295.

———. "Presentación." In *La defensa de los caucheros*, by Carlos Rey de Castro, Carlos Larrabure y Correa, Pablo Zumaeta, and Julio César Arana. Iquitos: CETA, 2005.

Chirif Tirado, Alberto, Pedro García Hierro, and Richard Chase Smith. *El indígena y su territorio son uno solo: Estrategias para la defensa de los pueblos y territorios indígenas en la Cuenca Amazónica*. Lima: Oxfam América, 1991.

Collier, Richard *Jaque al barón. La historia del caucho en la Amazonía*. Lima: Centro Amazónico de Antropología y Aplicación Práctica, 1981.

———. *The River that God Forgot: The Dramatic Story of the Rise and Fall of the Despotic Amazon Rubber Barons*. New York: E.P. Dutton, 1968.

Crevaux, Docteur J. *Voyages dans L'Amerique de Sud*. Paris: Librairie Hachette et Cie, 1883.

Crosby, Alfred W., Jr. *The Columbian Exchange: Biological and Cultural Consequences of 1492*. Westport, CT: Greenwood Pub. Co., 1982.

Curtin, Mary Ellen. *Black Prisoners and Their World, Alabama, 1865–1900*. Charlottesville: University Press of Virginia, 2000.

d'Ans, André-Marcel. *L'Amazonie Péruvienne indigène: Anthropologie écologique: Ethno-histoire: Perspectives contemporaines*. Paris: Payfatesot, 1982.

Davis, Wade. *One River: Explorations and Discoveries in the Amazon Rain Forest*. New York: Simon and Schuster, 1996.

Dean, Warren. *Brazil and the Struggle for Rubber: A Study in Environmental History*. Cambridge: Cambridge University Press, 1987.

Denevan, William M. "The Aboriginal Population of Amazonia." In *The Native Population of the Americas in 1492*, edited by William M. Denevan. Madison: The University of Wisconsin Press, 1976.

Dominguez, Camilo A., and Augusto Gómez. *La economía extractiva en la amazonía colombiana*. Bogotá: Tropenbos-Colombia, 1990.

———. *Nación y etnias: Conflictos territoriales en la Amazonía Colombiana 1750–1933*. Bogotá: Disloque Editores, 1994.

———. *La economía extractive en la Amazonia colombiana, 1850–1930*. Bogotá: Tropenbos-Colombia/Araracuara: Corporación colombiana para la Amazonia, 1990.

Dostoevsky, Fyodor. *The House of the Dead*. New York: Dell Publishing Company, 1959

Douglass, William A. "In Search of Juan de Oñate: Confessions of a Cryptoessentialist." *Journal of Anthropological Research* 56, no. 2 (2000): 137–162.

———. "Sabino's Sin: Racism and the Founding of Basque Nationalism." In *Ethnonationalism in the Contemporary World: Walker Connor and the Study of Nationalism*, edited by Daniele Conversi. London: Routledge, 2002.

———. *Basque Explorers in the Pacific Ocean*. Reno: Center for Basque Studies, University of Nevada, Reno, 2015.

———. *Casting About in the Reel World*. Oakland: RDR Books, 2002.

———. *Death in Murelaga: Funerary Ritual in a Spanish Basque Village*. Seattle: University of Washington Press, 1969.

———. "Witness in the Wilderness: The Tropical Tryst of Claude Lévi-Strauss and Theodore Roosevelt." *In Witness and Memory: The Discourse of Trauma*, edited by Ana Douglass and Thomas A. Vogler. New York: Routledge, 2003.

Douglass, William A., and Jon Bilbao. *Amerikanuak: Basques in the New World*. Reno: University of Nevada Press, 1975

Eagleton, Terry. *The Ideology of the Aesthetic*. Cambridge: Basil Blackwell, 1990.

Echeverri, Juan Álvaro. "Siete fotografías: una mirada obtusa sobre la Casa Arana." In *Imaginario e imágenes de la época del caucho: Los sucesos del Putumayo*, edited by M. Cornejo Chaparro and A. Chirif. Lima: CAAP, 2009.

———. "The People of the Center of the World: A Study in Culture, History, and Orality in the Colombian Amazon." PhD diss., New School for Social Research, 1997

Enock, Charles Reginald. *The Andes and the Amazon: Life and Travel in Peru*. New York: C. Scribner, 1907.

Enrique Espinar, "Viaje al Igara-Paraná, afluente izquierdo del río Putumayo, por el Capitán de Navío Enrique Espinar." In *Colección de leyes, decretos, resoluciones y otros documentos oficiales referentes al Departamento de Loreto*, edited by Carlos Larrabure y Correa. Vol. 4. Lima: Imprenta de "La Opinión Nacional," 1905.

Eustasio Rivera, José. *La vorágine*. Caracas: Biblioteca Ayacucho, 1993.

Fawcett, P. H., Lt. Col. *Exploration Fawcett: Arranged from his Manuscripts, Letters, Log-books, and Records by Brian Fawcett*. London; Hutchinson & Co., 1953.

Fernanda Sánchez, Luisa. "Paisanos en Bogotá." In *Perspectivas antropológicas sobre la Amazonia contemporánea*, edited by Margarita Chaves and Carlos del Cairo. Bogotá: Instituto Colombiano de Antropología e Historia, 2010.

Fifer, Valerie. "The Empire Builders: A History of the Bolivian Rubber Boom and the Rise of the House of Suarez." *Journal of Latin American Studies* 2, no. 2 (Nov. 1970): 121–125.

Flores Marín, José A. *La explotación del caucho en el Perú.* Lima: Concejo Nacional de Ciencia y Tecnología—CONCYTEC, 1987.

———. *La explotación del caucho en el Perú.* Lima: Concejo Nacional de Ciencia y Tecnología—CONCYTEC, 1987.

Foucault, Michel. *Power.* In *Essential Works of Foucault 1954–1984*, vol. 3, edited by James D. Faubion. New York: New York Press, 1994.

Fuentes, Hildebrando. *Loreto: apuntes geográficos históricos, estadísticos, políticos y sociales. 2 vols.* Lima: Imprenta La Revista, 1908.

Garate, Donald T. *Juan Bautista de Anza: Basque Explorer in the New World, 1693–1740.* Reno: University of Nevada Press, 2003.

García Altamirano, Alfredo. "FENAMAD 20 años después: Apuntes sobre el movimiento indígena amazónica en Madre de Dios." In *Los pueblos indígenas de Madre de Dios: historia, etnografía y conyuntura*, edited by Beatriz Huertas Castillo and Alfredo García Altamirano. Lima: Grupo Internacional de Trabajo sobre Asuntos Indígenas—IWGIA, 2003.

García Jordán, Pilar. "El oriente peruano territorio de confrontación social, económica, ideológica y política, 1821–1930." *In Fronteras, Territorios y Metáforas*, edited by in Clara Inés García. Medellín: Hombre Nuevo Editores, 2003.

———. "La misión del Putumayo (1912–1921): Religión, política, y diplomacia ante la explotación indígena." In *Memoria, creación e historia: Luchar contra el olvido/Memòria, creació I història: Lluitar contra l'oblit*, edited by Pilar García Jordán, Miguel Izard, and Javier Laviña. Barcelona: Universitat de Barcelona, 1994.

———. "Las misiones orientales peruanas: instrumento de pacificación, control y tutela indígena (1840–1915)." *Canadian Journal of Latin American and Caribbbean Studies/Revue canadienne des études latino-americaines et caraïbes* 13, no. 25 (1988): 89–105.

———. *Cruz y arado, fusiles y discursos: La construcción de los Orientes en el Perú y Bolivia, 1820–1940.* Lima: Instituto Francés de Estudios Peruanos, 2001.

García, María Elena, and José Antonio Lucero. "*Un País Sin Indígenas?* Rethinking Indigenous Politics in Peru." In *The Struggle for Indigenous Rights in Latin America*, edited by Nancy Grey Postero and Leon Zamosc. Brighton and Portland: Sussex Academic Press, 2004.

Geertz, Clifford. "Blurred Genres: The Refiguration of Social Thought." *The American Scholar* 29, no. 2 (Spring 1980): 165–179.

———. "Deep Play: Notes on the Balinese Cockfight." *Daedalus* 101, no. 1 (Winter 1972): 1–37.

———. *Available Light: Anthropological Reflections on Philosophical Topics.* Princeton: Princeton University Press, 2000.

Gewald, J. B. "The Herero and Nama Genocides, 1904–1908." In Encyclopedia of Genocide and Crimes against Humanity, edited by Dinah Shelton. New York: Macmillan Reference, 2004.

Gjelten, Tom. *Bacardi and the Long Fight for Cuba.* New York: Viking, 2008.
Goldman, Lawrence R., ed. *The Anthropology of Cannibalism.* Westport, CT: Bergin and Garvey, 1999.
Golob, Ann L. "The Upper Amazon in Historical Perspective." PhD Diss., City University of New York, 1982.
Gómez A., Ricardo. *La guarida de los asesinos: El relato histórico de los crímenes del Putumayo.* Pasto: Imp. "La Cosmopolita," 1933. First published in 1924.
Gómez López, Augusto Javier. *Putumayo: Indios, misiones, colonos, conflictos (1845–1970): Fragmentos para una historia de los procesos de incorporación de la frontera amazónica.* Bogotá: Imprenta Nacional, 1913.
———, ed. *Putumayo: La vorágine de las caucherías. Memoria y testimonio.* Bogotá: Centro Nacional de Memoria Histórica, 2014–2016.
———, ed. *Segunda parte, Documentos relativos a las violaciones del territorio colombiano en el Putumayo (1903–1910).* Bogotá: Centro Nacional de Memoria Histórica, 2014–2016.
Goodman, Jordan. *The Devil and Mr. Casement.* New York: Farrar, Straus and Giroux, 2010.
Gottlieb, Anthony. *The Dream of Reason: A History of Philosophy from the Greeks to the Renaissance.* New York and London: W. W. Norton & Company, 2000.
Goulard, Jean-Pierre. "Cruce de identidades: El Trapecio Amazónico colombiano." In *Fronteras: Territorios y Metáforas* edited by Clara Inés García. Medellín: Hombre Nuevo Editores, 2003.
———. "Un horizonte identitario amazónico." In *Perspectivas antropológicas sobre la Amazonia contemporánea,* edited by Margarita Chaves and Carlos del Cairo. Bogotá: Instituto Colombiano de Antropología e Historia, 2010.
Gould, Stephen Jay. *Rock of Ages: Science and Religion in the Fullness of Life.* New York: Ballantine Books, 2002.
———. *The Hedgehog, the Fox and the Magister's Pox: Mending the Gap between Science and the Humanities.* Cambridge, MA: The Belknap Press of Harvard University Press, 2011.
Grey Postero, Nancy, and Leon Zanozc, eds. *The Struggle for Indigenous Rights in Latin America.* Brighton and Portland: Sussex Academic Press, 2004.
Guillaume, H. *The Amazon Provinces of Peru as a Field for European Emigration.* London: Wyman and Sons, 1888.
Hardenburg, Walter E. *Mosquito Eradication.* New York: McGraw-Hill Book Company, Inc., 1922.
———. "Story of the Putumayo Atrocities 1." *The New Review* 1 (June 1913): 632.
———. "Story of the Putumayo Atrocities 2." *The New Review* 1 (July 1913): 689.
———. "The Devil's Paradise, A Catalogue of Crime." Bodleian Library MSS Brit. Emp. S-22 G335, 164.

———. *The Putumayo, the Devil's Paradise; Travels in the Peruvian Amazon and an Account of the Atrocities Committed upon the Indians Therein*, 2nd edition. Edited and with an Introduction by C. Reginald Enock. London: T. Fisher Unwin, 1912.

———. "The White Man's Burden." *The New Review* 1 (May 1913): 495.

———. *What Is Socialism? A Short Study of Its Aims and Claims*. Vancouver: Dominion Executive Committee [Socialist Party of Canada], circa 1912.

Herzog, Werner. *Fitzcarraldo: The Original Story*. San Francisco: Fjord Press, 1982.

Hildebrand, Martín, von, and Vincent Brackelaire. *Guardianes de la selva: Gobernabilidad y autonomía en la Amazonia colombiana*. Bogotá: Fundación Gaia Amazonas, 2012.

Hispano, Cornelio. *De París al Amazonas: Las fieras del Putumayo*. Paris: Librería Paul Ollendorff, n.d. (1913).

Hochschild, Adam. *King Leopold's Ghost*. Boston and New York: Houghton Mifflin Company, 1999.

Hughes, Robert. *The Fatal Shore: The Epic of Australia's Founding*. New York: Vintage Books, 1986.

Izaguirre, Bernardino. *Historia de las misiones frnaciscanas y narración de los progresos de la geografía en el Oriente del Perú, 1619–1921*. 12 vols. Lima: Taller Tipografía de la Penitenciaría, 1922–1929.

Jackson, Jean. "La política de la práctica etnográfica en el Vaupés colombiano." In *Perspectivas antropológicas sobre la Amazonia contemporánea*, edited by Margarita Chaves and Carlos del Cairo. Bogotá: Instituto Colombiano de Antropología e Historia, 2010.

Jaramillo Jaramillo, Efraín. "Colombia." In *The Indigenous World 2009*, edited by Kathrin Wessendorf. Copenhagen: International Work Group for Indigenous Affairs, 2009.

———. "Peru." In *The Indigenous World 2009*, edited by Kathrin Wessendorf. Copenhagen: International Work Group for Indigenous Affairs, 2009.

Jones, David S. "Death, Uncertainty, and Rhetoric." In *Beyond Germs: Native Depopulation in North America*, edited by Catherine M. Cameron, Paul Kelton, and Alan C. Swedlund.. Tucson: The University of Arizona Press, 2015.

Lacasta Zabalza, J. Ignacio. *La memoria histórica*. Iruña/Pamplona: Pamiela, 2015.

Lagos, Ovidio. *Arana, rey del caucho*. Buenos Aires: Emecé Editores, 2005.

Landaburu Jon, and Roberto Pineda Camacho. *Tradiciones de la gente del hacha: Mitología de los indios andoques del Amazonas*. Bogotá: UNESCO, 1984.

Larrabure y Correa, Carlos. *Colección de leyes, decretos, resoluciones y otros documentos oficiales referentes al Departamento de Loreto. 18 vols.*. Lima: Imprenta de "La Opinión Nacional," 1905.

———. *Peru y Colombia en el Putumayo*. Barcelona: Imprenta Viuda de Luis Tasso, 1913.

Las Corts, Estanislao de, Fray. "En busca de las tribus salvajes." *Revista de las Misiones* 20 (January 1924): 384.

Lausent Herrera, Isabelle. "Los inmigrantes chinos en la Amazonía peruana." *Bull. Inst. Fr. Et. And.* 15, nos. 3–4 (1986): 49–60.

Lissón, Carlos. *Breves apuntes sobre la sociología del Perú en 1886*. Lima: Imp. y Librerías de Benito Gil, 1887.

———. *Breves apuntes sobre la sociología del Perú en 1886*. Lima: Imp. y Librerías de Benito Gil, 1887.

Llanos Vargas, Hector, and Roberto Pineda Camacho. *Etnohistoria del Gran Caquetá*. Bogotá: Fundacion de Investigaciones Arqueológicas Nacionales, Banco de la República, 1982.

López Garcés, Claudia Leonor. "Etnicidad y nacionalidad en la frontera entre Brasil, Colombia y Perú: Los Ticuna frente a los procesos de nacionalidad." In *Fronteras, Territorio y Metáforas*, edited by Clara Inés García. Medellín: Hombre Nuevo Editores, 2003.

Lozano y Lozano, Juan. *Ensayos críticos y mis contemporáneos*. Bogotá: Biblioteca Colombiana de Cultura, 1978.

Madariaga, Salvador de. *Bolívar*. New York: Pellegrini & Cudahy, 1952.

Madley, Benjamin. *An American Genocide: The United States and the California Indian Catastrophe*. New Haven & London: Yale University Press, 2016.

Mallea-Olaetxe, José. "The Private Basque World of Juan Zumárraga, First Bishop of Mexico." *Revista de Historia de América* 114 (1992): 41–60.

Mariátegui, José Carlos. *Los siete ensayos de la interpretación de la realidad peruana*. Lima: Minerva, 1928.

Martínez Cuesta, Ángel O. A. R. "San Ezequiel Moreno, Misionero en Filipinas y Colombia." *Thesaurus* 52, nos. 1,2,3 (1997): 487–88.

Matthiessen, Peter. *At Play in the Fields of the Lord*. New York: Random House, 1965.

Maury, Matthew Fontaine. *The Amazon, and the Atlantic Slopes of South America: A Series of Letters Published in the National Intelligencer and Union Newspapers, under the Signature of "Inca"*. Washington, D.C.: Franck Taylor, 1853.

Mendieta, Eva. *In Search of Catalina de Erauso: The National and Sexual Identity of the Lieutenant Nun*. Reno: Center for Basque Studies, University of Nevada, 2009.

Micarelli, Giovanna. "Fortalecimiento del Consejo de Autoridades Tradicionales Indígenas del Resguardo Tikuna Uitoto Km. 6 y 11 para la construcción participativa de estrategias de gestión territorial y para su difusión. Leticia, Amazonas, Colombia)." Lima: PPD-ICAA, 2009.

———. "Pensar como un enjambre: redes interétnicas en las márgenes de la modernidad." In *Perspectivas antropológicas sobre la Amazonia contemporánea*, edited by Margarita Chaves and Carlos del Cairo. Bogotá: Instituto Colombiano de Antropología e Historia, 2010.

Michell, George B. *Report by His Majesty's Consul at Iquitos on His Tour in the Putumayo District.* London: His Majesty's Stationery Office, 1913.

Misiones católicas del Putumayo: Documentos oficiales relativos a esta Comisionaria. Bogotá: Imprenta Nacional, 1913.

Misrael Kuan Bahamón, S. J. "La misión capuchina en el Caquetá y el Putumayo 1893–1929." Master's thesis, Pontífica Universidad Javeriana (Bogotá, Colombia), 2013.

Mitchell, Angus. *Roger Casement.* Dublin: The O'Brien Press, 2013.

———, ed, *The Amazon Journal of Roger Casement.* London: Anaconda Editions, 1997.

———, introduction, commentary and footnotes. *Sir Roger Casement's Heart of Darkness: The 1911 Documents.* Dublin: Irish Manuscripts Commission, 2003.

Monnier, Marcel. *Des Andes au Para.* Paris: Chez Plon, 1890.

Neale-Silva, Eduardo. *Horizonte humano: Vida de José Eustasio Rivera.* Madison: University of Wisconsin Press, 1960.

Olarte Camacho, Vicente. *Los convenios con el Perú.* Bogotá: Impremta [sic] Eléctrica, 1911.

Páramo Bonilla, Carlos Guillermo. "Un mónstruo absoluto: Armando Normand y la sublimidad, del mal." *Maguaré* 22 (2008): 43–91.

Paternoster, G. Sydney. *The Lords of the Devil's Paradise.* London: S. Paul and Co., 1913.

Pearson, Henry C. *The Rubber Country of the Amazon.* New York: The India Rubber World, 1911.

Pennano, Guido. *La economía del caucho.* Iquitos: Centro de Estudios Teológicos de la Amazonía, 1988.

Pineda Camacho, Roberto. *Holocausto en el Amazonas: Una historia social de la Casa Arana.* Bogotá: Planeta Colombiana Editorial, 2000.

Pinell, Gaspar de, Fray. *Excursión apostólica por los rios Putumayo, San Miguel de Sucumbios, Cuyabeno, Caquetá y Caguán.* Bogotá: Imprenta Nacional, 1928.

———. *Un viaje por el Putumayo y el Amazonas. Ensayo de navegación.* Bogotá: Imprenta Nacional, 1924.

Portillo, Pedro. Las Montañas de Ayacucho y los Ríos Apurímac, Mantaro, Ene, Perené, Tambo y Alto Ucayali. London: Forgotten Books, 2013. Original work published in 1901.

Price, David W. *History Made, History Imagined: Contemporary Literature, Poiesis, and the Past.* Urbana: University of Illinois Press, 1999.

Ramsey, Robert D., III. *Savage Wars of Peace: Case Studies of Pacification in the Philippines 1900–1902.* Fort Leavenworth, KS: CSI Press, 2007.

Rathgerber, Theodore. "Indigenous Struggles in Colombia: Historical Changes and Perspectives." In *The Struggle for Indigenous Rights in Latin America*, edited by Nancy Grey Postero and Leon Zamosc. Brighton and Portland: Sussex Academic Press, 2004.

Restrepo López, P. José. *El Putumayo en el tiempo y el espacio*. Bogotá: Centro Editorial Bochica, 1985.

Restrepo, S. "Colombia and the Putumayo Question," *The Times*, May 27, 1913.

Rey de Castro, Carlos. *Antagonismos económicos, protección y librecomercio: Tratado de comercio entre Perú y el Brasil*. Barcelona: Viuda de Luis Tasso, 1913.

———. *Los escándalos del Putumayo: Carta abierta dirigida a Mr. Geo B. Mitchell, Consúl de S. M. B.* Barcelona: Imprenta Viuda de Luis Tasso, 1913.

———. *Los escándalos en el Putumayo: Carta al director del Daily News & Leader, de Londres.—Nuevos artículos alarmistas.—Plano de la zona sindicada.-Inglaterra en crisis*. Barcelona: Impenta Viuda de Luis Tasso, 1913.

———. *Los pobladores del Putumayo: Origen—nacionalidad*. Barcelona: Imprenta Viuda de Luis Tasso, 1914.

Reyes, Rafael, General. *The Two Americas*. New York: Frederick A. Stokes Company, 1914.

———. *Memorias, 1850–1885*. Bogotá: Fondo Cutural Cafetero, 1986.

Rice, Hamilton. "The River Vaupés. Further Explorations in the North-West Amazon Basin," *The Geographical Journal*, 44, no. 2 (1914): 148.

Robuchon, Eugenio. *En el Putumayo y sus afluentes: Edición oficial*. Lima: Imprenta La Industria, 1907.

Rocha, Joaquín. *Memorándum de viaje: Regiones amazónicas*. Bogota: Casa Editorial de 'El Mercurio', 1905.

Rodríguez Rodríguez, Isacio, OSA, and Álvarez Fernández, Jesús, OSA, eds. *Monumenta historico-augustiniana de Iquitos, Volumen Primero 1894–1902*. Valladolid: Centro de Estudios Teológicos de la Amazonía [CETA], 2001.

Rodríguez Rodríguez, Isacio, OSA, and Jesús Álvarez Fernández, OSA, eds. *Monumenta histórico-augustiniana de Iquitos. Volumen Tercero 1910–1915*. Valladolid: Centro de Estudios Teológicos de la Amazonía (CETA), 2001.

Röhlm, John C. G. *Wilhelm II.: Into the Abyss of War and Exile, 1900–1941*. Cambridge: Cambridge University Press, 2014.

Roldán Ortega, Roque. *Pueblos indígenas y leyes en Colombia: Aproximación crítica al estudio de su pasado y su presente*. Bogotá: Tercer Mundo Editores, 2000.

Roldán, Roque, and Ana María Tamayo. *Legislación y derechos indígenas en el Perú*. Lima: CAAAP, 1999.

Romero, Fernando. *Iquitos y la fuerza naval de la Amazonía (1830–1933)*. Lima: Dirección General de Intereses Marítimos, Ministerio de la Marina, 1983.

Rummenhoeller, Klaus. "Shipibos en Madre de Dios: La historia no escrita." In *Los pueblos indígenas de Madre de Dios: historia, etnografía y conyuntura*, edited by Beatriz Huertas Castillo and Alfredo García Altmirano. Lima: Grupo Internacional de Trabajo sobre Asuntos Indígenas—IWGIA, 2003.

Rumrrill, Roger, and Pierre de Zutter. *Amazonia y capitalismo: Los condenados de la selva*. Lima: Editorial Horizonte, 1976.
Sá, Lucía. "Introduction: Voicing Brazilian Imperialism: Euclides da Cunha and the Amazon." In *The Amazon: Land without History*, by Euclides da Cunha. Oxford: Oxford University Press, 2006.
Sáiz Díez, Fr. Félix, OFM. *Centenario de la Creación de las Prefecturas Apostólicas en el Perú. Después de tres siglos de Acción Misional en el Oriente Peruano*. Lima: Editorial Sin Fronteras, 2000.
Salamanca T., Demetrio. *Exposición sobre Fronteras Amazónicas de Colombia*. Bogotá: G. Forero Franco, 1905.
Salamanca T., Demetrio. *La Amazonia Colombiana: Estudio geográfico, histórico y jurídico del derecho territorial de Colombia*. Bogotá: La Imprenta Nacional, 1916.
San Román, Jesús Victor. *Perfiles históricos de la Amazonía peruana*. N.p.: Centro de Estudios Teológicos de la Amazonía; Centro Amazónico de Antropología Práctica; Instituto de Investigaciones de la Amazonía Peruana, 1994.
Santos Granero, Fernando, and Frederica Barclay. *La frontera domesticada. Historia económica y social de Loreto 1850–2000*. Lima: Pontificia Universidad Católica del Perú, Fondo Editorial, 2002.
———. *Tamed Frontiers: Economy, Society and Civil Rights in Upper Amazonia*. Boulder, CO: Westview Press, 2000.
Santos, Roberto. *História economica da Amazonia (1800–1920)*. São Paulo: T. A. Queiroz, 1980.
Serge, Margarita. "Fronteras carcelarias: Violencia y civilización en los territories salvajes y tierras de nadie en Colombia." In *Fronteras, Territorio y Metáforas*, edited by Clara Inés García. Medellín: Hombre Nuevo Editores, 2003.
Silva Bruno, Ernani. *História do Brasil, Geral e Regional*. Vol. 1, *Amazonia*. São Paulo: Ed. Cultrix, 1966.
Simon, W. O. "Frontier Life in South America." *The Wide World Magazine* (April 1913): 92.
Solzhenitsyn, Aleksandr. *One Day in the Life of Ivan Denisovich*. New York: Farrar, Straus and Giroux, 1991.
———. *The Gulag Archipelago, 1918–1956*. 3 vols. New York, Evanston, San Francisco, London: Harper & Row, 1974.
Souto Loureiro, António José. *A Grande Crise (1908–1916)*. Manaos: T. Loureiro & CIA., 1986.
Stanfield, Michael Edward. *Red Rubber, Bleeding Trees: Violence, Slavery, and Empire in Northwest Amazonia, 1850–1933*. Albuquerque: University of New Mexico Press, 1998.
Stevenson, Bryan. "A Presumption of Guilt." *The New York Review of Books* 64, no. 12: 8–10.
Stidsen, Stille, ed. *The Indigenous World 2006*. Copenhagen: International Work Group for Indigenous Affairs, 2006.

Taussig, Michael. *Shamanism, Colonialism, and the Wild Man.* Chicago: University of Chicago Press, 1987.
Tejedor, Senén, F. *Breve reseña histórica de la mission agustiniana de San León del Amazonas, Perú.* El Escorial: Imprenta del Real Monasterio de el Escorial, 1927.
Thomson, Norman. *Colombia and Peru in the Putumayo Territory.* London: N. Thomson & Co., 1913.
———. *The Putumayo Red Book.* London: N. Thomson & Co., 1913.
Tobón, Marco Alejandro. "Asentamientos multiétnicos: conflictos y distinciones en Leticia." In *Perspectivas antropológicas sobre la Amazonia contemporánea,* edited by Margarita Chaves and Carlos del Cairo. Bogotá: Instituto Colombiano de Antropología e Historia, 2010.
Tully, John. *The Devil's Milk: A Social History of Rubber.* New York: Monthly Review Press, 2011.
Turvasi, Francesco. *Giovanni Genocchi and the Indians of South America (1911–1913).* Roma: Editrice Pontifica Università Gregoriana, 1988.
United States Department of State. *Slavery in Peru: Message from the President of the United States, Transmitting Reports of the Secretary of State, with Accompanying Papers Concerning the Alleged Existence of Slavery in Peru.* Washington, DC: Government Printing Office, 1913.
Uribe Gaviria, Carlos. *La verdad sobre la Guerra.* 2 vols. Bogotá: Editorial Cromos, 1936.
Uribe Piedrahita, César. *Toá: Narraciones de caucherías.* Manizales: Arturo Zapata, 1933.
Valcárcel, Carlos A. *El proceso del Putumayo y sus secretos inauditos.* Lima: Imprenta "Comercial" de Horacio La Rosa & Co., 1915.
Valdez Lozano, Zacarías. *El verdadero Fitzcarrald ante la historia.* Iquitos: Imprenta El Oriente, 1944.
Vallvé, Frederic. "The Impact of the Rubber Boom on the Indigenous Peoples of the Bolivian Lowlands (1850–1920)." PhD diss., Georgetown University, 2010.
Vargas Llosa, Mario. *El hablador.* Barcelona: Seix Barral, 1987.
———. *The Storyteller.* New York: Farrar, Straus, Giroux, 1989.
Villegas, Jorge, and José Yunis. *Sucesos colombianos, 1900–1924.* Medellín: Universidad de Antioquia, 1976.
Vinding, Diana, ed. *The Indigenous World 2002–2003.* Copenhagen: International Work Group for Indigenous Affairs, 2003.
Vinding, Diana, and Sille Stidsen, eds. *The Indigenous World 2005.* Copenhagen: International Work Group for Indigenous Affairs, 2005.
Vonnegut, Kurt, Jr. *Slaughterhouse Five or the Children's Crusade.* New York: Delacorte Press, 1969.
Vuconic Wilkinson, Xenia. "Tapping the Amazon for Victory: Brazil's 'Battle for Rubber' of World War II." PhD Diss., Georgetown University, Washington, D.C, 2009.

Weinstein, Barbara. *The Amazon Rubber Boom 1850–1920*. Stanford, CA: Stanford University Press, 1983.
White, Hayden. *Metahistory: The Historical Imagination in Nineteenth Century Europe*. Baltimore: The Johns Hopkins University Press, 1973.
White, Hayden. *Tropics of Discourse: Essays in Cultural Criticism*. Baltimore: The Johns Hopkins University Press, 1978.
Wilson Odorio, John. "Prólogo." *Toá: Narraciones de caucherías*. Medellín: Colección Bicentenario de Antioquia, Universidad CES, 2013.
Woodroffe, Joseph F. *The Upper Reaches of the Amazon: Types of Brazilian Rubber Gatherers*. London: Methuen & Co., 1914.
Zulaika, Joseba, and William A. Douglass. *Terror and Taboo: The Follies, Fables and Faces of Terrorism*. New York: Routledge, 1998.
Zuluaga Gil, Ricardo. "Régimen jurídico de las entidades territoriales de frontera." In *Fronteras, Territorio y Metáforas*, edited by Clara Inés García. Medellín: Hombre Nuevo Editores, 2003.
Zumaeta, Pablo. *Las cuestiones del Putumayo: Segundo memorial*. Barcelona: Imprenta Viuda de Luis Tasso, 1913.

Index

La Acción Popular newspaper, 282
Acosta, Miguel Antonio, 120n358, 141, 157
Acre River, 32n64, 132, 318–19, 324
Acre War, 32n64, 51
African-American slavery, 522–23
African International Association, 170
Agrarian, Industrial and Mining Credit Bank (Caja), 496, 504
Agricultural Mortgage Bank (Colombia), 496
Aguero, Abelardo, 216, 217, 222, 244, 275, 363n72
Águila Hidalgo, Pedro de, 440
Aguirre, Julio (Basque-surnamed), 362
Alarco, Abel: and Arana, Alarco and Cia, 332–33; arrest of, 314n109; attacks on Hardenburg, 262; on the commission system, 322n144; at J. C. Arana y Hermanos, 70; letter on Acre River operations, 319; at PAC, 129, 190, 194, 235, 319, 332; responses to Casement Report, 262–63
Alarco, Germán: and Arana, Alarco and Cia, 332–33; and Arana y Alarco Company, 296; and David Cazes, 296; as Iquitos mayor, 160, 327; letter in *La Prensa*, 314n109; on public knowledge of atrocities, 327, 333
Alayza y Paz Soldán, Francisco: and Carlos Valcárcel's arrest warrants, 256–57; departure from Iquitos, 257; on foreign power in Iquitos, 535; Giovanni Genocchi on, 406; meeting with Roger Casement, 202; on the Putumayo, 257; on recruiting Spanish agriculturalists, 380n127; on reforms and rubber production, 346; reports on population drain, 213; security concerns, 214; and Stuart Fuller's reports, 265
Alcorta, Luis (Basque-surnamed), 234, 235n203, 361, 546
Álvaro Echeverri, Juan, 519n17
The Amazon, and the Atlantic Slopes of South America (Maury), 28–29
Amazon Colombian Rubber and Trading Company, 113
Amazonia, overview of: Borja, 20; Diego Vaca de Vega's incursion into, 20; European immigrants to, 381n129; European migration to, 16; first rail links, 31n62; first steamship expedition, 61; lack of infrastructure, 31; missionaries in, 43, 47–49; missionaries *versus caucheros*, 47–49; missionary settlements, 47–48; political disputes over, 29; remoteness of, 31; rubber industry in, 37; white-Indian dynamics in, 42–43
The Amazon Pact, 473
Anchieta, José de (Basque-surnamed), 18
Andrade, David Elías, 81
Andrés A Aramburú, 436n329
anthropologists *versus* missionaries, 527
anthropology: activist approaches, 527, 529–30; Alicia Dussan's, 528; Anglo-American, 527; author's, 527, 550–51; Clifford Geertz's, 551; Colombian,

527–29; Gerardo Reichel-Dolmatoff's, 528–29; Latin American, 527; Marcelino de Castellví's, 527–28; Michael Taussig's, 527; symbolic, 551; thick description, 551

Anti-Slavery and Aborigines Protection Society: booklet denouncing PAC, 314–15; calls for PAC investigation, 233; and Charles Roberts, 347; John Harris at, 181; letter from Edward Grey, 199; letter to Grey on PAC atrocities, 199; and PAC, 185; watchdog subcommittee, 347

Anti-Slavery and Aborigines Protection Society and Walter Hardenburg: advice on atrocity investigation and reporting, 200; lecture at, 343–44; letter on Putumayo missionary potential, 369–70; report summary given, 181; Select Committee testimony request, 325–26

Aquileo, Torres, 77, 89

Aragón, Arcesio, 427

Aramburú, Andrés A., 155–56

Arana, Alarco and Cia, 332–33

Arana, Benito, 66, 67, 71, 531

Arana, Eleonora (née Zumaeta, Basque surname): and Arana, 67, 84, 315; Casa Arana ownership, 226; children of, 440; illness of, 315, 320; impoverishment of, 441; PAC mortgage, 345; photographs of, 532; in *The River That God Forgot*, 533

Arana, Gregorio, 67

Arana, Julio César: absences from family, 68; acquisitions of Colombian properties, 433; André-Marcel d'Ans's assessment of, 538; anti-critic strategies of, 408n225; Arana, Vega y Cia, 85, 95; Arana, Vega y Larrañaga, 85; Arana y Vega, 70, 72n228; arrest warrant for, 280, 284; aspirations of, 53; as avid reader, 68; Basqueness of, 546–47; belief in own innocence, 541–42; and Benjamín Larrañaga, 79, 83, 87; in Biarritz, 546–47; biographies of, 423–24, 532; birth of, 66; Brazilian connections, 128, 534; on British/American Putumayo fact-finding mission, 266–68, 275, 298, 517; British directors' opinions of, 301, 305, 308–9; businesses of, non-PAC, 319; business strategies, 537; Calderón, Arana, and Company, 82, 87–88; Carlos Rey de Castro's defenses of, 156–58, 353, 354, 356–57; and Cecilio Hernández, 346–47; children of, 440; civilizing Indians discourses, 155, 277–78, 323; and David Cazes, 189–90, 197, 199, 200, 294–95; death of, 441; denial of selling Indians, 276; denunciation by Colombian government, 351; duplicity of, 536–37; education of, 67; and Eleonora Zumaeta (wife), 67, 84, 315; evilness question, 538–41; family of, 66–67, 531, 542–43; films commissioned by, 421; finances of, 535; guilt of, 288, 289, 301, 541–42; and Henry Read, 128–29; High Court of Justice appearance, 314–15; hiring of relatives, 86; historical vilification of, 525–26; Igaraparaná River control, 105; impoverishment of, 441; information on, lack of, 531; and Iquitos, 70, 71, 148–49, 160,

ARANA | 619

423, 535; Iquitos assets sale, 440; and Joaquín de Barros, murder of, 87; and John Russell Gubbins, 186; and John Whiffen, 187–89; and José Francisco Gómez, murder of, 87; and José Francisco Medina, 123; and José Gregorio Calderón, 82, 86, 87–88, 95; and Juan B. Vega, 70, 72n228, 85, 110; and Julio Egoaguirre, 362n71, 423; Larrañaga, Arana and Company, 82, 94; legacy of, 441–42, 542–43; Leguía government land grant, 425; letter to supporters, 312–13; linguistic abilities, 134n418, 188n25, 536; Lisbon interviews, 313; literacy of, 536; Manaos, importance in, 534; in Manaos, 124; meetings with Walter Hardenburg, 154; merchant banquet celebration, 278–79; and the military, 264n297; *The Observer*'s description of, 313; *La Opinión Nacional* on, 530; *versus* other rubber barons, 600; Ovidio Lagos on, 81, 105, 346, 441–42, 532; and Pablo Zumaeta, 68, 70, 84, 226, 347, 531–32; PAC atrocity scandal, as surprised by, 541; PAC atrocity scandal, business and life after, 345–47; patriotism of, 541–42; payments to Peruvian consul general, 124, 127; and Peru/Colombia border, 431; Peruvian recognition of, implicit, 84; photographs of, 532; political connections and influence, 128, 277, 535; positive characteristics of, 536; press defenses of, 283, 358–59; as private, 531–32, 535–36; as progressive legislator, 431; purchase of La Chorrera, 83, 87; purchase of Providencia, 79; references to PAC raids, 292n16; in *The River That God Forgot*, 532–33; as river trader, 67–68; rubber business in Yurimaguas, 68; and rubber export taxes, 423, 430; *versus* Sabino Arana Goiri, 548; Saldaña Rocca articles sent to London, 305; and Saldaña Rocca's murder, 333–34; and Salomón-Lozano Treaty, 432–433; and seizure of Leticia, 438; self-discipline of, 68; as Senator, 423–25, 430–31, 542; shipping business expansion, 84; station chief protection, 277; trading business in Tarapoto, 68; and *Truth* articles, 187–88; and Walter B. Perkins, 158; and Walter Hardenburg: blackmail accusations, 262, 309, 322, 330, 344; employment offer, 158; forgery accusations, 322, 323, 330, 344; meeting with, 154; writings about himself, 531; on Zapata report, 155–56. *See also* British Select Committee on PAC atrocities, Arana's appearance and testimony; Casa Arana; Casement, Roger and Julio César Arana; PAC

Arana, Julio César, attacks on: Benjamín Saldaña Rocca's, 126–27, 134, 318; in *La Felpa* newspaper, 127, 150–51, 164; Walter Hardenburg's, 345, 530–31

Arana, Julio César, PAC atrocity defense tactics: *ad hominem* attacks, 537; blackmail claims against Hardenburg, 262, 309, 322, 330, 344; blackmail claims against Saldaña Rocca, 330; blackmail claims against Valcárcel, 281; blackmail claims against

Whiffen, 188–89, 192, 309, 322, 328, 330; Carlos Rey de Castro's writings, 353, 354, 356–57; claims of defamation campaign, 356; denial, 312–13; forgery accusations against Hardenburg, 322, 323, 330, 344; formal statement, 329–30; ignorance claims, 316, 318, 332–33, 537–58; legislative, 272; open letter, 356, 573–77. *See also* British Select Committee on PAC atrocities

Arana, Julio César and PAC: Board meeting absences due to Casement, 229, 230; Board of Directors, 195, 196, 305n73, 537; debt collecting, 235; as liquidator, 232, 301, 303, 308–9, 314, 336, 345; stock holdings in, 535; valuation of, 302

Arana, Julio César and Walter Hardenburg: blackmail accusations, 262, 309, 322, 330, 344; employment offer, 158; forgery accusations, 322, 323, 330, 344; meeting with, 154

Arana, Julio César and the Putumayo: Carlos Zapata trip, 154–55; consolidation of holdings in, 85; defense for Peru, 115–16; first trip, 81; land ownership, 105; Pedro Portillo's support of, 82; rubber concession acquisitions, 86–88, 95, 112–13; second trip, 81; territory holdings, 132; trade with Colombians, 73. *See also* Aranalandia

Arana, Lizardo: Acre River operations, 319; and Casa Arana, 70; and PAC, 129, 332; and PAC Commission, 201–2; in the Putumayo, 93–94

Arana, Marie, 542–45
Arana, Martín, 67, 234, 234n202, 235n203, 332
Arana, Pedro Pablo, 542–43
Arana, Remigio, 292n16
Arana, Vega y Cia, 85, 95
Arana, Vega y Larraniaga, 85
Arana, Victor Manuel, 543–45
Arana Goiri, Sabino, 547–48
Aranalandia: Arana's control of, 84; Arana's desire to sell, 422; attacks on Colombians, 89, 91, 162–63; Banco Agrícola Hipotecario acquisitions, 496; Benjamin Larrañaga's exploration of, 102; Casa Arana dominance of, 104–5; Casa Arana rubber concession acquisitions, 86–89; Colombian control of, 439–40; Colombian expropriation of, 436; and the Colombian government, 89; Colombian land concessions, 110; Colombian properties in, 433; contemporary ownership of, 438; Crisóstomo Hernández in, 73; Indian attacks, 47; Indian-Colombian uprisings, 85–86; Indian population, 72; Indian population in, 72, 434; Indian resettlement, 434, 435; infrastructure in, 433; Leticia Trapezoid, 432, 436, 438–40, 493–95; mapping of, 516; maps of, 6–7; missionaries, lack of, 420; reservations created in, 496–97; rubber trees in, 81; rumors of foreign interest in, 110; sale to Colombian government, 440; sale to Victor Israel, 440; and Salomón-Lozano Treaty, 432–34; Sr. Mejía, 76; white settlement of, Colombians, 73–76. *See also* La Chorrera; Pinell, Gaspar de

ARANA | 621

Arana Ramírez, Luis, 441n343
Arana y Alarco Company, 296
Arana y Vega, 70, 72n228
Arédomi, 208, 236, 237, 517
Argentina, 12, 18, 111, 198, 233, 388, 447n2
Arriaga, José María, 377
artifactual conflicts, 517
Asociación Interétnica de Desarrollo de la Selva Peruana (AIDESEP), 451–452, 454, 458–60
Asociación Pro-Indígena, 210–11
La Unión assault: Ovidio Lagos on, 123
Astigarraga, José Luís (Basque), 459
Atenas station, 414
Augustinian order, 17, 383–85, 391, 410–11
Australia, 354
AZICATCH (La Asociación Zonal Indígena de Cabildos y Autoridades Tradicionales de La Chorrera), 497–98

Balcázar, Torres, 273–74
Bannister, Edward, 169, 170
Barbadians: Armando King, 244; atrocity testimonials, 202–3, 207, 209, 357; British Select Committee on, 357n55; Carlton Morris, 202; committing atrocities, 136, 186, 202–3, 218, 272, 282, 289, 524; as Commonwealth citizens, 148, 201; and David Cazes, 294, 295, 296n31; Frederick Bishop, 197n71, 202, 398; hiring of, 105; on Indian hunts, 295; John Brown, 187, 204; Joshua Dyall, 203; leaving the Putumayo, 207; and PAC Commission, 203–4; PAC employment of, 41, 136, 201; recruited by PAC, 105, 218, 357n55; and Roger Casement, 201–3, 205, 207, 289, 290
Barbados, 35, 105, 237, 244
Barchillón, F., 79
Barclay, Frederica, 520
Barnes, Louis Harding, 203, 299–301
Basque individuals: Arana's connections to, 546; in historical research, 545–46; in this history, 545–47; stereotypes of, 547
Basque surnames, use in research, 545n80
Belaunde, Luisa Elvira, 542
Belgium: Congo acquisition and reform, 176–77; diplomatic meetings regarding the Congo, 176; King Leopold, 40n94, 170–72, 175–76, 512; King Leopold II, 40–41; relationship with Britain, 173. See also the Congo and Belgium
Bell, Seymour, 203, 230, 254
Bellamy and Isaac, 129, 133
Benavides, Oscar R., 169
Betancourt, Alejandro, 58
Bidwell, Manuel, 387–88, 389, 397, 401
Billinghurst, Guillermo: attempts at Putumayo justice, 363; and Carlos Valcárcel, 283, 361n64; letters to *The Times* of London, 351; Putumayo investigative commission, 284; swearing-in as Peruvian president, 274
Bishop, Frederick, 197n71, 202, 398
Blanco, Aurelio, 158–59
Blue Book. *See* Casement Report
Bolívar, Simón de (Basque-surnamed), 14
Bolivia: Acre region, 51; Atlantic orientation post-Pacific War, 34;

Beni River, 597; border confusion, 14–15; Cachuela Esperanza, 597, 598, 599; Madeira-Mamoré railway, 51; quinine production, 29; rubber boom, 597; rubber collection in, 30; rubber production, 599; War of the Pacific, 50n125
Bonilla, Victor Daniel: on Capuchins, 394, 397, 466; career and education of, 470; on Catholic control of Colombia, 374; on Colombian DIA, 468; on Colombian *resguardos*, 467; *Servants of God or Masters of Men?*, 469–70
Brackelaire, Vincent, 473
Brazil: Acre region, 51; Acre War, 51; border agreement with Peru, 34; border confusion, 15; Empire of Brazil, 13–14; and the Granadine Federation, 372; independence movement, 13–14; Indian Protection Services, 211n126; Jesuits in, 18; Kingdom of Portugal, Brazil, and the Algarves, 13; leaf blight research, 55; Lower Putumayo control, 32; Madeira-Mamoré railway, 51; and MERCOSUR, 447n2; Peruvian extradition requests, 221–22; Peruvian river navigation agreements, 32, 34; Portuguese colonization, 12; and Putumayo Peru-Colombia conflicts, 166–67; refusal to extradite PAC employees, 255; rubber boom in, 37n88; rubber industry, 55–56; and Salomón-Lozano Treaty, 432. *See also* Manaos, Brazil
British/American Putumayo fact-finding mission: access problems, 265–66; Arana and supporters on, 266–68, 275, 298, 517; Carlos Rey de Castro's report, 275; George B. Michell on, 265–66, 268, 275; Germán Leguía y Martínez's article on, 274–75; John Brown as translator for, 265; Manaos Colombian consul's letter to, 266–67; photographs staged, 517; Stuart Fuller on, 265–66, 267–71
British Foreign Office: Colombian praise for, 250; and Eduardo Lembecke, 260; and Giovanni Genocchi's investigation, 391, 397–98; Loreto consulate, 343; Roger Casement's retirement from, 366; Thomas Whiffen's report to, 188; trust of *Truth* newspaper, 191–92. *See also* Grey, Edward; PAC atrocities and British Foreign Office
British Peruvian Corporation, 381
British Select Committee on PAC atrocities: on Barbadians, 357n55; and Casement Report, 512n1; Charles Roberts on, 287–88; and C. H. Rance, 334; David Cazes's testimony, 294–96; Douglas Hall on, 324–25; Douglas Hogg's defense of Arana, 330–32; Edward Morel's testimony, 334; findings of, 335–39; George Babington Michell's testimony, 298–99; G. S. Paternoster's testimony, 293–94; Harry Johnston's testimony, 334; Henry Lex Gielgud's testimony, 301–304, 409n227; Henry Read's testimony, 302n61, 309; Henry Samuel Parr's testimony, 310–11; John Lister-Kaye's testimony, 309, 326; John Russell Gubbins's testimony, 304–8, 310; José Medina's written brief, 308–9; laborer valuation debate, 302n61;

London PAC offices search, 287; Louis Harding Barnes's testimony, 299–301; Norman Thomson's testimony, 311–12; overview of, 287; Pablo Zumaeta's surprise letter, 305–6; problems and concerns, 334–35; proceedings of, 348; Swift MacNeill on, 320, 323; Thomas Whiffen's testimony, 322, 328–29, 357n53; Walter Hardenburg's testimony, 326–28

British Select Committee on PAC atrocities, and Roger Casement: collaboration with Charles Roberts, 287–88, 296, 314; concerns regarding, 296–97; Robuchon text, 288; testimony to, 288–93, 294n22; warning about Arana, 314

British Select Committee on PAC atrocities, Arana's appearance and testimony: on Acre River operations, 318–19, 324; on atrocity report verification, 324–25; attacks on Hardenburg, 322, 323; belief in atrocities, 324; on Board of Directors, 323–24; on Carlos Rey de Castro, 320; on civilized *versus* uncivilized regions, 323; claims of ignorance, 316, 318; on the commission system, 322; *conquistar versus correría* debate, 321; on desire to retire, 324; and Douglas Hogg, 315, 320, 325, 331–32; English use, 322; evidence and statement left behind, 329–30; evidence brought, 313, 315; as evidence of self-confidence, 537; on Indians as property, 325; learning of atrocities, 325; legal council, 315; on newspaper attacks, 333; as non-legally binding, 315, 343;

on PAC founding, 318, 319, 323; press reports of, 320–21, 536; reasons for, 313–14, 325, 331, 542; on La Reserva assault, 324; rhetorical tactics of, 320; on Robuchon's notes, 316–17; on Saldaña Rocca, 318; on slavery allegations, 317; use of translation, 536; on Zapata's investigation, 317–18

Brown, John, 187, 199, 204, 265
Browne, Percy, 401, 403, 407
Bryce, James, 250–51, 254
Buinaje, Marcelo, 506n167
Burga, Amadeo, 252
Buxton, Travers, 199
Buxton, Travis, 413

cabildos, 479n100, 490, 497, 498, 500–2
Caipe, Benjamín, 465
Caja de Crédito Agrario, Industrial y Minero, 496, 504
Cajiao, Leopoldo, 108–9
Calderón, Abel: on Benjamin Larrañaga, 77; and Colombians in Aranalandia, 89; Huitocia travel account, 75–76; Muñoz and Silva anecdote, 78–79; on Peruvian slavery practices, 77–78; political imprisonment of, 75n240; rubber enterprises stolen from, 88–89; on white settlement of Aranalandia, 76
Calderón, Arana, and Company, 82, 87–88, 95
Calderón, Gregorio, 292n16
Calderón, José Gregorio: in Aranalandia, 74; on Aranalandia Indian population, 72; Caraparaná settlement, 75; deceiving of Indians, 74–75; and Huitocia rebellion, 74; and the Larrañagas, 79;

partnership with Arana, 82, 86, 87–88, 95; Providencia sale to Arana, 79; rubber selling expedition, 75; on treatment of Indians, 89–90. *See also* Calderón, Arana, and Company
Calderón, Rafael, 159–60
Calderón, Teófilo, 74, 75, 79. *See also* Calderón, Arana, and Company
Calle, Bernardo, 384, 408–9
Camacho, Pineda, 44–46, 72, 414–15
Campuya River, 78–79
Canada, 454, 455
cannibalism: in Alayza Paz y Soldán's letter, 257; anthropological perspectives on, 240n220; Cashibo tribe, 67; cessation of, 417; designation of, 523–24; Huitotos thought to be, 64, 213; as justification for atrocities, 16–17, 336, 359, 523; Muinanes tribe, 103; in PAC Commission report, 223; in Paredes's reports, 240–42; rumors of, 102, 138; as threatening, 515, 522, 524; white belief in, 45, 47, 60, 64, 102, 515; white prohibition and punishment, 43, 193
Canseco, Ingoyen, 250
Capelo, Joaquín, 210–11
Capuchin order: abuses of Indians, 463–64, 467, 469; Apostolic Prefecture of, 463, 467; Bartolomé de Igualada, 363, 434, 463; in Colombia, 393, 472; Corts, Estanislao de Las, 422, 435; Güepí mission, 420; history of, 393n169; and INCORA, 469; Indian lawsuit against, 465; Marceliano de Vilafranca, 465, 467; Marcelino de Castellví, 466, 527–28; in Middle Putumayo, 25; mission at El Encanto, 439; and Pedro Juajibioy, 466; Putumayo expeditions, 394; reports on Indian treatment, 428n295; and *Servants of God or Masters of Men?*, 470; and Sibundoy Indians, 463; Sibundoy Valley mission, 137, 393–97, 463, 465. *See also* Montclar, Fidel de; Pinell, Gaspar de
Caquetá, Lower, Brazilian control of, 78
Caquetá Apostolic Prefecture, 394
Caraparaná River: Arana's interest and control, 81, 86, 105; Colombian administrative division, 107; Colombian administrator of, 118–19; Colombian claims to, 86; Colombian settlement of, 140; conflicts on, 78; Crisóstomo Hernández and, 73, 101–2; José Gregorio Calderón's settlement, 75; rumors of foreign interest in, 110. *See also* Aranalandia
Cárdenas, Abel, 93
Casa Arana (J. C. Arana y Hermanos): Abel Alarco at, 70, 84; acquiring land from Colombians, 106; arming of Teffé men, 117; audit by English accountant, 124; Barbadian employees, 41, 105, 136, 201; business strategies, 79; Calderón, Arana, and Company's debt to, 87–88; Caquetá expansion, 106; Eleonora Zumaeta's ownership, 226; encouragement of agriculture, 96; and Eugène Robuchon, 84–85, 124–25; government official interest in, 364; Henry Lex Gielgud's valuation of, 301–2; Indian ownership, 1920s, 435; international influences on, 95–96; international ties, 71;

Iquitos headquarters, 84; Iquitos prominence, 534–35; Lizardo Arana and, 70; Manaos headquarters, 128; and Middle Putumayo/Caquetá, 111; and missionaries, 420–21; newspaper articles on, 126–27; partners in, 70; Peruvian defense of, 91; as Peruvian naval force, 96; Peruvian political strategies, 128; and Peruvian Putumayo presence, 94; power of, 96; pre-PAC creation audit, 124; public offering reparations, 123–24; and the Putumayo, 94, 105, 423; the *Rápida* incident, 117; road construction, 95–96; rubber concession acquisitions, 86–89, 95, 112–13; sale to Peruvian Amazon Company, 128–29. *See also* El Encanto; PAC

Casa Arana attacks on Colombians: abductions, 77–78, 106, 118–19; and the Calderóns, 77–79; Caraparaná River, 95, 97, 105–6, 117–18; David Serrano, 118, 119; Gabriel Martínez, 118, 119; Helidoro Moreno, 118; Justino Hernández, 116; N. Hernández, 118; Peruvian support for, 119; reports of, 77–79, 116; at La Reserva, 97, 117–18, 144, 152, 154, 163, 288. *See also* La Unión assault

Casas, Bartolomé de las, 16

Casement, Roger: in Africa, 169–70, 172–73, 512; and Alice Strepford Green, 177; alleged blackmail attempts, 299; anti-British writings, 365, 366; anti-semitism of, 206; and Armando Normand, 206, 516; and Arthur Conan Doyle, 175n552, 367, 368n89, 401; attempts to change control of PAC, 230; awards given, 178, 229–30; Barbadian guilt, belief in, 289; Black Diaries controversy, 368; and British colonialism, 174, 177–78; British public opinion of post-execution, 369; Catholicism of, 401n199; and C. H. Rance, 334; on civilizing Indians discourses, 253; concerns regarding British/American Putumayo mission, 266, 267; in the Congo, 170, 172–73, 512; and Congo Reform Association, 175, 177; Congo *versus* Putumayo reports, 512; consulships held, 172, 178; David Cazes, opinion of, 294n22, 296n31; disappointment with the US, 226–27; and duchess of Hamilton, 389; eccentricities of, 237; and Edmund Dene Morel, 174–75; and Edward Bannister, 169, 170; and Edward Grey, 174, 244, 247, 248–49, 252–54; at Elder Dempster Shipping Company, 169; as English spy, 238; execution for treason, 367–68; exotica, use of, 517; family of, 169; Gerald Sydney Spicer, letters to, 228, 229–30, 250, 388–89, 403; and Germany, 364–66; and Giovanni Genocchi, 401; and Henry Spencer Dickey, 237, 368nn89–90; Henry Spencer Dickey on, 299n45; and Henry Stanley, 170; humanitarian reputation of, 177; invocations of Latin authoritarianism, 354; in Iquitos, 236–38, 242–43, 250; Irish identity and nationalism, 174, 177, 178, 229, 365, 366; and the Irish Volunteers, 366–67; on Isaac Escurra, 244; and James Bryce, 250–51; and John Harris, 405; and

John Russell Gubbins, 228–29; and Joseph Conrad, 170, 367, 518; in Manaos, 250; meeting with King Leopold, 172; on the Monroe Doctrine, 227; and Paredes Report, 245–46, 255; perspectives on, 513; on Peruvian government ineptness, 244–45, 252; photography use, 517; Putumayo mission advocacy, 387–89, 400–1, 403–4; and Putumayo Mission Fund, 400–1, 407; and Putumayo reform, 250–51; and Reginald Bertie, 218; religious tolerance of, 368; report on Belgian atrocities, 173–74; reputation of, 512; reservations toward government intervention, 228; retirement from Foreign Office, 366; return to Peru from England, 237; and Rómulo Paredes, 221, 245–46, 249; *versus* Sabino Arana Goiri, 547–48; as socialist crusader, 525; *El sueño del celta*, 518; and W. B. DuBois, 367; and William Cadbury, 226–28, 230, 389. *See also* British Select Committee on PAC atrocities, and Roger Casement; PAC Commission and Roger Casement

Casement, Roger and Julio César Arana: belief in guilt of, 288, 289; first meeting, 178; involvement in PAC financial difficulties, 232–33; letters from, 217–18; meeting requests from, 217–18, 250; telegram from, before execution, 367, 369, 542

Casement, Roger and PAC: anti-Arana strategies, 232–33; at Board meetings, 228–29, 230, 232; control transfer attempts, 230–31, 232, 235

Casement Report (Blue Book): Arana's awareness of, 220–21; Aurelio Rodríguez in, 219; Barbadians, 218, 272; British responsibility in, 272; control of publication, 232; control techniques, 220; in *The Devil's Paradise*, 297n34; and Edward Grey, 174, 252–54; Elías Martinengui in, 219; findings of, 223; and Genocchi's investigation, 398; in G. S. Paternoster's book, 347; Indian depopulation, 218–19; Indian enforcers, 219–20; missionary recommendations, 370, 387; and PAC directors, 222–23; and Peruvian government, 221; publication of, 254, 255, 260; responses to, American, 261, 264–65, 271; responses to, British, 260–61, 271; in *Slavery in Peru*, 512n1; use of Water Hardenburg's material, 511; and US government, 221; writing of, 215

Casement Report (Blue Book), Peruvian responses to: Abel Alarco's, 261–62; investigatory commissions, 274; newspaper articles, 261–62, 271–72, 274–75; parliamentary debates, 272–74

Castaneda, Carlos, 528n37

Castellví, Marcelino de, 466, 527–28

the Catholic Church: Augustinian order, 17, 383–85, 391, 410–11; and Casement Report, 387–88; Chapapoyas bishop, 377; Dominican order, 17, 383, 461; and European empire, 17; European power decline, 370; in Iquitos, 390; and Latin American generally, 370, 375, 377; liberation theology, 459–60, 469, 526; Maynas Region diocese, 376, 377; monitoring of treatment of indigenes, 446; Monsignor Quattrocchi, 389–90; and New World governments, 526; PAC atrocity investigations, 386–88, 391; and Paredes report, 390; and Peru, 389–90. *See also* Capuchin order; Genocchi, Giovanni; Jesuit order; missionaries; Pius X, Pope; Val, Merry del

the Catholic Church and Colombia: Caquetá Apostolic Prefecture, 394; Church control of, 374–75, 472; concordat of 1887, 393n169, 472; diplomatic relations, 25, 370–71; missionary authority, 374; power over, 25, 447, 472; religious wars, 371, 373–74

Cavalli, Aldo, 505

Cazes, David: Arana's letter to, 197; Arana's meeting request, 189–90; association with Arana, 199, 200, 294–95; and Barbadians, 294, 295, 296n31; as British consul in Iquitos, 148, 202, 294; and Carlos Zubiaur, 209; and G. S. Paternoster, 293; and John Russell Gubbins, 305; PAC Board communications, 186; and PAC Commission, 202, 208–9; on PAC Indian abductions, 244; Roger Casement's opinion of, 294n22, 296n31; Roger Casement's testimony on, 294n22; testimony on Saldaña Rocca's broadsheets, 154n484; testimony to Select Committee, 294–96; Whiffen's report to, 188

Centro Amazónico de Antropología y Aplicación Práctica (CAAAP), 450

Centro de Estudios Teológicos de la Amazonía (CETA), 450

Chile, 34, 50n125, 379, 381, 447n2, 481, 587n1

Chincha Islands War, 381

Chirif, Alberto: on La Cueva, 276n341; at House of Knowledge conference, 504, 506; on modernization *versus* civilization terms, 461–62; on Peruvian Indian treatment, 449; on Peruvian Lands' Law, 457; on Rey de Castro's ghostwriting, 353n38; works published, 519, 542

La Chorrera: Arana's purchase of, 83, 87; Benjamín Larrañaga at, 77, 78, 79, 83, 93, 104; British/American fact-finding mission at, 268–69; Calderón's attack on, 94; Captain Moya del Barco at, 282; Casa Arana headquarters, fate of, 497–98; English missionaries at, 409; first contact at, 46n109; Francisco Alayza y Paz Soldán on, 257; Henry Samuel Parr at, 310; Indian massacres at, 203, 262; Indian population of, 269–70; Indian torture at, 159; Juan Tizón at, 225; PAC Commission at, 300; PAC criminals working at, 245; Peruvian garrisons at, 94, 95; Rómulo Paredes at, 220, 240–41; and Salomón-Lozano Treaty, 432; in Stuart Fuller's report, 269–70; Victor Macedo at, 139; white exploration of, 74; wireless telegraph station, 364

La Chorrera in the 21st century: AZICATCH, 497–98; education in, 499; House of Knowledge conference, 504–7; Huitotos in, 498–99; infrastructure, 499; population, 488

La Chorrera Plan, 496–97

Ciceri, Agustin, 292n16

CILEAC (Center for Linguistic

and Ethnographical Research in Colombian Amazonia), 528
civilizing Indians discourses: Abel Alarco's use of, 262–63; and Amazon colonization, 381–82; Arana's use of, 155, 277–78, 323; and Cajiao concession, 108; and cannibalism, 45; church and state cooperation, 376, 378; Edmund Dene Morel's, 174; and *Indios Huitotos do Rio Putumayo* film, 360; John Russell Gubbins' use of, 301; as justification for exploitation, 16, 45; King Leopold's, 172; Lizardo Arana's use of, 201–2; *versus* modernizing, 461–62; and the OPFé, 377, 382; Pablo Zumaeta's use of, 305–6; Plan of Life's opposition to, 499; Roger Casement on, 253; ubiquity of, 540
civilizing Indians discourses and missionaries: as goal of, 372, 379, 413; overview of, 42–43; Pedro de Monclar's, 394–95, 412; power of, 395–96; rubber industry disrupting, 384, 414; state cooperation, 376, 378, 393n169, 421; Zamudio's skepticism of, 380
Cobo, Alfredo Vasquez, 114
COICA (Coordinadora de Organizaciones Indígenas de la Cuenca Amazónica), 451, 453, 474, 482–83
Collier, Richard: on Abel Alarco's firing, 194; on Arana and British Board, 305n73; on Arana in Iquitos, 70–71; on Hardenburg, 151, 344; invocations of Latin authoritarianism, 354; publications by, 519–20, 532; *The River That God Forgot*, 532–34, 542
Colombia: 1863 Constitution, 373; 1885 Constitution, 373; 1932 war with Peru, 436–39; 1970s-80s unrest, 471–72; 1991 Constitution, 446–47, 473–74; 2005 Constitutional changes, 479; Alfonso López Pumarejo's presidency, 464–65; as American ally, 447; Carlos Lleras Restrepo's presidency, 469; claims to Putumayo, 96–97; Compañía de Caquetá, 49–50; Compañía de Colombia, 49; contemporary government of, 446–47; denouncements of PAC atrocities, 154, 348; Departamento de Amazonas, 438; Enrique Olaya Herrera's presidency, 463–64; and the Granadine Federation, 372; internal divisions of, 438; internal military campaigns, 479; land concessions, 49–50, 108–9, 113; leaf blight research, 55; Leticia Trapezoid, 432, 436, 438–40, 493–95; Mocoa, 59–60, 63–65; Napo River jurisdiction, 81–82; nationalism in, 425; Natives' Law 89, 374; nineteenth century history: border confusion, 14–15; and the Catholic Church, 25, 370–71; Cuacan independence, 27n49; Federal War, 28; Jesuits in, 370–71; need for territorial defense, 33–34; *patronato* agreement, 371; quinine production, 29; religious order expulsions, 370–71; religious wars, 370–74; rubber boom, 71–72; War of the Cuaca, 373; War of the Pacific, 50n125; wars fought, 65, 81; paramilitary groups, 469n69; Pasto, 26–27; and Peru, contemporary similarities, 446–47; Peruvian oriente territorial disputes, 111; petroleum

boom, 492; Putumayo steamship service, 432; quinine exports, 64; Rafael Reyes's presidency, 107–8; Restrepo administration, 469–70; rubber collection in, 30; Sibundoy Valley, 467; Soratama enterprise, 57–58; tensions with Peru, 35–36, 80–81; U Party, 481; Uribe Vélez administration, 475–81, 486; War of a Thousand Days, 81, 107, 445, 463; War of Independence, 26–27; War of the Cuaca, 373. *See also* Capuchin order; the Catholic Church and Colombia; FARC; *modus vivendi* agreement; the Putumayo/Caquetá, Colombian-Peruvian conflicts in; Putumayo/Caquetá region; Putumayo/Caquetá region, Colombian activity in; Salomón-Lozano Treaty

Colombia, nineteenth century history: border confusion, 14–15; and the Catholic Church, 25, 370–71; Cuacan independence, 27n49; Federal War, 28; Jesuits in, 370–71; need for territorial defense, 33–34; *patronato* agreement, 371; quinine production, 29; religious order expulsions, 370–71; religious wars, 370–74; rubber boom, 71–72; War of the Cuaca, 373; War of the Pacific, 50n125; wars fought, 65, 81

Columbus, Christopher, 1, 11, 18, 382–83

El Comercio newspaper, 258–59, 284

Comisión Nacional de Pueblos Andinos, Amazónicos y Afroperuanos (CONAPA), 455–56

Confederación de Nacionalidades Amazónicas del Perú (CONAP), 451

the Congo: atrocities in fiction, 517–18; Belgian-British diplomatic incidents, 170; British hearings on atrocities, 173; population depletion, 41n95; Roger Casement in, 170, 172–73, 512; rubber production, 29, 30–31, 40–41; "rubber terror" articles, 171. *See also* Morel, Edmund Dene

the Congo and Belgium: acquisition and reform, 176–77; control of, 40n94, 42; King Leopold, 170n531, 171, 176, 512, 517; King Leopold II, 40–42

Congo Reform Association (CRA), 175–77

Conrad, Joseph, 170, 367, 518

Cornejo Chaparro, Manuel, 504, 519, 542

Cornjeo, Manuel, 506

correrías: debates over term meaning, 321, 333, 411; definition of, 20; Douglas Hogg on, 331; Francisco Irazola on, 421; Michael Taussig on, 515; Miguel San Román's account of, 410, 411; Percy Harrison Fawcett's account of, 381n130; Roger Casement's account of, 291–92; Santos Granero and Barclay on, 520–21; uses of, 523

Corts, Estanislao de Las, 422, 435

Cosmopolita (boat), 433

Crevaux, Jules, 43

La Crónica newspaper, 276, 281, 431

La Cueva, 276, 283

Cunha, Euclides da, 430

Curiel, Manuel, 418–19

The Daily News, 236, 357

Davidson, Randal, 400

debt peonage, 40, 256, 323, 331, 435, 520

Declaration of Machu Picchu, 455
Deew, John, 505
Del Aguila, Juan C., 69
Denevan, William M., 42n96
The Devil's Paradise, 297–98, 325, 326
Díaz, Paolino, 383–84, 385
Dickey, Henry Spencer, 193–94, 237, 239n167, 299n45, 368nn89–90
Dominican order, 17, 383, 461
El Dorado settlement, 162
Doyle, Arthur Conan: and Congo atrocities, 518; King Leopold opposition, 175–76; and Putumayo Mission Fund, 261, 401; and Roger Casement, 175n552, 367, 368n89, 401
Duarte, Fabio, 120, 140–41
Dumit, Miguel, 57–58
Dussan, Alicia, 528
Dyall, Joshua, 203

Eady, Swinfen, 314–15
Eagleton, Terry, 338–39
Easter Uprising, 367–68
Eberhardt, Charles B., 135–36, 148, 293–94, 312, 331
Echenique, José Rufino (Basque-surnamed), 378
Ecuador: border confusion, 14–15; boundary debate with Peru, 106; civil war, 32; independence of, 14; Oriente control, 32–33; quinine production, 29; rubber collection in, 30; War of the Cauca, 373
Egoaguirre, Julio (Basque-surnamed): Arana, collaborations with, 423; as Arana's attorney, 362n71; and Augusto B. Leguía, 191, 423; delay of Peruvian government investigation, 215, 284; denouncement of Carlos Valcárcel, 276; denouncement of Rómulo Paredes, 276; election of, 128; as foreign affairs minister, 432; and manifestation against Valcárcel, 283; and PAC governmental influence, 155, 191, 222, 272, 532; PAC's compensation demand from Colombia, 212; as Peruvian Senator, 423; and Salomón-Lozano Treaty, 433–34; on second Peruvian investigative commission, 255; and Walter Hardenburg, 149, 160, 191, 327–28
Elder Dempsey Shipping Company, 169, 171–72
El Encanto: Capuchin mission at, 439; Commandant Polack, 122; Gaspar de Pinnell at, 416–19; prisoners at, 122, 145–46; and Salomón-Lozano Treaty, 432; Walter Hardenburg at, 145–47
enconmiendas, 20
enganche systems, 37, 58, 66n197, 214n135, 381n130, 449, 521
English Booth Shipping Line, 148
Enock, Charles Reginald, 186, 290, 297, 354, 359
Entente Cordiale, 365
Erazo, Aurelio (Basque-surnamed), 159–60
Erazo, Manuel (Basque-surnamed), 163, 547
Escadón, Joaquín, 415
Escurra, Isaac (Basque-surnamed), 243–44
ethnic identity, 494n135
European atrocities, examples of, 354–55
evil, philosophical question of, 538–40

"faction" literary genre, 533–34

FARC (Fuerzas Armadas Revolucionarias de Colombia): 1970s-1980s, 471–72; attacks on Indigenous peoples, 474–75; casualties from conflict with, 486; exclusion from Constitutional Convention, 473; and Indigenous rights, 481; Middle Putumayo control, 491; overview of, 447n1; peace accords, 484–86; peace talks with Santos, 481, 482, 483, 484
Farquhar, Percival, 123, 137
Fawcett, Percy Harrison, 381n130
Federación de Comunidades Nativas del Corrientes (FECONACO), 452
La Felpa newspaper: attacks on Arana, 127, 150–51, 164; and Genocchi's investigation, 398; Orjuela on, 164n514; overview of, 126; *La Union* assault reporting, 150–51
Fernández, V. Romero, 261–62
film, 360
Firestone, Harvey, 54
First Moroccan Crisis, 365
Fitzcarrald, Carlos Fermín, 50–52
Fitzgerald, William, 50
Flores, Miguel, 97
Flórez, Miguel, 163, 427
La Florida, 95, 106, 118, 162, 426
"The Flower of the Selva: A Tale of the Upper Amazon" (Hardenburg), 565–72
Fonseca, José: accounts of atrocities committed, 216, 236, 427; attack on Agustin Ciceri, 292n16; Casement's arrest request, 237; flight to Brazil, 244
Ford, Henry, 54
Foucault, Michel, 548, 551
Fox, Walter, 203
France: atrocities committed by, 355;
Carlos Lissón on, 381; colonies post-WWI, 432; First Moroccan Crisis, 365; and Uruguay, 15–16
Franciscan order: accounts of Indian issues, 417; Antonio Jurado, 24; Colegio de Ocopa, 24; encounter with Gaspar de Pinell, 417; English province, 402; Irish province, 401–2; missionaries from England, 400; missionary work, 17, 24, 25, 417; and PAC, 417, 420–21. *See also* Genocchi, Giovanni
Fuentes, Hildebrando, 72n228, 147, 214n135, 250, 534n51
Fuller, Stuart J.: on Arana at merchant banquet, 279; and British/American fact-finding mission, 265–66, 267–71; on La Chorrera, 269–70; on continued atrocities, 264–65; fact-finding mission report, 268–71; and Francisco Alayza y Paz Soldán, 265; and George Babington Michell, 254; on Juan B. Vega's flight to Switzerland, 280n353; on Pablo Zumaeta, 257–58; Putumayo investigation, 254, 255–56; on Putumayo missions, 390; on reform potential, 255–56; rubber industry report, 256; and *Slavery in Peru* report, 312; South African assignment, 343

Gálvez, Miguel, 151
García Moreno, Gabriel, 373
García Pérez, Alan, 458
Gaviria, Carlos Uribe (Basque-surnamed), 439, 487
Geertz, Clifford, 549, 550–51
Genocchi, Giovanni: on Alayza y Paz Soldán, 406; audience with Pius X, 400; and Bernardo Calle,

408–9; and British Putumayo Mission Fund, 401; calls for Putumayo mission, 403–4; and English Franciscans, 402, 407–9; on English missionaries, 406; influence of, 400, 401; and Irish Franciscans, 401–2; and *Lacrimabili Statu Indorum* encyclical, 402; and Merry del Val, 402; return to Europe, second, 409; as socialist crusader, 525

Genocchi, Giovanni, Putumayo investigation: and British Foreign Office, 397–98; and Casement Report, 398; Father Alvarez's testimony to, 399; itinerary of, 392, 393; and Lucien Jerome, 391, 392; mandate of, 386–87; and prior reports, 511; purview of, 391; reports to Merry del Val, 391–92, 399–400; reports to the Vatican, 392–93, 398–99; secrecy precautions, 392; smallpox blanket anecdote, 388n152; writings on atrocities, 397

Germany: atrocities committed by, 355; Carlos Lissón on, 381n129; Casement's writings on, 364–65; foreign policy, 432; imperialist policies, 335; and the Irish Volunteers, 366; Moroccan Crises, 365–66; South American presence, 335, 364, 378

Gielgud, Henry Lex: letter to shareholders, 197; in Louis Barnes's testimony, 299; and Pablo Zumaeta's surprise letter, 306; and PAC atrocity articles, 190–91; on PAC Board, 307; and PAC Commission report, 301; on PAC investigative committee, 196–97; as PAC manager, 194–95; as PAC secretary, 194; PAC valuation, 301–3; Putumayo visits, 197n71; and Roger Casement, 203, 207; Select Committee testimony, 301–304, 409n227

Glass, Frederick C., 411
Glenny, Elliot T., 411, 412
Golob, Ann L., 19–22
Gómez, Cayetano, 117
Gómez, José Francisco, 426–27
Gómez, Julio, 117
Gómez A., Ricardo, 426–28
González, Ildefonso, 105
Goodman, Jordan, 532
Gottlieb, Anthony, 538–39
Goulard, Jean-Pierre, 493–94
Gould, Stephen Jay, 549n84
Goyeneche, Peruvian chargé d'affairs (Basque-surnamed), 405, 408
Granadine Federation, 372
Gran Colombia, 14–15
Gray, Andrew, 519–20
Graz, Charles Des, 243
Great Britain: and Belgium, 170; Boer War, 355; British Peruvian Corporation, 381; Carlos Rey de Castro's critiques of, 354–55; in Casement Report, 272; Casement Report responses, 260–61, 271; colonialism of, 174, 177–78; Congoese presence, 334; English Booth Shipping Line, 148; hearing on Congo atrocities, 173; immigrants from, to Peru, 380–81; Iquitos consulate, 148, 202, 294; Loreto consulate, 237; and Lucien Jerome, 391; PAC atrocity press coverage, 260–61, 271; and Peru, 15, 33, 34, 148, 381; petition for PAC oversight, 347; Portugal as protectorate of, 13; relationship with Belgium, 173; resolution on

the Congo, 173; Slave Trade Act, 337n188; and Uruguay, 15–16; and War of the Pacific, 381; and World War I, 432. *See also* British/American Putumayo fact-finding mission; British Foreign Office; British Select Committee on PAC atrocities; missionaries in the Putumayo from England
Green, Alice Strepford, 177
Grey, Edward: and Brown's deposition on Indian abuses, 199; and Casement Report, 174, 252–55; diplomatic pressure on Belgium, 176; and James Bryce, 254; John Harris meeting, 200; letter from Anti-Slavery and Aborigines Protection Society, 199; opinions of, 174, 178, 181; and Peurvian PAC investigation, 200–1; recommendations to Peruvian government, 222; and Roger Casement, 174, 244, 247, 248–49, 252–54; and the US government, 192, 226–27, 254, 260
Gubbins, John Russell: agreement with Casement Report, 304; Arana, trust in, 305, 308; credentials, 130; and David Cazes, 304–5; distrust of Pablo Zumaeta, 307; and Juan Tizón, 310; and PAC atrocity claims, 186, 192–93, 196, 301; and Roger Casement, 228–29, 307, 310; salary, 195, 197; and Saldaña Rocca's articles, 305; Select Committee testimony, 304–8, 310
Gubbins, John Russell, on PAC Board of Directors: as acting director, 129–30; as general manager, 196–97; information hidden from, 305, 307; and other British directors, 134

Guevara, Bartolomé (Basque-surnamed), 86, 146
Gutiérrez, Don Francisco, 98, 99, 100, 101

Hardenburg, Walter: Arana's testimony against, 322; articles on the Putumayo, 345; attacks on Arana, 345, 530–31; and Benjamín Saldaña Rocca, 125, 149–50, 333; blackmail accusations, 184–85, 262, 309, 322, 330, 344; in Canada, 195; characterizations of, 344; *The Devil's Paradise*, 297–98, 325, 326; equipment loss compensation from PAC, 162; and Fidel de Montclar, 137; "The Flower of the Selva: A Tale of the Upper Amazon," 565–72; forgery accusations, 322, 323, 330, 344; G. S. Paternoster's support of, 293, 344; and Guy T. King, 147, 148–49; information kept from Arana, 331; invocations of Latin authoritarianism, 354; in Iquitos, 149; and Jesús Orjuela, 119, 143–44; and Julio Egoaguirre, 149, 160, 191; in London, 181–82; and Mary Feeney, 181; motivations of, 293, 525; perspectives on, 513, 525; on Putumayo missionary potential, 369; on Robuchon's disappearance, 124n369; Select Committee's findings on, 337; Select Committee testimony, 326–28; socialist politics of, 344. *See also* Anti-Slavery and Aborigines Protection Society and Walter Hardenburg; Arana, Julio César and Walter Hardenburg
Hardenburg, Walter, and Walter B. Perkins: initial meeting of,

136–37; investment in David Serrano's business, 142; reunion post-abductions, 151–52; support from, 326

Hardenburg, Walter, PAC atrocity investigation and reporting: advice to Anti-Slavery Society, 200; and Aurelio Blanco, 158–59; and Aurelio Erazo, 159–60; book on, 182, 183; eyewitness accounts, 158–60; and Guy T. King, 160; and Julio Muriedas, 159, 161; London trip, 160–61; meeting with Julio César Arana, 154; and Miguel Gálvez, 151; origins of, 151; PAC monitoring of, 160–61; Peruvian responses to, 184–85; and Saldaña Rocca, 511; *Truth* articles on, 181–82

Hardenburg, Walter, Putumayo travels: abduction by PAC, 122, 143–46; at Argelia, 142, 143; Buenaventura to Mocoa, 137–38; and Carlos Zubiaur, 146; and David Serrano, 141–42; diary of, 137–38, 141; discussions with Jesús López, 139–40; at El Dorado, 142–43; El Encanto to Iquitos on *Liberal*, 145–47; and Octavio Materón, 138; La Sofía to Yaracaya rubber station, 138–40; stories of Peruvian abuses, 139–43; Yaracaya to La Unión, 140

Harris, John, 181, 189, 200, 352, 405

Heart of Darkness, 42, 518

Heath, Edwin, 597

Henson, Hensley, 260, 271, 400, 404

Henson, Herbert, 260, 271, 400, 404–5

Hernández, Cecilio, 69, 70, 128, 346–47

Hernández, Crisóstomo: about, 73; Aquileo Tobar's account of, 98–102; Caraparaná River exploration and control, 73, 101–2; and Don Francisco Gutiérrez, 100, 101; flight from authorities, 98–99; life with the Huitotos, 99–100; Rocha on, 75n237; rubber collecting and trading, 100–2; and Solís, 101–2

Hernández, Justino, 88

Higginson, Eduardo, 113–14

Hildebrand, Martín von, 472–73

Hispano, Cornelio, 353

historical narratives, 549–50

Hogg, Douglas, 315, 320, 325, 330–32

House of Knowledge conference, 504–7

Howard, Henry Clay, 251, 254

Howland, Tood, 505

Huitocia: Colombian settlement of, 76; Colombians naval base, 439; Huitotos in, 494, 496; Indians population, contemporary, 487–88; Joaquín Rocha's visit to, 72; overview of, 1, 445; remoteness of, 445–46. *See also* Aranalandia; Putumayo/Caquetá region

Huitoto Indians: Alto Cahuanarí uprising, 414–15; Aranalandia reservations given to, 496; Bogotá migrations, 489–90, 492; and Boras, 72n228; in La Chorrera Zone, 498–99; contemporary population of, 487, 488; coordinated killing of whites, 74; Crisóstomo Hernández and, 73, 99–100; dispersement of, 488–89; endangered people status, 488; enslavement of, 53; at Güepí, 420n266; in Huitocia, 494, 496; Huitoto Rebirth association, 501–52; Huitoto Regional Organization, 490; in Leticia,

501; in Leticia Trapezoid, 493–94; malaise of, 487; Nestor Reyes and, 64; NGOs aiding, 489; oral history of, 46n109, 47; Orejones, 60n174; outsider contact, 20th century, 489–90; perseverance of, 496; population estimates, 72; preferential endogamy, 492; in Puerto Leguízamo, 488, 489; and Resguardo Tijuna-Uitoto, 501; uprising of, 414–15; Yarocamena, 414–15

Humala Tasso, Ollanta Moíses, 460

Iferenanvique, 99, 100
Igaraparaná River: Arana's control of, 105; AZICATCH, 497–98; Benjamín Larrañaga's exploration of, 104; Colombian administrative division, 107; Crisóstomo Hernández's conquest of, 73; Enrique Espinar's mapping of, 82; Indian population on, 72; Indian uprisings, 86; Larrañaga, Arana and Company, 82
Ignatius of Loyola, Saint (Basque), 17–18
Igualada, Bartolomé de, 363, 434, 463
INCORA (Instituto Colombiano de la Reforma Agraria, 468–70, 496
Indians: Andoque tribe, 46; Aranalandia population, 72; authentic-false dichotomies, 500, 502–3; Boras, 47, 73, 304n69, 417, 501; *cabildos,* 479n100, 490, 497, 498, 500–2; *caciques* (leaders), 44; *capitanes,* 414, 494, 503, 521; Caripúnas, 598; Cashibo tribe, 67; collaboration with whites, 45–46; Colombian treatment of, 540–41; Colombian *versus* Peruvian treatment of, 89–91; contact concept, 523–24; contemporary isolated groups of, 49n120; contemporary struggles of, 552; cultural preservation programs, 500; cultural similarities between tribes, 43–44; culture of, as threatening, 522; and eco-ethno-tourism, 503; *enconmienda* raids, 20; epidemics experienced, 42n97, 46; ethnocide of, 488; exploitation of, 1933 account, 487; FARC assassinations, 530; fleeing whites, 48–49; Huarayos tribe, 50; Huitocia, contemporary, 487–88; Igaraparaná River, 72, 86; interpreters, 21; Jívaros rebellion of 1599, 20; Kaziya Narai *cabildo,* 503–4; Larrañaga, Arana and Company exploitation, 82; and Leticia war, 438; *malocas* (longhouses), 43–44, 503–4; *manguarés (large drums),* 44; Mascho tribe, 50; as minority, 490–91; and missionaries, early, 18–21; in mission settlements, 47–48; Multiétnico *cabildo,* 502–4; and national identities, 495; Nimaira Naimeki *cabildo,* 502, 504; and petroleum industry, 492; Pineda Camacho on, 45–46; popularity with intellectuals, 491; population declines, 24, 41–42, 292–93, 304, 310, 434–35; punishment of, *versus mestizos,* 521; racism against, ubiquity of, 540; resettlements of, 434–35; rural *versus* urbanite, 502–3; Sebúa tribe, 116–17; self-extinction reports, 435; "semi-civilized" *versus* "savages," 42–43; Shipibo tribe, 53, 55; Sibundoy valley, 393–94, 396–97; and steel axes, 19n13, 46; Ticunas, 493,

495, 500–1; tobacco use, 44, 45; and trade goods, 42–43, 46, 521; Waorani, 524; white collaboration, 523, 524; white culture, embracing of, 488, 525; white fear of, 44–45. *See also* cannibalism; civilizing Indians discourses; Huitoto Indians; indigenist policy; PAC atrocities

Indians, enslavement of: by Abelardo Aguero, 244; by Benjamin Larrañaga, 77; Brazilian, 26, 27–28; early history of, 18, 20–21; justifications for, 16–17; by Peruvians, 77–80, 259; by rubber barons, 53, 55

Indians, white treatment of: domination techniques, 45–46; Indian hunts, 295; Jesuit reductions, 18; Peruvian, 448; Portuguese, 21–23; raids against "cannibals," 16–17; Roger Casement's testimony on, 291–93; Spanish, 18, 20–21; theological debates over, 16. *See also* PAC atrocities

indigenist policy, 446, 455, 473

indigenist policy, Colombian: 1991 Constitution, 473–74; AATIs, 479; advocates for, 474; and Afro-Colombians, 474, 478; Aranalandia reservations, 496–97; Aurelio Iragorri Valencia and, 482–83; *cabildos,* 479n100, 490, 497, 498, 500–2; La Chorrera Plan, 496–97; and civil unrest, 471–72; and COICA, 474, 482–83; and Colombian Anthropological Institute, 472; constitutional rights, 447; Department of Native Affairs, 467–68; education, 479; environmental, 479–80, 481–82; and FARC, 474–75, 481, 482, 483, 485–86; favoring rural tribes, 503; Feliciano Valencia and, 484; health, 479; and ILO Convention 169, 490; INCORA, 468–70, 496, 497; Indian activism, 446, 482; Indigenous associations, 471, 472, 473; Indigenous leader assassinations, 471, 474–75; Indigenous protests, 478–80, 482, 484; Indigenous self-rule, 483; IWIGA reports, 474, 476, 478, 482–83; land policies, 468–69, 477, 484; Martín von Hildebrand and, 472–73; mineral concessions, 481–82; modernization *versus* preservation, 500; ONIC, 472, 476–77; and paramilitary groups, 476–77; *Resguardo Indígena Predio Putumayo,* 497; *resguardos,* 467, 469, 473, 496–97; Restrepo administration, 469–70; Rodolfo Stavenhagen's report on, 477; Roque Roldán and, 472; Santos administration, 480–86; Sibundoy Valley, 467; Uribe Vélez administration, 475–81, 486; and US military presence, 475; Vincent Brackelaire and, 473. *See also* Capuchin order

indigenist policy, Peruvian: 1993 constitution, 446–47, 454; and AIDESEP, 451–52, 454, 458–60; Alberto Chirif on, 449, 461–63; Alberto Pizango and, 459; Benavides administration, 448–49; bombing of Matsés tribe, 449; *campesino* concept, 450; and Canada, 454, 458; and the Catholic Church, 450; community recognition, 450–51; and CONAPA, 455–56; economic development prioritization, 456–57, 460–61; educational reforms, 449–50; Eliane Karp and, 455, 456; and environmentalism, 457, 458–49; First Summit of Indigenous Peoples of Peru, 457; freeing of enslaved and indentured peoples, 453; and Free Trade Agreement with US, 458–59; Fujimoro administration, 454; García Pérez administration, 458–59; Humala Tasso's administration, 460; and ILO Convention 169, 490; INDEPA, 456–58; Indian activism, 446, 449, 452; indigenous rights, 451–52, 454–57; *indio* category, 450, 451; international influences on, 455; and the IWGIA, 453, 456–58; José Carlos Mariátegui on, 448; Lands' Law, 457; and liberation theology priests, 459–60; Loretan Civic Front, 453–54; Mario Vargas Llosa on, 452; Miguel Palacín and, 454; modernization, 461; nature reserve creation, 452; and Oxfam, 453; protests of, 458, 459; Quechua language, 450, 451; racism hindering, 451; relocations, 449; and Shining Path rebellion, 454; Stefano Varese's work on, 449; Summer Institute of Linguistics, 461n47; territorial sovereignty, reductions of, 458; Terry administration, 449; Toledo Manrique administration, 455–57; tribe recognition, 449; and UN Declaration of the Rights of Indigenous Peoples, 456; universal democracy, 448; Velasco administration, 449, 450, 451; violence over, 459–61

Indios Huitotos do Rio Putumayo film, 360

International Labour Organization, 453

Iquitos: American consulate in, 113n343, 147–48; A. Morey and Company, 69–70; anti-centralist revolt, 424–25; Arana and, 70, 71, 148–49, 160, 423, 535; Arana's return to, post-scandal, 345; Augustinian missionaries in, 385; and bishop of Chachapoyas, 385; British influence in, 148; Casa Arana's prominence in, 534–35; the Catholic Church in, 390; College of Saint Augustine, 385; Colombian consul, 167n524; economic crisis of 1920, 423, 425; founding of, 23; growth of, 33, 34–35; hospital of, 430–31; judiciary colluding with PAC, 361–62; National College, 430–31; problems with, 213–14; river traffic to the Putumayo, 32; rubber export tax, 423, 430; rubber shipping cessation, 430; rumors of Colombian attacks, 122–23; separatist insurrection, 71; tensions with Colombia, 148, 164–65; threats to, 214

Irazola, Francisco (Basque-surnamed), 421–22

Isaacs, Robert, 271

Israel, Victor, 440

the Israelites, 494

Izaguirre, Benigno (Basque-surnamed), 50

Jacanamejoy, Tomás, 464

J. C. Arana y Hermanos. *See* Casa Arana

Jerome, Lucien, 222, 228, 391, 392

Jesuit order: abduction of Indians, 21; Amazonia, 20; Andean missions, 19; Audiencia of Quito, 17, 19; and Colombia's religious wars, 371, 372, 373; expulsion from South America, 23; founding of, 17; Indian reductions, 18, 19, 21–22, 23, 24; Iquitos mission, 23; José de Anchieta (Basque-surnamed), 18; missionary work, 17–18, 335, 391; overview of, 17–18; and Spanish *versus* Portuguese raiders, 23; Spiritual Exercises, 18. *See also* Ignatius of Loyola, Saint; Xavier, Francis, Saint

Jiménez, Augusto, 398
Johnston, Harry, 334
Juajibioy, Pedro, 466
Jurado, Antonio, 24

Katenere, 159, 289
King, Armando, 244
King, Guy T., 147–49
Knox, P. C., 312

Lagos, Ovidio: on Arana, 81, 105, 346, 441–42, 532; on Benjamín Larrañaga, 83; on Hardenburg at British Select Committee, 326; on La Unión assault, 123; on Martín Arana, 234n202; on Peruvian Putumayo sovereignty, 105
Lamis, Sofía Angulo, 59n169
Larrabure y Correa, Carlos (Basque-surnamed), 112, 350–51
Larrañaga, Arana and Company, 82, 94
Larrañaga, Benjamín (Basque-surnamed): Abel Calderón's businesses, taking of, 77, 89; Aquileo Tobar's account of, 102–4; and Arana, 79, 83, 87, 94; and Arana y Hermanos, 79; Arcesio Aragón on, 427; business partnerships, 79; and Calderón brothers, 74, 79, 94; Caraparaná conflict, 78; and La Chorrera, 79, 83, 93, 104; Igaraparaná exploration, 104; jailing of Abel Calderón's partners, 77, 89; Larrañaga, Arana and Company, 82, 94; Middle Caquetá exploration, 102–3; in Middle Putumayo, 65; Ovidio Lagos on, 83; Peruvian inquest into atrocities, 83; in Peruvian military, 77; in the Putumayo, 76–77, 93–94; and Rafael Reyes, 59–60; rubber trade with Indians, 102–4; *siringa* latex discovery, 76; and Solís, 102, 103–4; in Upper Iguaraparaná, 76–77; violence of, 77n247
Larrañaga, Rafael (Basque-surnamed), 76, 77n247, 83, 95, 102, 103
Larrea, Oscar R. Benavides, 438
latex, 29–30, 76, 80, 81n257. *See also* rubber
Latin American political swings, 553
Latin authoritarianism, 354
Lauri, Lorenzo, 421
leaf blight, rubber tree, 54–58
Leguía, Augusto B. (Basque-surnamed): imprisonment and death, 436; and John Russell Gubbins, 186; and Julio Egoaguirre, 191, 423; and Loreto boundary disputes, 431–32; Lucien Jerome's suspicions of, 228; meeting with Rómulo Paredes, 238; military coup, 423; PAC ownership interest, 364; political connections of, 128; Putumayo fugitive extradition, 222, 243
Leguía y Martínez, Germán, 243, 274–75
Lembcke, Eduardo, 184–85, 260, 350
Leticia, 56–57, 436–38, 500–1

Leticia Trapezoid, 432, 436, 438–40, 493–95
Liberal (boat): Arana's use of, post-scandal, 319n133, 425; and British/American fact-finding mission, 265–66, 274; Carlos Zubiaur's captaincy, 84, 209; loaned to Peruvian government, 433; overview of, 84; PAC Commission on, 203; PAC employees fleeing on, 222, 361; and de Pinell/Marquéz expedition, 416, 419–20; in Robuchon's book, 125; Roger Casement on, 203, 208; in *La Union* assault, 119–20, 121, 122, 143, 145, 150, 288; Walter Hardenburg on, 146
liberation theology, 459–60, 469, 526
Lissón, Carlos, 36, 379–81
Lister-Kaye, John: and Casement Report, 260n286; on PAC Board, 130, 196; and PAC Commission, 218; Select Committee testimony, 309, 326
literary representations of atrocities, 517–18
Llosa, Mario Vargas, 452
Loayza, Miguel: attacks on Colombians in Aranalandia, 162–63; attacks on David Serrano, 141, 144; and Gabriel Martínez, 117; in George Babington Michell's testimony, 298, 299; and Jesús Orjuela, 119, 145; letter regarding attacks, 199; and de Pinell/Marquéz expedition, 416–17, 419; on Putumayo rubber production, 346; as remorseless, 441n344; Walter Hardenburg abduction, 144–45
London Hearings. *See* British Select Committee on PAC atrocities
Lopez, Aurelio, 221

López, Jesús, 139
López Garcés, Claudia Leonor, 495
López Pumarejo, Alfonso, 464–65
Loreto region, Peru: anticentralist revolt, 424–25; British consul in, 237; British nationals in, 148; and central authority, 436n330; control of, 34–35; development plans, 378–80; economic growth, 466; governmental study of, 35n81; importance of, 65–66; Jesuit Reduction, 19; maritime and fluvial traffic, 35; migration to, 213–14, 379; population growth, 466; promotion of immigration to, 378, 380–81; Provincia Litoral de Loreto, 378; Puerto Leguízamo, 439–40, 467; security concerns, 214
Loyola, Iñigo de, 17–18

Macedo, Victor, 427; Abel Alarco's praise of, 191; Arana's admiration for, 190; arrest warrant for, 234, 235n203, 244, 361–62; and La Chorrera, 139; government protection of, 361–62; and Isaac Ezcurra, 243; and PAC atrocities, 198; and PAC Commission, 203, 207, 300; and PAC valuation, 302; resignation and flight, 215, 244
Madeira-Mamoré railroad, 123
Madeira River, 51
Magellan, Ferdinand, 12
magico-realism, 513
Mainas, 22, 375–77
Manaos, Brazil: accessibility of, 51, 73; Arana in, 124, 534; bishop of, 420; Casa Arana headquarters, 128; Colombian consul, 266–67; Pinell's expedition, 419; Putumayo connection, 51, 73; and

Rafael Reyes's voyages, 61; Roger Casement in, 250; and rubber exports, 37n88, 40, 49; trade with Pasto, 26

Manrique, Alejandro Toledo, 455

Manu/Madre de Dios River, 50–51 mapping war, 516–17

Mariátegui, José Carlos (Basque-surnamed), 448

Markham, Clements, 29

Marquéz, Thomas: career of, 425–26; expedition with Gaspar de Pinell, 416–20; and Fidel de Montclar, 416

Martinengui, Elías (Basque-surnamed): atrocities committed by, 398; flight from Putumayo, 361; and Ricardo Gómez A., 427; Roger Casement's report on, 219

Martínez, Antonio, 106

Martínez, Gabriel: in Iquitos, 164; as Middle Putumayo/Caquetá police inspector, 114; PAC abduction and imprisonment of, 118, 119, 122, 145, 146; PAC atrocity investigation, 116–18; report on Peruvian attacks, 164

Martínez, Vincent, 498–99

Maury, Matthew Fontaine, 28–29

Mayer, Dora, 275

Maynas Region, 22, 375–77

McNair, Stuart, 413

Medina, José Francisco: and Arana, 123; and PAC, 128–29, 134–35; Select Committee written brief, 308–9

Mejía, Carlos, 162–63

Menacho, Antonio, defense of PAC, 358–59

MERCOSUR, 447

mestizos: abuses of, 521; Aranalandia settlement, 445; Colombian paramilitary attacks on, 469n69; definition of, 493; and *enganche,* 37, 66n197; Huitoto-descended, 488; Leticia Trapezoid, 493; punishment of, *versus* Indians, 521; in the rubber industry, 37, 64, 520–21; rural, 450, 467; Santos Granero and Barclay on, 520–21

methods and theoretical framework, author's, 527, 550–52

Michell, George Babington: on Arana at merchant banquet, 279; and British/American fact-finding mission, 254, 265–66, 268, 275, 348; invocations of Latin authoritarianism, 354; as Loreto British consul, 237; Select Committee testimony, 298–99; and Stuart J. Fuller, 254

Middle Caquetá. *See* Aranalandia

missionaries in the Amazon: Augustinian, 383–85, 391, 410–11; Colombian requests for, 372; early history of, 22–23, 43, 47–49; Leo XIII's interest in, 382–83; Pius X's interest in, 386–87. *See also* Capuchin order; civilizing Indians discourses and missionaries; Franciscan order; Jesuit order

missionaries in the Putumayo: articles advocating, 378–79; Augustinian, 383–84, 391, 410–11; Bernardo Calle and, 384, 408–9; Carlos Lissón on, 379–80; *cauchero* resistance to, 399, 410; clergymens' attitudes toward, 382; Colombian concerns, 406; Colombian support, 412; and contemporary pentacostalism, 526; control of, 407–8; Diocese of Chachapoyas, 383; Dominicans, 383; foreign, 377n113; Franciscans, 379, 383,

400; Giovanni Genocchi as, 407, 408–9; Giovanni Genocchi's report on, 391–92; government support of, 376–77; in Iquitos, 385, 407; Jericho mission, 410–11; Jesuits, 391; José María Arriaga's plans for, 377; jurisdiction concerns, 406, 408; Marañon River, 384; Miguel San Román, 410–11; Napo River, 384; at Nazareth rubber station, 384; neglect of, 375; New Peru plan, 379; *La Obra de la Propagación de la Fé* (OPFé), 377, 382, 386, 410; Ocopa college for, 377; PAC obstruction of, 420–21; Paolino Díaz, 383–84, 385; Peruvians as preferred, 409; Protestant, 405, 526; Putumayo difficulties, 383–84; Roger Casement's advocacy for, 387–89, 400–1, 403–4; San León de Amazonas prefecture, 383, 421; Sotero Redondo, 421; *The Times*'s call for, 403–4; Vatican plans for, 383. *See also* Capuchin order; Montclar, Fidel de

missionaries in the Putumayo and PAC atrocities: Alemany on, 385–86, 387; Francisco Irazola on, 421–22; Monsignor Quattrochi on, 385; Prat on, 385

missionaries in the Putumayo from England: and Casa Arana, 413; Catholic, 402; at La Chorrera, 409–10; Elliot T. Glenny, 411, 412–13; false protests claim, 408; Father Sambrook, 409–10, 413–14; Franciscan, 400, 407–8, 420; Frederick C. Glass, 411; Peruvian resistance to, 407, 408; Peruvian support for, 405–6; Pius X's support of, 407; Protestant, 403; Protestant Evangelical Union of South America, 411, 412–13; Protestant *versus* Catholic schism, 405, 411; Putumayo Mission Fund, 400–1, 407

missionaries *versus* anthropologists, 527

Mitchell, Angus, 368

Mocoa, Colombia, 59–60, 63–65

modus vivendi agreement (Colombia and Peru): and Amazon Colombian Rubber and Trading concession, 113–14; as enabling PAC atrocities, 239, 272; as favoring Peru, 111; overview of, 111; Peru's disregard for, 111, 114, 139, 164; Peruvian misinformation regarding, 122–23, 157; Peruvian renegotiation promises, 165; Rafael Reyes's repudiation of, 115, 122

Molesworth Committee, 288n5

Monroy, Pablo J., 86, 95

Montclar (Monclar), Fidel de: as authoritarian, 397; and bishop of Manaos, 420; character of, 394; and Colombian government, 415–16; free river navigation goals, 416; influence of, 412; missionary work of, 394–96, 412; power of, 415; promotion of Capuchins, 415–16; return to Europe, 463; road building goals, 415; and Thomas Marquéz, 416; and Walter Hardenburg, 137

Montt, Alfredo, 216–17, 236, 243, 244

Morel, Edmund Dene: Aborigines Protection Society rally, 173; and Arthur Conan Doyle, 175–76; and Booker T. Washington, 175; and British colonialism, 174; on civilizing Indians, 174; Congolese rubber industry suspicions,

171–72; and Congo Reform Association, 175–77; and Mark Twain, 175; pressure on King Leopold, 175–76; reporting on Belgian atrocities, 172; and Roger Casement, 174–75
Morel, Edward, 334
Morey Arias, Adolfo, 69–70, 128
Morey Arias, Juan Abelardo, 69
Morey Arias, Luis Felipe, 69, 128
Morocco, 365–66
Morris, Carlton, 202
Mosquera y Arboleda, Tomás de, 371, 372–73
Mourrail, Charles, 65, 69, 70
Moya del Barco, Captain, 282
Muriedas, Julio, 159, 161

Napo River: Ecuador and open navigation of, 32; Ecuadorian-Peruvian conflicts, 106, 111; jurisdiction over, 81–82; missionaries in, early, 22–23; trail to, from Putumayo, 364
The Nation newspaper, 297–98
The New York Times, 271
Nofigageré, 103, 104
Normand, Armando: arrest of, 363; autobiographical account of, 516, 579–86; in Brazil, 250; English language ability, 105, 186; hiring of Barbadians, 105; and PAC atrocities, 206; and Ricardo Gómez A., 428; and Roger Casement, 206, 516; Salvador Olivares on, 259; and Thomas Whiffen, 186
Norzagaray (Basque-surnamed), 318–19, 422

Ochoa, Bernardino (Basque-surnamed), 86

Ochoa, Bernardo (Basque-surnamed), 138
O'Donnell, Andrés, 237, 243, 277
Oil Rivers Protectorate, 170
Olano, Luis, 420–21
Olartua, Miguel, 460
Olaya Herrera, Enrique, 463–64
Olivares, Salvador, 258, 259
Omarino, 208, 236, 237, 517
ONIC (Colombian National Indigenous Organizations), 472, 476
Ordóñez, Antonio: and El Remolino, 114, 139; and La Unión, 106, 119, 123n65, 140, 141, 144
Orient, search for sea route to, 11–12
El Oriente newspaper: ownership of, 126, 215; Rey de Castro's articles in, 156; Roger Casement's suspicion of, 209; Rómulo Paredes's editorials in, 123, 221; support of Arana, 221
orientes, political disputes over, 29
Orjuela, Jesús: abduction by Peruvians, 122, 143, 145, 146; as Columbian Caraparaná administrator, 118–19; discussion with Polack, 122; in Iquitos, 164; meeting with Peruvian prefect, 164; and Miguel Loayza's attack on David Serrano, 142; report to Colombian government, 165; La Unión assault, 119
Ortiz, Sergio Elías, 427
Osma y Pardo, Ramiro de, 142, 143, 145
Osorio, Alberto Salomón, 432
Oxfam, 453, 454

PAC (Peruvian Amazon Company): Abel Alarco at, 129, 190, 194, 235, 319, 332; accounting irregularities, 302, 303; Acre River asset

ARANA | 643

sale, 324; Arana family accounts, 300; Arana's early involvement, 128–30, 134; bankruptcy filing, 232; and Bellamy and Isaac, 129, 133; and Carlos Zapata, 154–55; Charles B. Eberhardt's report on, 135–36; Colombian government protest to, 129n390; compensation demand from Colombia, 212–13; creation of, 123, 128–29; debts of, 226, 230, 232, 236; development expenditures, 302; dissolution of, 314–15, 345n9, 423; dividend payments, 198; and Egoaguirre's governmental influence, 155, 191, 222, 272, 532; and Franciscan order, 417, 420–21; governmental influence of, 155, 191, 222, 257, 272, 532; government officials protecting, 418; Henry Read and, 128–30; Iquitos influence, 155; and J.C. Arana Y Hermanos, 226; John Lister-Kaye and, 130, 134; John Russell Gubbins and, 129–30, 186, 192; José Francisco Medina and, 128–29, 134–35; land holdings, 319; London incorporation of, 129, 130, 133–34, 526; media images of, 155–56; missionary obstruction, 420–21; Pablo Zumaeta and, 129; prospectus of, 131–33, 300, 323; public association with Arana, 345–46; public offering, 129, 130, 133–34, 535; Putumayo access control, 265–66; river fleet, 131; Roger Casement's desire to reform, 230–33; rubber export means, 357n53; unsold rubber problems, 422; winding-up order, 345; Zumaeta lien against, 226. *See also* Arana, Julio César;

Casa Arana; El Encanto; PAC atrocities; PAC atrocities, PAC responses to accusations of; PAC Board of Directors

PAC atrocities: Alfredo Montt and, 216–17, 236, 237, 243, 244; and Arana's family, 332–33; Arana's guilt of, 288, 289, 301, 541–42; and Asociación Pro-Indígena, 210–11; Atenas station retribution, 414; Augusto Jiménez, 398; Barbadians committing, 136, 186, 202–3, 218, 272, 282, 289, 524; *versus* Belgian, 40–42; Carlos A. Valcárcel's legal responses, 212; and Casa Arana's IPO, 521–22; the Catholic Church's investigation into, 386–88, 391; La Chorrera Indian massacre, 262; Colombian land, seizures of, 427; and the commission system, 160, 191, 192, 225, 322; as common, 540–41; Las Delicias assault, 418; economic rationales for, 521–22; Elias Martinengui's, 398; employee flights to Brazil, 215, 221–22, 243, 244, 255, 275; employees committing, as minority, 523; in European press, 258; evidence in company documents, 198–99; examples of, 140, 145, 158–60; Hardenburg's book on, 183–84; historical contexts of, 526; incentives for, 160; internal investigative commission report, 223–26; Isaac Escurra's disclosure attempts, 243–44; John Brown's deposition on, 199; John Whiffen's witnessing of, 186–87; and missionaries, 383, 385–87, 421–22; and *modus vivendi* agreement, 239; *modus vivendi* enabling, 239,

272; Pablo Zumaeta's ignorance of, 187; and Peruvian military authorities, 352–53; publications disputing, 350–51, 353–54; raids acknowledged by Arana, 292n16; Roger Casement's letter to Louis Mallet, 215–18; shareholder meeting discussions, 193, 214–15; slavery, 256, 275–76; support for accused employees, 250; suspects feeling to Brazil, 215; and the United States, 250–51; and Victor Macedo, 198. *See also* British Select Committee on PAC atrocities; Casa Arana attacks on Colombians; Casement, Roger; Genocchi, Giovanni; Hardenburg, Walter; missionaries in the Putumayo and PAC atrocities; PAC atrocities, publications covering; PAC Commission; the Putumayo/Caquetá, Colombian-Peruvian conflicts in; La Reserva; Serrano, David; La Unión assault

PAC atrocities, abductions and imprisonments: Colombians, 77, 106; Gabriel Martínez, 118, 119, 122, 145, 146; Indians, 244; Jesús Orjuela, 122, 143, 145, 146; Walter B. Perkins, 144–46, 152–54, 315n114; Walter Hardenburg, 122, 143–46. *See also correrías*

PAC atrocities, accounts of: Alfredo Montt's, 236; Arédomi's, 236; from Casement Report, 398; Fray De las Corts', 422; Indian women provided to the military, 282; José Fonseca's, 216, 236, 427; Manaos Colombian consul's, 266–67; Manuel Quintana's, 423; Omarino's, 236; oral histories from descendants, 414; Prefecture of Loreto documentation, 414; problems with, 511; Robert Isaacs's, 271; V. Romero Fernández's, 262

PAC atrocities, arrest warrants for: issued by Carlos Valcárcel, 234–35, 242–43, 256–57, 280; issued by Herrera, 363; for Juan B. Vega, 234, 280, 284; for Julio César Arana, 280, 284; for Pablo Zumaeta, 234, 235n203; Victor Macedo, 244; for Victor Macedo, 234, 235n203, 244, 361–62

PAC atrocities, PAC responses to accusations of: *ad hominem* attacks, 184–85, 189, 537; Alarco's letter to management, 190; alleged circulated letter, 305–6; attempt to bribe Horace Thorogood, 182–83, 194; and British/American investigative committee, 266–68, 275, 298, 517; denials, 184–86, 189, 192, 193–94, 196, 197; denouncements of Hardenburg, 191, 262; fake blackmail letter, Whiffen's, 188–89, 192; internal investigative commission, 196, 197, 200, 223–26; investigation obstructions, 215; legal obstructions, 277; letter to British Foreign Office, 196; letter to shareholders, 192; management changes, 190; Philips and Dickey letters, 193–94

PAC atrocities, Peruvian government responses to: calls for investigation, 272–73, 274; cessation claims, 274; inaction, 248, 252, 361; investigations, 191, 215, 254–55; minister of external affairs's, 274; ongoing collaboration with PAC, 245; parliamentary debates, 272–74; protecting perpetrators, 243–45; as stock market ploy,

263–64. See also Paredes, Rómulo
PAC atrocities, publications covering: academic, 518–20; *La Amazonía Colombiana*, 353; American newspaper articles, 261, 271; Arana, historical vilification of, 525–26; Armando Normand's autobiography, 516; British newspaper articles, 260–61, 271; cannibalism concept in, 523–24; Colombian *versus* Peruvian, 518; contact concept in, 523; Cornelio Hispano's, 353; *The Devil's Paradise*, 297–98, 325, 326; epistemic murk in, 514, 516; and European humanitarian guilt, 526; G. Sidney Paternoster's book, 347–48; Guido Pennano's, 519; Horace Thorogood's articles, 183; journalistic, 518–19; on merit and blame systems, 522–23; overview of, 347–48; Pablo Valcárcel's book, 360–62; Peruvian press, 209, 211–12, 213; problematic terms, 524–25; problems with, 511–12; protagonists of, 512–13; *The Putumayo Red Book*, 348–51; Ricardo Gómez A.'s book, 426–28, 518; Spanish press, 209–10; *El sueño del celta*, 518; *Toá: Narraciones de caucherías*, 518; *The Upper Reaches of the Amazon*, 352–53; use of stereotypes in, 522. *See also* PAC atrocities, reports on; *Shamanism, Colonialism, and the Wild Man: A Study in Terror and Healing*; *Truth* newspaper articles on PAC atrocities

PAC atrocities, reports on: British Directors' awareness of, 305; Carlos Zapata's, 155–56, 158, 317–18; Charles B. Eberhardt's, 135–36, 148, 293–94; Eugène Robuchon's, edited by Rey de Castro, 124–25, 288, 316–17, 517; Francisco José Urrutia's, 165–66; Giovanni Genocchi's, to Merry del Val, 391–92, 399–400; Giovanni Genocchi's, to the Vatican, 392–93, 398–99; internal investigative commission's, 223–26; PAC board, to British Foreign Office, 226; PAC Commission, 221, 301; *Slavery in Peru*, 312, 343, 348, 512n1; Stuart J. Fuller's, 268–71; Thomas Whiffen's, to British Foreign Office, 188. *See also* Casement Report; Hardenburg, Walter; Paredes Report

PAC atrocities against Colombians: attacks on, 139–44; Caraparaná River conflicts, 95, 97, 105–6, 117–18; Gabriel Martínez, 116–18, 119, 122, 145, 146

PAC atrocities and British Foreign Office: belief in allegations, 191–92; demands for investigation, 195, 196; PAC board report to, 226; PAC board responses to, 192, 196; requests for intervention from, 199–201. *See also* Grey, Edward

PAC atrocities and the Board of Directors: and Casement's report, 222–23; concerns of, 186, 190, 192; denial letter to shareholders, 197; and internal investigative committee, 196; investigation request to Peruvian government, 290; meetings with PAC Commission, 218; reports to, 198; report to British Foreign Office, 226; support for Arana, 195, 196

PAC atrocity scandal aftermath:

Anglophone *versus* Spanish rhetoric, 354–355; Anti-Slavery Society watchdog group, 347; British oversight petitions, 347; Colombia's capitalization on, 348–52; International Board, humanitarian, 364; PAC employees indicted but free, 361–62; Peruvian claims of British conspiracy, 363–64; Putumayo border dispute resolution, 364; *The Times* of London exchanges, 349–51; warrants issued, 363

PAC atrocity scandal aftermath, PAC defenses and public relations: Antonio Menacho's letter, 358–59; Arana's open letter, 356, 573–77; Carlos Larrabure y Correa's writings, 350–51; films and photography, 360; South American campaign, 358–59; South American *versus* International, 360; Spanish texts defending Arana, 354–58

PAC Board of Directors: and Arana, 195, 196, 305n73, 537; awareness of atrocity reports, 305; and British Foreign Office, 192, 196, 226; British Select Committee's finding on, 336–38; calls for prosecution of, 271; and Casement Report, 222–23; Collier on Arana's relationship to, 305n73; and David Cazes, 186; Henry Lex Gielgud, 194–97, 301–3, 307; information hidden from, 300, 305, 307; John Lister-Kaye, 130, 196, 218, 260n286, 309, 326; John Russell Gubbins, 129–30, 186, 192, 307; Juan Tizón's report to, 198; meetings, Arana's absences from, 229, 230; meetings, Roger Casement at, 228–29, 230, 232; Pablo Zumaeta, 305; and Roger Casement, 228–29, 230, 232, 289, 290; status of, 134; *The Times* of London on, 271

PAC Board of Directors and PAC atrocities: concerns of, 186, 190, 192; concerns over, 186, 190, 192; denial letter to shareholders, 197; and internal investigative committee, 196; investigation request to Peruvian government, 290; meetings with PAC Commission, 218; reports to, 198; report to British Foreign Office, 226

PAC Commission: Arana's objection to, 307; Barbadian testimonies, 203–4; and Carlton Morris, 202; at La Chorrera, 300; company reform goals, 204; and David Cazes, 202, 208–9; findings of, 223; and Frederick Bishop, 202–3; and Henry Lex Gielgud, 301; and John Russell Gubbins, 307; Joshua Dyall's testimony to, 203; and Juan Tizón, 204, 205, 207–8, 307; and Lizardo Arana, 201–2; Louis Harding Barnes on, 203, 299–300; members of, 203; need for secrecy, 204–5; Peruvian officials on, 300; Reginald Bertie on, 196, 203, 218; report of, 221, 301; Roger Casement's testimony on, 290; Seymour Bell on, 203, 230; and Vicara, 203; Walter Fox on, 203

PAC Commission and Roger Casement: Arana's objection to, 307; and Arédomi and Omarino, 208; and Barbadians, 203–4, 207, 208; and David Cazes, 208–9; descriptions of PAC employees, 206; and Gielgud, 203, 207; Gubbins's and

Read's approval of, 307; Indian enslavement evidence, 201; and Iquitos prefect, 209; and Juan Tizón, 205, 207–8, 307; letter to Grey's secretary, 215–17; mandate, 201, 205, 290; meetings with informants, 202–3; on need for secrecy, 204–5; opinion of Carlos Valcárcel, 212; and other commission members, 203, 207; personal safety fears, 205, 208; Peruvian government authorization, 200–1, 307; translator for, 203; zeal of, 205–6
Pacelli, Eugenio, 405–6
Palacín, Miguel, 454
Panama Canal, 168
Panama and the Granadine Federation, 372
Pan-American Conference, 351–52
Pan American Union, 437
Paraguay, 18, 447n2
Pardo, José: and Arana, 84, 105; and Casa Arana, 104; as Peruvian president, 104–5
Paredes, Rómulo: on abuses of Colombians by Peruvians, 97; "A Samarem" poem, 369, 587–92; belief in reforms made, 265; book, unpublished, 363; on Bora Indians, 304n69; on Carlos Rey de Castro, 156; on Casa Arana's political clout, 278; flight from Iquitos and death, 369; history of the Putumayo, 93–97; investigation into Arana, 93, 97–98; letter to Travers Buxton, 363; and Macedo's arrest warrant, 244; meeting with Augusto B. Leguía, 238; nationalism of, 234; and *El Oriente*, 123, 126, 156, 221, 234; PAC employment, 277n344; perspectives on, 513; press attacks on, 276, 277n344, 283, 354; press defenses of, 281; retraction of credentials, 222; and Roger Casement, 221, 245–46, 249; skepticism of, 209, 221; ties to PAC supervisors, 221; on La Unión assault, 123; and V. Romero Fernández, 262; writing abilities, 233n196
Paredes, Rómulo and Peruvian government investigation of PAC: and Aurelio Lopez, 221; findings of, 233–35, 238–42; on lack of action, 248; as leader of, 215; overview of, 220; second, 249, 265n300, 284
Paredes Report: Arana in, 280; against British intervention, 280; burying of, 252; edits by Peruvian government, 255; excerpts from, 238–42; Isaac Ezcurra's seconding of, 243; Juan Tizón in, 280; press attacks on, 283; and Roger Casement, 245–46, 255; use of prior reports, 511; version sent to US Secretary of State, 279–80
Parr, Henry Samuel, 310–11
Pasos, J. M., 113–14
Pastrana, Antonio, 418
Paternoster, G. Sidney, 181, 293–94, 344, 347–48
Patiño, Bákula, 23–24, 436n330
Patriotic Board, 436–37
Paul III, Pope, 16
Pearson, Henry, 45, 202n86
La Pedrera, 168–69
Pennano, Guido, 519
Pérez, Hipólito, 88
Perkins, Walter B.: abduction by PAC, 144–46, 152–54, 315n114; Arana encounter, 158; and La Reserva attack, 152. *See also*

Hardenburg, Walter, and Walter B. Perkins
The Permanent Coordinator of the Indigenous People of Peru (COPPID), 455
Peru: 1930-31 presidents, 436; 1932 war with Colombia, 436–39; as American ally, 447; American diplomatic pressure on, 343; Carlos Lissón, 36; and the Catholic Church, 374–75, 447; Chinese laborers and immigrants, 83n266; Colombia, tensions with, 35–36, 80–81, 111; and Colombia, contemporary similarities, 446–47; colonization by Spain, 12; constitution of, 446–47; contemporary government of, 446–47; Departamento de Amazonas, 376, 378; Ecuadorian boundary dispute, 106, 111; and foreign missionaries, 421; and Great Britain, 15, 33, 34, 381; immigrant nationality preferences, 380–81; José Rufino Echenique's presidency, 378; leaf blight research, 55; military presence in Putumayo, 62; Moyobamba, 34–35, 66n198; Napo River, 81–82; *La Obra de la Propagación de la Fé* (OPFé), 377, 382, 386, 410; oriente, foreign investment in, 535; Oscar R. Benavides Larrea's presidency, 438; parliamentary debates on Casement Report, 272–74; population depletion, 42n96; and Putumayo/Caquetá disputes, 35–36; Putumayo steamship service, 432; Quechuan-descended presidents, 447; Rioja, 66n197; rubber industry protection, 346n13; Shining Path rebellion, 447n1, 454; suspicions of the US, 526; Yurimaguas, 68–69. *See also* Casement Report, Peruvian responses to; indigenist policy, Peruvian; Iquitos; Loreto region, Peru; *modus vivendi* agreement; PAC atrocities, Peruvian government responses to; the Putumayo/Caquetá, Colombian-Peruvian conflicts in; Salomón-Lozano Treaty

Peru, nineteenth century history of: agricultural promotion initiatives, 378; border agreement with Brazil, 34; Brazilian river navigation agreements, 32, 34; and British navigation company, 34; Chincha Islands War, 381; Oriente of, 32–34, 378, 380n126, 382; quinine production, 29; Reduction of the Loreto, 19; river boat acquisitions, 33, 34; rubber collection in, 30; War of the Pacific, 34, 50n125, 379

Peruvian Amazon Company. *See* PAC

photography: Eugène Robuchon's, 124n369, 517; in Hardenburg and Enock's book, 359–60; uses of, 360, 517

Pinell, Gaspar de: aiding Huitotos, 434; Aranalandia expedition of 1926, 434–35; on Arana's desire to sell Aranalandia, 422; and Bartolomé de Igualada, 363, 434; death of, 467; at Sibundoy mission, 463

Pinell, Gaspar de, Aranalandia expedition of 1918: and Arana, 418, 419–20; book on, 425–26; Colombians encountered, 418; El Encanto, 416, 417; El Encanto, forced return to, 418–19;

encounter with Franciscans, 417; goals of, 416; Indian encounters, 416; and Miguel Loyaza, 416–17, 419; outcomes of, 420; PAC interference, 418–20; on PAC ship *Liberal*, 420; and Peruvian officials, 418; return to Manaos, 419
Pinheiro, Manuel, 60
Pius X, Pope: Amazon missionary support, 386–87, 407; call for Indian protections, 261; and Giovanni Genocchi, 400; Lacrimabili Statu Indorum encyclical, 402; and Peter Hickey, 402
Pizango, Alberto, 459
Pizarro, Pedro Antonio, 77n247
Plan of Life, AZICATCH's: overview of, 498–99; "Words of the Elders of Tobacco, Coca, and Sweet Yucca," 593–96
Platform for Amazonian Indigenous Peoples United in Defence of Their Territories (PUINAMUNT), 460
Porras, Peruvian foreign minister, 389
Portillo, Pedro, 82, 83, 384
Portugal, 11–14
post-modernism, 516, 548–49
Prado y Ungarteche, Javier (Basque-surnamed), 254–55
La Prensa newspaper, 126, 211, 213, 259, 274–76, 284, 314n109
Prezet, Federico Alfonso, 265
prison literature, 522n31
Pro-Indígena Society, Lima, 245–46
Protestant Evangelical Union of South America, 411
Puerto Asís, 467, 492
Puerto Leguízamo, 439–40, 467
Pulido, Primitivo M., 162–63
Pupiales, Fray Basilio, 47

Putumayo (ship), 77–78
the Putumayo: Benjamin Larrañaga in, 76; Brazilian raids in, 26; cannibals in, 64, 67; Colombian-Peruvian agreement, 111–12; Cotuhé customs house, 111; Indians in, 45–46, 89–91, 104, 148; map showing Peruvian ownership, 78; "no man lands," 36n85; Paredes on Colombian claims to, 96–97; Putumayo River, 23, 32; reform in, 231–32; rubber production statistics, 132; as Spanish, 26
the Putumayo, Middle: accessing from Iquitos, 84; Brazil's relinquishing of claim to, 37; Colombian activity in, 107, 109, 439–40; Colombian-Peruvian disputes over, 35–36; end of nineteenth century state of, 36–37; European presence as ephemeral, 36; FARC control of, 491; Indigenous leader corruption in, 491; missionaries in, 25, 420; Pedro Portillo in, 82–83; Peruvian military expeditions, 82. *See also* Aranalandia
the Putumayo, white settlement histories: of Abel Calderón, 75–79, 88–89; of Aquileo Tobar, 98–104; of Eugène Robuchon, 317; of Joaquín Rocha, 73, 85–86; of José Gregorio Calderón, 74–75, 87–88; Peruvian, 91–94; of Rómulo Paredes (Peruvian), 93–97
the Putumayo/Caquetá, Colombian-Peruvian conflicts in: attacks on Colombians, 89, 91, 162–63; Benjamín Larrañaga and, 77; boundary marker destruction, 516; and Brazil, 166–67; Colombian military activity, 95; Colombian military activity rumors, 122–23,

140, 164; Colombian spies, 252; Francisco José Urrutia's report on, 165–66; historical roots of, 35–36; and the International Board, humanitarian, 364; International Commission on, 166–67; mediation attempts, 148, 165n519, 198, 279; missions as recognitions of authority, 406; PAC raids, 292n16; and PAC scandal aftermath, 364; at La Pedrera, 168–69, 251; Peru's ploy for Vatican recognition, 407–8; Peruvian military presence, 252, 257; La Reserva attack, 97, 117–18, 144, 152, 154, 163, 288; territorial claims, 78, 112–13; troop massing, 197–98. *See also* Casa Arana attacks on Colombians; Hardenburg, Walter; *modus vivendi* agreement; Orjuela, Jésus; PAC; Perkins, Walter B.; Salomón-Lozano Treaty; Serrano, David; La Unión assault

Putumayo/Caquetá region: Amazon Colombian Rubber and Trading Company, 113; Brazilian slavers in, 26, 27; Brazil's forfeiture of claims to, 32; *coca* industry, 491; Colombia, value to, 109n328; Colombian authority in, 85, 167–68; European interest in, 109; Pedro F. Urrutia as prefect, 59; petroleum industry, 492; and the Portuguese, 25–26; El Remolino, 114, 116; remoteness of, 29, 65; and the rubber boom, 65, 71–72; rubber tree destruction, 65; rubber trees in, 80; violence in, 85

Putumayo/Caquetá region, Colombian activity in: early nineteenth century, 26–28; infrastructure improvements, 114; infrastructure petitions, 109; investigation of 1924, 428–29; Reyes's plans for sovereignty over, 107–8; rubber concession sale, 86–87; settlement strategy, 65

Putumayo/Caquetá region, Lower, 432, 436, 438–40, 493–95

Putumayo/Caquetá region, Upper, 24, 25, 27, 65, 415

Putumayo exploration and settlement: Rafael Reyes's, 59–62, 516; Spanish, 24–25

Putumayo Mission Fund, 261, 400–1, 407

The Putumayo Red Book (Thomson), 348–51

quinine, 29, 49
Quintana, Manuel, 423

Ramírez, J. M. Moris, 79
Rance, C. H., 334
Read, Henry: approval of Casement investigation, 307; opinion of Arana, 309; and PAC, 128–30; Select Committee testimony, 302n61, 309
Reátegui, Manuel (Basque-surnamed), 266
Redondo, Sotero, 421
Reichel-Dolmatoff, Gerardo, 528–29
La Reserva: assault on, 97, 117–18, 144, 152, 154, 163, 288; and Casa Arana holdings, 105–6; Gabriel Mártinez at, 116–17; Walter Hardenburg at, 141
Resguardo Tijuna-Uitoto Km. 6–11, 501
Restrepo, Carlos Eugenio, 412
Restrepo, Carlos Lleras, 469
Restrepo, S., 279, 349–50

Rey de Castro, Carlos: on Anglophone moralizing, 354–55; Arana, relationship to, 128, 154, 531–32; and Arana's open self-defense letter, 356; on British/American fact-finding mission, 266, 267–68, 275, 517; defenses of Arana, 156–58, 353, 354, 356–57; denouncement of Carlos Valcárcel, 276; denouncement of Rómulo Paredes, 276; denouncements of Great Britain, 354–55, 358; denouncements of the US, 354–55, 358; media manipulation, 156–58; on PAC payroll, confirmation of, 320; on Peruvian-Brazil trade agreements, 357n53; as Peruvian consul in Manaos, 124, 155, 158; and Peruvian nationalism, 354; Putumayo history, 357; report on fact-finding mission, 275; Robuchon report edits, 124–25, 288, 316–17, 517; and Salomón-Lozano Treaty, 433; texts ghostwritten by, 351, 353
Reyes, Andrés, 59
Reyes, Carlos, 108
Reyes, Enrique, 65
Reyes, Ernesto, 59
Reyes, Florentino Calderón, 108–9
Reyes, Nestor, 64
Reyes, Rafael: and Abel Calderón's book, 75n240; and Benjamín Larrañaga, 59–60; and Brazil-Peru border markers, 60; business travels, 59; and cannibals, 64; and Chief Chua, 60; La Concepción rubber station, 59; concessions given to foreigners, 108, 110; Elías Reyes y Hermanos quinine firm, 58, 59, 63; *versus* Enrique Olaya Herrera, 463; fiancée of, 59; and Manuel Pinheiro, 60; in Mocoa, Colombia, 59–60; notebooks of, 58n166, 64; *versus* other rubber barons, 600; in Pará, 60, 61; and Pedro F. Urrutia, 59; presidency of, 107–8, 110, 115, 138, 166, 168, 428; Putumayo/Amazona expeditions, 59–62, 516; river boat fleet project, 62; steamship plans, 59, 60, 62; treatment of Indians, 63; on treatment of Indians, 90–91; and the United States, 168n526; Zacarías incident, 62–63
Rice, Cecil Spring, 343
Rijoa, Peru, 66n197
Rio de Janeiro, 18
Rio de la Plata region, 15–16
Rivera, José Eustasio, 428–29
Roberts, Charles, 287–88, 296, 314, 347
Robuchon, Eugène: and Casa Arana, 84–85, 516; and David Cazes, 295; death of, 260, 517; disappearance of, 124, 126–27; photos of atrocities, 124n369, 517; on Putumayo settlement, 317; report of, edited by Rey de Castro, 124–25, 288, 316–17, 517; use in propaganda, 124–25; writings of, 288, 316–17
Rocha, Joaquín, 72–74, 84n271, 85, 321n140
Rodríguez, Aurelio, 219
Rodríguez, Maximo, 52–53, 55
Rojas, Miguel A., 128, 191n42, 272
Roosevelt, Franklin D, 54, 465
Roosevelt, Theodore, 355
rubber, 29–30
rubber barons: Alejandro Betancourt, 58; common patterns, 600; Morey Arias brothers, 69; Nicolás Suárez, 51–53, 597–600; overview

of, 49; Vaca Diez, 598. *See also* Arana, Julio César; Fitzcarrald, Carlos Fermín; Rodríguez, Máximo; Suaréz, Nicolás

Rubber Development Corporation, 54

rubber industry: Asian, 230; Compañía de Caquetá, 49–50; Compañía de Colombia, 49; Congolese, 171; *correrías,* 520–21; debt peonage, 40, 256, 323, 331, 435, 520; deforestation, 38, 81; end of South American, 436; foreign interests in, 535n55; globalization of, 31; and the Huallaga, 66; Iberia rubber station, 52–53, 55, 56, 57; Indians, fear of, 47; labor shortages, 30–31; leaf blight, 54–58; market busts, 134; *The Nation* article on, 231; Peruvian export tax removal, 423, 430; Peruvian reforestation, 346n13; plantation production, 53, 134; Putumayo/Caquetá deforestation, 65; Sánchez brothers operation, 58n163; Soratama enterprise, 57–58; South American crises, 53, 55, 214n135, 230, 238, 315; sustainability of, 38; synthetic rubber production, 56n156; terror, role of, 521; United States, 54–56; and World War II, 55. *See also* rubber barons

rubber industry labor system: atrocities against indigenes, 40–42, 58n163; *aviadores* (river traders), 37, 38, 39–40, 49; *caucheros* (merchants), 40, 47, 48; *caucheros* (route builders), 37, 38, 39; changes in, 81; collection, 20; control methods, 48; debt as capital in, 40; as debt slavery, 38–39, 40; *estrada* routes, 37, 38; fortified headquarters buildings, 48; Fuller's report on, 256; and Indian attacks, 45, 47; indigene disinterest in, 40; *mestizos,* 37; *peones* (manual laborers), 37, 38, 40, 45, 48, 525n34; processing, 38; as slavery, 381–82; terminology of, 37n88; tree tapping, 37–38

Rubber Reserve Company, 54

Rumrrill, Roger, 433, 441n343, 449n7

Sagols, Francisco, 378–79

Salamanca, Demetrio, 112

Salamanca T., Demetrio, 353

Saldaña Rocca, Benjamín: about, 125; articles by, and Arana, 305, 318, 333; attacks on Arana, 126–27, 134, 318; blackmail claims against, false, 330; criminal charges against Casa Arana, 125, 212, 333; death of, 259, 333; defense of Indians, 126; departure from Iquitos, 149–50; informants on Arana, 149–50; newspapers of, 125–26, 398; perspectives on, 512–13; Peruvian reactions to, 154; public opinion of, 294; and Ricardo Gómez A., 427; *La Sanción* newspaper, 125, 398; as socialist crusader, 525. *See also La Felpa* newspaper

Salomón-Lozano Treaty: and Arana, 432–34, 542; and Aranalandia, 432–33; Article 9, 436; breaking and reinstatement of, 436–38; Colombian books influencing, 425–27; and Enrique Olaya Herrera, 464; fears of land loss to Colombia, 424–25; Indian migrations, 434; Indian resettlements after, 434–35; initiation of, 423; Leticia Trapezoid, 2n1, 432,

436, 440, 493–95; Patriotic Board response, 436–37; Peruvian approval of, 433; Peruvian debates over, 432; Peruvian protests, 433; physical presence reinforcement during, 432; President Leguía and, 423, 431–32; Ricardo Gómez A.'s support for, 428; territories given to Colombia, 432

Sánchez, Alfonso, 119, 122

Sánchez, Ramón, 295

Sánchez Rangel, Hipólito, 377

San Martín, José de (Basque descendant), 14

San Román, Miguel, 410–11

Santos, Silvino, 266, 360, 517

Santos Calderón, Juan Manuel: career and education of, 481; FARC peace talks and accord, 481, 482, 483, 484–86; at House of Knowledge conference, 506; as national defense minister, 479; Nobel Peace Prize, 487; presidential campaign, 480; second term, 483

Santos Granero, Fernando, 520

São Paolo, 18

Scapardini, Angelo, 263–64, 363–64, 387n147, 405, 406–7, 409

Schultes, Richard, 56–58

scientific methodology, 549n84

Seibert, Russell, 57

Select Committee. *See* British Select Committee on PAC atrocities

Serrano, David: Casa Arana attacks and murder, 117–18, 119, 141–42, 163; employment Indians fleeing Casa Arana, 117; flight from PAC, 144; and Francisco Gutiérrez, 101; and Jesús Orjuela, 142; and Walter Hardenburg, 141–42. *See also* La Reserva

Servants of God or Masters of Men? (Bonilla), 469–70

Shamanism, Colonialism, and the Wild Man: A Study in Terror and Healing (Taussig): Andrew Gray on, 519–20; *versus* Anglo-American anthropology, 527; on *correrías*, 515; deconstructions of, 520; on history and memory, 515; on Indian-white co-construction, 513–15; overview of, 513; on PAC atrocity origins, 515

Shaw, George Bernard, 367, 518

Simon, W. O., 213

Sjöblom, E. V., 171

"Sketch of the Putumayo": authorship of, 131n403; full text of, 555–58; on PAC, 131n403; on rubber production, 133; on white settlement in the Putumayo, 91–93

Slavery in Peru report, 312, 343, 348, 512n1

slavery *versus* debt peonage, 331

Solís, Paulino, 75

Solzhenitsyn, Aleksander, 539–40

Sorensen, Hans, 55, 56, 57

South America, history of: Catholic orders, 17; eighteenth century, 12–13; fifteenth century, 12; mixed-race populations, 31; nineteenth century, 14–15; Spanish treatment of Indians, 19–21

Souza, Rafael de, 52

Spain: Catholic power in, 370; Chincha Islands War, 381; eighteenth century history, 12–13; ethnic differences within, 545–46; fifteenth century history, 11–12; immigrants from, to Peru, 381; imperial policies of, 14; nineteenth century history, 13

Spicer, Gerald Sydney, 228, 229–30, 236

Stanfield, Michael Edward, 36, 61, 80, 81n257, 111–12, 125
Stimson, Harry, 437
Suárez, Francisco, 597–98
Suárez, Gregorio, 597–98
Suárez, Luis, 540–41
Suaréz, Nicolás, 51–53, 597–600
Suárez, Pedro, 598, 599
Suárez, Rómulo, 597, 598
Suarez Hermanos & Co, 597–99
Sublimus Dei 1537, Pope Paul III's, 16
El sueño del celta (Llosa), 518

Tabatinga, 13
Taussig, Michael. See *Shamanism, Colonialism, and the Wild Man: A Study in Terror and Healing*
Terry, Fernando Belaúnde, 449
Thomson, Norman: and British Select Committee, 311–12; Colombian government connections, 351; Colombian sympathies, 312, 350; and International Board, humanitarian, 364; letters in *The Times* of London, 350, 351; White Book, 351. See also *The Putumayo Red Book*
Thorogood, Horace, 182–83, 194
The Times of London: calls for Putumayo missions, 403–4; Hensley Henson sermon, 404–5; on PAC Board of Directors, 271; summary of Casement Report, 403
Tizón, Juan, 190, 198; blackmail accusation against Hardenburg, 285; desire for reform, 233; and English missionaries, 409; in George Babington Michell's testimony, 298, 299; improvements made, 311; and John Russell Gubbins, 310; and PAC Commission, 204, 205, 207–8, 307; in Paredes report, 280; replacement of Macedo, 190, 215, 225; report to PAC Board of Directors, 198; Roger Casement's testimony on, 290
Toá: Narraciones de caucherías (Uribe Piedrahita), 518
Tobar, Aquileo, 98–104
Torralbo, José, 348–49
Torre, Carlos de la, 273
Torres, Aquileo, 77
Treaty of San Ildefonso, 12–13, 26
Treaty of Tordesillas, 11
Triana, Leopoldo, 167–68
Truth newspaper: Eduardo Lembcke's response to Hardenburg, 184–85; G. S. Paternoster and, 181–82; verification of Hardenburg's claims, 293; and Walter Hardenburg, 181–82, 195, 293n18
Truth newspaper articles on PAC atrocities: Abel Alarco's disavowal of, 190; British Foreign Office's trust in, 191–92; "The Devil's Paradise," 182; Eduardo Lembcke's responses, 184–85; in G. S. Paternoster's book, 347; Henry Lex Gielgud and, 191, 197; Henry Read on, 309; and Horace Thorogood, 182; impact of, 331; international spread of, 189; as Iquitos public knowledge, 327, 333; and PAC Commission creation, 290; PAC Commission's verification of, 225; PAC directors' disavowals of, 185–86, 196–97; Peruvian Senator Ward and, 191; sent to US government, 192; shareholder meeting discussions of, 193, 214
Turvasi, Francesco, 265n302, 374–75, 391, 402n205

Ucayali river, 50
Último Retiro, 310–11
La Unión assault: Carlos Rey de Castro (PAC) account, 157; Carlos Valcárcel's view of, 157–58; Casement's testimony on, 288; Colombian citizen denouncement, 164; Colombian government denouncement, 154; Felipe Cabrera M.'s account, 559–63; Gómez López's account, 119–23, 162; Hardenburg's account, 140, 144, 150–51, 157n494; and Jesús Orjuela, 119; Roger Casement's testimony on, 288
the United States of America: atrocities committed by, 354–55; Carlos Lissón on, 381n129; Carlos Rey de Castro's critiques of, 354; and Colombia, 447, 475, 478; House resolution on slavery documentation, 264; imperialism of, 355–56, 526; Iquitos consulate, 113n343, 147, 251, 254, 343; Knickerbocker Crisis, 133–34; Monroe Doctrine, 15, 227, 526; and PAC atrocity claims, 197–98, 312; and the Panama Canal, 358; Peruvian alliance, 447; and Peruvian seizure of Leticia, 437–38; Plan Colombia, 475; and Putumayo reforms, 232, 343; rubber industry, 54–56; slavery in, 522–23; *Slavery in Peru* report, 312, 343, 348, 512n1; treatment of Indians, 354–55; and World War I, 432
Upper Amazon. *See* Amazonia
The Upper Reaches of the Amazon (Woodroffe), 352–53
Uribe, Alredo White, 435
Uribe, Rafael (Basque-surnamed), 107, 151, 425

Uribe Gaviria, Carlos (Basque-surnamed), 487
Uribe Piedrahita, César (Basque-surnamed), 518
Uribe Vélez, Álvaro (Basque-surnamed), 475–81, 483, 485–86, 498
Urrutia, Francisco José (Basque-surnamed), 165–66
Urrutia, Pedro F., 59
Urubamba river, 50
Uruguay, 15–16, 447n2

Vaca Diez, Antonio, 52
Val, Merry del: and Angelo Scarpadini, 263–64, 387n147, 409; communications regarding Putumayo mission, 406–8; on English missionaries, 409; Genocchi investigation reports to, 391–92; and Giovanni Genocchi, 402; and Monsignor Quattrocchi, 389–90; and Peru's ploy for Vatican recognition, 407–8
Valcárcel, Carlos: Arana's false blackmail claims, 281; on Arana's literacy, 536; arrest warrants issued, 234–35, 242–43, 280; book on PAC atrocities, 360–62; firing of, 242; and Guillermo Billinghurst, 283, 361n64; health issues, 222, 283–84; on Julio Egoaguirre, 128, 222; legal proceedings against PAC, 212; on PAC deception of Peruvian government, 157–58; on PAC investigations, 154–55; and Paredes report, 511; personal investigation requests, 222; perspectives on, 513; on Peruvian patriotism, 362; President Billinghurst's protection order, 283; press attacks on, 276, 283, 354; press defenses of, 281; public

manifestation against, 281, 283; reinstatement of, 276; return to Iquitos, 222; treatment in New York, 215
Valderrama, Alberto, 79
Valderrama, Gabriel, 144–45
Valencia, Aurelio Iragorri (Basque-surnamed), 482–83
Varese, Stefano, 449
Vega, Juan B.: Arana, Vega y Cia, 85; Arana and Vega, 70; arrest warrant for, 234, 280, 284; as Colombian viceconsul in Iquitos, 85; flight to Switzerland, 280n353; and Miguel A. Rojas, 128; PAC money owed to, 302; political positions, 110; Vega, Morey, and Company, 69
Velasco, Lisímaco, 114, 116–17
Velez, Germán, 112–13
Venezuela, 14, 380, 447n2, 473n81, 481
Vilafranca, Marceliano de, 465, 467
Villa, Veliz de, 244–45
La vorágine novel (Rivera), 429–30, 518

War of the Oranges, 13
Whiffen, Thomas: and Arana, 187–89; blackmail claims against, false, 188–89, 192, 309, 322, 328, 330; Douglas Hogg's attack on, 332; lawsuit against Arana, 328; Putumayo visits, 197n71; report to British Foreign Office, 188; Select Committee's findings on, 337–38; Select Committee testimony, 322, 328–29, 357n53; witnessing of PAC atrocities, 186–87
White, Hayden, 550
Williams, Morgan, 193, 214–15
Wilson, Edmund O., 549n84

Woodroffe, Joseph F., 351–52
World War I, 365–66, 432

Xavier, Francis, Saint (Basque), 18

Yapurá River, 111
Yaracaya rubber station, 139
Yeats, William Butler, 367, 518
Yurimaguas, Peru, 68–69

Zamudio, Manuel Patiño (Basque-surnamed), 380
Zapata, Carlos: Arana's statement on, 329–330; and PAC, 154–55, 317–18; PAC atrocity report, 155–56, 158, 317–18
Zubiaur, Carlos (Basque-surnamed), 84, 120, 146, 147, 209, 244, 546
Zumaeta, Angélica Arana, 440, 441
Zumaeta, Bartolomé: characterization of, 332; death of, 159, 288–89; and Manuel Erazo, 547; murder of Justino Hernández, 88; and La Reserva assault, 97, 117–18, 155
Zumaeta, Eleonora. *See* Arana, Eleonora
Zumaeta, López, 361
Zumaeta, Luis Arana, 436–37, 437n331, 438, 440–41
Zumaeta, Marcial, 131n403, 266n301, 303, 313, 329
Zumaeta, Miguel, 435
Zumaeta, Nestor, 86
Zumaeta, Pablo: and Arana, 68, 70, 84, 226, 347, 531–32; arrest warrant for, 234, 235n203; attacks on Rómulo Paredes, 277n344; and British Board members, 305; Casement's request to replace, 230; civilizing Indians discourse, 305–6; financial claims against

Colombia, 212; investigation obstructions, 215; as Iquitos mayor, 347; and John Brown, 187; John Russell Gubbins's distrust of, 307; legal charges against, 212, 234–35; letters to London board on atrocities, 246–47; letter to Select Committee, 305–6; lien filed against PAC, 226; at PAC, 129, 332; PAC atrocities, ignorance of, 187; on PAC Board of Directors, 305; in the Putumayo, 93, 224; request to buy out section chiefs, 192n50; Roger Casement's testimony on, 288, 289; special legal treatment, 242–43; Stuart fuller on, 257–58; writings of, 353–54